Proceedings of the 9th International Symposium on

FOUNDATIONS OF QUANTUM MECHANICS IN THE LIGHT OF NEW TECHNOLOGY

ISQM — TOKYO '08

Proceedings of the 9th International Symposium on

FOUNDATIONS OF QUANTUM MECHANICS IN THE LIGHT OF NEW TECHNOLOGY

ISQM — TOKYO '08

Advanced Research Laboratory
Hitachi, Ltd., Hatoyama, Saitama, Japan

25–28 August 2008

Edited by

Sachio Ishioka
Advanced Research Laboratory
Hitachi, Ltd., Japan

Kazuo Fujikawa
Institute of Quantum Science
Nihon University, Japan

World Scientific

NEW JERSEY · LONDON · SINGAPORE · BEIJING · SHANGHAI · HONG KONG · TAIPEI · CHENNAI

Published by

World Scientific Publishing Co. Pte. Ltd.
5 Toh Tuck Link, Singapore 596224
USA office: 27 Warren Street, Suite 401-402, Hackensack, NJ 07601
UK office: 57 Shelton Street, Covent Garden, London WC2H 9HE

British Library Cataloguing-in-Publication Data
A catalogue record for this book is available from the British Library.

FOUNDATION OF QUANTUM MECHANICS IN THE LIGHT OF NEW TECHNOLOGY

ISBN-13 978-981-4282-12-3
ISBN-10 981-4282-12-X

Printed by Fulsland Offset Printing (S) Pte Ltd, Singapore

The Ninth International Symposium on Foundations of Quantum Mechanics (ISQM-Tokyo '08)

PREFACE

The Ninth International Symposium on Foundations of Quantum Mechanics in the Light of New Technology (ISQM-TOKYO '08) was held on August 25–28, 2008 at the Advanced Research Laboratory, Hitachi, Ltd. in Hatoyama, Saitama, Japan. The symposium was organized by its own scientific committee under the auspices of the Physical Society of Japan, the Japan Society of Applied Physics, and the Advanced Research Laboratory, Hitachi, Ltd. A total of 137 participants (28 from abroad) attended the symposium, and 32 invited oral papers, 7 contributed oral papers, and 47 poster papers were presented.

Just as in the previous eight symposia, the aim was to link the recent advances in technology with fundamental problems in quantum mechanics. It provided a unique interdisciplinary forum where scientists from different disciplines, who would otherwise never meet each other, convened to discuss basic problems of common interest in quantum science and technology from various aspects and "in the light of new technology."

Quantum Coherence and Decoherence was chosen as the main theme for this symposium because of its importance in quantum science and technology. This topic was reexamined from all aspects, not only in terms of quantum computing, quantum information, and mesoscopic physics, but also in terms of the physics of precise measurement, spin related phenomena, and other fundamental problems in quantum physics.

We are now very happy to offer the fruits of the symposium in the form of the proceedings to a wider audience. As shown in the table of contents, the proceedings include a special lecture by Professor C.N. Yang and 75 refereed papers in ten sections: cold atoms and molecules; spin-Hall effect and anomalous Hall effect; magnetic domain wall dynamics and spin related phenomena; Dirac fermions in condensed matter; quantum dot systems; entanglement and quantum information processing, qubit manipulations; mechanical properties of confined geometry; precise measurements; novel properties of nano-systems; and fundamental problems in quantum physics.

We will mention just some of the important keywords here to give the flavor of the proceedings: Bose-Einstein condensate, quantum computation and communication, qubits, quantum dots, spintronics, graphene, rattling, atomic clocks, and vortices in high-temperature superconductors. We hope that the proceedings will not only be the record of the symposium but also serve as a good reference book for experts on quantum coherence and decoherence and as an introductory book for newcomers in this field.

In conclusion, we thank the participants for their contribution to the symposium's success. Thanks are also due to all the authors who prepared manuscripts and to the referees who reviewed the papers. We also thank the members of the Advisory Committee and Organizing Committee; without their invaluable assistance, the symposium would not have been a success. Finally, we express our deepest gratitude to the Advanced Research Laboratory, Hitachi, Ltd. and Dr. Nobuyuki Osakabe, who was its General Manager at that time, for providing us with financial support and an environment that was ideal for lively discussion. We also thank his staff members, in particular Mr. Yoshimichi Yamamoto and Ms. Tamae Nagamine, for their efforts in making the symposium enjoyable as well as productive.

March 2009

Sachio Ishioka
Kazuo Fujikawa

COMMITTEES

Chair:

H. Fukuyama, Tokyo University of Science

Advisory Committee:

J. E. Mooij, Delft University of Technology
S. Nakajima, Superconductivity Research Laboratory,
International Superconductivity Technology Center
N. Osakabe, Advanced Research Laboratory, Hitachi, Ltd.
C. N. Yang, Tsinghua University
A. Zeilinger, University of Vienna

Organizing Committee:

K. Fujikawa, Nihon University
Y. Iye, University of Tokyo
N. Nagaosa, University of Tokyo
Y. A. Ono, University of Tokyo
F. Shimizu, University of Electro-Communications
H. Takayanagi, Tokyo University of Science
S. Ishioka, Advanced Research Laboratory, Hitachi, Ltd.

Sponsors:

The Physical Society of Japan
The Japan Society of Applied Physics
Advanced Research Laboratory, Hitachi, Ltd.

Secretariat:

S. Ishioka, Advanced Research Laboratory, Hitachi, Ltd.

CONTENTS

Spin Hall Effect and Anomalous Hall Effect

Magnetic Domain Wall Dynamics and Spin Related Phenomena

Quantum Dot Systems

Entanglement and Quantum Information Processing, Qubit Manipulations

OBITUARY: PROFESSOR SADAO NAKAJIMA

Professor Sadao Nakajima, the founder of International Symposium on the Foundation of Quantum Mechanics in the Light of New Technology (ISQM), passed away on December 14, 2008, at the age of 85.

Professor Nakajima was born in Shizuoka prefecture and graduated from Tokyo University in September, 1945. After spending some years (1950-1961) in Nagoya University, he had been in the Institute for Solid State Physics (ISSP) of Tokyo University till his retirement in 1984. In the last 3 years in ISSP, he had been the director of ISSP, when he started this ISQM together with Dr. Takeda, then General Manager of the Central Research Laboratories of Hitachi. The first ISQM was held in 1983 to be followed by those in 1986, 1989, 1992, 1995, 1998, 2001, 2005 and 2008. Professor Nakajima chaired till 1995.

The scope of ISQM, which Professor Nakajima and Dr. Takeda had made clear, is to search for ever deeper understanding of the implication of quantum mechanics with the help of frontier of technology. It will be fair to say that they had a great foresight in identifying this paradigm a quarter of century ago in view of the fact that broad scientific and technological interest are present in these days in properties of condensed matter, as frequently termed as nano-science and nano-technology, where quantum effects play crucial roles.

The identification of this scientific paradigm is based on many successful scientific experiences of Professor Nakajima who was always positive to new findings and had a remarkable record of scientific achievements such as the deep understanding of electron-phonon interaction in solids contributing to the creation of BCS theory for superconductivity and various aspects of low temperature physics including the introduction of paramagnon theory in He_3 superfluidity. Besides such scientific activities Professor Nakajima had played crucial roles in encouraging and fostering young generations by creating a very free and warm atmosphere around him, whose memories will live long in the heart of many researchers.

Hidetoshi Fukuyama
Chair of ISQM-TOKYO'08

OPENING ADDRESS

HIDETOSHI FUKUYAMA
Chair of the Organizing Committee, ISQM-TOKYO'08
Faculty of Science, Department of Applied Physics, Tokyo University of Science, Tokyo 102-0073, Japan

Good morning, ladies and gentlemen, and dear colleagues;

On behalf of the Organizing Committee, I would like to welcome all of you to the International Symposium on the Foundation of Quantum Mechanics in the Light of New Technology, ISQM-TOKYO'08.

This is the 9th of the series. It was started in 1983, a quarter century ago, when Professor Sadao Nakajima, then the director of ISSP, and Dr. Yasutsugu Takeda, then General Manager of the Central Research Laboratories of Hitachi, resonated in the idea to bridge basic science and technology, namely to discuss subjects on the frontiers of science which have become possible with the help of new technology. Unfortunately both of them can not join the meeting this time.

By now this series has a history. Actually it is impressive to observe that, during last 25 years, there has been tremendous progress in both science and technology.

Behind the decision made 25 years ago by Nakajima and Takeda, there was a beautiful experimental verification of the Aharonov-Bohm effect by Dr. Akira Tonomura, who received strong inspirations and warm encouragements from Professor C.N. Yang to carry out this very difficult experiment. Professor Yang is expected to join this Symposium tomorrow.

By the way, Dr. Tonomura, Dr. Fujikawa, one of the organizers, and myself were together in the same class in the days of undergraduate student in the University of Tokyo.

The experiment by Dr. Tonomura is so beautiful and appealing as you may all know that I now introduce his experimental results in the first class hour of quantum mechanics for the undergraduate students by using power points which Dr. Tonomura kindly edited for me to be used originally for the public lecture. No doubt, the experiment by Tonomura is symbolic in demonstrating the fact that science and technology are hand in hand - technological developments can lead to exploration of mysteries of nature and, in return, new understanding of scientific facts leads to new technology

In the last ISQM held three years ago I said the following "in this conference many interesting experimental questions associated with the quantum coherence of not only electron waves but also that of spins has been addressed. This trend will grow since scientific and technological interests are moving more toward to materials in nano-scales, where obviously the quantum coherence really manifests itself." This trend persists, and the importance of transport phenomena in magnetic systems has been recognized last year as Nobel Prize in Physics. Very clear feature of the symposium this time is that there is much more interest in the active roles played by this spin degree of freedom.

It is our hope that this Symposium could contribute to the deeper understanding of the implications of quantum mechanics. We, organizers, thank you for your joining this Symposium. Please enjoy discussions.

Thank you very much for your attention.

WELCOMING ADDRESS

NOBUYUKI OSAKABE

General Manager, Advanced Research Laboratory, Hitachi, Ltd.
Hatoyama, Saitama 350-0395, Japan

Good morning, distinguished guests, ladies and gentlemen. On behalf of our laboratory, let me welcome you to Hatoyama and ISQM-TOKYO '08. It is a great pleasure for us to host this event for you all here today.

This symposium was initiated in 1983, as Professor Fukuyama explained. The motivation for Hitachi to host and sponsor the first ISQM came from Dr. Akira Tonomura's successful verification of the Aharonov-Bohm effect. In his study, new micro-fabrication technology developed in the semiconductor industry and an electron microscope technology developed at Hitachi were successfully combined to verify the important concept of the gauge field.

In addition, through the strong thrust of Professor Chen Ning Yang, we decided to hold the first ISQM. The aim of the symposium has been to discuss the foundation of quantum mechanics achieved by using new technologies based on industrial innovation and to contribute to the scientific community in general.

Twenty five years have past since then. Many new areas in physics have been discussed over the years. Today those fundamental quantum phenomena are now being used in the world of technology: Spin instead of charge can now be controlled to make new memory or logic chips to breakthrough the barrier of modern device performance. Quantum entanglement will be used to secure future communication. Macroscopic quantum tunneling and coherence will be the basis for quantum computing. The foundation of quantum mechanics will leverage industrial companies. Thus, because of all these possibilities, we found enormous value in hosting and sponsoring the symposium here again.

I hope all of you find the symposium rewarding. Before concluding, I would like to thank all the members of the Advisory and Organizing Committees for putting together such an exciting program. I also would like to thank all the invited speakers for coming to share their latest findings with the participants here. I very much look forward to hearing them.

Thank you very much for your attention.

PSEUDOPOTENTIAL METHOD IN COLD ATOM RESEARCH

C. N. YANG

Tsinghua University, Beijing, China and Chinese University of Hong Kong, Hong Kong

Keywords: Cold atom; pseudopotential method.

More than fifty years ago, the dilute hard sphere problem (in 3 dimensions) was studied by Bogoliubov[1], and Lee, Huang, Yang[2]. In particular, the latter arrived at the ground state energy (per particle) of the system in an asymptotic expansion:

$$\frac{E_0}{N} = 4\pi a\rho\left[1+\frac{128}{15\sqrt{\pi}}\sqrt{\rho a^3}+\cdots\right] \quad (1)$$

where ρ=density and a=scattering length. Amazingly in the last few years it had become possible to test the validity of equation (1) in cold atom research.

The method used in reference 2 to obtain equation (1) was a very *physical* pseudopotential method. Because of recent experimental developments it is worthwhile to review this method in 3D and generalize it to other dimensions.

The key idea of the method is to replace the boundary condition that φ=0 when two spheres touch with a pseudopotential.

How to construct this pseudopotential in 3D? We notice that for 2 hard spheres in the CM system, the wave function is

$$\varphi = 1-\frac{a}{r}. \quad (2)$$

Furthermore

$$\nabla^2\frac{1}{r} = -4\pi\delta^3(\mathbf{r}). \quad (3)$$

If we take

$$V' = 4\pi a\delta^3(\mathrm{r}), \quad (4)$$

then

$$\left[-\nabla^2+V'\right]\left[1-\frac{a}{r}\right] = 0\left(a^2\right), \quad (5)$$

suggesting V' as a candidate for the pseudopotential.

Using V' as the potential, a first order perturbation calculation yields the first term $4\pi a\delta$ of equation (1) for the energy of the ground state. However, the second order perturbation gives infinity. This divergence was examined in reference 2 and was understood to be due to a $\frac{a}{r}$ term in the first order wave function.

This understanding also generates the cure for the divergence: replace V' with

$$V = 4\pi a\delta^3(\mathrm{r})\frac{\partial}{\partial r}r \quad (6)$$

which is the same as V' when operating on a regular function, but gives zero when operating on $\frac{1}{r}$. Using this V equation (1) was obtained in reference 2.

To generalize to dimensions 2 and 4, we need to first generalize (3):

$$\text{Dim 2}\quad \nabla^2(-\ell n r) = -2\pi\delta^2(\mathbf{r}). \tag{7}$$

$$\text{Dim 4}\quad \nabla^2 \frac{1}{r^2} = -4\pi^2\delta^4(\mathbf{r}). \tag{8}$$

Starting from these basic formulae one can easily construct the generalizations of the pseudopotential to these dimensions. Calculation can then proceed for E_0/N, resulting in Table 1 below which is explained in detail in a paper to appear shortly in the *European Physics Letters*.

References

1. Bogoliubov N. N., *J. Phys. USSR*, II (1947) 23.
2. Lee T. D., Huang K. and Yang C. N., *Phys. Rev.* **106**, 1135 (1957). This paper used the pseudopotential method: See Huang K. and Yang C. N., *Phys. Rev.* **105**, 767 (1957); Huang, Yang and Luttinger, *Phys. Rev.* **105**, 776 (1957). The ground state energy E_0 was actually first obtained by a method called binary collision method: T. D. Lee and C. N. Yang, *Phys. Rev.* **117**, 12 (1960). Cf. also additional remarks about the binary collision method by C. N. Yang, p.38-398, p.225-235, *Selected Papers 1945-1980 With Commentary*.
3. Yang C. N., *Rev. Mod. Phys.* **34**, 694 (1962), Section 4.

Dimension	Pseudopotential for fixed scatterer	E_0 / N	Comments
1	—	$\pi^2\rho^2/3(1-\rho a)^2$	No BEC
2	$\dfrac{\sim 4\pi\delta^2}{\lvert \ell n(\rho a^2)\rvert}$	$\sim\dfrac{4\pi\rho}{\lvert \ell n(\rho a^2)\rvert}+\cdots$	Almost BEC
3	$8\pi a\delta^3 \dfrac{\partial}{\partial r} r$	$4\pi\rho a(1+\dfrac{128\sqrt{\rho a^3}}{15\sqrt{\pi}}+\cdots)$	BEC. BG= 1 Boson
4	$8\pi^2 a^2\delta^4 \dfrac{\partial^2}{2(\partial r)^2} r^2$	$4\pi^2\rho a^2(1+4\pi^2\rho a^4\lvert \ell n(\rho a^4)\rvert+\cdots)$	BEC. BG = 1 Boson
5		?	?

Table 1: Comparison of BEC in dilute interacting Boson gases for different dimensions. (BG = basic group)

SYMMETRY BREAKING IN BOSE-EINSTEIN CONDENSATES

MASAHITO UEDA

Department of Physics, University of Tokyo, Tokyo, 113-0033, Japan
**E-mail: ueda@phys.s.u-tokyo.ac.jp*

Bose-Einstein condensates offer an ideal testing ground for symmetry breaking. I will explain the reasons why and briefly overview our recent work on the related subject.

Keywords: Bose-Einstein condensate; symmetry breaking; topological excitation.

A gaseous Bose-Einstein condensate (BEC) offers a cornucopica of symmetry breaking, because a trapped BEC system is in a mesoscopic regime, and situations exist under which symmetry breaking may or may not occur. Investigating this problem sheds new insights into why mean-field theories have been so successful in describing gaseous BEC systems and when many-body effects play a significant role.

By way of introduction, we begin by discussing where we should look for new phenomena. First, the system is extremely dilute with the atomic density five orders of magnitude more dilute than that of the air. This implies a very long collision time of the order of milliseconds and kinetics becomes important. In particular, the dynamics of phase transitions and defect nucleation can be investigated in real time.

Second, energy and angular momentum of the system are conserved. In condensed matter physics, the system is usually in contact with a heat bath such as container walls which allow exchange of energy and angular momentum with the system. In contrast, a system of ultracold atoms is usually suspended in a vacuum chamber by means of an electromagnetic potential. Since it does not have microscopic rugosities, no energy exchange occurs between the system and the trapping potential. Moreover, if the potential is isotropic or axisymmetric, the total angular momentum or its projection on the symmetry axis is conserved. Thus the thermodynamics of the system looks more like that of a stellar object rather than a terrestrial one. Those conservation laws impose special constraints on time evolution of the system, leading, among other things, to spontaneous structure formation.

Third, spin-exchange and magnetic dipole-dipole interactions play crucial roles in determining magnetism, spin textures, and topological excitations. The energy scales of these interactions are much smaller than the spin-independent interaction and temperture of the system. However, they give rise to observable consequences because of Bose-Einstein condensation, as discussed below.

What makes this system a conucopia of symmetry breaking is the fact that the system involves energy scales that differ by five orders of magnitude. Apart from the Zeeman energy which fixes the total magnetization, the largest energy scale is the spin-independent Hartree interaction, which together with a trapping potential determines the distribution of particle density. Three orders of magnitude smaller than this is the spin-dependent interaction which determines the magnetism or the symmetry of the spinor order parameter. By far the smallest is the magnetic dipole-dipole interaction. Albeit its minuteness, it plays a dominant role in determining a spin texture which is a spatial distribution of spin orientation. Because these interactions are different in their orders of magnitude, they can play distinct roles in determining the characters of BECs and thus making the physics of atomic-gas BECs very rich.

Now the crucial question is: Why can these very small, nano-Kelvin physics be observed at micro-Kelvin temperature? The answer is bosonic stimulation. Once the system undergoes BEC, all Bose-condensed atoms respond to an external perturbation in a collective manner. In other words, the off-diagonal long-range order enhances the underlying interaction by a factor of the number of Bose-condensed atoms. For example, the magnetic dipole-dipole interaction is about 0.1nK for Rb87 and 10nK for Cr52. Once the system undergoes BEC, the collective energy amounts to 1 μK for Rb and 100μK for Cr. Therefore we cannot ignore dipolar effects when considering the dynamics longer than 100ms for Rb87 and 1ms for Cr52. This is similar to the situation in superfluid helium-three, where the energy scale of the nuclear dipole interaction is of the order of 10^{-7}K, yet it plays a dominant role in determining the properties of the system at 10^{-3}K.

We have substantiated these ideas in several distinct situations: namely, soliton formation in attractive BECs [1], topological winding/unwinding in metastable BECs [2], spontaneous magnetization in spinor BECs, and spin texture formation in spinor-dipolar BECs. As an illustration, let us consider an attractive BEC confined in a quasi one-dimensional potential. It is known that that the ground state of an infinite attractive BEC spontaneously breaks a space-translation symmetry and forms a bright soliton. This is because the attractive BEC is unstable against long-wavelength density fluctuations of the system. However, when the system is confined in a finite system, the zero-point energy of the system introduces a cut-off for low-lying excitations. Conseqeuently, a quantum phase transition between a uniform BEC and a bright soliton occurs at the critical strength of attractive interaction which is equal to the zero-point energy [1]. This is a prime example in which the mesoscopic nature of the system yields a new type of symmetry breaking.

For the case of repulsive BECs, the density distribution is fixed by the spin-independent contact interaction and the trapping potential. When the potential is an optical trap, the spin degrees of freedom give rise to several symmetry breakings unique to atomic-gas BECs. For example, a spin-1 Bose-Einstein condensate with ferromagnetic interaction exhibits chiral-symmetry breaking when an external magnetic field is rapidly quenched. In this case, local spins grow randomly and spin vortices are created spontaneously [3]. This prediction was subsequently vindicated experimentally by the Berkeley group [4]. We have discussed the underlying mechanism [3] and ramifications of this phenomenon such as the Kibble mechanism [5]. Recently, it has been predicted that knot excitations can be realized in a spinor BEC [6]. It has been demonstrated that dipolar collapse in a chromium BEC exhibits a cloverleaf pattern due to the anisotropic nature of the interaction[7]. These examples illustrate a rich variety of symmetry breakings in gaseous BECs which merit further study.

References

1. R. Kanamoto, H. Saito and M. Ueda, *Phys. Rev. Lett.* **94**, 090404 (2005).
2. R. Kanamoto, L. Carr and M. Ueda, *Phys. Rev. Lett.* **100**, 060401 (2008).
3. H. Saito, Y. Kawaguchi and Ueda, *Phys. Rev. Lett.* **96**, 065302 (2006).
4. L. Sadler, et al.@*Nature* **443**, 312 (2006).
5. H. Saito, Y. Kawaguchi and Ueda, *Phys. Rev.* **A76**, 043613 (2007).
6. Y. Kawaguchi, M. Nitta and Ueda, *Phys. Rev. Lett.* **100**, 180403 (2007).
7. T. Lahye, et al. *Phys. Rev. Lett.* **101**, 080401 (2008).

QUANTIZED VORTICES IN ATOMIC BOSE-EINSTEIN CONDENSATES

M. TSUBOTA
Department of Physics, Osaka City University, Osaka, 558-8585, Japan
**E-mail: tsubota@sci.osaka-cu.ac.jp*

We discuss theoretically several important topics of atomic Bose-Einstein condensates. We address the dynamics of vortex lattice formation in a rotating condensate, and various types of vortex patterns in rotating two-component Bose condensates. Finally we discuss quantum turbulence in trapped condensates, followed by proposing how to make turbulence and showing that the energy spectra obey the well-known Kolmogorov law.

Keywords: Quantized vortex; Atomic Bose-Einstein condensate; Quantum turbulence.

1. Introduction

Quantized vortices are topological defects created through macroscopic quantum condensation. Since their discovery about a half-century ago, they have been studied extensively in superfluid helium, superconductors etc. The realization of atomic Bose-Einstein condensation in 1995 has opened a new powerful field of studying quantized vortices. The Bose-condensed system exhibits the macroscopic wave function $\Psi(\mathbf{r},t) = |\Psi(\mathbf{r},t)|e^{i\theta(\mathbf{r},t)}$ as an order parameter. The superfluid velocity field is given by $\mathbf{v}_s = (\hbar/m)\nabla\theta$ with boson mass m. Since the macroscopic wave function should be single-valued for the space coordinate $\mathbf{r}$, the circulation $\Gamma = \oint \mathbf{v} \cdot d\ell$ for an arbitrary closed loop in the fluid is quantized by the quantum $\kappa = h/m$. A vortex with such quantized circulation is called a quantized vortex. Any rotational motion of a superfluid is sustained only by quantized vortices.

Physics of quantized vortices has been studied thoroughly in superfluid helium[1]. Since the discovery of atomic Bose-Einstein condensates (BECs) in 1995[2], the system has become another important field for quantized vortices. The characteristics of trapped BECs are as follows: (i) a BEC system is weakly interacting and can be easily treated theoretically, (ii) many physical parameters of BECs are experimentally controllable, and (iii) various physical quantities such as the density and phase of BECs can be directly observed, which is in stark contrast to superfluid helium systems.

In this article we describe several important topics of quantized vortices in atomic BECs. The first topic is the dynamics of vortex lattice formation in a rotating condensate. The second topic is various types of vortex patterns appearing in rotating two-component Bose condensates. The last topic is quantum turbulence in trapped condensates.

2. Vortex lattice formation in a rotating BEC

After the realization of Bose-Einstein condensation in dilute atomic gases, one of the main interests was whether this system shows superfluidity or not. It is impossible to apply some flow to the system as was done experimentally in superfluid helium. However, superfluid systems make some characteristic response when it is subjected to rotation: the system makes a lattice of quantized vortices to mimic a solid body rotation with the same angular velocity as the rotation. This idea was applied to atomic Bose-Einstein condensates, and some groups have succeeded in making and observing vortex lattice formation [3,4].

Madison *et al.* directly observed nonlinear processes such as vortex nucleation and lattice formation in a rotating condensate [3]. By sudden rotational application on the trapping potential, the initially axisymmetric condensate undergoes a collective quadrupole oscillation with elliptically deformed condensates. This oscillation continues for a few hundred milliseconds with gradually decreasing amplitude. Then, the axial symmetry of the condensate is suddenly recovered and concurrently the vortices go into the condensate through its surface, eventually settling into a lattice configuration.

This observation has been well reproduced by a simulation of the Gross-Pitaevskii (GP) equation in two-dimensional [5,6] and three-dimensional [7] spaces. The corresponding GP equation in a frame rotating with a frequency $\mathbf{\Omega} = \Omega\hat{\mathbf{z}}$ is given by

$$(i-\gamma)\hbar\frac{\partial\Psi(\mathbf{r},t)}{\partial t} = \left[-\frac{\hbar^2\nabla^2}{2m} + V_{\rm ex} + g|\Psi(\mathbf{r},t)|^2 - \Omega L_z\right]\Psi(\mathbf{r},t). \quad (1)$$

Here $V_{\rm ex} = (1/2)m\omega^2 r^2$ is a trapping potential, and $L_z = -i\hbar(x\partial_y - y\partial_x)$. The interparticle potential $V_{\rm int}$ is approximated by a short-range interaction $V_{\rm int} \simeq g\delta(\mathbf{r}-\mathbf{r}')$, where $g = 4\pi\hbar^2 a/m$ is a coupling constant, characterized by the s-wave scattering length a. The term γ indicates phenomenological dissipation.

A typical numerical simulation of Eq. (1)[5,6] is shown in Fig.1. After a few hundred milliseconds, the boundary surface of the condensate becomes unstable and generates ripples that propagate along the surface, identified as invisible "ghost" vortices in the low-density surface region. The ripples develop into vortex cores, which enter the condensate and eventually form a vortex lattice. The numerical results agree quantitatively with the observations.

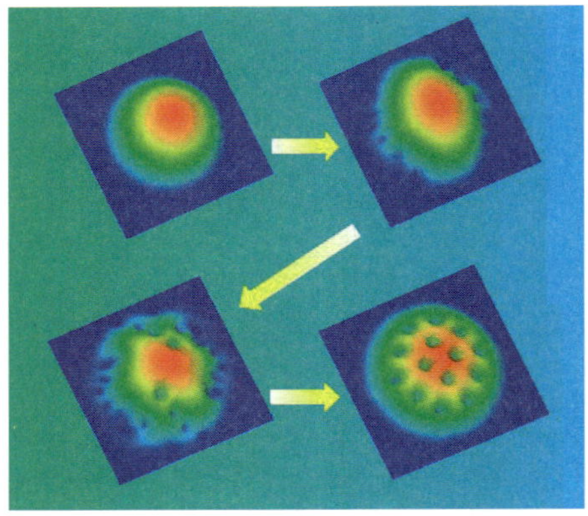

Fig. 1. Dynamics of vortex lattice formation in a rotating BEC (http://matter.sci.osaka-cu.ac.jp/ bsr/tsubotag/tsubotasim-e.html) .

3. Various vortex lattices in rotating two-component BECs

A system described by a multicomponent order parameter can excite various topological defects. One of the well-known systems is superfluid ^{3}He, where many kinds of vortices, solitons and texture have been theoretically predicted and observed experimentally[8]. Multicomponent BECs[9] are also a nice system for this issue.

This section describes briefly vortex lattices in rotating two-component BECs represented by the wave functions $\Psi_i (i = 1,2)$[10]. The theoretical analysis is performed for the normalized coupled GP equations

$$\mu_i\Psi_i = \left[-\frac{m_{12}}{m_i}\nabla^2 + V_i - \frac{\Omega}{\bar{\omega}}L_z + \sum_{j=1,2} C_{ij}|\Psi_j|^2\right]\Psi_i \quad (i=1,2) \quad (2)$$

Here the reduced mass m_{12} is given by $m_1m_2/(m_1+m_2)$ and the trapping potential V_i is given by $(m_i\omega_i^2/4m_{12}\bar{\omega}^2)r^2$. Length and energy are normalized by $b_{ho} = \sqrt{\hbar/2m_{12}\bar{\omega}}$ and $\hbar\bar{\omega} = \hbar(\omega_1 + \omega_2)$, respectively. The intracomponent- and intercomponent-coupling constants are C_{ii} and C_{ij} $(i \neq j)$. For simplicity, the number of parameters are reduced by assuming $C_{11} = C_{22} = C$, $m_1 = m_2$ and $\omega_1 = \omega_2 = \bar{\omega}$, so that the remaining parameters are $\delta = C_{12}/C$ and $\bar{\Omega} = \Omega/\bar{\omega}$.

Here we confine ourselves to the case in which all coupling constants are repulsive. Then the essence of the system is the following. When δ is smaller than unity, two condensates manage to overlap inside the trapping potential. If δ exceeds unity, however, the intercomponent interaction makes them phase-separated[11].

Under some realistic condition, the equilibrium states of the vortex structures are given in the $\delta - \bar{\Omega}$ phase diagram (Fig. 2). Depending on the parameters, various structures appear. The dependence on δ discussed above appears apparently. For $\delta = 0$, where two components are not interacting, thereby triangular vortex lattices are formed in each component. As δ increases, positions of the vortex cores in one component gradually shift from those of other component and triangular lattices are distorted. Eventually vortices in each component form a square lattice rather than a triangular one, as shown in Fig. 3. Such square lattice was actually observed [12] after this work. When δ exceeds unity, two condensates themselves become phase-separated, yielding various structures such as stripe or double-core vortex lattices and vortex sheets. The details are described in Ref. 10.

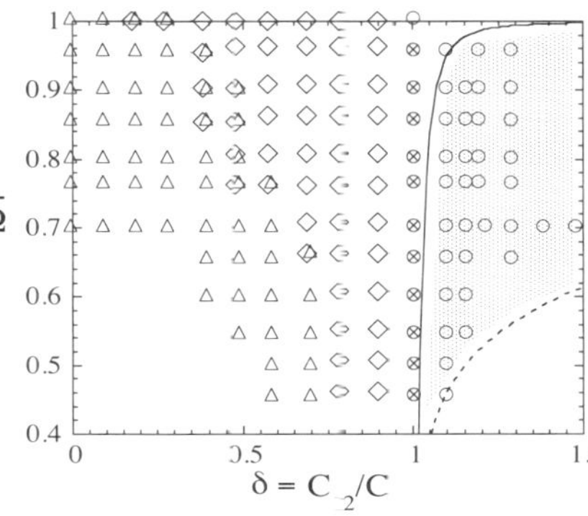

Fig. 2. Phase diagram of vortex structures in rotating two-component BECs. The symbol △ refers to triangular lattice; ◇, square lattice; ⊗, stripe or double-core lattice; ○, vortex sheet. [Kasamatsu, Tsubota and Ueda, *Phys. Lev. Lett.* **91**, 150406-2 (2003), reprodiced with permission. Copyright(2003) by the American Physical Scociety.]

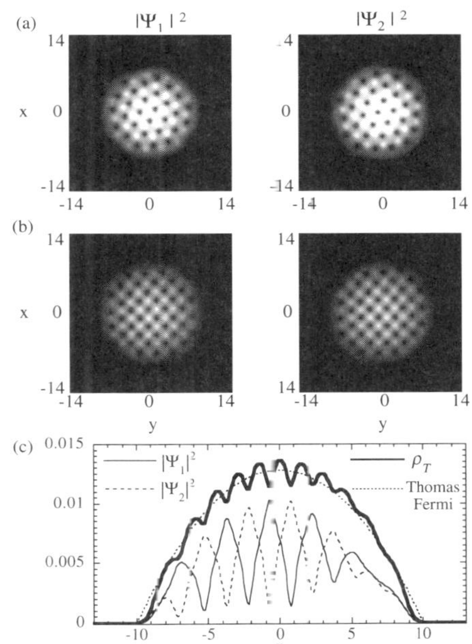

Fig. 3. Triangular lattice (a) for $\bar{\Omega} = 0.6$ and square lattice (b) for $\bar{\Omega} = 0.75$. Both cases have $\delta = 0.7$. (c) Cross section of (b) along the $y = 0$ line. Here $\rho_T = |\Psi_1|^2 + |\Psi_2|^2$. [Kasamatsu, Tsubota and Ueda, *Phys. Lev. Lett.* **91**, 150406-2 (2003), reprodiced with permission. Copyright(2003) by the American Physical Scociety.]

4. Quantum turbulence in atomic BECs

Quantum turbulence (QT) is turbulence of superfluid component, being comprised of quantized vortices. The QT was discovered in superfluid ^{4}He in the 1950s, and research has tended toward a new direction since the mid 90s[13]. The similarities and differences between quantum and classical turbulence have become an important area of research. Recent numerical studies show that QT obeys the Kolmogorov law which is one of the most important statistical laws in classical turbulence[14,15]. Since QT is comprised of quantized vortices that are definite topological defects, it is expected to yield a model of turbulence that is much simpler than the classical model.

The long history of research into superfluid helium tells us two main cooperative phenomena of quantized vortices: one is vortex lattice under rotation and the other is vortex tangle in QT. Both have been thoroughly studied in superfluid helium. However, almost all studies of quantized vortices in atomic BECs have been limited to vortex lattices. This section briefly discusses QT in atomic BECs.

An important point is how to make QT in trapped BECs: the method should be experimentally accessible. Kobayashi and Tsubota proposed an easy and powerful method to make a steady QT in trapped BEC, namely using precession[16]. The dynamics of the wave function is described by the GP equation with dissipation. First, we trap a BEC cloud in a weakly elliptical harmonic potential:

$$U(\mathbf{r}) = \frac{m\omega^2}{2}[(1-\delta_1)(1-\delta_2)x^2 + (1+\delta_1)(1-\delta_2)y^2 + (1+\delta_2)z^2], \quad (3)$$

where the parameters δ_1 and δ_2 exhibit elliptical deformation in the xy- and zx-planes. Second, to develop the BEC to a turbulent state, we apply a rotation along the z-axis followed by a rotation along the x-axis, as shown in Fig. 4 (a). The rotation vector $\Omega(t)$ is given by $\Omega(t) = (\Omega_x, \Omega_z \sin \Omega_x t, \Omega_z \cos \Omega_x t)$, where Ω_z and Ω_x are the frequencies of the first and second rotations, respectively. Note that this is precession. We consider the case where the spinning and precessing rotational axes are perpendicular to each other. Hence the two rotations do not commute, and thus cannot be represented by their sum. Actually this sort of rotation is used for turbulence in water [17].

Starting from a stationary solution without rotation and elliptical deformation, Kobayashi *et al.* numerically calculated the time development of the GP equation by turning on the rotation $\Omega_x = \Omega_z = 0.6$ and elliptical deformation $\delta_1 = \delta_2 = 0.025$. By monitoring the kinetic energy and some anisotropic parameters, the system is found to become almost steady and isotropic eventually. For the steady state the spectrum $E^i_{kin}(k,t)$ of the incompressible kinetic energy per unit mass was calculated to be consistent with the Kolmogorov law (Fig. 4 (b)). The inertial range which sustains the Kolmogorov law is determined by the Thomas-Fermi radius $R_{\rm TF}$ and the coherence length ξ. The application of combined rotation around three axes enables us to obtain more isotropic QT[18].

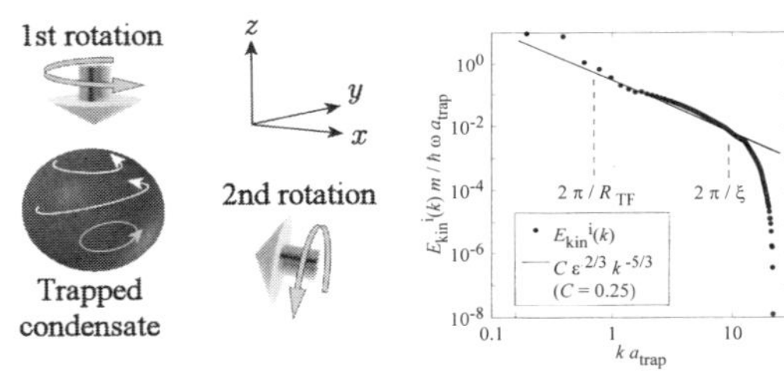

(a) (b)

Fig. 4. QT in atomic BECs. (a)Rotating the condensate around two axes. (b)Energy spectrum of a steady QT. The dots refer to the numerically obtained spectrum, while the solid line is the Kolmogorov spectrum. Here, $a_{\rm trap} = \sqrt{\hbar/m\omega}$ is the characteristic scale of the trap.

There are several advantages of studying QT in atomic BECs over superfluid helium. First, we can observe the vortex configuration, probably even the Richardson cascade process of vortices. The research along this direction is quite important, because we can study directly the relation between the real-space Richardson cascade and the wavenumber-space cascade (the Kolmogorov spectrum). Second, we can control the transition to turbulence by changing the rotation frequencies or other parameters. For example, we know that rotation along one axis forms a vortex lattice. When we apply another rotation, it may just rotate the lattice if the frequency is low. If the frequency is high enough, however, the second rotation should melt the lattice towards a vortex tangle. It would be possible to investigate in detail the transition to turbulence. Third, by changing the shape of the trapping potential, we can study the effect of the anisotropy on turbulence; a typical question is how the Kolmogorov spectrum is changed when the BEC becomes anisotropic. This interest should lead to the transition between 2D and 3D turbulence. Fourth, we can create certain interesting QTs in multicomponent BECs by controlling the interaction parameters.

5. Conclusions

This article described theoretically and numerically several interesting problems on quantized vortices in atomic BECs. First, we revealed the dynamics of vortex lattice formation in a rotating BEC; the results agree quantitatively with the typical experimental results. Second, we investigated rotating two-component BECs and found various vortex structures: especially, square lattices predicted by us were later observed experimentally. Third, we discussed quantum turbulence in trapped atomic BECs. We showed that combining two rotations enabled us to make turbulence. The energy spectra were found to obey the Kolmogorov law that is the most important statistical law in turbulence. This method should be confirmed experimentally. However, there are still many unresolved and interesting problems. Research of topological defects in cold atoms will make important contributions into many fields such as superfluid ^{4}He, ^{3}He, superconductors, quantum Hall systems, nonlinear optics, nuclear physics, and cosmology[19].

Acknowledgments

The author would like to thank M. Kobayashi, K. Kismets and M. Ueda for the collaboration on the works described here.

References

1. R. J. Donnelly, *Quantized Vortices in Helium II* (Cambridge University Press, Cambridge, 1991).
2. C. J. Pethick and H. Smith, *Bose-Einstein Condensation in Dilute Gases* (Cambridge University Press, Cambridge, 2002).
3. K. W. Madison, F. Chevy, W. Wohlleben and J. Dalibard, *Phys. Rev. Lett.* **84**, 806 (2000).
4. J. R. Abo-Shaeer, C. Raman, J. M. Vogels and W. Ketterle, *Science* **292**, 476 (2001).
5. M. Tsubota, K. Kasamatsu and M. Ueda, *Phys. Rev.* **A65**, 023603 (2002).
6. K. Kasamatsu, M. Tsubota and M. Ueda, *Phys. Rev.* **A67**, 033610 (2003).
7. K. Kasamatsu, M. Machida, N. Sasa and M. Tsubota, *Phys. Rev.* **A71**, 063616 (2005).
8. G. E. Volovik, in *The Universe in a Helium Droplet* (Clarendon Press, Oxford, 2003).
9. K. Kasamatsu, M. Tsubota and M. Ueda, *Int. J. Mod. Phys.* **B19**, 1835 (2005).
10. K. Kasamatsu, M. Tsubota and M. Ueda, *Phys. Rev. Lett.* **91**, 150406 (2003).
11. P. Ao and S. T. Chui, *Phys. Rev.* **A58**, 4836 (1998).
12. V. Schweikhard *et al.*, *Phys. Rev. Lett.* **93**, 210403 (2004).
13. M. Tsubota, *J. Phys. Soc. Jpn.* **77**, 11111 (2008).
14. M. Kobayashi and M. Tsubota, *Phys. Rev. Lett.* **94**, 065302 (2005).
15. M. Kobayashi and M. Tsubota, *J. Phys. Soc. Jpn.* **74**, 3248 (2005).
16. M. Kobayashi and M. Tsubota, *Phys. Rev.* **A76**, 045603 (2007).
17. S. Goto, N. Ishii, S. Kida and M. Nishioka, *Phys. Fluids* **19**, 061705 (2007).
18. M. Kobayashi and M. Tsubota, *J. Low Temp. Phys.* **150**, 587 (2008).
19. Y. M. Bunkov and H. Godfrin eds., *Topological Defects and the Non-Equilibrium Dynamics of Symmetry Breaking Phase Transitions* (Kluwer Academic Publishers, Les Houches, 1999).

QUANTUM DEGENERATE GASES OF YTTERBIUM ATOMS

S. UETAKE*
CREST, Japan Science and Technology Agency, Kawaguchi, Saitama 332-0012, Japan
**E-mail: uetake@scphys.kyoto-u.ac.jp*

A. YAMAGUCHI, S. KATO, T. FUKUHARA, S. SUGAWA, K. ENOMOTO, Y. TAKASU
Department of Physics, Graduate School of Science, Kyoto University, Kyoto 606-8502, Japan

Y. TAKAHASHI
Department of Physics, Graduate School of Science, Kyoto University, Kyoto 606-8502, Japan
CREST, Japan Science and Technology Agency, Kawaguchi, Saitama 332-0012, Japan

We have performed high-resolution spectroscopy of quantum degenerate gases of bosonic and fermionic ytterbium atoms using ultra-narrow intercombination transitions to probe quantum properties of the gases. The mean field interaction of the Bose-Einstein condensation and the energy distribution characteristic of the Fermi degeneracy were observed in the spectra.

Keywords: Bose-Einstein condensation; Fermi degeneracy; ytterbium

1. Introduction

The first observation of Bose-Einstein condensation (BEC) in a weakly interacting atomic gas in 1995 [1] and a Fermi degenerate gas (FDG) in 1999 [2] opened up a new frontier in physics. Many important studies have been performed using alkali atoms. Extending atomic species beyond alkali atoms is an important step for a future investigation.

In this respect, atoms of a two-electron system such as an alkaline-earth and ytterbium (Yb) atom, a rare earth element with the electronic structure similar to the alkaline-earth atoms, offer many possibilities for fundamental research and applications. The ground state $^1S_0$ and the metastable states $^3P_0$ and $^3P_2$ are connected by extremely narrow intercombination transitions, which has received considerable attention as a frequency and time standard with unprecedented precision [3]. The simple structure of the spin-zero ground state $^1S_0$ with no nuclear spins results in spinless BEC, which is considered as a dilute gas analogue of the well-known superfluid ^{4}He. The spinless BEC is insensitive to an external magnetic field, which is invaluable in precision coherent atom interferometry. In addition to these distinct features connected to the two-electron systems, Yb has another great advantage due to the existence of rich varieties of stable isotopes of five bosons and two fermions, which allows us to study various interesting quantum degenerate gases of Yb atoms.

It should be noted, however, that the rich variety of isotopes does not always mean the rich variety of inter-atomic interaction. The most important parameter to describe a quantum degenerate gas is a s-wave scattering length that characterizes an interatomic interaction at ultracold temperatures. Previously information on the inter-atomic interaction of Yb atoms was not known theoretically and experimentally, which introduces great difficulty in working with Yb atoms. However, our recent experiments together with theoretical works have resulted in an accurate determination of scattering lengths of all pairs of Yb isotopes [4]. As

Table 1. Calculated s-wave scattering lengths in nm.

	^{168}Yb	^{170}Yb	^{171}Yb	^{172}Yb	^{173}Yb	^{174}Yb	^{176}Yb
^{168}Yb	13.33	6.19	4.72	3.44	2.04	0.13	-19.0
^{170}Yb		3.38	19.3	-0.11	-4.30	-27.4	11.08
^{171}Yb			-0.15	-4.46	-30.6	22.7	7.49
^{172}Yb				-31.7	22.1	10.61	5.62
^{173}Yb					10.55	7.34	4.22
^{174}Yb						5.55	2.88
^{176}Yb							-1.28

one can readily find out from Table 1 that Yb atoms in fact offer a rich variety of interatomic interactions, and now this has turned out to be a great advantage of working with Yb atoms. In this paper, we will describe in detail creation of various quantum degenerate gases of Yb atoms and recent experiments on high-resolution laser spectroscopy of the Yb quantum gases. We also discuss prospects for several interesting applications.

2. Cooling to Quantum Degeneracy

Up to now, we successfully obtained two Yb BECs (^{174}Yb and ^{170}Yb), and one FDG (^{173}Yb). Since there is no electron spins in the ground state of Yb atoms, we cannot use a magnetic trapping method, and thus we created quantum degenerate gases of Yb atoms by an optical method of evaporative cooling in an optical trap.

A BEC transition of ^{174}Yb was confirmed by observation of an anisotropic expansion and a bimodal distribution [5]. From the time evolution of the horizontal expansion widths, we found the scattering length of ^{174}Yb to be 6^{+10}_{-5} nm. This value is consistent with the value of 5.55 nm obtained from the photoassociation experiment [4,6].

A BEC transition of ^{170}Yb was observed with a similar procedure and setup for the ^{174}Yb BEC transition [7]. The number of atoms was about 10^4. The scattering length was measured to be 3.6 ± 0.9 nm from the expansion of ^{170}Yb BEC after release from the trap. The measured value of the scattering length was again consistent with the value of 3.38 nm obtained in the photoassociation experiment [4].

We also performed evaporative cooling with unpolarized ^{173}Yb in which the populations are equally distributed in six spin states [8]. After evaporative cooling, we obtained ultracold 1×10^4 ^{173}Yb atoms at 75 nK, which corresponds to 0.37 T/T_F, where T_F is the Fermi temperature. The velocity distribution was well fitted by Fermi-Dirac distribution. We also observed a decrease of evaporative cooling efficiency at small T/T_F because of a Fermi pressure and Pauli blocking. The rather rapid thermalization rate of 4 ms was measured, which is consistent with a large scattering length of 10.55 nm obtained in the photoassociation experiment [4].

In addition, it is also possible to trap mixture of Yb isotopes with almost the same procedure as that for a single isotope. Contrary to the already realized alkali atom mixtures, Yb atom mixture is ideal in that the isotope mass ratio is almost unity, which simplifies experimental situations. To create a quantum degenerate gas mixture, we performed sympathetic evaporative cooling. For the pairs of ^{171}Yb - ^{174}Yb, ^{173}Yb - ^{174}Yb, and ^{174}Yb - ^{176}Yb, the sympathetic evaporative cooling worked well. The bosonic ^{176}Yb atoms were cooled down to the BEC transition temperature in the presence of the ^{174}Yb BEC[9]. This BEC-BEC mixture is interesting in that one of the BEC is attractive and the other is repulsive. Fermionic ^{171}Yb atoms were cooled down below a Fermi temperature[9], but in the final stage of the sympathetic evaporative cooling, no ^{174}Yb atoms remained in the trap. By using ^{173}Yb - ^{174}Yb pairs, a BEC-Fermi degenerate gas mixture was successfully formed. The temperature of ^{173}Yb was slightly below the Fermi temperature when bosonic ^{174}Yb atoms reached the BEC transition[9].

3. High-Resolution Laser Spectroscopy of Quantum Degenerate Yb Gases

Quite recently we have demonstrated high-resolution laser spectroscopy of a ^{174}Yb BEC by using the ultranarrow $^1S_0$ - $^3P_2$ optical transition [10]. The lineshape of BEC is written as

$$I(\nu) = -\frac{15h(\nu - \nu_0 - \nu_{\rm rec})}{4n_0\Delta U} \times \sqrt{1 - \frac{h(\nu - \nu_0 - \nu_{\rm rec})}{n_0 \Delta U}}, \quad (1)$$

where h is Planck's constant, ν_0 is the resonant frequency of free atoms, $\nu_{\rm rec}$ is the recoil frequency, and n_0 is the peak density in the BEC. ΔU is given by $h^2(a_{12} - a_{11})/\pi m$, where a_{12} is the scattering length between ground state and excited state atoms, a_{11} is the scattering length between ground state atoms, and m is the mass. The large mean field shift of the $^1S_0$ - $^3P_2$ transition enabled us to successfully observe not only the frequency shift and broadening but also the distortion of the lineshape due to the mean field energy of a BEC in a harmonic trap, which has never been observed

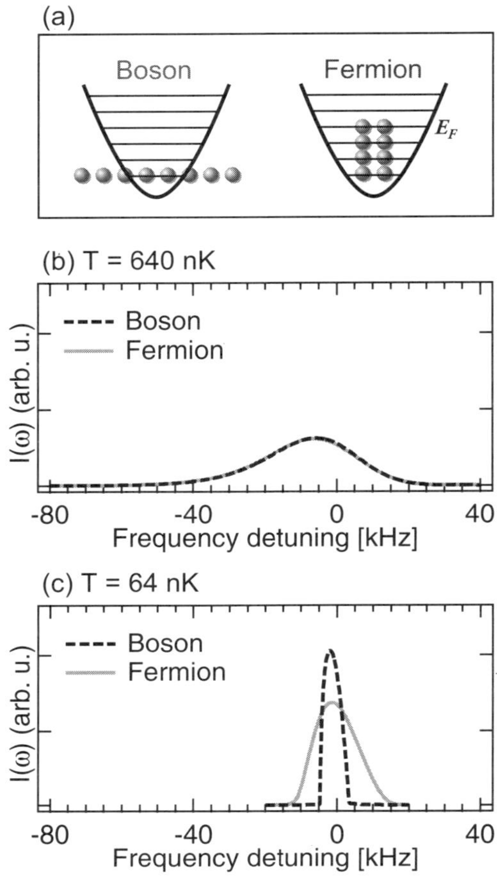

Fig. 1. (a) Difference of energy distribution between boson and fermion at zero temperature. (b) Calculated lineshape at $T = 640$ nK, above the critical temperature of BEC and the Fermi temperature. At this high temperature, both spectra are almost the same, and thus the lines for boson and fermion completely overlap. (c) Calculated lineshape at $T = 64$ nK. At this very low temperature, the difference between the spectra of BEC and Fermi degenerate gas is significant.

in the radio-frequency spectroscopy of alkali atoms[11]. To highlight the role of the interatomic interaction, we have also performed high-resolution laser spectroscopy of a BEC in a one-dimensional optical lattice and successfully detected the on-site interaction of atoms in an unambiguous way.

In addition, also quite recently, we have successfully performed high-resolution laser spectroscopy of a ^{174}Yb BEC and a ^{173}Yb

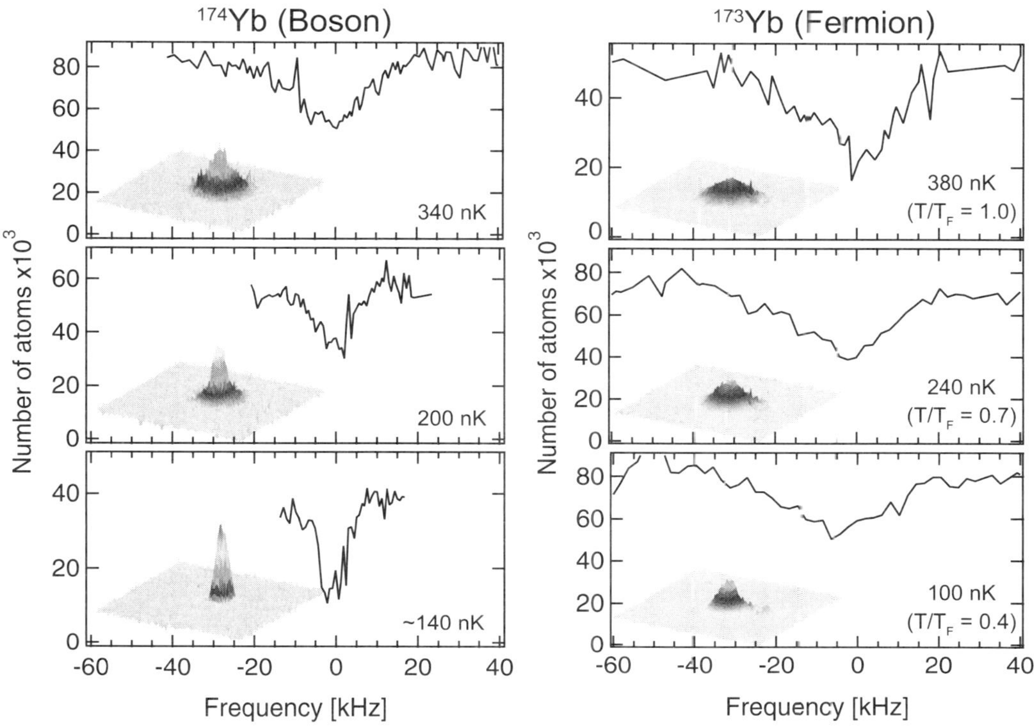

Fig. 2. Results of high-resolution laser spectroscopy of ^{174}Yb (boson) and ^{173}Yb (fermion) by using the ultranarrow $^1S_0$ - $^3P_0$ transition. Inset of each figure shows the absorption image corresponding to each spectrum.

Fermi degenerate gas by using the ultranarrow $^1S_0$ - $^3P_0$ optical transition[12]. Figure 1 shows numerical calculations of lineshape for bosons and fermions. In this calculation, the BEC transition temperature (T_c) is 330 nK and the Fermi temperature (T_F) is 640 nK. For thermal atoms (at $T \sim 2T_c$, T_F) both spectra are almost the same. However, when atoms are further cooled, the shape difference of the spectra becomes significant. While the spectrum for fermions is still broad, the spectrum for bosons has narrowed down and has a distorted parabolic lineshape. This shows that fermions have a high kinetic energy even at low temperatures. On the other hand, the narrow but finite linewidth and distortion in the boson spectrum reflect the atom-atom interaction due to its high density. Thus the difference between the spectra of the BEC and Fermi degenerate gas is attributed to the difference of quantum statistics, which is schematically shown in Fig. 1(a) for T=0.

Figure 2 shows the experimental results of the high-resolution spectroscopy of ^{174}Yb BEC and ^{173}Yb Fermi degenerate gas. The distinct difference between the spectra of the BEC and Fermi degenerate gas was clearly observed. At temperature above 300 nK, two spectra are approximately the same. The spectra of Bose gas change significantly at lower temperatures, whereas the spectra of Fermi gas do not change much. At very low temperatures ($T \sim 140$ nK for Bose gas, condensate fraction is ~ 0.7 and $T \sim 100$ nK for Fermi gas), the difference of spectra between the two gases is apparent. As expected, the lineshape of BEC becomes a deformed parabolic shape.

4. Possible Applications of Quantum Degenerate Yb Gases

In this section, we discuss in detail various future possibilities of Yb quantum degenerate gases. A magnetically induced Feshbach resonance for tuning the scattering length is a key technique for an investigation of quantum degenerate gases of alkali atoms. The Feshbach resonance occurs when a scattering state in a potential (open channel) is resonantly coupled to a bound state in another potential (closed channel). This resonance modifies the scattering wavefunction. While a rich variety of Yb isotopes offers a rich variety of inter-atomic interaction as is shown in Table 1, it is not still possible to tune the scattering length continuously. Because of the lack of a hyperfine structure in the ground state of Yb atoms, the already established method of a magnetically induced Feshbach resonance does not work for Yb atoms. However, essentially the same effect occurs when a laser field couples a scattering state and a molecular bound state, which is often called an optical Feshbach resonance. Recently it has been pointed out [15] that the optical Feshbach resonance effect is quite efficient in the case of a weak intercombination transition which exists in a Yb atom. The narrow linewidth of the intercombination transition allows us to use an excited molecular level close to the dissociation limit, which provides a strong optical coupling because of a large Franck-Condon overlap, while keeping the atom loss small due to the photon scattering relevant to the atomic resonance. Quite recently Enomoto *et al.* have successfully demonstrated a large effect of the optical Feshbach resonance [16] using the $^1S_0-^3P_1$ transition for ^{172}Yb and ^{176}Yb isotopes. The observed scattering phase shift corresponds to the change of the scattering length by about 30 nm in the zero energy limit.

Ultracold polar molecules will offer fascinating possibilities such as a new quantum phase of the quantum degenerate matters and a scalable quantum computation. An especially interesting polar molecule is composed of one alkali atom and one two-electron atom since this molecule has an electron spin degree of freedom in the ground molecular state, contrary to the spin singlet molecular ground state for alkali dimers. This is an important requirement for demonstration of a quantum simulation of quantum spin physics using polar molecules in optical lattices. Okano *et al.* have already proposed to create a YbLi molecule as a promising candidate of such a system [13], and the experiment towards this goal is now underway. The large ratio of the mass of a Yb atom to that of a Li atom is expected to cause suppression of inelastic collisions [14]. This novel behavior is also interesting to explore.

5. Conclusion

We have performed high-resolution spectroscopy of quantum degenerate gases of bosonic and fermionic ytterbium atoms using ultra-narrow intercombination transitions to probe quantum properties of the gases. We successfully observed the mean field interaction of the Bose-Einstein condensation and the energy distribution of the Fermi degeneracy in the spectra. The difference between the spectra of the BEC and Fermi degenerate gas is attributed to the difference of quantum statistics. We have also discussed the future possibilities of quantum degenerate gases of Yb atoms.

Acknowledgments

We acknowledge experimental assistance by M. Sugimoto, H. Kakiuchi, T. Tsujimoto, H. Wayama, S. Tojo, A. Wasan, M. Kitagawa, and K. Kasa of Kyoto University. The studies were partially supported by Grant-in-Aid for Scientific Research of Japan Society of the Promotion of Science(JSPS) (Grant Nos. 18043013, 18204035) and Global COE Program "The Next Generation of Physics, Spun from Universality and Emergence" from Ministry of Education, Culture, Sports, Science and Technology of Japan. S. Sugawa and S. Kato acknowledge support from JSPS.

References

1. M. H. Anderson, J. R. Ensher, M. R. Matthews, C. E. Wieman, and E. A. Cornell, *Science* **269**, 198(1995).
2. B. DeMarco and D. S. Jin, *Science* **285** 1703(1999) .
3. M. Takamoto, F. Hong, R. Higashi, and H. Katori, *Nature* **435**, 321(2005) .
4. M. Kitagawa, K. Enomoto, K. Kasa, Y. Takahashi, R. Ciuryło, P. Naidon, and P. S. Julienne, *Phys. Rev. A***77**, 012719(2008).
5. Y. Takasu, K. Maki, K. Komori, T. Takano, K. Honda, M. Kumakura, T. Yabuzaki, and Y. Takahashi, *Phys. Rev. Lett.* **91**, 040404(2003).
6. K. Enomoto, M. Kitagawa, K. Kasa, S. Tojo, and Y. Takahashi, *Phys. Rev. Lett.* **98**, 203201(2007).
7. T. Fukuhara, S. Sugawa, and Y. Takahashi, *Phys. Rev. A* **89**, 051604(R) (2007) .
8. T. Fukuhara, Y. Takasu, M. Kumakura, and Y. Takahashi, *Phys. Rev. Lett.* **98**, 030401(2007).
9. T. Fukuhara, S. Sugawa, Y. Takasu, and Y. Takahashi, *Proc. of the Int. School of Physics "Enrico Fermi"*, ed. M. Inguscio, W. Ketterle, and C. Salomon (IOS Press, Amsterdam 2008), p. 857.
10. A. Yamaguchi, S. Uetake, S. Kato, H. Ito, and Y. Takahashi, presented at ICAP 2008, Int. Conference of Atom Physics 2008.
11. A. Görlitz, T. L. Gustavson, A. E. Leanhardt, R. Löw, A. P. Chikkatur, S. Gupta, S. Inouye, D. E. Pritchard, and W. Ketterle, *Phys. Rev. Lett.* **90** 090401(2003).
12. S. Uetake *et al.*, in preparation.
13. M. Okano, H. Hara, Y. Takasu, and Y. Takahashi, presented at the 21st Century COE Symp. *Diversity and Universality in Physics* at Kyoto University, Japan (2007).
14. B. Marcelis, S. J. J. M. F. Kokkelmans, G. V. Shlyapnikov, and D. S. Petrov, *Phys. Rev. A* **77**, 032707(2008).
15. R. Ciurło, E. Tiesinga, and P. S. Julienne, *Phys. Rev. A.* **71**, 030701(2005).
16. K. Enomono, K. Kasa, M. Kitagawa, and Y. Takahashi *Phys. Rev. Lett.* **101**, 203201(2008).

SUPERFLUID PROPERTIES OF AN ULTRACOLD FERMI GAS IN THE BCS-BEC CROSSOVER REGION

YOJI OHASHI[1,2,*] AND NAOKI FUKUSHIMA[3]

[1] *Department of Physics, Keio University, 3-14-1, Hiyoshi, Yokohama, 223-8522, Japan*
[2] *CREST(JST), 4-1-8 Honcho, Saitama, 332-0012, Japan*
[3] *CANON INC, 4202, Fukara, Susono-shi, Shizuoka, 410-1196, Japan*
[*] *E-mail: yohashi@rk.phys.keio.ac.jp*

We investigate superfluid properties of a two-component Fermi gas in the BCS-BEC crossover region. Including strong-coupling effects within a Gaussian fluctuation approximation, we self-consistently determine the superfluid order parameter and chemical potential at finite temperatures. Using these self-consistent solutions, we calculate single-particle excitations, as well as collective mode, over the entire BCS-BEC crossover region. As one increases the strength of a pairing interaction, we show that, while the single-particle excitation gap becomes large, the velocity of collective Goldstone mode becomes small. As a result, the origin of temperature dependence of a physical quantity continuously changes from single-particle excitations to collective excitations, as one passes through the BCS-BEC crossover region. To see this, we examine the superfluid density.

Keywords: BCS-BEC crossover; superfluid Fermi gas; superfluid density.

1. Introduction

Since the realization of superfluid state in ^{40}K and ^{6}Li Fermi gases[1,2,3,4], the BCS-BEC crossover phenomenon[5,6,7] has become one of the most exciting topics in cold atom physics[8,9]. In this phenomenon, the character of fermion superfluidity continuously changes from the weak-coupling BCS type to the Bose-Einstein condensation (BEC) of tightly bound molecules, as one increases the strength of a pairing interaction between Fermi atoms. In ^{40}K and ^{6}Li Fermi gases, a tunable interaction can be realized by using a Feshbach resonance[10,11]. In a sense, the BCS-BEC crossover enables us to study Fermi superfluids and Bose superfluids in a unified manner, so that this phenomenon is useful for the study of superfluidity on general viewpoint. We also note that, in ^{40}K and ^{6}Li Fermi gases, the superfluid phase transition temperature reaches $T_c \sim 0.2T_F$ in the strong-coupling BEC regime[1,2,3,4] (where T_F is the Fermi temperature). Thus, the superfluid Fermi gases are also expected to be useful for the study of high-T_c superconductivity.

In this paper, we investigate superfluid properties of a cold Fermi gas. Extending the strong-coupling (Gaussian fluctuation) theory for T_c developed by Nozières and Schmitt-Rink (NSR)[6] to the superfluid phase below T_c[12], we calculate the superfluid order parameter Δ and chemical potential μ over the entire BCS-BEC crossover region. Using these self-consistent solutions, we consider two fundamental excitations in the superfluid phase, namely, single-particle excitations and collective Goldstone mode. We show that, while quasi-particle excitations are crucial for temperature dependence of physical quantities in the weak-coupling BCS regime, collective excitations become dominant in the strong-coupling BEC regime. We also show that this continuous change of important excitations can be seen in the behavior of superfluid density ρ_s in the BCS-BEC crossover.

In the following discussions, we take $\hbar = k_B = 1$. We also set the system volume $V =$

1, so that the number of particle N and the particle density ρ are the same.

2. BCS-BEC crossover below T_c

We consider a uniform Fermi gas with two atomic hyperfine states described by pseudo-spin $\sigma =\uparrow,\downarrow$. So far, all the current experiments on superfluid Fermi gases are using a broad Feshbach resonance[1,2,3,4]. In this case, we can safely study this system by using the BCS model described by the Hamiltonian

$$H = \sum_{\mathbf{p},\sigma} \xi_{\mathbf{p}} c^\dagger_{\mathbf{p}\sigma} c_{\mathbf{p}\sigma} - U \sum_{\mathbf{p},\mathbf{p}',\mathbf{q}} c^\dagger_{\mathbf{p}+\mathbf{q}/2\uparrow} c^\dagger_{-\mathbf{p}+\mathbf{q}/2\downarrow} \times c_{-\mathbf{p}'+\mathbf{q}/2\downarrow} c_{\mathbf{p}'+\mathbf{q}/2\uparrow}. \qquad (1)$$

Here, $c^\dagger_{\mathbf{p}\sigma}$ is the creation operator of a Fermi atom with the kinetic energy $\xi_{\mathbf{p}} \equiv \varepsilon_{\mathbf{p}} - \mu = p^2/2m - \mu$, measured from the chemical potential μ. U describes a tunable pairing interaction associated with a Feshbach resonance.

To study the superfluid properties in the BCS-BEC crossover region, we extend the strong-coupling theory for T_c developed by Nozières and Schmitt-Rink (NSR)[6] to the superfluid phase below T_c[12]. The NSR theory consists of the equation for T_c and the equation for the number of fermions. The T_c-equation is derived from the Thouless criterion, which states that the superfluid phase transition occurs when the particle-particle scattering matrix develops a pole at $\omega = q = 0$. Within the t-matrix approximation, this T_c-equation has the same form as the ordinary BCS gap equation at T_c. However, in contrast to the weak-coupling BCS theory (where one can safely set the chemical potential μ to be equal to the Fermi energy ε_F), NSR pointed out that μ deviates from ε_F as one approaches the strong-coupling BEC regime. This strong-coupling effect is taken into account by solving the equation for the number of fermions including pairing fluctuations beyond the mean-field approximation. The NSR theory includes the pairing fluctuations within a Gaussian fluctuation approximation[7].

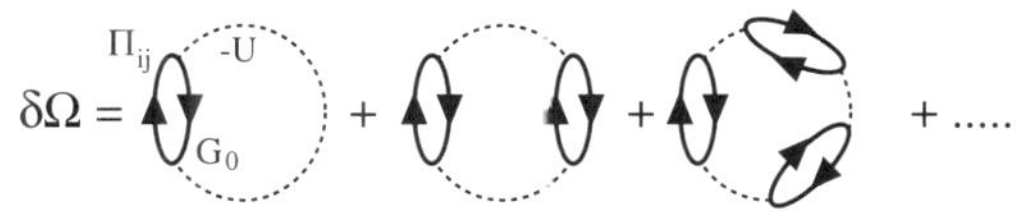

Fig. 1. Fluctuation contribution to the thermodynamic potential. Solid lines and dashed lines show the single-particle Green's function G_0 and pairing interaction $-U$, respectively.

When we extend the NSR theory to the superfluid phase below T_c[12], the T_c-equation is replaced by the BCS gap equation. As usual, since it involves the ultraviolet divergence originating from the contact type pairing interaction, we need to regularize the equation by introducing the two-body s-wave scattering length a_s. The regularized gap equation is given by[7]

$$1 = -\frac{4\pi a_s}{m} \sum_{\mathbf{p}} \Big[\frac{1}{2E_{\mathbf{p}}} \tanh \frac{E_{\mathbf{p}}}{2T} - \frac{1}{2\varepsilon_{\mathbf{p}}} \Big], \qquad (2)$$

where $E_{\mathbf{p}} = \sqrt{\xi^2_{\mathbf{p}} + \Delta^2}$ is the single-particle excitation spectrum. a_s is related to the interaction $-U$ as

$$\frac{4\pi a_s}{m} = -\frac{U}{1 - U\sum_{\mathbf{p}} \frac{1}{2\varepsilon_{\mathbf{p}}}}. \qquad (3)$$

We solve the gap equation (2), together with the equation for the number of Fermi atoms given by[12]

$$\rho = \rho^0_F - \frac{T}{2}\frac{\partial}{\partial\mu} \sum_{\mathbf{q},i\nu_n} \ln\det\Big[1 - \frac{4\pi a_s}{m}\big[\Xi(\mathbf{q}, i\nu_n) + \sum_{\mathbf{p}} \frac{1}{2\varepsilon_{\mathbf{p}}}\big]\Big]. \qquad (4)$$

Here, ρ is the number density of Fermi atoms. $\rho^0_F = \sum_{\mathbf{p}}[1 - (\xi_{\mathbf{p}}/E_{\mathbf{p}})\tanh(E_{\mathbf{p}}/2T)]$ is the number density of Fermi atoms in the mean-field approximation. The last term in Eq. (4) ($\equiv \delta\rho$) describes effects of pairing fluctuations, which is obtained from the identity $\delta\rho = -\partial\delta\Omega/\partial\mu$, where $\delta\Omega$ is the fluctuation contribution to the thermodynamic potential described by the diagrams shown in Fig. 1.

In Eq. (4), $\Xi(\mathbf{q}, i\nu_n)$ is given by[12,13]

$$\Xi = \frac{1}{4} \times \begin{pmatrix} \Pi_{11}+\Pi_{22}+2i\Pi_{12} & \Pi_{11}-\Pi_{22} \\ \Pi_{11}-\Pi_{22} & \Pi_{11}+\Pi_{22}-2i\Pi_{12} \end{pmatrix}, \tag{5}$$

$$\Pi_{ij} = T\sum_{\mathbf{p},i\omega_m} \mathrm{Tr}[\tau_i G_0(\mathbf{p}+\mathbf{q}/2, i\omega_m+i\nu_n) \times \tau_j G_0(\mathbf{p}, i\omega_m)], \tag{6}$$

where τ_j $(j=1,2,3)$ are Pauli matrices. ω_m and ν_n represent fermion and boson Matsubara frequency, respectively. $G_0^{-1}(\mathbf{p}, i\omega_m) = i\omega_m - \xi_{\mathbf{p}}\tau_3 + \Delta\tau_1$ is the single-particle Green's function in the Nambu representation[14]. In Eq. (6), Π_{11} and Π_{22}, respectively, describe amplitude and phase fluctuations of the order parameter. Π_{12} describes coupling between amplitude and phase fluctuations[13].

Figure 2 shows the self-consistent solutions of the coupled equations (2) and (4). In panel (a), as one increases the strength of the pairing interaction, while T_c approaches a constant value ($= 0.218T_F$[8]), the magnitude of Δ at $T = 0$ continues to increase. As a result, the ratio $2\Delta(T=0)/T_c$ becomes larger than the BCS universal constant ($= 3.54$) in the strong-coupling BEC regime.

In Fig.2(b), we find that μ remarkably deviates from ε_F to be negative in the BEC region. When $\mu < 0$, the energy gap E_g in the single-particle excitation spectrum $E_{\mathbf{p}} = \sqrt{(\varepsilon_{\mathbf{p}}-\mu)^2+\Delta^2}$ is not equal to Δ, but given by $\sqrt{\mu^2+\Delta^2} > \Delta$. In particular, since $|\mu| \gg \Delta$ in the BEC limit, one finds $E_g \simeq |\mu|$. In Fig. 3, we show E_g in the BCS-BEC crossover.

We briefly note that the first-order behavior can be seen in the BEC regime of Figs.2 and 3, which is, however, an artifact of the present crossover theory[12]. The same problem has been also known in the theory of Bose gas BEC. From the knowledge of the Bose gas theory, we find that, to recover the expected second-order phase transition in the BEC regime, we need a more sophisticated treatment of interaction between bound pairs beyond the present Gaussian fluctuation approximation[15]. Although this is a crucial problem, leaving it as a future problem, we will discuss collective excitations and superfluid density within the Gaussian fluctuation approximation in the following sections.

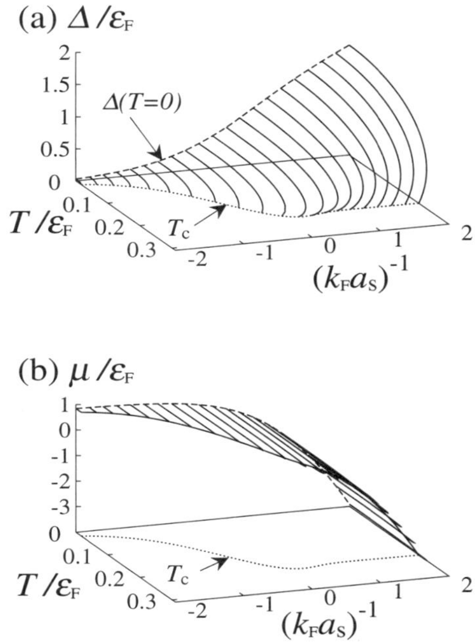

Fig. 2. Self-consistent solutions of the coupled equations (2) and (4) in the BCS-BEC crossover. The interaction is measured in terms of the inverse scattering length $(k_F a_s)^{-1}$, where k_F is the Fermi momentum. In this scale, the BCS region and BEC region are, respectively, given by $(k_F a_s)^{-1} \lesssim -1$ and $(k_F a_s)^{-1} \gtrsim 1$.

3. Collective excitations

Besides the single-particle excitations, superfluid Fermi gases also have collective excitations with linear dispersion $\omega = v_\phi q$ in the long wave length limit. This is the Goldstone mode associated with spontaneous breakdown of continuous $U(1)$-symmetry. The collective excitations are obtained from poles of

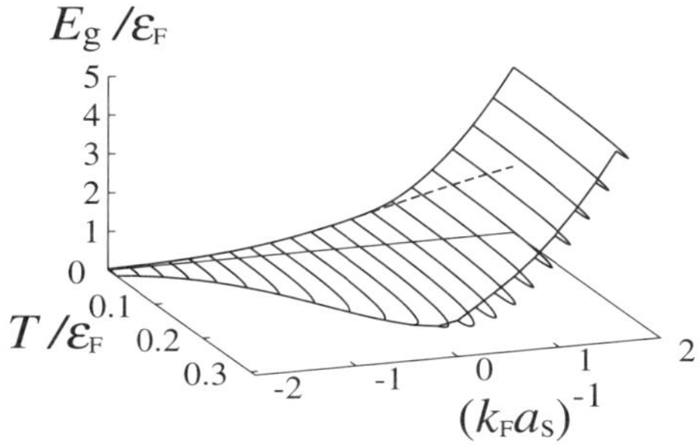

Fig. 3. Single-particle excitation gap E_g. The dashed line shows $\Delta(T=0)$.

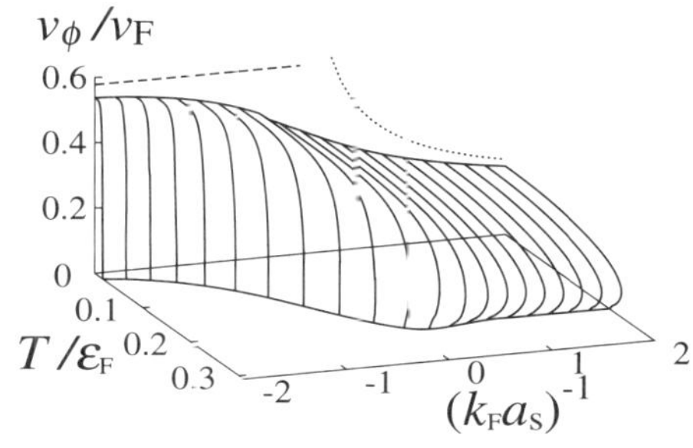

Fig. 4. Velocity v_ϕ of the collective Goldstone mode in the BCS-BEC crossover The dashed line and dotted line show the velocity $v_\phi = v_F/\sqrt{3}$ of the Anderson-Bogoliubov mode in the weak-coupling BCS theory and that of Bogoliubov phonon in a weakly-interacting molecular BEC given by Eq. (8), respectively.

phase-correlation function[13]. In the present approximation, they are determined by the equation[12],

$$0 = \det\Big[1 - \frac{4\pi a_s}{m}\big[\Xi(\mathbf{q}, \omega + i\delta) + \sum_{\mathbf{p}} \frac{1}{2\varepsilon_{\mathbf{p}}}\big]\Big]. \tag{7}$$

We briefly note that a similar expression to the right hand side of Eq. (7) can be seen in the last term of Eq. (4). In the weak-coupling BCS regime, Eq. (7) at $T = 0$ gives the well known Anderson-Bogoliubov mode with $v_\phi = v_F/\sqrt{3}$ in the long wave length limit (where v_F is the Fermi velocity). In the strong-coupling BEC limit, one obtains

$$v_\phi = \sqrt{\frac{U_B \rho_B}{M}} \quad (T = 0), \tag{8}$$

where $M = 2m$, $\rho_B = \rho/2$, and $U_B = 4\pi(2a_s)/M$. Equation (8) is just the velocity of Bogoliubov phonon in a superfluid molecular Bose gas where the number of molecules and molecular mass are given by ρ_B and M, respectively. U_B is interpreted as a *repulsive* molecular interaction characterized by the molecular scattering length $a_B \equiv 2a_s$.

Figure 4 shows the velocity v_ϕ of the Goldstone mode in the BCS-BEC crossover. As one increases the strength of the pairing interaction, $v_\phi(T = 0)$ continuously changes from the velocity of the Anderson-Bogoliubov mode ($v_\phi = v_F/\sqrt{3}$) to that of the Bogoliubov phonon in Eq. (8). Since the Goldstone mode only exists below T_c, v_ϕ vanishes at T_c.

In Fig.4, we find that the collective mode becomes slow in the BEC regime, reflecting that the system reduces to a non-interacting ideal molecular Bose gas. This tendency can be also seen in Eq. (8). Namely, since the *effective* molecular scattering length ($a_B = 2a_s$) vanishes in the BEC limit ($(k_F a_s)^{-1} \to \infty$), v_ϕ approaches zero in this limit.

4. Superfluid density

In Secs.2 and 3, we have shown that the single-particle excitation gap E_g becomes large while collective mode velocity v_ϕ becomes small as one approaches the BEC regime (see Figs.3 and 4). These results indicate that the relative importance of these two kinds of excitations continuously changes in the BCS-BEC crossover. As schematically shown in Fig.5, only single-particle excitations are important in the weak-coupling BCS regime, except for a small momentum region around $q = 0$, where $\omega = v_\phi q \lesssim 2\Delta$. On the other hand, low-energy excitations are dominated by collective mode in the

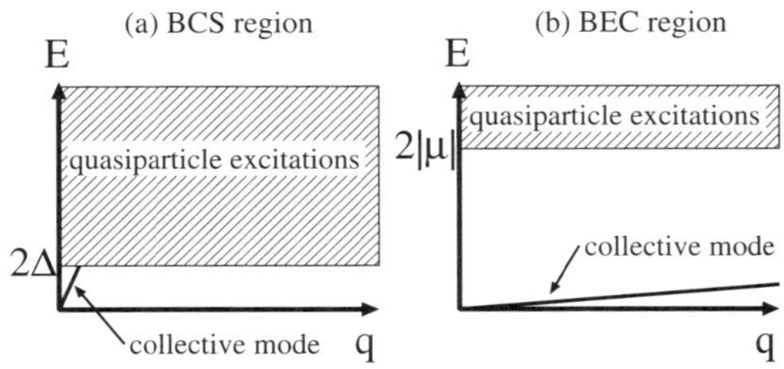

Fig. 5. Schematic picture of quasi-particle and collective excitations in the superfluid phase. In this figure, the threshold energy $2E_g$ of number-conserving quasi-particle excitations is shown.

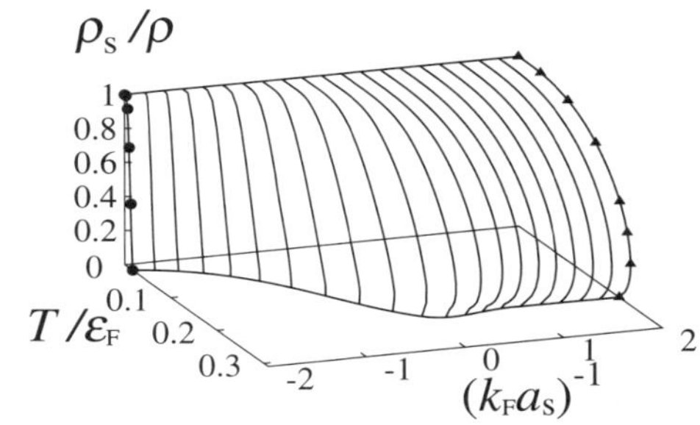

Fig. 6. Calculated superfluid density ρ_s in the BCS-BEC crossover. The solid circles show the weak-coupling result $\rho_s = \rho - \rho_n^F$. The solid triangles show $\rho_s = \rho - \rho_n^B$, where Eq.(12) is used.

strong-coupling BEC regime. This continuous change of dominant excitations is expected to directly affect temperature dependence of various physical quantities. To see this, in this section, we consider the superfluid density ρ_s as a typical example.

We consider the superfluid phase with finite superflow in the z-direction. This state is described by the current-carrying order parameter $\Delta(z) = \Delta e^{iQz}$. The superfluid density ρ_s is related to the supercurrent density $J_s = \sum_{\mathbf{p},\sigma}(p_z/m)\langle c^\dagger_{\mathbf{p}\sigma}c_{\mathbf{p}\sigma}\rangle$ as $J_s = \rho_s v_s$ (where $v_s = Q/2m$) when Q is small. Using this, we find that the normal fluid density $\rho_n \equiv \rho - \rho_s$ is given by

$$\rho_n = -2T\sum_{\mathbf{p},i\omega_m} p_z\Big[\frac{\partial}{\partial Q}\mathrm{Tr}[G_s(\mathbf{p},i\omega_m)]\Big]_{Q\to 0}. \tag{9}$$

Here, G_s is the single-particle Green's function in the supercurrent state under the Nambu representation. The Green's function G_s which is consistent with the present strong-coupling theory is obtained so that the equation $\rho = \sum_{\mathbf{p}} 1 + T\sum_{\mathbf{p},i\omega_m}\mathrm{Tr}[\tau_3 G_s(\mathbf{p},i\omega_m)]$ can reproduce the number equation (4) when $Q = 0$. The resulting G_s has the form $G_s = G_{s0} + G_{s0}\Sigma_s G_{s0}$, where $G_{s0}^{-1} = i\omega_m - v_s p_z - \xi_{\mathbf{p}}\tau_3 + \Delta\tau_1$. The self-energy Σ_s involves effects of pairing fluctuations. (For the expression for Σ_s, we refer to Ref.[12].) Substituting the detailed expression for G_s into Eq. (9), one finds that the normal fluid density can be written as $\rho_n = \rho_n^F + \rho_n^B$, where

$$\rho_n^F = -\frac{2}{3m}\sum_{\mathbf{p}} p^2\frac{\partial f(E_{\mathbf{p}})}{\partial E_{\mathbf{p}}}, \tag{10}$$

$$\rho_n^B = -2Tm\Big(\frac{\partial^2}{\partial Q^2}\ln\det\Big[1 - \frac{4\pi a_s}{m}[\Xi + \sum_{\mathbf{p}}\frac{1}{2\varepsilon_{\mathbf{p}}}]\Big]\Big)_{Q\to 0}. \tag{11}$$

Equation (10) is just the well-known expression for the normal fluid density in the weak-coupling BCS theory, where $f(E_{\mathbf{p}})$ is the Fermi distribution function. Equation (11) describes the fluctuation correction to ρ_n, where G_0 in Ξ is replaced by G_{s0}. (See Eqs. (5) and (6).) In the strong-coupling BEC limit, ρ_n^B reduces to the normal fluid density in a molecular Bose superfluid characterized by the collective Bogoliubov excitation spectrum $E^B_{\mathbf{q}} \equiv \sqrt{\varepsilon^B_{\mathbf{q}}(\varepsilon^B_{\mathbf{q}} + 2U_B\rho_B)}$ (where $\varepsilon^B_{\mathbf{q}} = q^2/2M$ and $M = 2m$) as[12],

$$\rho_n^B = -\frac{2}{3M}\sum_{\mathbf{q}} q^2\frac{\partial n_B(E^B_{\mathbf{q}})}{\partial E^B_{\mathbf{q}}}. \tag{12}$$

In the weak-coupling BCS regime, since pairing fluctuations are weak, $\rho_n \simeq \rho_n^F$ is realized. In this case, Eq. (11) indicates that the decay of the superfluid density $\rho_s \simeq \rho - \rho_n^F$ originates from single-particle excitations. As one increases the strength of the pairing interaction, the contribution from ρ_n^F becomes weak due to the increase of the magnitude of single-particle excitation gap E_g. Instead, the fluctuation part ρ_n^B becomes important in the BEC regime ($\rho_n \simeq \rho_n^B$). In the BEC limit, we find from Eq. (12) that ρ_n comes from collective Bogoliubov modes excited thermally.

Figure 6 shows the superfluid density in the BCS-BEC crossover. The superfluid density dominated by single-particle excitations in the BCS regime are found to continuously change into that dominated by collective excitations, as one passes through the BCS-BEC crossover.

5. Summary

In this paper, we have investigated superfluid properties of a two component Fermi gas in the BCS-BEC crossover region. We have self-consistently determined Δ and μ based on the extended NSR theory. We have also calculated the single particle excitation gap E_g, as well as the velocity v_ϕ of the collective Goldstone mode, over the entire BCS-BEC crossover below T_c. To see their effects on physical quantities, we have examined the superfluid density.

The temperature dependence of ρ_s is dominated by single-particle excitations in the weak-coupling BCS regime. On the other hand, the contribution from collective excitations becomes crucial, as one approaches the BEC regime. In the strong-coupling BEC limit, the decrease of superfluid density at finite temperatures is completely dominated by thermal excitations of Bogoliubov mode. Since single-particle excitations and collective excitations are both fundamental excitations in Fermi superfluids, our results would be useful in considering temperature dependence of various physical quantities in the BCS-BEC crossover region.

Acknowledgments

The author would like to thank A. Griffin and E. Taylor for useful discussions. This work was supported by a Grant-in-Aid for Scientific research from MEXT (18043005) and CTC program of Japan.

References

1. C. A. Regal, M. Greiner and D. S. Jin, *Phys. Rev. Lett.* **92**, 040403 (2004).
2. M. Bartenstein, A. Altmeyer, S. Riedl, S. Jochim, C. Chin, J. Denschlag and R. Grimm, *Phys. Rev. Lett.* **92**, 120401 (2004).
3. M. W. Zwierlein, C. A. Stan, C. H. Schunck, S. M. F. Raupach, A. J. Kerman and W. Ketterle, *Phys. Rev. Lett.* **92**, 120403 (2004).
4. J. Kinast, S. Hemmer, M. Gehm, A. Turlapov and J. Thomas, *Phys. Rev. Lett.* **92**, 150402 (2004).
5. A. Leggett, in *Modern Trends in the Theory of Condensed Matter*. ed. A. Pekalski and J. Przystawa (Springer Verlag, Berlin, 1980), p. 14.
6. P. Nozières and S. Schmitt-Rink, *J. Low. Temp. Phys.* **59**, 195 (1985).
7. M. Randeria, in *Bose-Einstein Condensation*, ed. A. Griffin, D. W. Snoke and S. Stringari (Cambridge University Press, N.Y., 1995), p. 355.
8. Y. Ohashi and A. Griffin, *Phys. Rev. Lett.* **89**, 130402 (2002).
9. For a review, see Q. Chen, J. Stajic, S. Tan and K. Levin, *Phys. Rep.* **412**, 1 (2005).
10. E. Timmermans, K. Furuya, P. W. Milonni and A. K. Kerman, *Phys. Lett.* **A285**, 228 (2001).
11. M. Holland, S. J. J. M. F. Kokkelmans, M. L. Chiofalo and R. Walser, *Phys. Rev. Lett.* **87**, 120406 (2001).
12. N. Fukushima, Y. Ohashi, E. Taylor and A. Griffin, *Phys. Rev.* **A75**, 033609 (2007).
13. Y. Ohashi and S. Takada, *J. Phys. Soc. Jpn.* **66**, 2437 (1997).
14. J. R. Schrieffer, *Theory of Superconductivity* (Addison-Wesley Publishing Company, N.Y., 1964) chapter 7.
15. Y. Ohashi, *J. Phys. Soc. Jpn.* **74**, 2659 (2005).

FERMIONIC SUPERFLUIDITY AND THE BEC-BCS CROSSOVER IN ULTRACOLD ATOMIC FERMI GASES

M. W. ZWIERLEIN
MIT-Harvard Center for Ultracold Atoms, Research Laboratory of Electronics, Department of Physics, Massachusetts Institute of Technology, Cambridge, MA 02139, USA

Keywords: BEC-BCS Crossover.

In our recent experiments at MIT, superfluidity has been observed in a strongly interacting atomic Fermi gas. These systems constitute a novel form of matter with model character: One can vary the temperature, density and dimensionality, the number of "spin up" versus "spin down" fermions, and, most remarkably, the interactions can be precisely controlled over an enormous range. This allows to study the crossover of a Bose-Einstein condensate of tightly bound molecules to a Bardeen-Cooper-Schrieffer superfluid of long-range Cooper pairs.

Superfluidity in this crossover regime is demonstrated by setting the gas under rotation and observing ordered lattices of quantized vortices [1]. Thanks to its strong interactions, the gas is a high-temperature superfluid: Scaled to the density of electrons in a metal, superfluidity would occur already far above room temperature.

A new regime is entered when the number of spin up versus spin down atoms is imbalanced. In this case, not every spin up atom can find a spin down partner. The ground state of such a system has been under debate for over 40 years. We observe the breakdown of superfluidity at a critical imbalance, the Chandrasekhar-Clogston limit [2]. The superfluid gives way to an intriguing, strongly-interacting Fermi gas with unequal spin populations [3, 4].

Bosonic vs Fermionic Superfluids

It is a remarkable fact of history that the first fermionic superfluid was created using a bosonic one as the coolant. In 1908, Heike Kamerlingh-Onnes liquefied helium. In these experiments, he already reached the critical temperature for superfluidity of helium at 2.2 K, the onset of frictionless flow, although these and other remarkable properties of superfluids remained unnoticed at the time. He moved on to use helium to cool down mercury. In 1911 he observed that at the critical temperature of 4.2 K the resistance of the metal dropped suddenly to non-measurable values, it became "superconducting". Tin (at 3.8 K) and lead (at 6 K) showed the same remarkable phenomenon. This was the discovery of superfluidity in an electron gas.

The fact that bosonic superfluidity and fermionic superfluidity were first observed at very similar temperatures, is due to purely technical reasons (because of the available cryogenic methods) and rather obscures the very different physics behind these two phenomena.

Bosonic superfluidity occurs via Bose-Einstein condensation at the degeneracy temperature, i.e. the temperature T at which the spacing between particles $n^{-1/3}$ at density n becomes comparable to the thermal de Broglie wavelength $\lambda = h/mv_{\text{th}}$, where m is the particle mass and $v_{\text{th}} \sim (k_B T/m)^{1/2}$ is the thermal velocity. The predicted transition temperature of $T_{\text{BEC}} \sim h^2\, n^{2/3}/m \approx 3$ K for liquid helium at a typical density of $n = 10^{22}$ cm^{-3} coincides with the observed lambda point.

In contrast, the degeneracy temperature (equal to the Fermi temperature $T_F = E_F/k_B$) for conduction electrons is higher by the mass ratio $m_{4\text{He}}/m_e$, bringing it up to several ten-thousand degrees. It was only in 1957 when it became clear why in fermionic systems, superfluidity occurs only at temperatures much smaller than the degeneracy temperature.

Of course, the main difference to Bose gases is that electrons, being fermions, cannot be in one and the same quantum state but instead must arrange themselves in *different* states. An obvious scenario for superfluidity might be the formation of tightly bound pairs of electrons that can act as bosons and could form a condensate. But apart from the problem that the condensation temperature would still be on the order of E_F/k_B, there is no known interaction which could be sufficient to overcome the strong Coulomb repulsion and form tightly bound electron pairs. The idea itself of electrons forming pairs was indeed correct, but the conceptual difficulties were so profound that it took several decades from the discovery of superconductivity to the correct physical theory.

In 1950, it became clear that there was indeed an effective attractive interaction between electrons, mediated by the crystal lattice vibrations (phonons), that was responsible for superconductivity. The lattice vibrations left their mark in the characteristic variation of the critical temperature with the isotope mass of the crystal ions, the isotope effect. But the scale of the lattice vibrations, the Deby-temperature T_D, was still too large to explain the observed low critical temperatures for superconductivity.

A breakthrough came in 1956, when L. Cooper realized that fermions interacting via an arbitrarily weak attractive interaction on top of a filled Fermi sea can form a bound pair. In other words, the Fermi sea is unstable towards pair formation.

However, unlike the tightly bound pairs considered before, the "Cooper"-pair is very large, much larger than the interparticle spacing. That is, a collection of these pairs necessarily needs to overlap very strongly in space. In this situation, it was far from obvious whether interactions between different pairs could simply be neglected. But it was this simplifying idea that led to the final goal: Bardeen, Cooper and Schrieffer (BCS) developed a full theory of superconductivity starting from a new, stable ground state in which pair formation was included in a self-consistent way. The pair binding energy was found to be exponentially suppressed compared to the Debye energy-scale, $T_C \sim T_D \, exp(-1/\rho(E_F)V)$, with $\rho(E_F)$ the density of states at the Fermi energy and V the effective phonon-mediated electron-electron interaction. Hence, the critical temperature is proportional to the Debye-temperature, in accord with the isotope effect, but the fragile nature of Cooper pairing suppresses T_C by an exponentially small factor.

The BEC-BCS Crossover

It appears that there are vastly different scenarios for bosonic and fermionic superfluidity. Bose-Einstein Condensation of point-like bosons and Condensation of long-range Cooper pairs, often depicted as pairing in momentum space. On the other hand, a ^{4}He atom can be viewed as a tightly bound fermion

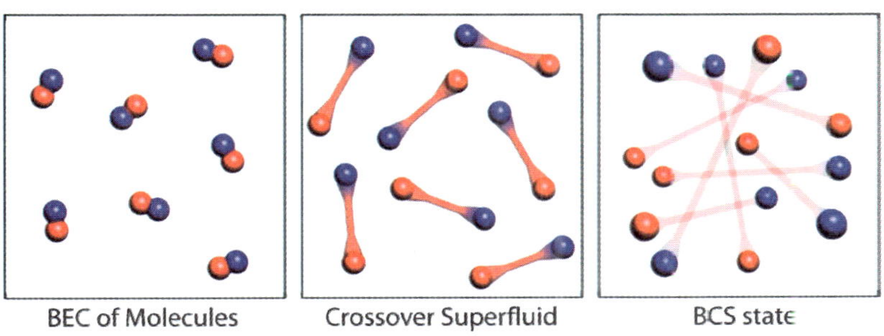

Fig. 1 The BEC-BCS crossover in a mixture of spin up (red) and spin down atoms (blue) connects the limit of a Bose-Einstein condensate of tightly bound molecules (left) to the limit of a Barden-Cooper-Schrieffer superfluid of long-range Cooper pairs (right). In between, one deals with a strongly interacting "soup" of pairs whose pairsize is on the order of the interparticle distance.

pair, consisting of a fermionic ion bound to an electron. From the BCS viewpoint, one can ask what will happen when Cooper pairs are more and more strongly bound as the interactions or the density are increased.

It was Popov, Keldysh and Eagles who realized in different contexts that the BCS formalism and its Ansatz for the ground state wavefunction provides not only a good description for a condensate of Cooper pairs, but also for a Bose-Einstein condensate of tightly bound pairs. Leggett showed in 1980 in a beautifully simple model that the two limits of fermionic superfluidity, tightly bound molecules and long-range Cooper pairs, are connected in a smooth crossover. Nozières and Schmitt-Rink extended Leggett's model to the normal state at finite temperatures and verified that the critical temperature for superfluidity varies smoothly from the BCS limit, where it is exponentially small, to the BEC limit where one recovers the value for the Bose-Einstein condensation of tightly bound molecules (see Fig. 3).

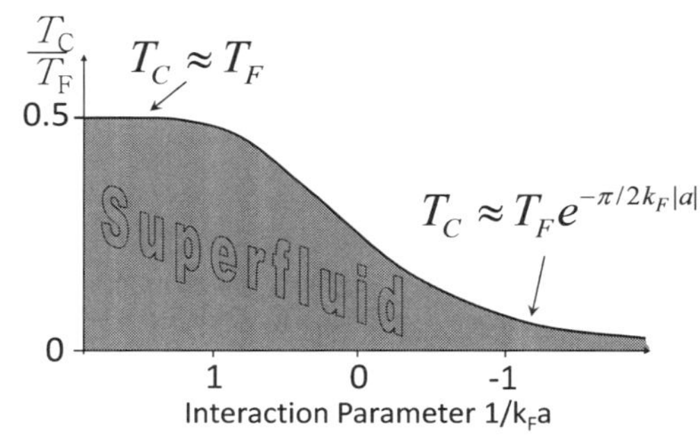

Fig. 2 Sketch of the critical temperature in the BEC-BCS crossover as a function of the interaction parameter $1/k_Fa$.

The interest in strongly interacting fermions and the BCS-BEC crossover increased with the discovery of high-temperature superconductors in 1986 by Bednorz and Müller. The physics of the BEC-BCS crossover in a gas of interacting fermions does not directly relate to the complicated phenomena observed in High-T_C materials. However, the two problems share several features: A pair size (or correlation length) comparable to the interparticle distance and a normal state above T_C which is far from being an ordinary Fermi gas. In the normal state, correlations are still strong enough to form uncondensed pairs at finite momentum. In High-T_C materials, this region in the phase diagram is referred to as the "Nernst regime", part of a larger region called the "Pseudo-gap".

The Unitarity Limit

One point in the BEC-BCS crossover is of special interest: When the interparticle potential is just about strong enough to bind two particles in free space, the bond length of this molecule tends to infinity (unitarity regime). In the medium, this bond length will not play any role anymore in the description of the many-body state. The only length scale of importance is then the interparticle distance $n^{-1/3}$, the corresponding energy scale is the Fermi energy E_F. In this case, physics is said to be universal. The average energy content of the gas, the binding energy of a pair, and (k_B times) the critical temperature must be related to the Fermi energy by universal numerical constants. The size of a fermion pair must be given by a universal constant times the interparticle distance.

It is at the unitarity point that fermionic interactions are at their strongest. Further increase of attractive interactions will lead to the appearance of a bound state and turn fermion pairs into bosons. As a result, the highest transition temperatures for fermionic superfluidity are obtained around unitarity and are on the order of the degeneracy temperature. Finally, almost 100 years after Kamerlingh Onnes, it is not just an accidental coincidence anymore that bosonic and fermionic superfluidity occur at similar temperatures!

High-Temperature Superfluidity

The crossover condensates realized in the experiments on ultracold Fermi gases are a new type of fermionic superfluid. This superfluid differs from ^{3}He, conventional and even High-T_C superconductors in its high critical temperature T_C when compared to the Fermi temperature T_F. Indeed, while T_C/T_F is about $10^{-5}...10^{-4}$ for conventional superconductors, 5×10^{-4} for ^{3}He and 10^{-2} for High-T_C superconductors, the strong interactions induced by the Feshbach resonance allow atomic Fermi gases to enter the superfluid state already at $T_C/T_F \approx 0.2$, as summarized in Table 1. It is this large value which allows us to call this phenomenon "*high-temperature superfluidity*". Scaled to the density of electrons in a metal, this form of superfluidity would already occur far above room temperature (actually, even above the melting temperature).

System	T_C	T_F	T_C/T_F
Metallic Super-conductors (typ.)	1-10 K	50000 – 150000 K	$10^{-4} - 10^{-5}$
^{3}He	2.6 mK	5 K	$5 \cdot 10^{-4}$
High-T_C Super-conductors	35-140 K	2000 – 5000 K	$(1-5) \cdot 10^{-2}$
Neutron stars	10^{10} K	10^{11} K	10^{-1}
Stronlgly interacting Fermi Gases	200 nK	1 μK	0.2

Table 1 Transition temperatures, Fermi temperatures and their ratio T_C/T_F for a variety of fermionic superfluids or superconductors.

Realization in Ultracold Atomic Fermi Gases

After the accomplishment of quantum degeneracy in bosons in 1995, one important goal was the study of quantum degenerate fermions. A difficulty in cooling fermions arises directly from the Pauli exclusion principle. In evaporative cooling, the technique that allowed reaching the degeneracy temperature in bosons, the most energetic atoms are forced to escape from the trap, while the rest of the sample rethermalizes at a lower temperature. However, spin-polarized fermions can no longer thermalize at temperatures below the threshold for p-wave collisions, as s-wave (head-on) collisions are forbidden by the Pauli principle. Experiments on ultracold fermions thus have to employ a mixture of either two hyperfine states of the same atom or a second species of atoms to guarantee thermalization. The first degenerate Fermi gas was created by B. DeMarco and D. Jin in 1999 at JILA using a spin-mixture of fermionic ^{40}K [5].

Until the end of 2003, six more groups had succeeded in producing ultracold degenerate Fermi gases, one more using ^{40}K [6], and five using ^{6}Li [7-11]. Between 1999 and 2001, the ideal Fermi gas and its collisional properties were studied.

At first sight, the prospects for reaching superfluidity in these alkali Fermi gases seemed dim. The van-der-Waals interactions in these gases are typically weak, resulting in a very low binding energy of Cooper pairs and thus a critical temperature for superfluidity far too low to be achievable in experiments. Interactions at low temperatures are mediated exclusively via s-wave collisions between unlike spins. They are characterized by the scattering length a, which is typically in the range of the van der Waals-length, about 100 a_0. This is much smaller than the interparticle distance $\sim 1/k_F \sim 3000\ a_0$ between particles. The ratio $1/k_F a$ is a measure for the interaction strength. The critical temperature for BCS-type s-wave superfluidity is given by

$$T_C \approx E_F \exp(-\pi/2k_F a)$$

It scales as the Fermi energy, essentially the degeneracy temperature, which experimenters knew how to reach since 1995. However, it is suppressed with respect to that energy scale by a factor that depends exponentially on the interaction strength. This factor is easily 10^{-20}. However, the ^{6}Li atom was early on thought to be a good candidate for a fermionic superfluid due to its large and negative (triplet) scattering length $a \sim -2000\ a_0$. This spurred hope that one might be able to actually reach T_C in experiments. In addition, already from work with Bose-Einstein condensates, it was known that the scattering length can be tuned near a so-called Feshbach resonance.

In a Feshbach resonance, a bound state of the intermolecular potential (say, a singlet with total spin S=0) is tuned in resonance with the energy of two colliding atoms (colliding, say as a triplet, S=1) (see Fig. 3). Due to the different Zeeman shifts of molecular state (essentially zero magnetic moment) and the triplet state (magnetic moment typically $2\mu_B$), this tuning can be achieved by simply applying a bias magnetic field. For the field where the free-atom and the molecular states cross, a scattering resonance occurs. One one side of the resonance, atom pairs can form stable molecules, while on the other side, atom pairs are, at least in free space, unstable. Only the presence of a surrounding Fermi sea can stabilize fermion pairs in this regime ("BCS-side" of the Feshbach resonance). Diatomic molecules can be formed by either adiabatically sweeping the magnetic field from the "free atom" to the "molecular" side or by three-body collisions of free atoms directly on the molecular side.

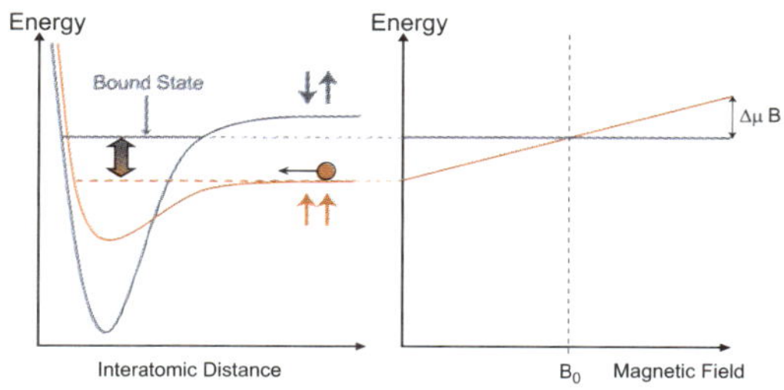

Fig. 3 Origin of Feshbach resonances. Atoms entering f. ex. in the triplet potential are coupled to a singlet bound molecular state. By tuning the external magnetic field, this bound state can be brought into resonance with the incoming state (at B_0 in the graph on the right).

In bosons, the resonant enhancement of the elastic collision rate is unfortunately accompanied by resonantly enhanced three-body losses. The hope for achieving superfluidity in Fermi gases truly intensified in the year 2003, when C. Salomon's group at the ENS in Paris found that *fermionic atoms*, in contrast to their bosonic counterparts, are extremely stable close to such a Feshbach resonance. It thus became clear that one could directly cool the Fermi mixture at a Feshbach resonance, induce strong interactions so that rather small and stable fermion pairs (Feshbach molecules) would form, which should then condense. The race was on to form the first Bose-Einstein condensate of molecules.

In November 2003, three groups reported the realization of Bose-Einstein condensates of molecules (see Fig. 4). In fact, these experiments already realized crossover condensates. They operated in the strongly interacting regime with $k_F a > 1$, where the size of the pairs is not small with respect to the interparticle spacing. When the interparticle spacing $\sim 1/k_F$ becomes smaller than the scattering length $\sim a$, the two-body molecular state is not relevant anymore and pairing is a many-body affair. Soon after these first experiments on fermion pair condensates, their observation was extended throughout the whole BEC-BCS crossover region by employing a rapid ramp to the "BEC"-side of the Feshbach resonance [12][13].

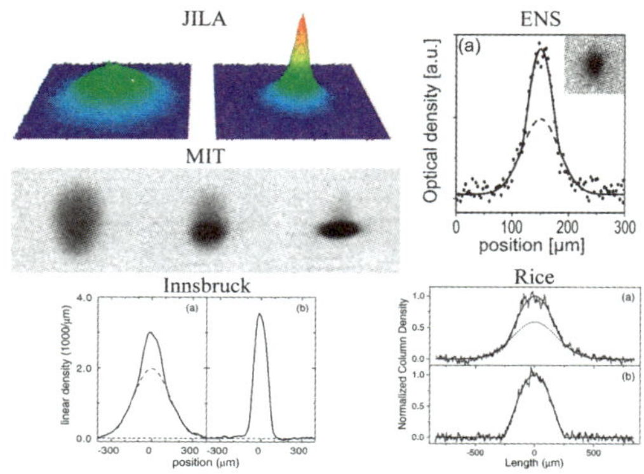

Fig. 4 Observation of Bose-Einstein condensation of molecules. The gallery shows bimodal density distributions observed after expansion and molecule dissociation at JILA, after expansion, dissociation and zero-field imaging at MIT [14] and at the ENS [15], and in-situ profiles from Innsbruck [16] and Rice [17].

In the following years, properties of this new crossover superfluid were studied in thermodynamic measurements, collective excitations, and radiofrequency spectroscopy revealing the formation of pairs [18]. However, while the experiments were consistent with superfluid behavior, they did not address properties unique to superfluids, i.e. hydrodynamic excitations can reflect superfluid or classical hydrodynamics, and the RF spectrum shows no difference between the superfluid and the normal state [18]. What was needed was a proof of superfluidity in atomic Fermi gases.

Observation of High-Temperature Superfluidity

In April 2005, fermionic superfluidity and phase coherence was finally demonstrated at MIT through the observation of quantized vortices. Superfluids are described by a macroscopic wavefunction $\psi(\boldsymbol{r})$, that is zero in the normal state and non-zero in the superfluid state. As a wavefunction, it is a complex quantity with a magnitude and phase ϕ:

$$\psi(\boldsymbol{r}) = |\psi(\boldsymbol{r})| \exp(i\,\phi)$$

The velocity of the superfluid is the gradient of the phase, $\boldsymbol{v} = \hbar\,\nabla\phi\,/\,m^*$, where m^* is the mass of the boson or fermion pair forming the superfluid. Integrating the velocity around a closed loop inside the superfluid, we immediately arrive at the Onsager-Feynman quantization condition for the circulation

$$\oint \boldsymbol{v}\cdot\mathbf{d}\boldsymbol{l} = n\,h/m^*$$

with integer n. If the superfluid wave function has no nodal lines and the loop lies in a simply connected region of space, we must have n=0. However, we can have $n\neq0$ if the wave function contains a *vortex*, that is, a flow field that depends on the vortex core distance r like $v\sim1/r$. At the location of the vortex, it has a nodal line. This is the way a superfluid carries angular momentum. Circulation is quantized in units of h/m^*. Note that vortices are a property of the superfluid in the ground state at given angular momentum. This is in marked contrast to *classical* vortices that exist only in metastable or non-equilibrium situations. Vortex patterns will ultimately decay into rigid body rotation whenever the viscosity is non-zero.

Vortices of equal charge repel each other. This immediately
follows from kinetic energy considerations. Two vortices on top of each other double the velocities and quadruple the energy. Two vortices far separated have only twice the energy of a single vortex. As a result, vortices with charge $n > 1$ will quickly decay into singly charged vortices. If many vortices are created, they minimize the total kinetic energy of the cloud by arranging themselves into a regular hexagonal lattice, called Abrikosov lattice.

How can quantized vortices nucleate? Vortices cannot suddenly appear within the condensate, as the angular momentum contained within a closed loop inside the condensate cannot abruptly jump. Rather, the nodal lines have to enter the condensate from a *surface*, where the condensate's wave function is zero. This surface can also be the surface of a stirrer, if it fully expels the condensate.

In the MIT experiment, the stirrer is formed by two tightly focused, green laser beams that were symmetrically rotated around the trapped atomic cloud (Fig. 5). The laser light is blue-detuned with respect to the atomic resonance, and thus creates a repulsive potential for the atoms. The optical trap, formed by an infrared laser beam, has to be cylindrically symmetric and as smooth as possible. In other words, one has to "sand off the bumps of the bucket" in order to make the cloud rotate.

In the trap, the vortex size is on the order of the healing length (for an atomic or a molecular BEC) or of the inverse Fermi wavevector (for a strongly interacting Fermi gas), about 200 nm. This small size is prohibitive for in-situ detection using optical techniques. Fortunately, angular momentum conservation allows vortices to survive the expansion of the condensate, which we can thus use as a "magnifying glass".

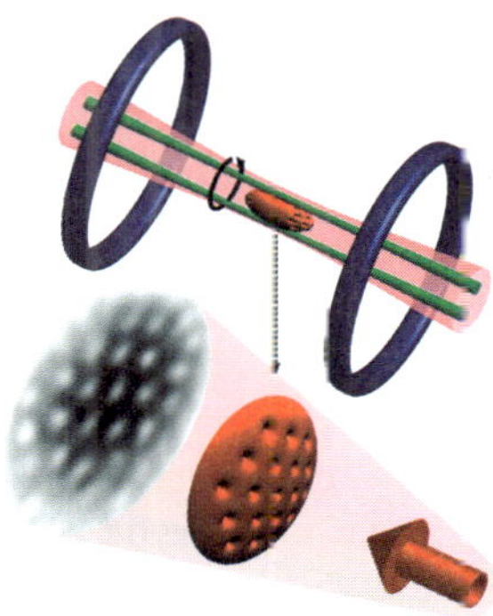

Fig. 5 Experimental Setup for the observation of Vortex lattices in ultracold Fermi gases. The gas is held in an optical trap (pink) in a magnetic field created by magnet coils (blue). Two additional laser beams (green) set the cloud in rotation. An absorption image of the expanded cloud shows the vortex lattice.

An additional complicatoin in the detection is due to the nature of the condensate in the BEC-BCS crossover: As the interactions are tuned from the BEC-side of the crossover towards the BCS-side, an ever smaller fraction of the gas is found in the condensate. In fact, in the BCS-limit, only an exponentially small number of atom pairs contributes to the coherent condensate wavefunction. The vortex contrast decreases dramatically. This can be overcome

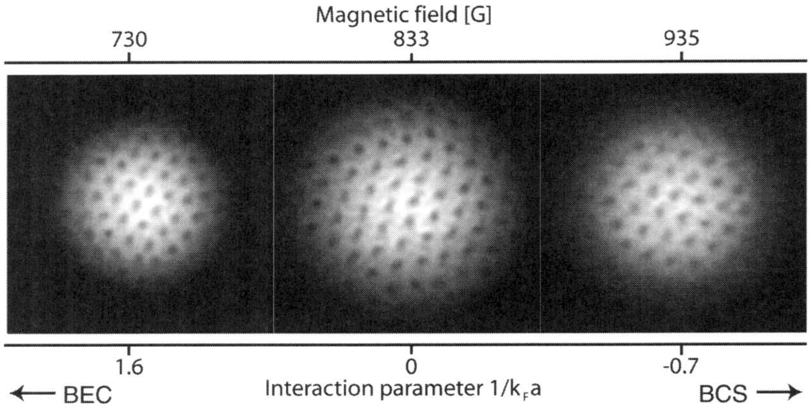

Fig. 6 Vortex lattices in the BEC-BCS crossover of an ultracold Fermi gas of atoms. The images show the gas clouds after expansion from the trap, for different values of the interaction parameter $1/k_F a$ within the BEC-BCS crossover. From [1].

by ramping the interaction strength during expansion toward the BEC-side of the resonance, where the vortex contrast is large and the vortices are easy to detect.

Fig. 6 shows the resulting vortex lattices in the BEC-BCS crossover of an ultracold Fermi gas of atoms [1]. These images demonstrate superfluidity and phase coherence in gases of molecules and of fermion pairs. The regularity of the lattice proves that all vortices have the same vorticity. From their number, the size of the cloud and the quantum of circulation $h/2m$ for each vortex, we can estimate the rotational frequency of the lattice. For an optimized stirring procedure, we find that it is close to the stirring frequency.

Apart from demonstrating superfluidity, vortices have been used to map out the phase diagram of imbalanced Fermi mixtures (see next section) and to study the breakdown of superfluid flow during expansion on the BCS-side of the resonance [18].

Fermionic Superfluidity with Imbalanced Spin Populations

Whether it occurs in Superconductors, 3He or inside Neutron stars, fermionic superfluidity requires pairing of fermions. What, however, happens if we deliberately imbalance the populations in the two spin states? In this case, not every spin up (majority) fermion can find a spin down (minority) partner. Immediately, several questions arise: Can the gas still be superfluid? If so, are the excess fermions tolerated inside the gas of pairs or are they expelled from the superfluid? If superfluidity breaks down, what is the nature of such a strongly interacting, imbalanced normal mixture?

These questions cannot be answered in typical superconductors. To imbalance spin up versus spin down electron densities would require a magnetic field, but that field cannot enter due to the Meissner effect, a direct consequence of the electron charge. It can only enter in the form of quantized magnetic flux lines, the vortices.

Ultracold fermionic gases of atoms are neutral, so that effects on the spatial wavefunction, like rotation and vortices, are decoupled from their internal spin degrees of freedom. They thus allow us to freely choose the number of spin up versus spin down atoms used in the experiment, offering access to a new and unexplored part of the phase diagram of Fermi mixtures.

We studied this phase diagram in our strongly interacting Fermi mixture by varying the spin imbalance, temperature and interaction strength. Superfluidity was found to be robust against population imbalance in the strongly interacting regime [2]. We could show that, below a certain temperature, the superfluid state requires equal spin densities, and phase separates from the partially polarized normal state [19]. At a critical population imbalance, we observed the final breakdown of the

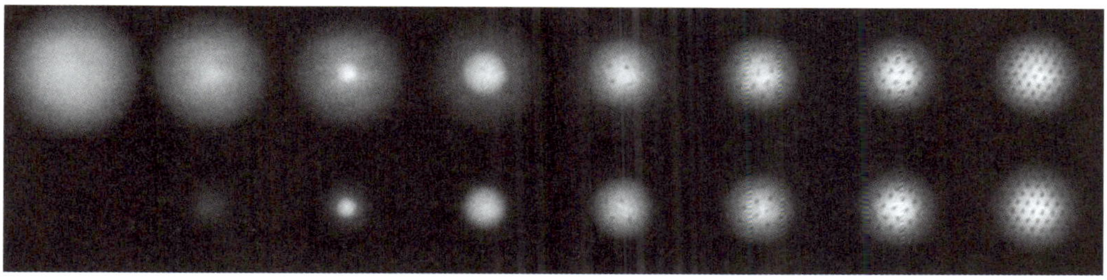

Fig. 7 Fermionic Superfluidity in imbalanced Fermi mixtures. The top and bottom row show absorption images of atoms in "spin up" and "spin down" at a magnetic field around the Feshbach resonance.

superfluid state, the Clogston or Pauli limit of superfluidity. Studying imbalanced Fermi mixtures enabled us to directly observe the superfluid transition in situ, without any magnetic field ramps into the molecular regime [20]. RF spectroscopy [18] allowed us to study the nature of pair correlations and interactions in the normal state in the imbalanced regime [3].

While the physics of an equal Fermi mixture with varying interaction strength was a smooth BEC-BCS crossover, imbalanced mixtures provide us with a much richer phase diagram, including zero temperature phase transitions, which challenges present many-body theories.

Fermi Polarons in a Tunable Fermi Liquid of Ultracold Atoms

An intriguing limit of imbalanced Fermi mixtures is reached when one is left with just a few mobile "impurities" in a Fermi sea of "environment" atoms. In such a highly imbalanced mixture of fermionic atoms, we have observed so-called Fermi polarons using tomographic RF spectroscopy [4]. Feshbach resonances again allow to freely tune the interactions between the two spin states involved. A single spin down atom immersed in a Fermi sea of spin up atoms can do one of two things (see Fig. 8): For strong attraction, it can form a molecule with exactly one spin up partner, but for weaker interaction it will spread its attraction and surround itself with a collection of majority atoms. This spin down atom "dressed" with a spin up cloud constitutes the spin-polaron. We have observed a striking spectroscopic signature of this quasi-particle

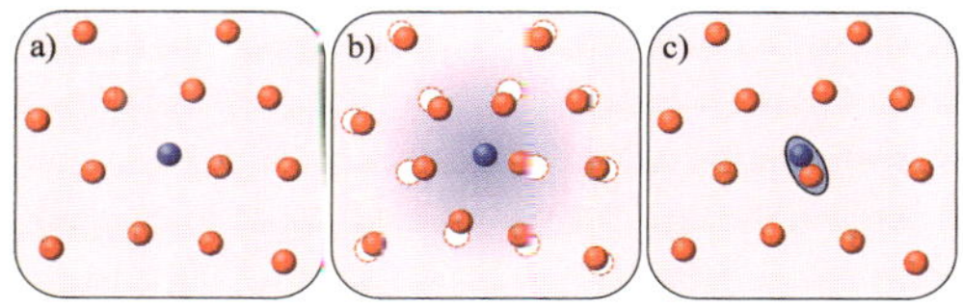

Fig. 8 From polarons to molecules. a) For weak attraction, an impurity (blue) experiences the mean field of the medium (red). b) For stronger attraction, the impurity surrounds itself with a localized cloud of environment atoms, forming a polaron. c) For strong attraction, molecules form despite Pauli blocking from the environment.

for various interaction strengths, a narrow peak in the spin down spectrum that emerges above a broad background. The narrow width signals a long lifetime of the Fermi polaron, much longer than the collision rate with spin up atoms, as it must be for a proper quasi-particle. The peak position allows to directly measure the polaron energy. The broad pedestal at high energies reveals physics at short distances and is thus "molecule-like": It is exactly matched by the spin up spectra. The comparison with the area under the polaron peak allows to directly obtain the quasi-particle weight Z.

We observe a smooth transition from polarons to molecules. At a critical interaction strength of $1/k_F a = 0.7$, the polaron peak vanishes and spin up and spin down spectra exactly match, signalling the formation of molecules. This is the same critical interaction strength found earlier to separate a normal Fermi mixture from a superfluid molecular Bose-Einstein condensate.

The spin-polarons determine the low-temperature phase diagram of imbalanced Fermi mixtures. In principle, polarons can interact with each other and might, at low enough temperatures, form a superfluid of p-wave pairs. The question of the ground state of imbalanced Fermi mixtures is still open.

Outlook

Ultracold atomic Fermi gases allow us to explore a very unusual richness of phenomena, from low-viscosity hydrodynamic behavior, fermionic superfluids, Fermi mixtures with population imbalance, zero temperature phase transitions etc. Many patches of this phase diagram have not yet been explored experimentally, and there remain controversies on the ground state of imbalanced Fermi gases in the different limits of interaction strength. Most importantly, the phase diagram of attractively interacting Fermi mixtures in optical lattices will directly impact the fermionic Hubbard model with repulsive interactions. Indeed, there is a remarkable mapping between these two major unsolved problems of condensed matter physics. What is found in one regime (for attractive interactions) will directly impact the other (repulsive interactions). At this point, any prediction of what physics ultracold Fermi gases might uncover in the future will soon be outdated. It is fascinating that Nature has given us this amazing system to work with, where essentially every parameter can be tuned at will. She has indeed been very kind to us.

References

[1]M. W. Zwierlein *et al.*, Nature **435**, 1047 (2005).
[2]M. W. Zwierlein *et al.*, Science **311**, 492 (2006).
[3]C. H. Schunck *et al.*, Science **316**, 867 (2007).
[4]A. Schirotzek *et al.*, preprint arXiv:0902.3021 (2009).
[5]B. DeMarco, and D. S. Jin, Science **285**, 1703 (1999).
[6]G. Roati *et al.*, Phys. Rev. Lett. **89**, 150403 (2002).
[7]A. G. Truscott *et al.*, Science **291**, 2570 (2001).
[8]F. Schreck *et al.*, Phys. Rev. Lett. **87**, 080403 (2001).
[9]S. R. Granade *et al.*, Phys. Rev. Lett. **88**, 120405 (2002).
[10]Z. Hadzibabic *et al.*, Phys. Rev. Lett. **88**, 160401 (2002).
[11]S. Jochim *et al.*, Science **302**, 2101 (2003).
[12]C. A. Regal, M. Greiner, and D. S. Jin, Phys. Rev. Lett. **92**, 040403 (2004).
[13]M. W. Zwierlein *et al.*, Phys. Rev. Lett. **92**, 120403 (2004).
[14]M. W. Zwierlein *et al.*, Phys. Rev. Lett. **91**, 250401 (2003).
[15]T. Bourdel *et al.*, Phys. Rev. Lett. **93**, 050401 (2004).
[16]M. Bartenstein *et al.*, Phys. Rev. Lett. **94**, 103201 (2004).
[17]G. B. Partridge *et al.*, Phys. Rev. Lett. **95**, 020404 (2005).
[18]W. Ketterle, and M. W. Zwierlein, in *Ultracold Fermi Gases, Proceedings of the International School of Physics "Enrico Fermi", Course CLXIV, Varenna, 20 - 30 June 2006*, edited by M. Inguscio, W. Ketterle, and C. Salomon (IOS Press, Amsterdam, 2008).
[19]Y. Shin *et al.*, Phys. Rev. Lett. **97**, 030401 (2006).
[20]M. W. Zwierlein *et al.*, Nature **442**, 54 (2006).

KIBBLE-ZUREK MECHANISM IN MAGNETIZATION OF A SPINOR BOSE-EINSTEIN CONDENSATE

HIROKI SAITO

Department of Applied Physics and Chemistry, The University of Electro-Communications, Tokyo 182-8585, Japan

YUKI KAWAGUCHI

Department of Physics, University of Tokyo, Tokyo 113-0033, Japan

MASAHITO UEDA

Department of Physics, University of Tokyo, Tokyo 113-0033, Japan
Macroscopic Quantum Control Project, ERATO, JST, Bunkyo-ku, Tokyo 113-8656, Japan

Spontaneous magnetization of a spin-1 Bose-Einstein condensate with a ferromagnetic interaction is investigated. Polar-core vortices are shown to be created through the Kibble-Zurek mechanism. The number of created spin vortices obeys the system-size scaling and quench-time scaling predicted by Zurek.

Keywords: Bose-Einstein condensate; Kibble-Zurek mechanism; spin vortex.

1. Kibble-Zurek mechanism

Kibble[1] pointed out that topological defects such as magnetic monopoles and cosmic strings were created in the phase transition of the vacuum in the early Universe. The basic idea is schematically shown in Fig. 1. In the early stage of a phase transition, domains of an order parameter grow independently. If the broken symmetry is multiply connected, topological defects are created with some probability when the domains merge.

Zurek[2] extended this idea to superfluid transition in helium and proposed scaling properties. For simplicity, let us consider a phase transition on a 1D ring with a radius R, in which U(1) symmetry is broken. The size of each domain of superfluid state is characterized by a coherence length ξ, and the number of domains on the ring is proportional to $\sim R/\xi$. Since the phase randomly changes domain by domain, the net change of the phase in circumnavigating the ring is just like the random walk of the phase and

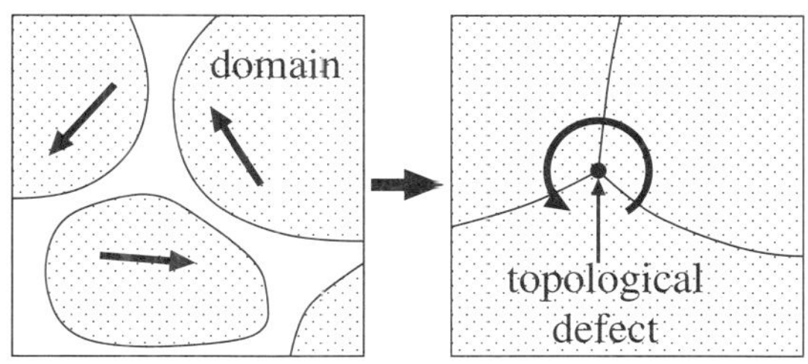

Fig. 1. Schematic illustration of topological defect formation through the Kibble-Zurek mechanism. Directions of arrows represent e.g., U(1) phases of an order parameter.

given by

$$w \sim (R/\xi)^{1/2}. \qquad (1)$$

Thus, the winding number is proportional to the square root of the size of the closed path[2]. When a phase transition is driven by a change of an external parameter, the winding number also depends on the speed of the change. The external parameter q is assumed to be changed as

$$\frac{q_c - q}{q_c} = \frac{t}{\tau_Q} \equiv \varepsilon, \qquad (2)$$

where q_c is the critical value and τ_Q is called a quench time. In the vicinity of the critical point, the time scale τ and the coherence length ξ of the system are assumed to diverge as $\tau(t) \propto |\varepsilon|^{-\alpha}$ and $\xi(t) \propto |\varepsilon|^{-\beta}$, where α and β are critical exponents. The "freeze out" time $\hat{t}$ at which the phase of the order parameter is frozen is estimated by $\tau(\hat{t}) = \hat{t}$, giving $\hat{t} \propto \tau_Q^{\alpha/(\alpha+1)}$. From Eq. (1), the quench time scaling of the winding number reads[2]

$$w \propto \xi^{-1/2}(\hat{t}) \propto \tau_Q^{-\beta/(2\alpha+2)}. \qquad (3)$$

Thus, such kind of mechanism for topological defect formation is called the Kibble-Zurek (KZ) mechanism, which has been applied to a variety of systems[2].

2. Spontaneous magnetization of a spinor BEC

Recently, spontaneous magnetization of a spin-1 ^{87}Rb BEC was observed in a nondestructive manner by the Berkeley group[3], in which polar-core vortices were frequently observed in complicated magnetization patterns. In this experiment, a BEC in the $m = 0$ state is prepared in a strong magnetic field applied in the direction of the quantization axis, where m denotes the magnetic sublevel of hyperfine spin. In a strong magnetic field, the $m = 0$ state is stable due to the quadratic Zeeman effect[4]. At $t = 0$, the magnetic field is suddenly "quenched" to below the critical value, and the system spontaneously magnetizes due to the ferromagnetic interaction. The direction of the magnetization is perpendicular to that of the magnetic field, and the magnetization breaks the axisymmetry.

Because of the axisymmetry breaking in this magnetic phase transition in a spinor BEC, we expect that topological defects are generated through the KZ mechanism. However, the polar-core vortex formation observed in the experiment is not due to the KZ mechanism but due to dynamical instabilities[5], since the system is not so large compared with the size of domains, and domains are not independent. The aim of the present paper is to show that the KZ mechanism does emerge for a sufficiently large system, in which domains far apart form each other are independent. Recently, several authors have studied this problem [6,7,8,9].

3. System size scaling

We employ the mean-field approximation, and numerically solve the three-component Gross-Pitaevskii equation describing dynamics of macroscopic wave functions ψ_m of a spin-1 BEC. We assume that the system is tightly confined in the z direction and the 2D approximation is valid. We also assume that the system is confined in the x-y direction by a hard wall with radius $R_w = 100$ μm that is much larger than the spin healing length $\simeq 2$ μm. The atomic density is $n = 2.8 \times 10^{14}$ cm^{-3} and the number of atoms is $N = 1.6 \times 10^8$.

The initial state is the $m = 0$ ground state prepared by the imaginary-time propagation method. We add a small white noise to the $m = \pm 1$ components in order to simulate quantum fluctuations, which triggers growth of the transverse magnetization. The strength of the noise is chosen to reproduce the Bogoliubov theory when ensemble average is taken.

Figure 2 shows time evolution of the transverse magnetization $F_+ = F_x + iF_y = \sqrt{2}(\psi_1^* \psi_0 + \psi_0^* \psi_{-1})$. At $t = 0$, the quadratic Zeeman energy is suddenly quenched to $q < q_c$ for which the $m = 0$ state is dynamically unstable against magnetization in the x-y direction. At $t \simeq 50$ ms, we can see that magnetic domains gradually emerge and at the same time the direction of the magnetization begins to fix. The magnetic domains then merge and form topological structures, which are polar-core vortices. An example is indicated by the black circle in Fig. 2. The

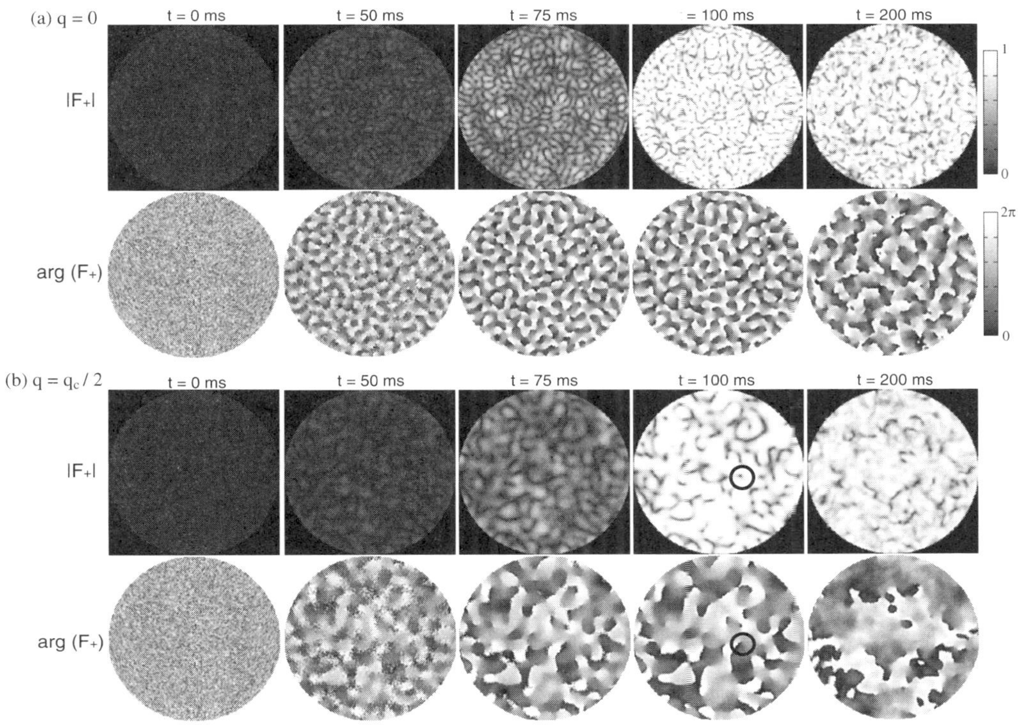

Fig. 2. Time evolution of the strength $|F_+|$ and direction $\arg F_+$ of the transverse magnetization. The quadratic Zeeman energy is suddenly quenched from $q > q_c$ to (a) $q = 0$ and (b) $q = q_c/2$, where q_c is the critical value for magnetization. The black circle indicates an example of the position of a spin vortex. Reprinted from Ref. 6.

number of spin vortices is larger for $q = 0$, since the domain size is smaller.

In order to confirm that this spin-vortex formation is due to the KZ mechanism, we investigate the system-size scaling for the winding number in Eq. (1). The winding number w is defined by the net number of rotation of the spin in the x-y plane along the circumference a circle with a radius $R < R_w$. Figure 3 shows the variance of w as a function of R. The variance is taken for 1000 runs of simulations with initial noises produced by random numbers. The winding number is taken at $t = 100$ ms. For large R, w is proportional to $R^{1/2}$ on average, in agreement with the system size scaling (1) in the KZ theory. For small R, on the other hand, w is proportional to R. This is because R is comparable to the size of the domain and the

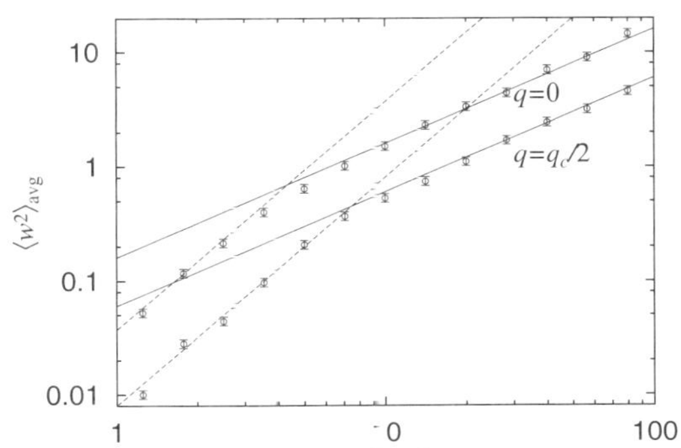

Fig. 3. Plots show variance of the spin winding number w for a circular path with a radius R. The error bar represents the 95% confidence interval of the χ^2 distribution for 1000 samples. The solid and dashed lines are proportional to R and R^2, respectively. Reprinted from Ref. 6.

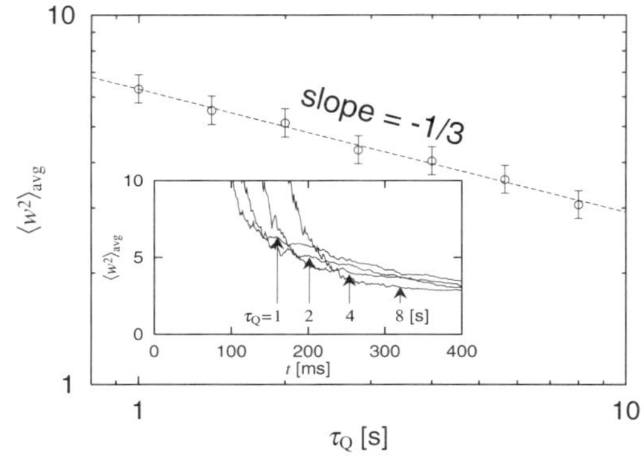

Fig. 4. Plots show variance of the spin winding number w as a function of the quench time τ_Q of the quadratic Zeeman energy. The inset shows time evolution of $\langle w^2 \rangle_{\rm avg}$, where the arrows indicate the time at which each plot in the main panel is taken. The error bar represents the 95% confidence interval of the χ^2 distribution for 1000 samples. Reprinted from Ref. 6.

KZ assumption to derive Eq. (1) is invalid. In fact, the crossover region of R between $w \propto R$ and $w \propto R^{1/2}$ roughly agrees with the size of the domain for each q.

4. Quench time scaling

We next study the quench time dependence of the spin winding number. The quadratic Zeeman energy q is decreased in time as Eq. (2). According to the Bogoliubov analysis in the uniform system[6], the critical exponent for the time scale is $\alpha = 1/2$ and that for the coherence length is $\beta = 1/2$. Thus, from Eq. (3) the quench time scaling of the winding number is predicted to be $w \propto \tau_Q^{-1/6}$.

Figure 4 plots the variance of the winding number for various quench times τ_Q. We can clearly see that w is proportional to $\tau_Q^{-1/6}$, in good agreement with the KZ theory. The inset of Fig. 4 shows time evolution of the variance of w. As time proceeds, the direction of magnetization is gradually fixed as in Fig. 2, and w decreases in time. This time is estimated to be the "freeze out" time $\hat{t}$, which is in this case proportional to $\tau_Q^{1/3}$. We therefore take the winding number at $t \propto \tau_Q^{1/3}$ in Fig.4.

5. Conclusions

We have studied the spontaneous magnetization of a spin-1 ^{87}Rb BEC prepared in the $m = 0$ state. We found that for a large system compared with the spin coherence length, magnetic domains merge and generate spin vortices through the KZ mechanism. We showed that the spin winding numbers follow the scaling laws with respect to the system size and the quench time, and the powers are in excellent agreement with those predicted by the KZ theory.

In the present paper, we have considered an almost uniform system. The effect of a harmonic trapping potential will be an important problem in connection with future experimental study.

Acknowledgments

This work was supported by the Ministry of Education, Culture, Sports, Science and Technology of Japan (Grants-in-Aid for Scientific Research, No. 17071005 and No. 20540388) and by the Matsuo Foundation.

References

1. T. W. B. Kibble, *J. Phys. A* **9**, 1387 (1976).
2. W. H. Zurek, *Nature (London)* **317**, 505 (1985); *Phys. Rep.* **276**, 177 (1996).
3. L. E. Sadler, J. M. Higbie, S. R. Leslie, M. Vengalattore, and D. M. Stamper-Kurn, *Nature (London)* **443**, 312 (2006).
4. K. Murata, H. Saito, and M. Ueda, *Phys. Rev. A* **75**, 013607 (2007).
5. H. Saito, Y. Kawaguchi, and M. Ueda, *Phys. Rev. Lett.* **96**, 065302 (2006); *Phys. Rev. A* **75**, 013621 (2007).
6. H. Saito, Y. Kawaguchi, and M. Ueda, *Phys. Rev. A* **76**, 043613 (2007).
7. A. Lamacraft, *Phys. Rev. Lett.* **98**, 160404 (2007).
8. M. Uhlmann, R. Schützhold, and U. R. Fischer, *Phys. Rev. Lett.* **99**, 120407 (2007).
9. B. Damski and W. H. Zurek, *Phys. Rev. Lett.* **99**, 130402 (2007).

QUASIPARTICLE INDUCING JOSEPHSON EFFECT IN A BOSE-EINSTEIN CONDENSATE

S. TSUCHIYA* and Y. OHASHI
Department of Physics, Keio University, 3-14-1 Hiyoshi, Kohoku-ku, Yokohama, Japan
CREST(JST), 4-1-8 Honcho, Saitama 332-0012, Japan
**E-mail: tsuchiya@rk.phys.keio.ac.jp*

We investigate tunneling properties of Bogoliubov phonons in a Bose-Einstein condensate. We find that the quasiparticle current of Bogoliubov phonon is remarkably enhanced near the potential barrier at low energies. This leads to the increase of the transmission probability in the low energy region. We show that the quasiparticle current twists the relative phase of the condensate wavefunctions and induces a Josephson supercurrent through the barrier. The Josephson critical current is estimated by calculating the induced supercurrent and phase difference.

Keywords: Bose-Einstein condensate; Bogoliubov phonon; anomalous tunneling; Josephson effect.

1. Introduction

In the pioneering work by Bogliubov in 1947,[1] it was shown that Bose-Einstein condensates (BECs) of weakly interacting bosons have phonon excitations called Bogoliubov phonons. Bogoliubov phonons are in a Goldstone mode, which is a manifestation of broken U(1) symmetry. Bogoliubov phonons also play a crucial role to connect BEC and superfluidity.[2] Since the realization of BECs in ultracold atomic gases, Bogoliubov phonons have been investigated extensively.[2,3]

Recently, it has been predicted by Kovrizhin and co-workers[4,5,6] that Bogoliubov phonons exhibit striking tunneling properties. They showed that the transmission probability of Bogoliubov phonons through a potential barrier increases as the incident energy decreases and Bogoliubov phonons exhibit a perfect transmission in the low energy limit, irrespective of the height of the potential barrier. This phenomenon is now called *anomalous tunneling*[6] and has attracted much attention. The origin of the anomalous tunneling is under intense investigation.[6,7,8,9,10] Anomalous tunneling has been also extended to various cases.[7,8,11,12]

In this paper, we study tunneling properties of Bogoliubov phonons to investigate anomalous tunneling properties. We show that the quasiparticle current carried by Bogoliubov phonons is enhanced near the potential barrier at low energies. This leads to the anomalous increase of the transmission probability at low energies first found by Kovrizhin *et al.*[4] We find that a finite phase difference arises between the condensates across the barrier due to the backreaction effect of tunneling Bogoliubov phonons on condensates. This phase difference induces the counter flow of Josephson supercurrent through the potential barrier. The Josephson critical current is estimated by calculating the supercurrent and phase difference. The role of Bogoliubov phonons for the fluctuation of the phase difference is briefly discussed.

2. Formalism

We study the tunneling problem of Bogoliubov phonons through a potential barrier at $T = 0$ using the Bogoliubov mean-field theory.[2] We consider a BEC in the presence of a potential barrier $U(x)$. For simplicity, we treat a one-dimensional problem along the x-direction and neglect y and z dependence. The field operator for bosons can be written as a sum of the condensate wavefunction $\Psi_0(x)$ and the non-condensate part $\delta\hat{\psi}(x)$, as $\hat{\psi}(x) = \Psi_0(x) + \delta\hat{\psi}(x)$. The non-condensate part consists of the operators for Bogoliubov excitations as $\delta\hat{\psi}(x) = \sum_j [u_j(x)\hat{\alpha}_j - v_j(x)^*\hat{\alpha}_j^\dagger]$. Here, $\hat{\alpha}_j^\dagger$ is a creation operator of an excitation in the j-th state. The coefficients (u_j, v_j) are wavefunctions for Bogoliubov excitations.

The condensate wavefunction $\Psi_0(x) = \langle\hat{\psi}(x)\rangle$ satisfies the static Gross-Pitaevskii (GP) equation,[2] given by (hereafter we set $\hbar = 1$)

$$\left(-\frac{1}{2m}\frac{d^2}{dx^2} + U(x) + g|\Psi_0|^2\right)\Psi_0 = \mu\Psi_0, \quad (1)$$

where m is the mass of a boson, μ is the chemical potential, and $g(> 0)$ is a repulsive interaction between bosons. In this paper, we assume a simple rectangular potential barrier $U(x) = U_0\theta(d/2 - |x|)$ ($U_0 > 0$). We only consider the ground state solution of Eq. (1).

The wave function of Bogoliubov excitations (u_j, v_j) can be obtained by solving the Bogoliubov equations[2] given by

$$\begin{pmatrix} \hat{h} & -g\Psi_0^2 \\ -g(\Psi_0^*)^2 & \hat{h} \end{pmatrix}\begin{pmatrix} u_j \\ v_j \end{pmatrix} = E_j\tau_3\begin{pmatrix} u_j \\ v_j \end{pmatrix}, \quad (2)$$

where $\hat{h} \equiv -\frac{1}{2m}\frac{d^2}{dx^2} + U(x) + 2g|\Psi_0|^2 - \mu$, τ_3 is the Pauli matrix, and E_j is the excitation energy.

Substituting the ground state solution of Eq. (1)[6] into Eq. (2), we solved Eq. (2) numerically using the finite element method. We impose the boundary condition that Bogoliubov excitations with energy E come from the left side of the barrier.

3. Enhancement of quasiparticle current

The transmission probability W and phase shift δ are numerically calculated. They are shown in Fig. 1 as functions of the incident energy E for several values of U_0 when $d = \xi$, where $\xi = 1/\sqrt{2m\mu}$ is the healing length. The anomalous tunneling behavior[6] is clearly seen in Fig. 1: The W increases at low energies as E decreases, while the δ decreases at low energies as E decreases. Moreover, the perfect transmission ($W \to 1$ and $\delta \to 0$) can be seen in the low energy limit ($E \to 0$). This tunneling property of Bogoliubov phonons is in contrast with that of free particles i.e. free particles are perfectly reflected by a potential barrier in the low energy limit.

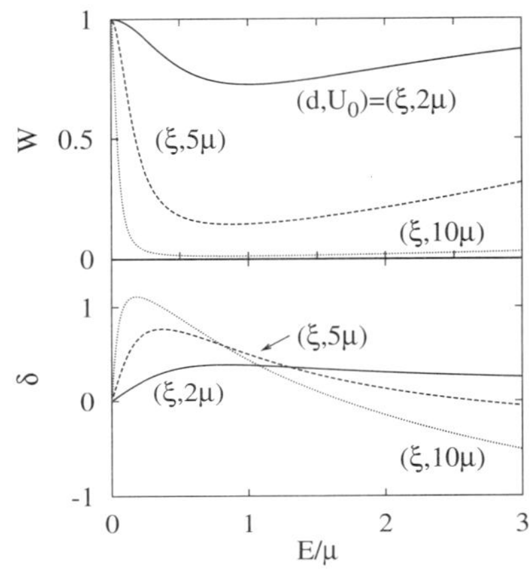

Fig. 1. Transmission probability W (upper panel) and phase shift δ (lower panel) of Bogoliubov excitations as functions of the incident energy E/μ.

To investigate anomalous tunneling properties of Bogoliubov phonons in more detail, we calculate the quasiparticle current carried by Bogoliubov excitations. The quasiparticle current is given by $J_q = J_u - J_v$[9], where $J_u = (1/m)\mathrm{Im}(u^*\partial_x u)$ and $J_v = (1/m)\mathrm{Im}(v^*\partial_x v)$. From Eq. (2), we obtain the equation of continuity for quasiparticles as $\partial_x J_q = S$, where $S = -4g\mathrm{Im}(\Psi_0^2 u^* v)$ is the source term. This equation indicates that the number of quasiparticles is *not* conserved when S is finite. In a uniform system, $S = 0$ and J_q is conserved. However, S is finite and quasiparticle number is not conserved in our tunneling problem.[9]

The excess quasiparticle current defined by $\Delta J_{\rm q}(x) = J_{\rm q}(x) - J_{\rm q}(x = -\infty)$ can be calculted from the relation $\Delta J_{\rm q}(x) = \int_{-\infty}^{x} S(x')dx'$. In Fig. 2, $\Delta J_{\rm q}(x)$ is plotted for several incident energies when $(d, U_0) = (\xi, 10\mu)$. One finds that $\Delta J_{\rm q}$ is supplied and enhanced near the potential barrier at $|x|/\xi < 1/2$. This enhancement of $\Delta J_{\rm q}$ becomes remarkably larger when the incident energy gets smaller. The increase of transmission probability of Bogoliubov phonons in Fig. 1 results from this quasiparticle current enhancement. In fact, by comparing Figs. 1 and 2, one finds that the energy at which W starts increasing is the same as the energy at which $J_{\rm q}$ becomes remarkably large ($E \simeq 0.1\mu$).

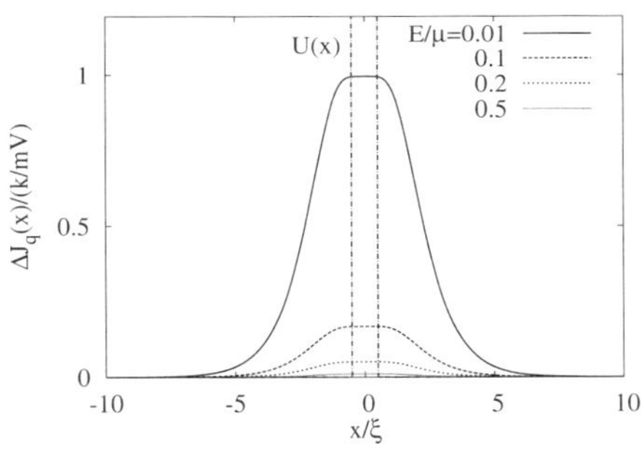

Fig. 2. Excess quasiparticle current $\Delta J_{\rm q}(x)$ for $(d, U_0) = (\xi, 10\mu)$. V is the volume of the system. The dash-dot line indicates the barrier region.

4. Induced Josephson effect

The origin of the excess quasiparticle current $\Delta J_{\rm q}$ was identified by studying the number conservation law.[9] It was pointed out that the Bogoliubov mean-field theory based on Eqs. (1) and (2) do not satisfy the conservation of total current, because the backreaction effect of Bogoliubov excitations on condensates is neglected.[9] The modified GP equation with a backreaction term originating from quasiparticles was proposed in order to eliminate this inconsistency.[9] As a result, the conservation of the total current requires a finite supercurrent $\Delta J_s = -\sum_j \Delta J_{{\rm q}j}\langle \hat{\alpha}_j^\dagger \hat{\alpha}_j\rangle$ induced by quasiparticles in the barrier region. Thus, the excess quasiparticle current $\Delta J_{\rm q}$ is supplied from the condensate, which is canceled out by the induced supercurrent ΔJ_s.

When we assume that the condensate wavefunction is changed by a phase shift $\theta(x)$ from the solution of Eq. (1) by the backreaction of quasiparticles, $\theta(x)$ is given by[9]

$$\theta(x) = -m\sum_j \left(\int_{-\infty}^{x} dx' \frac{\Delta J_{{\rm q}j}(x')}{n_{\rm s}(x')}\right)\langle \hat{\alpha}_j^\dagger \hat{\alpha}_j\rangle. \tag{3}$$

The normalized phase shift $N_0\theta(x)$ (N_0 is the number of condensate atoms) is plotted in the inset of Fig. 3. Here, $\theta(x)$ varies sharply at the barrier region. One clearly sees the phase difference ϕ across the barrier. This indicates that the tunneling Bogoliubov phonons twist the relative phase of the condensates.

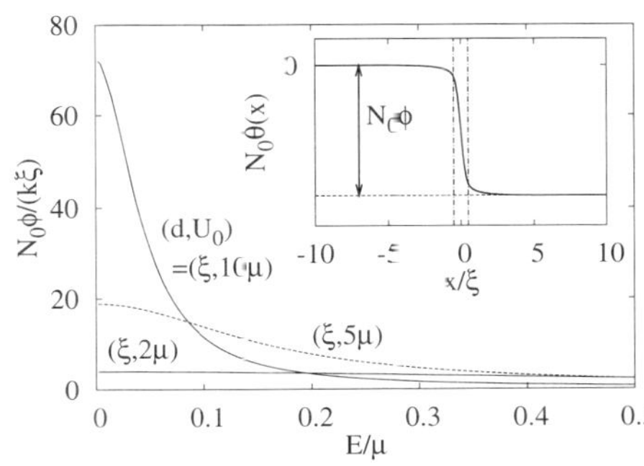

Fig. 3. Phase difference $N_0\phi$ as a function of the energy E/μ. The spatial variation of the condensate phase $N_0\theta(x)$ for $(d, U_0) = (\xi, 10\mu)$ and $E = 0.01\mu$ is shown in the inset. The dash-dot line in the inset indicates the barrier region.

In Fig. 3, the normalized phase difference $N_0\phi$ is plotted as a function of E/μ. It shows that ϕ increases as E decreases. This implies that at low energy the coupling effect of Bogoliubov phonons with condensates becomes large. Thus, low energy Bogoliubov phonons create a large phase difference.

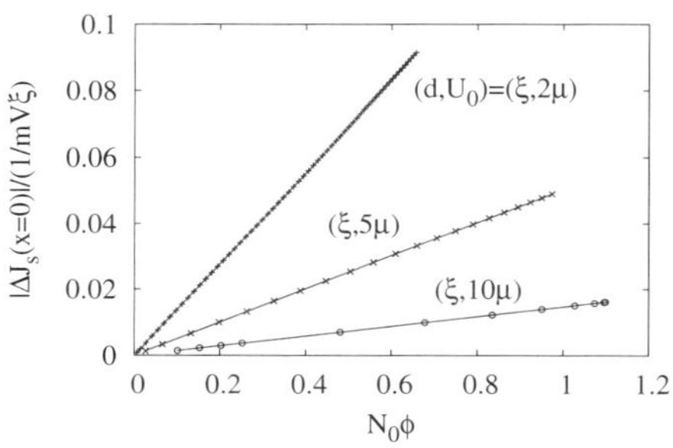

Fig. 4. The induced supercurrent $\Delta J_s(x=0)$ as a function of the phase difference ϕ. The slopes of the lines are $\gamma = 0.14$, 0.050, and 0.015 for $(d, U_0) = (\xi, 2\mu)$, $(\xi, 5\mu)$, and $(\xi, 10\mu)$, respectively.

The induced supercurrent ΔJ_s can be regarded as the Josephson current originating from the relative phase ϕ. In Fig. 4, the supercurrent in the barrier region $\Delta J_s(x=0)$ is plotted as a function of ϕ at low energies. One sees that $\Delta J_s(x=0)$ is proportional to ϕ. Josephson current is usually given by $I_J \sin\phi$,[2] where I_J is the Josephson critical current. Since ϕ is inversely proportional to N_0 and negligibly small, the Josephson current is proportional to ϕ in the present case. From the slope of the line, γ, in Fig. 4, the Josephson critical current can be obtained as $I_J = \gamma(\frac{n_0}{m\xi})$, where $n_0 = N_0/V$ is the condensate density. Thus, the first-principle calculation of the Josephson critical current I_J is possible from the Bogoliubov equations.

Finally, we remark that the tunneling property of Bogoliubov excitation discussed above suggests an important role of quasiparticles on the fluctuation of the relative phase at finite temperatures which was recently observed.[13] When quasiparticles are excited on both sides of the barrier at finite temperatures, they tunnel the potential barrier twisting the relative phase and induce the fluctuation of the phase difference. In particular, since the population of Bogoliubov phonon is dominant at low temperatures and it creates a large phase difference, strong phase fluctuations may be induced by Bogoliubov phonons at low temperatures.

5. Conclusion

We discussed the tunneling properties of Bogoliubov phonons in a BEC at $T = 0$. It was found that the anomalous enhancement of the quasiparticle current near a potential barrier leads to the increase of the transmission probability of Bogoliubov phonons at low energies. We have shown that tunneling Bogoliubov phonons twist the relative phase of the condensates across the barrier that induces the Josephson supercurrent. In addition, we estimated the Josephson critical current by calculating the supercurrent and phase difference.

S.T. thanks I. Danshita, K. Kamide, S. Inouye, and F. Dalfovo for helpful discussions.

References

1. N. N. Bogoliubov, *J. Phys. USSR* **11**, 23 (1947).
2. L. Pitaevskii and S. Stringari *Bose-EinsteinCondensation* (Oxford, New York, 2003).
3. R. Ozeri *et al.*, *Rev. Mod. Phys.* **77**, 187 (2005).
4. D. L. Kovrizhin and L. A. Maksimov, *Dokl. Phys* **46**, 328 (2000).
5. D. L. Kovrizhin, *Phys. Lett. A* **287**, 392 (2001).
6. Yu Kagan *et al.*, *Phys. Rev. Lett.* **90**, 130402 (2003).
7. I. Danshita *et al.*, *New J. Phys.* **8**, 44 (2006).
8. Y. Kato *et al.*, *J. Phys. Soc. Jpn.* **77**, 013602 (2008).
9. S. Tsuchiya and Y. Ohashi, *Phys. Rev. A* **78**, 013628 (2008).
10. Y. Ohashi and S. Tsuchiya, *Phys. Rev. A* **78**, 043601 (2008).
11. I. Danshita *et al.*, *Phys. Rev. A* **72**, 053611 (2005).
12. I. Danshita and S. Tsuchiya, *Phys. Rev. A* **75**, 033612 (2007).
13. R. Gati *et al.*, *Phys. Rev. Lett.* **96**, 130404 (2006).

STABILITY OF SUPERFLUID FERMI GASES IN OPTICAL LATTICES

YOSHIHIRO YUNOMAE*
Department of Physics, Waseda University, Okubo, Shinjuku-ku, Tokyo 169-8555, Japan
**E-mail: yunon@kh.phys.waseda.ac.jp*

IPPEI DANSHITA
Department of Physics, Faculty of Science, Tokyo University of Science, Kagurazaka, Shinjuku-ku, Tokyo 162-8601, Japan

DAISUKE YAMAMOTO
Department of Physics, Waseda University, Okubo, Shinjuku-ku, Tokyo 169-8555, Japan

NOBUHIKO YOKOSHI
Nanotechnology Research Institute, AIST, Tsukuba 305-8568, Japan
CREST(JST), Saitama 332-0012, Japan

SHUNJI TSUCHIYA
Department of Physics, Keio University, Hiyoshi, Kohoku-ku, Yokohama 223-8522, Japan
CREST(JST), Saitama 332-0012, Japan

Critical velocities of superfluid Fermi gases in optical lattices are theoretically investigated across the BCS-BEC crossover. We calculate the excitation spectra in the presence of a superfluid flow in one- and two-dimensional optical lattices. It is found that the spectrum of low-lying Anderson-Bogoliubov (AB) mode exhibits a roton-like structure in the short-wavelength region due to the strong charge density wave fluctuations, and with increasing the superfluid velocity one of the roton-like minima reaches zero before the single-particle spectrum does. This means that superfluid Fermi gases in optical lattices are destabilized due to spontaneous emission of the roton-like AB mode instead of due to Cooper pair breaking.

Keywords: superfluidity; critical velocity; Anderson-Bogoliubov mode.

1. Introduction

According to Landau criterion, superfluid flow of bosonic particles propagates without dissipation, unless its group velocity exceeds a certain critical value and spontaneous creation of excitations is induced.[1] The direct experimental verifications of the criterion have been attempted mainly in superfluid ^{4}He [2] and dilute atomic Bose gases.[3] These studies have inspired various discussions and contributed to understanding of the superfluid phenomenon itself.

The recent realization of the BCS-BEC crossover in ultracold Fermi gases[4] has raised a new question, i.e., how does the critical velocity for bosonic superfluids change in the BCS regime? Recently, Miller *et al.* investigated critical velocities of superfluid Fermi gases across the BCS-BEC crossover in a moving one-dimensional (1D) optical lattice.[5] They determined the critical velocities at which the abrupt decrease of fermionic pairs in the condensate occurs, when the velocity of the lattice potential is varied.

Theoretically, critical velocities in the BCS-BEC crossover have been investigated in a uniform system.[6] It was claimed that the instability of superfluid flow in a uniform system is induced by Cooper pair breaking in the BCS regime.[6] In this paper, motivated by the experiment done by Miller *et al.*,[5] we investigate critical velocities of superfluid Fermi gases in the BCS-BEC crossover in optical lattices. We calculate excitation spectra and determine critical velocities in 1D and 2D optical lattices. We find that the excitation spectrum of the low-lying Anderson-Bogoliubov (AB) mode[6,7,8] has a characteristic roton-like structure, and the instability of superfluid flow in the BCS regime is induced by a different mechanism from that in a uniform system: the spontaneous emission of AB mode destabilizes the superfluidity.

2. Formalism

We consider two-component atomic Fermi gases with equal populations. We assume that the lattice potential is sufficiently deep that the tight-binding approximation is valid. Thus, the system can be described by a single-band Hubbard model

$$H = -J \sum_{\langle i,j \rangle,\sigma} \left(c^{\dagger}_{i\sigma} c_{j\sigma} + h.c. \right) + U \sum_{i} c^{\dagger}_{i\uparrow} c^{\dagger}_{i\downarrow} c_{i\downarrow} c_{i\uparrow}, \qquad (1)$$

where $c_{j\sigma}$ is an annihilation operator of an atom with mass m and pseudospin $\sigma = \uparrow, \downarrow$ on the j-th site, J is a hopping energy between nearest-neighbor sites, and $U(<0)$ is an on-site attractive interaction. We set $\hbar = 1$ throughout the paper.

We assume that a superfluid flows with the velocity $\boldsymbol{v}$ in the coordinate system fixed with respect to the lattice potential, i.e. Cooper pairs have the center of mass quasi-momentum $\boldsymbol{q} = 2m\boldsymbol{v}$. The corresponding order parameter is given by

$$\Delta_{\boldsymbol{v}} = -\frac{U}{M} \sum_{\boldsymbol{k}} \langle c_{-\boldsymbol{k}+m\boldsymbol{v}\downarrow} c_{\boldsymbol{k}+m\boldsymbol{v}\uparrow} \rangle, \qquad (2)$$

where M is the number of lattice sites. By diagonalizing the Hamiltonian within the BCS mean-field approximation, we obtain the single-particle excitation energy as

$$E^{\pm}_{\boldsymbol{v}}(\boldsymbol{k}) = \eta_{\boldsymbol{v}}(\boldsymbol{k}) \pm \sqrt{(\varepsilon_{\boldsymbol{v}}(\boldsymbol{k}) - \mu)^2 + |\Delta_{\boldsymbol{v}}|^2}, \qquad (3)$$

where $\eta_{\boldsymbol{v}}(\boldsymbol{k}) = 2J\Sigma_\nu \sin(mv_\nu d)\sin(k_\nu d)$ and $\varepsilon_{\boldsymbol{v}}(\boldsymbol{k}) = 2J\Sigma_\nu(1 - \cos(mv_\nu d)\cos(k_\nu d))$ with d being the lattice constant. The energy gap $\Delta_{\boldsymbol{v}}$ and the chemical potential μ are determined by solving the gap equation and the number equation self-consistently. This scheme interpolates the BCS and BEC regimes at $T = 0$.[9]

To determine the critical velocities, it is necessary to calculate the excitation spectra in the presence of a superfluid flow. For neutral fermionic superfluids, there exists a low-lying AB mode[6,7,8] in addition to the single-particle excitation. The spectrum of the AB mode can be obtained from the pole of the density response function $\chi(\boldsymbol{q}, \omega)$.[10] We calculate $\chi(\boldsymbol{q}, \omega)$ by employing the generalized random phase approximation (GRPA) developed by Côté and Griffin.[10]

3. Results

Conditions for Cooper pair breaking are determined from the single-particle excitation energy Eq. (3). When the absolute value of the superfluid velocity $\boldsymbol{v}$ exceeds a certain critical value $v_{\rm pb}$, one finds that $E^{+}_{\boldsymbol{v}}(\boldsymbol{k})$ becomes negative. The negative energy of the single-particle excitation indicates the occurrence of spontaneous Cooper pair breaking. Thus, the critical velocity for Cooper pair breaking is given by $v_{\rm pb}$.[5,6,11]

In order to compare the instabilities induced by two kinds of excitations, i.e., the single-particle excitation and AB mode excitation, it is necessary to monitor both the excitation spectra in parallel. For that purpose, it is useful to plot the dynamical structure foctor $S(\boldsymbol{q}, \omega) = -{\rm Im}\chi(\boldsymbol{q}, \omega)/\pi$. First, we investigate the case of 1D optical lattices. In Fig. 1, $S(q, \omega)$ is shown for (a) current-free and (b) current-carrying cases. In Fig. 1,

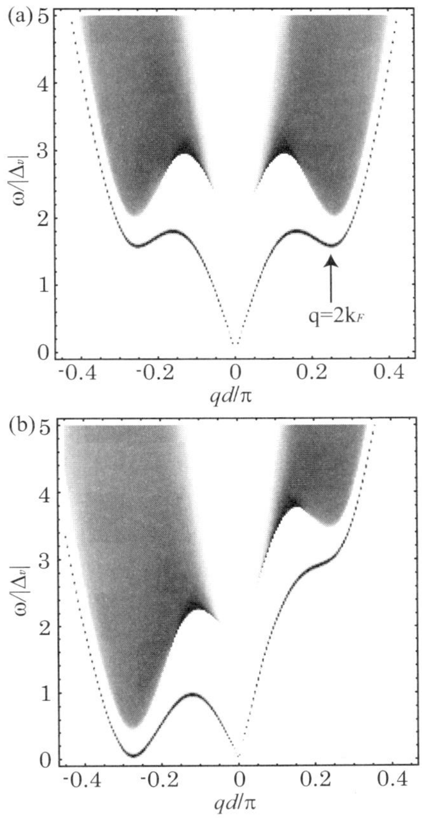

Fig. 1. Dynamical structure factors $S(q,\omega)$ in 1D optical lattices are shown for (a) current-free ($v=0$) and (b) current-carrying cases ($v=0.095/md$), when $n=0.25$ and $U=-1.0J$. The dark region indicates large $S(q,\omega)$. In each panel, the upper gray region corresponds to the particle-hole continuum, and the lower dashed curve is the spectrum of the AB mode.

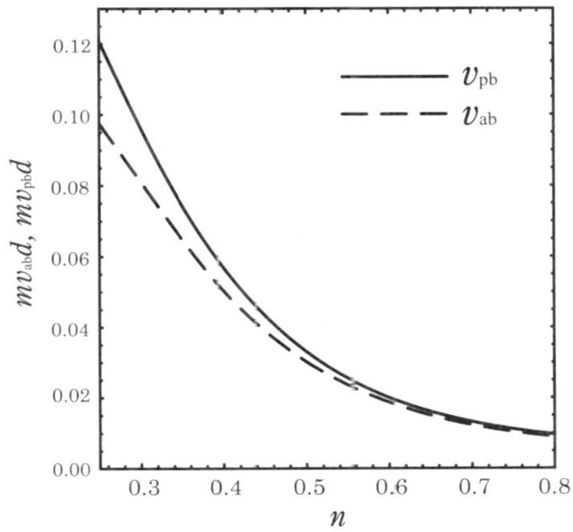

Fig. 2. Critical velocities v_{pb} and v_{ab} as functions of the filling n when $U=-1.0J$. The solid line represents v_{pb} and the dashed line represents v_{ab}.

the upper gray regions are the particle-hole continuums, and the lower boundaries of the particle-hole continuums correspond to $\omega = \min\left[E_{\boldsymbol{v}}^{+}(\boldsymbol{k}+\boldsymbol{q}) - E_{\boldsymbol{v}}^{-}(\boldsymbol{k})\right]$. The lower curves in Fig. 1 are the spectrum of AB mode, which becomes phonon-like when $qd \ll 1$.[7] It should be noted that the spectrum of the AB mode has roton-like minima at $|q| \simeq 2k_F$ where k_F is the Fermi wavevector. Such a roton-like structure also appears in the spectrum of AB mode in uniform 1D Fermi gases.[12] The energies of the roton-like minima decrease as one approaches the half-filling when $v=0$. They eventually become zero at $k=\pi/d$ at the half-filling. This suggests that exciting density fluctuations with $k=\pi/d$ costs zero energy and the superfluid ground state is in competition with the charge density wave (CDW) state. Thus, the roton-like minima reflect the strong CDW fluctuations.

In the presence of a superfluid flow with $v>0$, the spectrum slants towards the left side (see Fig. 1(b)). The critical velocity due to pair breaking v_{pb} coincides with the velocity at which the particle-hole continuum touches the line $\omega=0$. Since the AB mode spectrum locates below the particle-hole continuum, one of the roton-like minima reaches zero at the velocity v_{ab} which is smaller than v_{pb}. According to the Landau criterion,[1] this means that the spontaneous emission of the AB mode is induced before the Copper pair breaking occurs. In Fig. 2, the critical velocities v_{pb} and v_{ab} are plotted as functions of the filling n. One clearly sees that v_{ab} is always smaller than v_{pb}, and that both v_{ab} and v_{pb} decrease monotonically with increasing filling. The difference between the two velocities diminishes near the half-filling ($n \simeq 1$).

We proceed to discuss Fermi gases in 2D

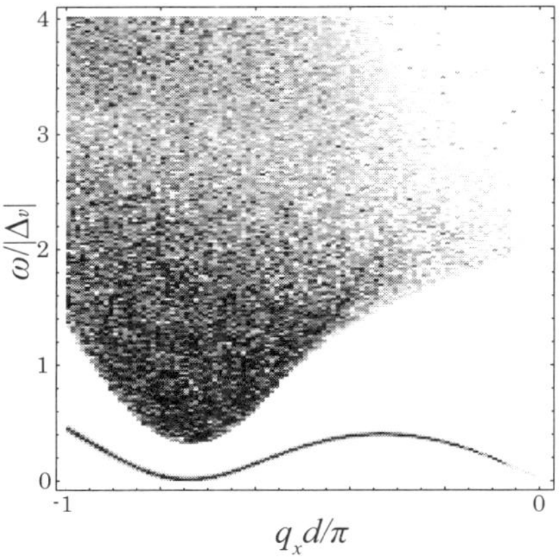

Fig. 3. Dynamical structure factor $S(\boldsymbol{q}, \omega)$ in 2D optical lattices for $q_x = q_y$. We set $U = -4.5J$, $n = 0.5$, and $\boldsymbol{v} = (v, v)/\sqrt{2}$ with $v = 0.467/md$. The dark region indicates large $S(\boldsymbol{q}, \omega)$. The upper gray region corresponds to the particle-hole continuum, and the lower curve is the spectrum of AB mode.

optical lattices. In Fig. 3, we plot the dynamical structure factor $S(\boldsymbol{q}, \omega)$ for $q_x = q_y$, when $\boldsymbol{v} = (v/\sqrt{2}, v/\sqrt{2})$. Similarly to the 1D case, the spectrum of AB mode has a roton-like minimum in Fig. 3. It reaches zero before the particle-hole continuum does when v is increased. Thus, $v_{\rm ab}$ is smaller than $v_{\rm pb}$, and the instability induced by emission of the AB mode also dominates in 2D optical lattices. Furthermore, it is found that the direction of the superfluid velocity $\boldsymbol{v}$ affects the spectrum of AB mode. The energy of the roton-like minimum decreases more rapidly for $\boldsymbol{v} = (v/\sqrt{2}, v/\sqrt{2})$ than for $\boldsymbol{v} = (v, 0)$. This implies that the CDW fluctuation is enhanced when the supercurrent is flowing in the direction of (π, π).

4. Conclusion

In summary, we have studied critical velocities of superfluid Fermi gases in 1D and 2D optical lattices using the attractive Hubbard model. We calculated the excitation spectra of single-particle excitation and AB mode by employing the GRPA. We have shown that the spectrum of AB mode has a roton-like structure at $|q| \simeq 2k_F$, and it always lies below the particle-hole continuum due to the strong CDW fluctuations in optical lattices. As the velocity of the superfluid flow is increased, the energy of a roton-like minimum decreases and the instability due to the emission of roton-like excitations was shown to occur before Cooper pair breaking occurs. We note that Burkov and Paramekanti have proposed another type of current-induced instability of superfluid Fermi gases in optical lattices.[13]

Acknowledgments

We would like to thank K. Kamide, S. Kurihara, and T. Nikuni for valuable comments and discussions. I.D. and D.Y. are supported by a Grant-in-Aid from JSPS.

References

1. L. D. Landau, J. Phys. (USSR) **5**, 71 (1941).
2. D. R. Tilley and J. Tilley, *Superfluidity and Superconductivity*, (Bristol, Hilger 1991).
3. R. Onofrio *et al.*, Phys. Rev. Lett. **85**, 2228 (2000).
4. C. A. Regal *et al.*, Phys. Rev. Lett. **92**, 040403 (2004); M. W. Zwierlein *et al.*, *ibid.* **92**, 120403 (2004).
5. D. E. Miller *et al.*, Phys. Rev. Lett. **99**, 070402 (2007).
6. R. Combescot *et al.*, Phys. Rev. A **74**, 042717 (2006).
7. P. W. Anderson, Phys. Rev. **112**, 1900 (1958).
8. L. P. Pitaevskii *et al.*, Phys. Rev. A **71**, 053602 (2006).
9. A. J. Leggett in *Modern Trends in the Theory of Condensed Matter*, edited by A. Pekalski and R. Przystawa (Springer-Verlag, Berlin, 1980).
10. R. Côté and A. Griffin, Phys. Rev. B **48** 10404 (1993).
11. M. Rodriguez and P. Törmä, Phys. Rev. A **69**, 041602(R) (2004).
12. T. Alm and P. Schuck, Phys. Rev. B **54**, 2471 (1996).
13. A. A. Burkov and A. Paramekanti, Phys. Rev. Lett. **100**, 255301 (2008).

Z_2 SYMMETRY BREAKING IN MULTI-BAND BOSONIC ATOMS CONFINED BY A TWO-DIMENSIONAL HARMONIC POTENTIAL

MASAHIRO SATO[1], AND AKIYUKI TOKUNO[2]

[1]*Condensed Matter Theory Laboratory, RIKEN, Wako, Saitama 351-0198, Japan*
[2]*Department of Applied Physics, Hokkaido University, Sapporo 060-8628, Japan*

When a two-dimensional harmonic potential is applied to bosonic cold atoms, the atoms become a cigar shape and the radial energy level is discretized. If the strength of the potential or the total number of atoms is tuned, the atoms can occupy not only the lowest radial band but also higher bands. This paper shows that when both the lowest band and the second lowest doubly-degenerate bands are filled, the repulsive interaction among atoms induces a ground state circulating along the trap axis, in which the Z_2 reflection symmetry is spontaneously broken.

Keywords: Cold Atoms, Spontaneous Symmetry Breaking, Effective Field Theories

1. Introduction

Cold-atom systems[1] have attracted much interest for more than a decade. Thanks to the development of techniques of controlling them, we can obtain cold atoms confined in various kinds of geometries and lattices. Furthermore, we can change the spatial dimensionality. This variability of dimensions and geometries is a significant feature of the cold-atom field, while it is absent in electron systems in solids.

In this paper, we focus on one-dimensional (1D) single-species Bose gases[2,3] such as polarized atoms ^{7}Li, ^{23}Na, ^{39}K, ^{87}Rb. In experiments of cold atoms, 1D systems are fabricated by applying a 2D trap potential, and the atoms take a cigar shape as shown in Fig. 1. Due to the trap potential, the radial energy level is quantized. When cold atoms are all confined in the lowest energy level, the dynamics for the radial direction is negligible and the atoms can be regarded as a purely 1D system. In general, the low-energy properties of 1D repulsively interacting bosons can be described by the gapless Tomonaga-Luttinger liquid (TLL) theory.[4] Actually, the TLL behavior has been observed in recent experiments of 1D cold Bose gases.[5,6]

If we change the strength of the potential and interactions, and the number of the atoms, the momentum distribution of the atoms also varies and then they can occupy higher radial levels as well as the lowest one. The effects of the higher energy bands have not been considered yet. We study them in the present paper by considering a harmonically trapped 1D Bose gas with both the lowest and the second lowest doubly-degenerate bands occupied. It is shown that for bosons on the second lowest degenerate bands, interband interactions yield spontaneous rotation around the confinement axis as shown by arrows in Fig. 1.[7] In the following sections, we will explain this mechanism.

2. 1D Multi-band Bose Gas

We start from a 3D Bose gas system in the presence of a 2D harmonic potential, whose

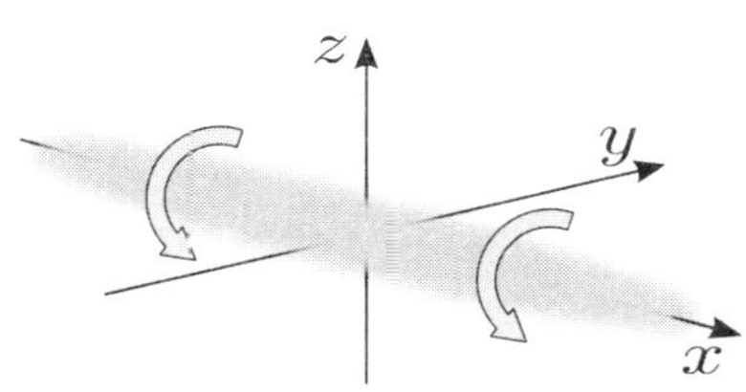

Fig. 1. The 1D atomic gases. The arrow indicates the spontaneous rotation explained in this paper.

Hamiltonian is given by

$$H = \int d\vec{r}\Big[\psi^\dagger\Big(-\frac{\hbar^2\nabla^2}{2m} - \mu + V(\vec{r})\Big)\psi + \frac{U}{2}\mathcal{D}^2 + \cdots\Big], \quad (1)$$

where $\psi(\vec{r})$ is the annihilation field and $\mathcal{D} = \psi^\dagger\psi$ is the density operator. Positive parameters m, μ and U denote the mass of the bosons, the chemical potential and the repulsive coupling constant, respectively. The 3D s-wave scattering length a is related to U via $U = 4\pi\hbar^2 a/m$. Three or more-body interactions are assumed to be quite small compared with U. The 2D confinement potential is represented by $V(\vec{r}) = \frac{1}{2}m\omega_0^2(y^2+z^2) - \hbar\omega_0$, where the axial direction is equal to the x axis (see Fig. 1).

For $U = 0$, the Hamiltonian (1) is reduced to a one-particle problem. The dynamics in the y-z plane is described by a 2D harmonic oscillator. Using its eigenstates $u_{n,l}(y,z)$, we can expand $\psi(\vec{r})$ as

$$\psi(\vec{r}) = \sum_{n=0}^{\infty}\sum_{l} u_{n,l}(y,z)\phi_{n,l}(x), \quad (2)$$

where $l = -n, -n+2, \ldots, n-2, n$ for a given n. It is important to note that the quantum number l is the angular momentum around the trapped axis. The one-particle energy spectrum is given by $\epsilon_{n,l}(k) = \frac{\hbar^2k^2}{2m} + \hbar\omega_0 n$ as shown in Fig. 2, and the bands with a fixed n are $(n+1)$-fold degenerate (k is the wavenumber for x direction). Since we concentrate on the case that the lowest three bands ($\phi_{0,0}$ and $\phi_{1,\pm1}$) are filled, the chemical potential is fixed in the region $\hbar\omega_0 < \mu < 2\hbar\omega_0$ as in Fig. 2. In addition, we impose the following conditions:

$$a/a_\perp \ll 1, \qquad (\hbar\omega_0 \gg U/a_\perp^3), \quad (3)$$

$$2\hbar\omega_0 - \mu \gg U/a_\perp^3. \quad (4)$$

Here we have defined $a_\perp = \sqrt{\hbar/m\omega_0}$ which corresponds to the width of the wave function $u_{n,l}(y,z)$. The condition (3) means that the interaction U is much smaller than the gap between the neighboring bands. As we will see later, these conditions allow us to neglect higher-band bosons $\phi_{n>1,l}$.

Substituting Eq. (2) into the Hamiltonian (1) and then integrating over variables of the 2D-plane coordinate (y,z), we obtain the effective Hamiltonian H_{1D} with fields $\phi_{n,l}(x)$, which describes the dynamics of x direction. Furthermore, we may trace out the boson fields $\phi_{n>1,l}$ in the empty bands when the low-energy properties of the Hamiltonian (1) are considered. This procedure is performed, for example, via the cumulant expansion with respect to the gapful bosons $\phi_{n>1,l}$ in the action of H_{1D}. The resulting Hamiltonian, including only occupied boson fields $\phi_{0,0}$ and $\phi_{1,\pm1}$, is expressed by

$$H_{\text{eff}} = \int dx\ \mathcal{H}_0 + \mathcal{H}_{\text{int}}, \quad (5)$$

$$\mathcal{H}_0 = \frac{\hbar^2}{2m}\sum_\alpha \partial_x\phi_\alpha^\dagger\partial_x\phi_\alpha - \mu_0\rho_0 - \mu_1(\rho_1+\rho_{-1}), \quad (6)$$

$$\mathcal{H}_{\text{int}} = \frac{U}{8\pi a_\perp^2}\Big[2\Gamma_0\rho_0^2 + \Gamma_1(\rho_1^2+\rho_{-1}^2) + 4\Gamma_{01}\rho_0(\rho_1+\rho_{-1}) + 4\Gamma_2\rho_1\rho_{-1} + 2\Gamma_3(\phi_0^\dagger\phi_0^\dagger\phi_1\phi_{-1} + h.c.) + \cdots\Big], \quad (7)$$

where we have defined $\phi_{0,0} = \phi_0$, $\phi_{1,\pm1} = \phi_{\pm1}$, and the density operator of each band $\rho_\alpha = \phi_\alpha^\dagger\phi_\alpha$. Chemical potentials are $\mu_0 = \mu > 0$ and $\mu_1 = \mu - \hbar\omega_0 > 0$. Coupling

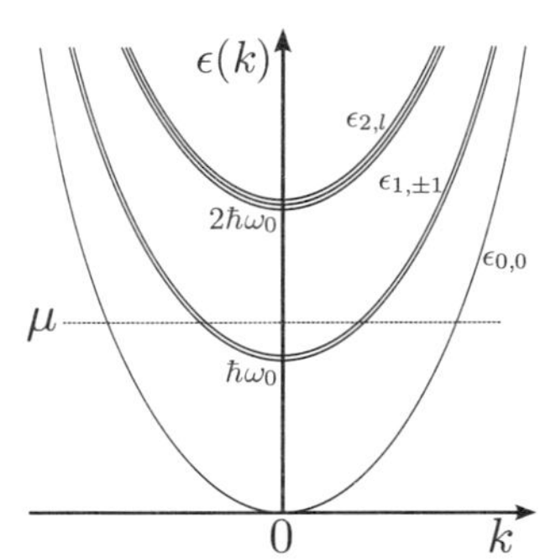

Fig. 2. The band structure of a free boson under the 2D potential $V(\vec{r})$.

constants $\Gamma_\alpha = 1$ if we entirely omit the effects of the higher-bands fields $\phi_{n>1,l}$. Using Eqs. (3) and (4), we can verify that the corrections to Γ_α are at most $O(a/a_\perp)$. This result justifies our approximation of neglecting the higher bands.

3. Spontaneous Rotation

Let us discuss a possibility of the spontaneous rotation, analyzing the effective Hamiltonian (5). For this purpose, we define $\rho_s = \rho_1 + \rho_{-1}$ and $\rho_a = \rho_1 - \rho_{-1}$. Here, ρ_s and ρ_a denote the total density and the angular-momentum density in the second lowest bands, respectively. Then, a solution with $\langle\rho_a\rangle \neq 0$, if possible, indicates a ground state rotating around the x axis.

To determine the density profiles, we first apply a mean-field approach for the potential part, i.e., the chemical potential terms and the interaction $\mathcal{H}_{\rm int}$. Before starting the mean-field analysis, we should note the following properties: (i) couplings Γ_{01} and Γ_3 combine the bosons in the lowest band and those in the second lowest bands and (ii) the distance between the neighboring bands, $\hbar\omega_0$, is much larger than the strength of the interaction $U/a_\perp^3$ [see Eq. (3)]. These suggest that effects of Γ_{01} and Γ_3 are considerably smaller than those of other intra-band couplings. Therefore, we replace Γ_{01} and Γ_3 with effective small constants $\tilde{\Gamma}_{01} \ll 1$ and $\tilde{\Gamma}_3 \ll 1$, respectively. Moreover, for a while, let us thoroughly ignore the tunneling term $\tilde{\Gamma}_3$. After these approximations, the mean-field free energy $\mathcal{F}_{\rm MF}$ is given by

$$\mathcal{F}_{\rm MF} = -\mu_0\rho_0 - \mu_1\rho_s + \frac{U}{8\pi a_\perp^2}\Big[2\Gamma_0\rho_0^2 + \Gamma_s\rho_s^2 + \Gamma_a\rho_a^2 + 4\tilde{\Gamma}_{01}\rho_0\rho_s\Big], \quad (8)$$

where $\Gamma_{s,a} = (\Gamma_1 \pm 2\Gamma_2)/2$. The mean-field equations $\partial\mathcal{F}_{\rm MF}/\partial\rho_\alpha = 0$ provide the density profiles $\langle\rho_\alpha\rangle_{\rm MF}$. We can easily confirm that an unphysical solution $\langle\rho_s\rangle < 0$ appears when we use the original $\Gamma_{01} \approx 1$ instead of $\tilde{\Gamma}_{01}$. This shows that the replacement $\Gamma_{01,3} \to \tilde{\Gamma}_{01,3}$ is reasonable. Remarkably, the negative coupling constant Γ_a indicates a finite angular momentum $\langle\rho_a\rangle_{\rm MF} \neq 0$. This negative sign originates from the fact that the inter-band interaction $4\Gamma_2$ is stronger than the intra-band one Γ_1 in the second lowest bands. The mean-field theory thus concludes that the ground state has an angular momentum and the Z_2 reflection symmetry breaking ($\rho_a \to -\rho_a$) occurs.

Next, we should consider quantum fluctuation around the mean-field result. However, such a procedure is difficult in the present case since the free energy $\mathcal{F}_{\rm MF}$ has no minimum and leads to the unphysical expectation value $\langle\rho_a\rangle_{\rm MF} \to \pm\infty$. To obtain a physical solution, we here add a phenomenological quartic interaction $\frac{U}{8\pi a_\perp^2}\frac{\Gamma_4}{4}\rho_a^4$ with $\Gamma_4 > 0$ to $\mathcal{F}_{\rm MF}$. The Γ_4 term can originate from the higher-order cumulants or multi-body interactions neglected in the Hamiltonian (1). This modified free energy has two minimums, i.e., physical solutions $\langle\rho_a\rangle_\pm = \pm[-2\Gamma_a/\Gamma_4]^{1/2}$. We can thus take into account effects of quantum fluctuation around one solution of $\langle\rho_a\rangle_\pm$. The bosonization[4] is a powerful method of treating such effects. Applying it to Eq. (5), we arrive in the bosonized Hamiltonian

$$H_{\rm TLL} \approx \int dx \sum_{q=0,s,a} \frac{v_q}{2\pi}\Big\{K_q(\partial_x\theta_q)^2 + K_q^{-1}(\partial_x\varphi_q)^2\Big\} + g_1\partial_x\theta_s\partial_x\theta_a + g_2\partial_x\varphi_0\partial_x\varphi_s + g_3\cos(2\theta_0 - 2\theta_s), \quad (9)$$

where the phase-field pair $(\varphi_\alpha, \theta_\alpha)$ is defined via the bosonization formula $\phi_\alpha \sim e^{-i\theta_\alpha}$ and $\rho_\alpha \sim \langle\rho_\alpha\rangle + \partial_x\varphi_\alpha/\pi$, and we have defined $\varphi_{s,a} = \varphi_1 \pm \varphi_{-1}$ and $\theta_{s,a} = (\theta_1 \pm \theta_{-1})/2$. In the derivation of Eq. (9), we have restored the $\tilde{\Gamma}_3$ term. The quantities v_q, K_q, and $g_{1,2,3}$ are functions of parameters in Eq. (5). In the Hamiltonian (9), the first part proportional to $v_{0,s,a}$ describes decoupled three TLLs, and the remaining terms with $g_{1,2,3}$ denote interactions among TLLs. Since the vertex operator $\cos(2\theta_0 - 2\theta_s)$ is usually relevant in the renormalization-group framework[4], a gap opens in single sector of three TLLs and the low-energy physics is described by a two-component TLL. It is worth noting that Eq. (9) does not involve any vertex operator of θ_a. It means that the quantum fluctuation of the angular momentum density $\langle \rho_a \rangle \neq 0$ is sufficiently small. Therefore, the spontaneous rotation still survives after introducing quantum fluctuation.

If Γ_4 is fairly small, $\langle \rho_a \rangle_\pm$ approaches $\pm\infty$. For this situation, singular mean-field solutions $(\langle \rho_1 \rangle, \langle \rho_{-1} \rangle) = (\mathrm{const}, 0)$ or $(0, \mathrm{const})$ should be adopted instead of $\langle \rho_a \rangle_\pm$. In the case of $\langle \rho_{-1} \rangle = 0$ $[\langle \rho_1 \rangle = 0]$, neglecting the degree of freedom of ϕ_{-1} $[\phi_1]$ in Eq. (5) and bosonizing it, we obtain a low-energy effective Hamiltonian including only two phase-field pairs $(\varphi_{0,1}, \theta_{0,1})$ $[(\varphi_{0,-1}, \theta_{0,-1})]$. It leads to a two-component TLL again. Thus we can predict that for the Hamiltonian (1), the Z_2 symmetry is spontaneously broken in the ground state and the low-energy excitations are expressed by the two-component TLL.

4. Discussions

We have discussed the effects of the second lowest degenerate bands in the Bose gas system (1) confined by a 2D harmonic potential $V(\vec{r})$. It is explained that due to the repulsive interaction between the bosons on the second lowest bands, the ground state has a finite angular momentum around the trapped axis, and the Z_2 reflection symmetry is spontaneously broken (see Fig. 1).

The key point of the mechanism of this rotation is simple: The inter-band interaction $4\Gamma_2$ is larger than intra-band interaction Γ_1 in Eq. (7). This situation and a Z_2 symmetry breaking also occur in various kinds of multi-species quantum systems[8,9,10,11]. For instance, in frustrated spin tubes,[11] the Z_2 symmetry breaking corresponds to the emergence of a vector chiral order $\langle (\vec{S}_j \times \vec{S}_k)^\alpha \rangle$.

To correct the mean-field result and to treat quantum fluctuation around it, we have introduced an artificial term with Γ_4. However, constructing more sophisticated methods for 1D multi-species bosons are necessary. Quite recently, field theory techniques of analyzing 1D two-component bosons in the strong-coupling limit have been developed.[12,13] These must be an important reference point when we develop methods for multi-species weakly interacting bosons such as the present model (5).

M.S. was supported by Grant-in-Aid for Scientific Research from MEXT (17071011) and A.T. was supported by JSPS and 21st Century COE programs at Hokkaido University 'Topological Science and Technology'.

References

1. C. J. Pethick and H. Smith, *Bose-Einstein Condensation in Dilute Gases* (Cambridge University Press, 2001); L. Pitaevskii and S. Stringari, *Bose-Einstein Condensation* (Oxford University Press, 2003).
2. M. Olshanii, *Phys. Rev. Lett.* **81**, 938 (1998).
3. M. A. Cazalilla, *At. Mol. Opt. Phys.* **37**, S1 (2004).
4. T. Giamarchi, *Quantum Physics in One Dimension* (Oxford University Press, New York, 2004).
5. T. Kinoshita, T. Wenger and D. S. Weiss, *Science* **305**, 1125 (2004); *Phys. Rev. Lett.* **95**, 190406 (2005).
6. B. Paredes, *et. al.*, *Nature (London)* **429**, 277 (2004).
7. A. Tokuno and M. Sato, *Phys. Rev. A* **78**, 013623 (2008).
8. K. Yang, *Phys. Rev. Lett.* **93**, 066401 (2004).
9. A. Kolezhuk and T. Vekua, *Phys. Rev. B* **72**, 094424 (2005).
10. M. A. Cazalilla and A. F. Ho, *Phys. Rev. Lett.* **91**, 150403 (2003).
11. M. Sato and T. Sakai, *Phys. Rev. B* **75**, 014411 (2007).
12. M. B. Zvonarev, V. V. Cheianov, and T. Giamarchi, *Phys. Rev. Lett.* **99**, 240404 (2007).
13. K. A. Matveev and A. Furusaki, *Phys. Rev. Lett.* **101**, 170403 (2008).

RECENT ADVANCES IN ANOMALOUS HALL EFFECT AND SPIN HALL EFFECT

N. NAGAOSA
Department of Applied Physics, University of Tokyo, Tokyo 113-8656, Japan

Cross-correlated Materials Research Grouyporrelated Electron Research Center, National Institute of Advanced Industrial Science and Technology, 1-1-1, Higashi, Tsukuba, Ibaraki 305-8562, Japan

Recent advances in the studies of anomalous Hall effect (AHE) and the spin Hall effect (SHE) are described with the emphasis on the (i) universal scaling of AHE, (ii) the metallic SHE, and (iii) the noncentro-symmetric superconductors as an analogue of quantum spin Hall system.

Keywords: Anomalous Hall effect; spin Hall effect; non-centrosymmetric superconductor.

1. Introduction

The dissipationless currents in condensed matter have been attracting intensive interests over decades. A representative example is the superconducting current, which flows in the thermal equilibrium without any dissipation. Anomalous Hall effect (AHE) and spin Hall effect (SHE) are the two representative phenomena where this dissipationless current plays an important role *in the normal state*, and hence are of fundamental importance. However, the microscopic mechanism of AHE has been a controversial issue of debates over 50 years [1,2,3,4]. The essential problem is if the anomalous velocity can contribute to the d.c. current in the steady state, which is attained by the dissipative processes. Therefore, one needs to consider the effect of the relaxation to treat the AHE. This issue has been attacked recently in terms of the non-equilibrium Green's function method, i.e., Keldysh formalism, and the universal behavior of the AHE is now revealed by considering the anomalous Hall conductivity (AHC) as a function of the conductivity σ_c [5]. This is the topic discussed in Section 2. With this scaling behavior, it is now possible to know which mechanism is the dominant one for the AHE. Extending this idea to the SHE, which can be regarded as the two copies of AHE, SHE in metallic systems is considered in Section 3 [6]. Another focus related to the SHE is the topological insulator and the quantum spin Hall system. In this system, the gap opens for the bulk states, while the gapless edge modes are protected by the topology. We will consider the extension of this idea to the superconductors in section 4 [7]. Namely, in a certain superconductor, the superconducting gap opens for the bulk states, while the gapless Andreev bound states appear due to the topology analogous to the quantum spin Hall system. We found that the non-centrosymmetric superconductors are a promising candidate for this novel state, and the spin transport properties of this system are examined. Secrion 5 is devoted to conclusions.

2. Universal scaling of anomalous Hall effect

Usually, the quantum transport properties of solids are characterized by the two energy scales, i.e., (i) the Fermi or kinetic energy of electrons ε_F, (ii) the inverse of the transport lifetime τ, i.e., $\hbar/\tau$. On the other hand, the conductance has the natural unit of e^2/h. In d-dimensions, the conductivity σ_c has the unit of $e^2/(ha^{d-2})$ with a being the lattice constant. For example, the conductivity in

3D due to the Boltzmann transport theory is given by $\sigma_c^B \cong (e^2/(ha))(\varepsilon_F \tau/\hbar)$ and the quantum correction due to the weak localization is an order of $\hbar/(\varepsilon_F \tau) << 1$ smaller than σ_c^B. Namely, the expansion parameter is $\hbar/(\varepsilon_F \tau) << 1$ for the metallic conducting region. In the problem of the AHE, the spin-orbit interaction (SOI) Δ enters as a new energy scale. If one assumes that the perturbative expansion in Δ is justified, the dominant term of AHC σ_H is the first order in Δ, and the following generic expansion can be obtained.

$$\sigma_H \cong (e^2/(ha))(\Delta\tau/\hbar)(c_1 + c_2(\hbar/\varepsilon_F \tau) + \cdots), \tag{1}$$

with c_1, c_2 being constants of the order of unity. The first term corresponds to the extrinsic skew scattering contribution while the second term to the intrinsic contribution and the extrinsic side jump contribution. Therefore, in the metallic region $\hbar/\varepsilon_F \tau << 1$, it is expected that the skew scattering term is always the dominant contribution and one can not identify the intrinsic mechanism for the AHE since when $\hbar/\varepsilon_F \tau \sim 1$ the metallicity itself is questionable due to the localization effect and the metal-insulator transition. However, the good agreement with the first-principles band calculations for the intrinsic AHC and the experimental results strongly suggest the metallic region where intrinsic contribution is the dominant source of the AHE [8,9]. A clue to this puzzle is the band crossings behaving as the sources/sinks of the gauge field in the momentum space, which leads to the dominant contribution to the AHC. In this case, the expansion in Δ is not justified and the nonperturbative topological effect arises due to the SOI. A band crossing can be modeled by a Dirac fermion with a mass term m. For simplicity we consider the two-dimensional model given by

$$H_0(k) = \sigma_x k_x + \sigma_y k_y + \sigma_z \Delta \tag{2}$$

with Δ being the mass term due to the SOI. The idea is that the up- and down-spin bands are independent of each other without the SOI, and hence can cross accidentally along a curve in k-space. The SOI introduces the mixing between up- and down-spins to open the gap. We expand the dispersion around the k-point where this gap is minimum to obtain eq.(2). One can calculate the Hall conductance of this model as

$$\sigma_H = (e^2/2h)\text{sign}(\Delta). \tag{3}$$

Since the infinitesimal m leads to the finite σ_H, one can not expand with respect to Δ. In the generic cases, the Fermi energy is usually not in the gap, and hence there are bands crossing the Fermi energy. This situation is mimicked by introducing the quadratic dispersion in addition to eq.(2), i.e.,

$$H(k) = H_0(k) + \frac{k^2}{2m}. \tag{4}$$

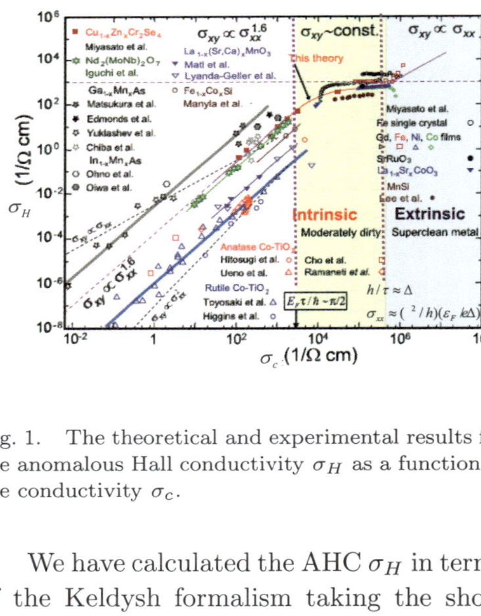

Fig. 1. The theoretical and experimental results for the anomalous Hall conductivity σ_H as a function of the conductivity σ_c.

We have calculated the AHC σ_H in terms of the Keldysh formalism taking the short range impurity potential by the T-matrix approximation. The obtained results together with the various experimental results are shown in Fig. 1 [5,10]. There are three regimes as the disorder strength increases. The most clean case, where the longitudinal conductivity σ_c is larger than $(e^2/h)(\varepsilon_F/\Delta)$, the skew scattering contribution is dominant and the relation $\sigma_H \propto \sigma_c$ holds. In the intermediate region $10e^2/h < \sigma_c < (e^2/h)(\varepsilon_F/\Delta)$, the intrinsic contribution due to Berry curvature dominates and σ_H stays more or less constant with the values of the order of e^2/h. In this regime, σ_H can be estimated by the band structure calculation. The reason for this constancy is the robustness of the intrinsic contribution due to its topological nature. It can be regarded as the "hidden quantum Hall effect due to the metalicity". In the strongly disordered regime $\sigma_c < 10e^2/h$, the new scaling law $\sigma_H \propto \sigma_c^{1.6}$ is found. These theoretical calculations are compared with the experimental data for various systems as shown in Fig.1. Here, the translation from the model calculation of 2D model to the experimental results in 3D systems have been done by replacing e^2/h by e^2/ha with $a = 5A$ is the typical lattice constant. It is evident that the predicated crossovers are clearly seen, and the new scaling $\sigma_H \propto \sigma_c^{1.6}$ is also in good agreement with the theory. From this global behavior, one can now identify the dominant mechanism of the AHE from the values of σ_c and σ_H.

3. Spin Hall effect in metals

Up to now the studies of the spin Hall effect (SHE) have been done mainly in semiconductor systems [11,12,13] Recently, SHE and the inverse spin Hall effect (ISHE) in metals are the topic of intensive interest [14,15,16]. In spite of the recent experimental advances in metallic SHE, its theoretical understanding is still lacking and is urgent. Among metallic systems, Pt shows remarkably large SHE surviving even up to room temperature [14,15], whereas aluminum and copper show relatively tiny SHE [16]. The SHC in Pt at room temperature is 240 $(\hbar/e)(\Omega\text{cm})^{-1}$, ten times larger than that of aluminum at 4.2K.

We have done an *ab initio* calculation for the SHC in Pt, and its analysis based on an effective Hamiltonian [6]. The band structure of Pt is calculated using a fully relativistic extension of the all-electron linear muffin-tin orbital method based on the density functional theory with local density approximation. The lattice constants for Pt and Al used are 3.92 and 4.05 Å, respectively. The basis functions used are s, p, d and f muffin-tin orbitals for Pt but s, p, and d muffin-tin orbitals for Al. In the self-consistent band structure calculations, 89 k-points in the fcc irreducible wedge (IW) of the BZ were used in the BZ integration. The SHC is evaluated by the Kubo formula.

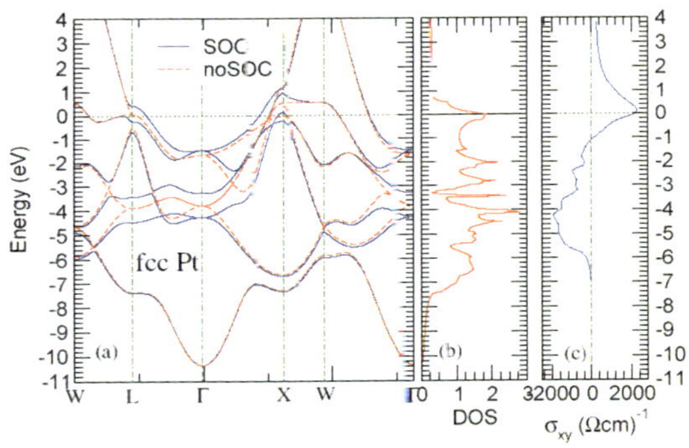

Fig. 2. (a) Relativistic band structure, (b) Density of states, and (c) spin Hall conductivity of fcc Pt. The Fermi energy is $E_F = 0$ specified by the dotted line. The dashed curves in (a) are the scalar-relativistic band structure without the spin-orbit interaction.

Figure 2 shows (a) the relativistic band structure of Pt, (b) the density of states, and (c) the SHC (σ_{xy}) as a function of E_F. The Fermi level for Pt is $E_F = 0eV$, and the rigid band approximation predicts the SHC for other 5d transition metals. The SHC has the peak at $E_F = 0$ with a large value of2200 $(\hbar/e)(\Omega\text{cm})^{-1}$, which is orders of magnitude larger than the corresponding value in p-type semiconductors Si, Ge, GaAs and AlAs [17,18] and that in simple metal Al -17 $(\hbar/e)(\Omega\text{cm})^{-1}$. We have found by analyzing the k-resolved contribution to the SHC that the region near the L and X points are dominant, where the degeneracies of the bands occur which are lifted by the SOI. Also it is found that SHC decreases monotonically as the temperature (T) is raised. This rather strong temperature dependence is also due to the near-degeneracies since the small energy scale is relevant to the SHC there. Nevertheless, the SHC $\sigma_{xy} = 240(\hbar/e)(\Omega\text{cm})^{-1}$ at $T = 300$K is still large, and is close to the measured value (240) [15]. The SHC for Al at 4 and 300 K is -17 and $-6(\hbar/e)(\Omega\text{cm})^{-1}$, respectively. The former value is similar to the experimental values (-27, -34) at 4.2 K [16].

Recently even larger SHE has been reported in FePt/Au devices at room temperature [19]. The resistivity (ρ) of Au at room temperature is $2.7\mu\Omega$cm . With this small value of ρ for Au, the main mechanism of the SHE/ISHE is considered to be the extrinsic skew scattering. They also measured both SHE and ISHE, and the obtained value of the spin Hall angle α_{SH} is 0.113, which is very large compared with the values obtained so far. This amazingly large spin Hall angle is unexpected from the anomalous Hall angle in ferromagnets. This suggests that the SHE is not the simple two copies of AHE. A possibility is that the spin fluctuation and Kondo effect can lead to the large SHE due to the skew scattering [20]. Namely, the Kondo temperature scale reduced from the hybridization energy scale compete with the SOI. Therefore, the renormalization of the energy scales due to the electron correlation is an important issue in the study of the SHE especially 3d impurities are included in the sample.

4. Non-centrosymmetric superconductors - an analogue to quantum spin Hall system

Another development in the study of SHE is the discovery of the quantum spin Hall system analogous to the quantum Hall system (QSHS). QSHS corresponds to the generalization of the topological concept to the systems with the time-reversal symmetry T, where the Z_2 topological number is defined. This number is related to the Kramers's doublet due to the T-symmetry [21,22], and is the number mod 2 of the helical mode pairs at the edge of the sample. When the number is odd, there always remain at least one pair, which is protected from the disorder scattering by the Kramers's theorem. The existence of the helical edge mode has been experimentally demonstrated for the quantum well of HgTe system by the measurement of the charge conductance [23].

An interesting issue is the superconducting analogue of QSHS. We will discuss below that the non-centrosymmetric superconductors [24,25,26,27] are a promising candidate for it. Our aim is to look for the spin transport phenomenon enhanced by the superconductivity. We will show below that the Andreev reflection associated with the helical edge modes produces the large spin polarization of the superconducting current for angled incident electrons to the normal to the interface and/or the weak external magnetic field. We consider the following Hamiltonian [7]

$$H = \sum_k c_k^\dagger (\xi_k + \vec{\lambda}(k) \cdot \vec{\sigma}) c_k + H_{\text{pairing}}, \quad (5)$$

where $c_k =^t (c_{\uparrow k}, c_{\downarrow k})$ is the two-component spinor operator of the electrons, and $\vec{\sigma}$ is the three component Paili matrix vector. The Rashba type spin-orbit interaction term is characterized by the vector $\vec{\lambda}(k)$, which satisfy $\vec{\lambda}(k) = -\vec{\lambda}(-k)$ due to the time-reversal symmetry. If the inversion symmetry is there in addition, $\vec{\lambda}(k) = \vec{\lambda}(-k)$ must be satisfied, and hence $\vec{\lambda}(k) = \vec{0}$. Therefore, nonzero $\vec{\lambda}(k)$ is allowed only for the non-centrosymmetric system. H_{pairing} is the pairing interaction term, the explicit form of which is not specified here.

First, it is seen that this superconductor belongs to the Z_2 nontrivial class as long as $\Delta_s < \Delta_p$, where Δ_s (Δ_P) is the gap value for the s- (p-)wave pairing. In the Z_2 nontrivial case, we expect the existence of the helical edge modes protected by the topology. This is realized by the Andreev bound state in the superconductor. Special to the superconductor is that the fermion operator for negative energy is related to that of the positive energy by the Hermite conjugation, i.e., the Majorana fermion. We have confirmed by explicit calculation that this is indeed the case, and there appear linear dispersing edge modes with the opposite direction which are related to the time-reversal symmetry.

We have studied the Andreev reflection associated with the helical edge modes, and calculated the incident angle dependence of the spin polarization of the superconducting current through the junction between the normal metal and the non-centrosymmetric superconductor. Let ϕ be the angle between the incident electron wavevector and the normal to the junction, and ε be its energy. When the The wavenumber parallel to the junction is given by $k = k_F \sin\phi$ (k_F: Fermi wavenumber), and when $\varepsilon = \pm vk$ is satisfied the spin conductance is enhanced and the supercurrent with large spin polarization can be produced. This is due to the Doppler effect, i.e., the Andreev reflection is sensitive to the relative velocity between the incident electron and that of the edge mode. Since the direction of the velocities are opposite to each other between up- and down-spins, leading to the large spin polarization of the supercurrent.

5. Conclusions

We have discussed the recent advances in the studies of anomalous Hall effect and spin Hall effect. Both of these phenomena are essentially topological in nature, leading to the stability against the disorder and thermal agitation. Quantum spin Hall effect is the most representative to reveal this topological nature, but also the quantized anomalous Hall effect is waiting to be discovered yet. The interplay between the electron-electron interaction and the topology will be the most important and challenging problem to be studied in the future, and we have shown that the extension to the superconductors is a fruitful direction even from the viewpoint of the applications. Recently, we have considered the possible quantum spin Hall system realized in a 5d transition metal oxide, where the strong spin-orbit interaction and the electron correlation coexist. In this case, the magnetism breaks the time-reversal symmetry, but in a nontrivial way. Therefore, the topologically nontrivial magnets will be also an interesting direction to be studied in the future.

Acknowledgments

The results reported here have been obtained in the collaboration with S. Onoda, N. Sugimoto, G.Y. Guo, S. Murakami, T.-W. Chen, S. Maekawa, and the author acknowledge them for the fruitful discussion. This was partly supported by the Grant-in-Aids from under the Grant No. 15104006, No. 16076205, and No. 17105002) and NAREGI Nanoscience Project from the Ministry of Education, Culture, Sports, Science, and Technology.

References

1. R. Karplus and J. M. Luttinger, Phys. Rev. **95**, 1154 (1954).
2. M. Onoda, and N. Nagaosa, J. Phys. Soc. Jpn. **71**, 19 (2002).
3. J. Smit, Physica **21**, 877 (1955).
4. L. Berger, Phys. Rev. B **2**, 4559 (1970).
5. S. Onoda, N. Sugimoto, N. Nagaosa, Phys. Rev. Lett. **97**, 126602 (2006).
6. G. Y. Guo, S. Murakami T.-W. Chen, N. Nagaosa, Phys. Rev. Lett. **100**, 096401 (2008).

7. Y. Tanaka et al., arXiv:0806.4639.
8. Z. Fang *et al.*, Science **302**, 92 (2003).
9. Y. Yao *et al.*, Phys. Rev. Lett. **92**,037204 (2004).
10. T. Miyasato *et al.*, Phys. Rev. Lett. **99**, 086602 (2007).
11. S. Murakami, N. Nagaosa, and S.-C. Zhang, Science **301**, 1348 (2003).
12. J. Sinova *et al.*, Phys. Rev. Lett. **92**, 126603 (2004).
13. Y. Kato *et al.*, Science **306**, 1910 (2004).
14. E. Saitoh, M. Ueda, H. Miyajima, and G. Tatara, Appl. Phys. Lett. **88**, 182509 (2006).
15. T. Kimura *et al.*, Phys. Rev. Lett. **98**, 156601 (2007).
16. S. O. Valenzuela, M. Tinkham, Nature **442**, 176 (2006).
17. G.Y. Guo, Y. Yao, and Q. Niu, Phys. Rev. Lett. **94**, 226601 (2005).
18. Y. Yao and Z. Fang, Phys. Rev. Lett. **95**, 156601 (2005).
19. T. Seki et al., Nature Materials **7**, 125 (2008).
20. G. Y. Guo et al., arXiv:0809.2921
21. C. L. Kane and E. J. Mele, Phys. Rev. Lett. **95**, 146802 (2005); C. L. Kane and E. J. Mele, Phys. Rev. Lett. **95**, 226801 (2005).
22. B. A. Bernevig, and S. C. Zhang, Phys. Rev. Lett. **96**, 106802 (2006).
23. B. A. Bernevig, *et. al.*, Science, **318**, 766 (2007).
24. E. Bauer, *et. al.*, Phys. Rev. Lett. **92**, 027003 (2004).
25. P. A. Frigeri, *et. al.*, Phys. Rev. Lett. **92**, 097001 (2004).
26. M. Sato, Phys. Rev. B **73**, 214502 (2006).
27. R. Roy, arXiv:cond-mat/0608064v1

TOPOLOGICAL INSULATORS AND THE QUANTUM SPIN HALL EFFECT

C.L. KANE
Department of Physics and Astronomy, University of Pennsylvania, Philadelphia, PA 19104, USA
E-mail: kane@physics.upenn.edu

A topological insulator is a material with a bulk excitation gap generated by the spin orbit interaction that is topologically distinct from an ordinary insulator. This distinction, characterized by a Z_2 topological invariant, necessitates the existence of gapless electronic states on the sample boundary, which have important implications for electronic transport. In two dimensions, the topological insulator is a quantum spin Hall insulator, which is a close cousin of the integer quantum Hall state. Here we will outline the theory of this phase and describe two recent experiments in which its signatures have been observed. (1) Transport experiments on HgTe/HgCdTe quantum wells have demonstrated the existence of the edge states predicted for the quantum spin Hall insulator. (2) Photoemission experiments on the semiconducting alloy $Bi_{1-x}Sb_x$ have observed the gapless surface states predicted for a three dimensional topological insulator. We will close by arguing that the proximity effect between an ordinary superconductor and a topological insulator leads to a non-trivial interface state that supports Majorana fermions and may provide a new venue for realizing proposals for topological quantum computation.

Keywords: Topological insulator; spin orbit interaction; Majorana fermion.

1. Introduction: Topological Band Theory

The insulating state is the most basic electronic phase of matter. An insulator has a band structure in which occupied and empty bands are separated by an energy gap. The energy gap, however, does not guarantee a material is a simple insulator. A counterexample is the two dimensional integer quantum Hall state, which has an energy gap between the filled and empty Landau levels in a magnetic field. Despite the gap, this state is not a conventional insulator, but rather has a quantized Hall conductivity. The classification of distinct insulating band structures was pioneered by Thouless, et al. (TKNN) [1], who showed that the quantized Hall conductivity defines an integer topological invariant. This invariant characterizes the mapping from the crystal momentum $\mathbf{k}$ (defined in the magnetic Brillouin zone - a torus) to the Bloch wavefunction $|u_\mathbf{k}\rangle$ characterizing the valence band. Due to $U(1)$ gauge in variance, $|u_\mathbf{k}\rangle$ defines a $U(1)$ fiber bundle over the torus, which is classified by its first Chern class - or Chern number,

$$n = \frac{1}{2\pi i}\int d^2\mathbf{k}\langle\nabla_\mathbf{k}u_\mathbf{k}| \times |\nabla_\mathbf{k}u_\mathbf{k}\rangle. \qquad (1)$$

The integer n classifies topologically distinct gapped band structures. n is insensitive to continuous changes in the Hamiltonian, provided the energy gap remains finite. It can only change at a phase transition, where the gap goes to zero. TKNN showed that n is precisely the quantized Hall conductivity, $\sigma_{xy} = ne^2/h$. $n = 0$ is the conventional insulator, while $n \neq 0$ is a quantum Hall state.

An important consequence of this topological structure is the existence of gapless edge states at spatial interfaces between topological classes. This is easiest to see by imagining a smooth limit where the band structure slowly interpolates between the two sides. Somewhere along the way the gap has to vanish, since otherwise the two sides would be in the same class. Gapless states are thus bound to the interface. For the interface between the integer quantum Hall state and the vacuum ($n = 0$), these are precisely the well known chiral edge states.

n is odd under time reversal, so only the trivial class $n = 0$ is allowed for time reversal invariant systems. Recently, however, a new topological invariant has been discovered for systems with time reversal symmetry[2] . This new invariant has two values $\nu = 0, 1$ in Z_2. The nontrivial "topological insulator" state, requires a spin orbit interaction. As in (1), ν can be computed from the valence band's Bloch wavefunctions. However, the mathematical formulation is somewhat more involved[2,3,5]. In order to physically motivate the Z_2 structure it is useful to consider the edge states that occur at a translationally invariant interface with the vacuum (or trivial insulator). These states disperse as a function of momentum k in a one dimensional Brillouin zone indicated in Fig 1(a). According to Kramers' theorem, time reversal invariance requires that energy eigenstates come in time reversed pairs. At two special momenta $k = \Lambda_{a=1,2}$, for which $\pm\Lambda_a$ are related by a reciprocal lattice vector, the edge states must be twofold degenerate. At other points, the degeneracy will be lifted by spin orbit interactions. There are two possibilities, shown in Fig. 1(c,d), for how the states at $\Lambda_{1,2}$ connect. In Fig. 1(c) they connect pairwise. Though edge states may be present at E_F, they are not guaranteed, and there must be an even number of Fermi points between $\Lambda_{1,2}$. In contrast in Fig. 1(d) the edge states "switch partners", and can not be eliminated. There must be an odd number of Fermi level crossings between $\Lambda_{1,2}$. Which of these alternatives occurs is clearly a topological property, which cannot change by smoothly varying the bulk Hamiltonian. It is determined precisely by the bulk Z_2 invariant ν.

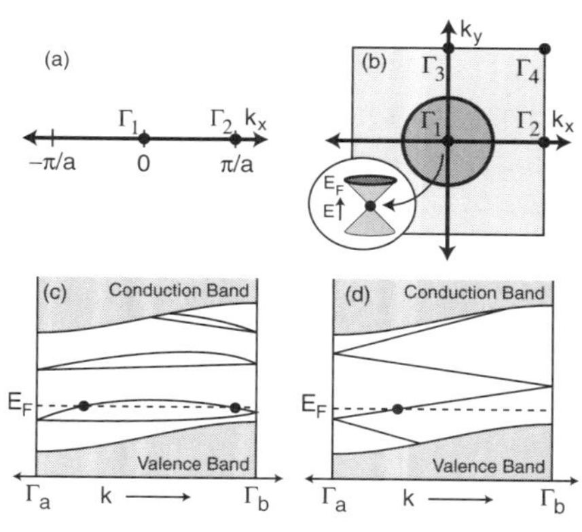

Fig. 1. Brillouin zone for (a) edge of 2D topological insulator or (b) surface of 3D topological insulator, identifying the two time reversal invariant points Γ_a. (c,d) Surface state dispersion showing two possible connections for spin orbit split surface states.

Topological insulators are also possible for three dimensional crystals[4,5,6]. It is again most useful to consider the states at a translationally invariant surface. Now, the states are labeled by momentum in a two dimensional Brillouin zone, as in Fig. 2(b). There are four distinct time reversal invariant points, where the surface states are Kramers degenerate. The surface states between any pair can resemble Fig. 2(c) or 2(d), which is determined by the topological class of the bulk crystal. It turns out that the surface state connections are determined by four Z_2 invariants, $(\nu_0; \nu_1\nu_2\nu_3)$. Three of these invariants characterize a 3D state which is essentially a stacked version of the 2D topological insulator (stacked in a direction determined by $\nu_1\nu_2\nu_3$. The fourth invariant ν_0 defines an entirely new 3D state, which we will refer to as a strong topological insulator, and characterize below in section 3.

2. 2D: Quantum Spin Hall Insulator

The 2D topological insulator is also known as a quantum spin Hall insulator (QSHI). This is because the edge states have a unique "spin filtered" property, which can not occur in an isolated purely 1D system. The edge states which cross the Fermi energy are spin polarized (though not necessarily 100%). In the case where a single band crosses E_F between $\Gamma_{1,2}$ the edge state (including its Kramers partner) have the property that opposite spins propagate in opposite directions. Such edge states have the important property that elastic backscattering is forbidden by time reversal symmetry - essentially because it requires a spin flip. The edge states are robust in the presence of disorder and weak interactions, forming a novel one dimensional state which is immune to Anderson localization.

In a special limit where the spin S_z is conserved, this state can be understood as two copies of the integer quantum Hall state[7], with Hall conductivity $\sigma_{xy} = \pm e^2/h$ for the up and down spins. There are two Chern numbers, $n_{\uparrow,\downarrow}$. Time reversal symmetry requires $n_\uparrow + n_\downarrow = 0$. $n_\sigma = (n_\uparrow - n_\downarrow)$, however, defines an integer "spin Chern number". It defines a quantized spin Hall conductivity $\sigma^s_{xy} = n_\sigma e/4\pi$, which relates the spin current $J^s = (J_\uparrow - J_\downarrow)\hbar/2e$ to the electric field: $J^s_x = \sigma^s_{xy} E_y$. The edge states follow naturally in this limit. They are simply the chiral edge states associated with the independent quantum Hall states for spin $\uparrow, \downarrow$, which propagate in opposite directions.

While the S_z conserving limit is simple to understand, it is important to emphasize that it is not general. With spin orbit interations there is no fundamental conservation law for S_z, which makes it impossible to define n_σ. Nonetheless, in the S_z conserving limit, it can be shown that the Z_2 invariant is $\nu = n_\sigma$ mod2. When S_z nonconserving terms are added, n_σ looses its meaning, but ν remains well defined. In this sense, the quantum spin Hall effect transends the S_z conserving limit.

The QSHI was introduced and characterized in graphene by E.J. Mele and the author[2]. We showed that spin orbit interactions open a gap at the Dirac point. Unfortunately the spin orbit induced gap is quite small, and may not be observable. More recently, Bernevig, Hughes and Zhang[8] proposed that the QSHI state could occur in a quantum well structure in which a layer of HgTe is sandwiched between CdTe, and that by varying the well thickness, it should be possible to tune between the QSHI state and the conventional insulator.

This prediction led to a beautiful transport experiment by the Wuerzburg group[9], which found convincing evidence of the edge states in the HgCdTe quantum well structures. For thin wells, in the conventional insulator, the transport was strongly suppressed when E_F was in the gap. In contrast, in the QSHI state they observed a conductance as large as $2e^2/h$ independent of the sample width, which is expected for the edge states.

3. 3D: Strong Topological Insulator

In a three dimensional topological insulator the four Z_2 invariants determine the structure of the surface Fermi surface for any crystal face in that it determines the number of Fermi level crossings between each pair of time reversal invariant momenta Λ_{1-4}. The strong topological invariant ν_0 specifies whether the Fermi surface encloses an even or an odd number of the Λ_a on any face[4]. When this number is odd, the surface exhibits a nontrivial metallic phase.

The simplest case is shown in the inset to Fig. 1b. The Kramers degeneracies at Γ_a are Dirac points in the 2D surface state spectrum. In the strong topological insulator, the surface Fermi surface encloses an odd number of Dirac points. This is impossible for an ordinary 2D Fermi system, where the number of Dirac points must be even. Any 2D Fermi surface may be characterized by a Berry's phase acquired by an electron when it adiabatically circles the Fermi surface. There are two time reversal invariant values for the Berry's phase: 0 and π. Since circling a Dirac point is like rotating a spin by 360°, the Berry's phase at the surface of a strong topological insulator is π. This minus sign could have important consequences for transport. For instance, the weak localization correction to the conductivity will be positive, leading to antilocalization. In fact, it can be shown that at least in the absence of electron interactions it is impossible to localize the states at the surface, even for arbitrarily strong disorder (provided the bulk mobility gap remains intact).

In general it is a nontrivial problem to determine the Z_2 topological invariants characterizing a given insulating band structure. In 3D, the trick of considering spin conserving models is of no use. Liang Fu and the author found that the situation is considerably simpler in crystals that have inversion and time reversal symmetry[6]. We showed that the Z_2 invariants are determined by the parity eigenvalues $\xi_n(\Gamma_a)$ of the occupied bulk energy bands n at the eight parity (and time reversal) invariant points Γ_a in the 3D Brillouin zone. In particular, the strong topological invariant is simply given by

$$\nu_0 = \prod_{n,a} \xi_{2n}(\Gamma_a). \qquad (2)$$

Each band is Kramers degernerate at Γ_a, and $\xi_{2n+1} = \xi_{2n}$, so each pair is to be included only once. Since the parity eigenvalues are tabulated in the band theory literature, it is feasible to use this technique to identify topological insulators.

Using this technique we predicted that the semiconducting alloy $Bi_{1-x}Sb_x$ is a strong topological insulator. Pure Bi and Sb are semimetals, which have a finite direct gap, but a negative indirect gap due to band overlap. When they are alloyed there is a range of concentration where the indirect gap becomes positive. Moreover, when x is varied from 0 to 1 there is an inversion transition where the *direct* gap goes to zero at a 3D Dirac point. For the range of x where the crystal is semiconducting, the system is on the Sb side of this inversion transition. By looking up the known band structure of Sb and Bi, we deduced that Bi has the trivial topological invariant, while Sb has the non trivial one. From this we concluded that the alloy is a strong topological insulator.

A recent experiment has verified this prediction. Hsieh et al. [10] performed a detailed angle resoloved photoemission spectrocopy (ARPES) study of $Bi_{.9}$ $Sb_{.1}$. By mapping the dispersion of the surface energy bands along lines connecting the Γ_a they were able to establish that there are an odd number of level Fermi level crossings along the lines connecting the $\bar{\Gamma}$ point ($\mathbf{k} = 0$) to the three equivalent $\bar{M}$ points. This proves that the metallic surface states of the alloy have the topologically protected properties described above.

4. Superconducting Proximity Effect and Topological Quantum Computation

Now that the strong topological insulator state is known to exist, we are faced with the question of what we can do with it. In this section we will explore a tantalizing possibility, which is to consider the interface between an ordinary superconductor and a topological insulator[11]. We will argue that this leads to a highly nontrivial state, which might be useful for realizing schemes for topological quantum computation.

When a superconductor is in contact with the surface of a topological insulator, Cooper pairs from the superconductor can tunnel into the surface states. This leads to the proximity effect, where the surface states themselves acquire a superconducting gap which, though it may be smaller than that of the bulk superconductor, has the same phase. We will consider a simple model of the surface states by considering a Fermi surface which encloses a single Dirac point. This may be described by a four component theory, which combines a two component Dirac spinor with a two component Bogoliubov particle-hole spinor. The Bogoliubov de Gennes Hamiltonian is[11]

$$\mathcal{H} = -iv\tau^z\sigma\cdot\nabla - \mu\tau^z + \Delta_0(\tau^x\cos\phi + \tau^y\sin\phi), \quad (3)$$

where Pauli matrices $\vec{\sigma}$ operate on the Dirac indices and $\vec{\tau}$ operate on the particle-hole indices. v is the velocity of the Dirac electrons, and μ is the chemical potential, which need not be at the Dirac point $E = 0$. Δ_0 defines the magnitude of the induced superconducting gap, and ϕ is the phase, inherited from the bulk superconductor. For a homogeneous system, the energy spectrum is $E(\mathbf{k}) = \sqrt{(v|\mathbf{k}| - \mu)^2 + \Delta_0^2}$. This is different from an ordinary superconductor because the surface states do not have a spin degeneracy. There is similarity to a *spinless* electron superconductor. However, in that case the Pauli principle requires the order parameter to be p wave, and it must be $p_x + ip_y$ to avoid nodes in the gap. Unlike the $p_x + ip_y$ superconductor, Eq. (3) does not violate time reversal symmetry.

An important feature of $p_x + ip_y$ superconductors is that they have bound states at exactly zero energy in the cores of a vortex. We showed that this is also the case for (3) by solving the BdG equation $\mathcal{H}\xi = E\xi$ with $\Delta(r,\theta) = \Delta_0 e^{i\theta}$. The zero energy states are nontrivial because they are really "half" a state. In the Bogoliubov theory, the Hilbert space is doubled, so that positive and negative energy states are redundant. The spectrum pairs every positive energy state with a negative energy state. The pair defines a single fermionic degree of freedom. A *pair* of vortices defines a single zero energy fermionic state, which can be either empty or occupied. The ground state of a pair of vortices thus has a twofold degeneracy. The fact that the two vortices can be spatially separated means that the quantum information in that two level system is stored in a highly nonlocal way. Similar physics has been proposed to occur in the $\nu = 5/2$ quantum Hall effect[12] - though it has not yet been observed.

This groundstate degeneracy has been proposed as a basis for topological quantum computation[13]. $2N$ vortices lead to a 2^N fold ground state degeneracy, which encodes N qubits of quantum information. Moreover, processes where the vortices are adiabatically rearranged (or braided) perform nontrivial unitary operations on the states. Though this system does not have the structure necessary to produce a universal quantum computer, the quantum information present in the system is topologically protected from local forms of decoherence.

In order to engineer the presence of vortices and to allow for them to be created, manipulated and measured we consider junctions between superconductors mediated by the surface of a topological insulator. Consider first a linear junction, shown in Fig. 2(a). To model this, we consider the two dimensional theory with $\Delta(x,y) = \Delta_0$ for $y < -W/2$, $\Delta(x,y) = \Delta_0 \exp i\phi$for $y > W/2$ and $\Delta_0 = 0$ for $|y| < W/2$. Of interest is the spectrum of Andreev bound states in the channel, shown as a function of momentum k_x along the channel for various values of the phase difference ϕ across the junction. For $\phi = \pi$ we find that the spectrum is gapless, and has a single pair of right and left moving bands crossing $E = 0$. This resembles an ordinary spinless 1D fermi gas, however, it must be remembered that the positive and negative energy states are redundant. It is thus "half" a 1D fermi gas, or a non chiral "Majorana quantum wire". Away from $\phi = \pi$ a gap opens in the spectrum.

Next consider a tri-junction where three linear junctions meet. When the phases are 0, $2\pi/3$ and $4\pi/3$ it is similar to having a vortex at the junction, and by solving the BdG equation in this geometry it can be verified that a zero energy Majorana bound state is present. When the phases are changed, the bound state can not disappear until the gap vanishes on one of the three arms. As a function of the phases, there is thus a region of stability for the Majorana bound states, shown in Fig. 2c.

Consider next a pair of tri junctions connected by single linear junction, shown in Fig 2d. By controling the phases it is in principle possible to create, manipulate and measure the states of the Majorana fermions. In the top panels, the phases are adjusted so that no Majorana boundstates are present. If the phase across the linear junction is adiabatically tuned through π, then Majorana states appear at both tri junctions, in the state which is connected to the vacuum. On the other hand, if we start with two Majoranas and change the phase through π then the Majoranas disappear, and depending on the state of the Majoranas, the system either returns to the groundstate or a state with one extra quasiparticle. In principle, the difference these two states can be distinguished by measuring the current across the junction. Finally, in the phases are adjusted

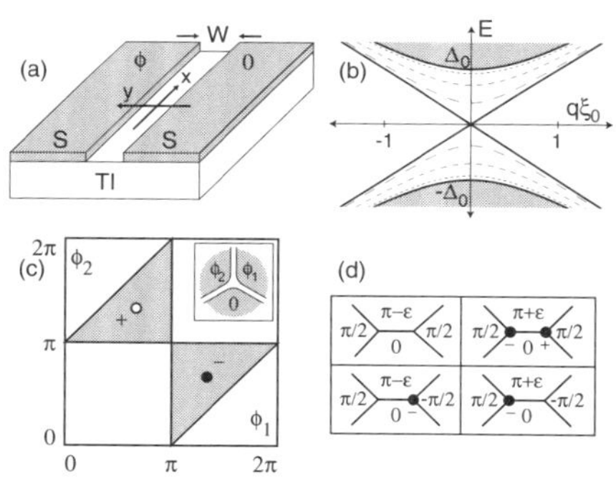

Fig. 2. (a) A STIS line junction. (b) Energy levels of the junction as a function of momentum for various phases ϕ. Solid lines show bulk quasiparticles and the Andreev bound states for $\phi = \pi$. The dashed lines are for $\phi = 3\pi/4$, $\pi/2$ and $\pi/4$. (c) Phase diagram for the tri-junction (shown in inset). In the shaded regions there is a Majorana at the junction. (d)Two tri-junctions joined by a linear junction allow Majorana states to be created, transported and fused.

so that one Majorana is present, then by tuning the phase through π the Majorana can be moved from one junction to the other. One can imagine larger arrays, where these operations can be used to create, braid and fuse Majorana states.

5. Conclusion

A new electronic phase of matter - the topological insulator - has been predicted and observed. This opens the door to experiments probing the electronic, magnetic and superconducting properties of this state. We hope that this work stimulates further experimental and theoretical interest in this subject.

Acknowledgments

We thank Gene Mele, Liang Fu and Jeffrey Teo for collaboration in this work. This work was supported by NSF grant DMR-0605066 and ACS PRF grant 44776-AC10.

References

1. D.J. Thouless, M. Kohmoto, M.P. Nightingale and M. den Nijs, *Phys. Rev. Lett* **49**, 405 (1982).
2. C.L. Kane and E.J. Mele, *Phys. Rev. Lett.* **95**, 2 (2)6801 (2005); *Phys. Rev. Lett.* **95**, 146802 (2005).
3. L. Fu and C.L. Kane, *Phys. Rev.* **B74**, 195312 (2006).
4. L. Fu, C.L. Kane and E.J. Mele *Phys. Rev. Lett.* **98**, 106803 (2007).
5. J. Moore and L. Balents, *Phys. Rev.* **B75**, 121306 (2007).
6. L. Fu and C.L. Kane, *Phys. Rev.* **B76**, 045302 (2007).
7. F.D.M. Haldane, *Phys. Rev. Lett.* **61**, 2015 (1988).
8. A. Bernevig, T. Hughes and S.C. Zhang, *Science* **314**, 1757 (2006).
9. M. König, et al., *Science* **318**, 766 (2007).
10. D. Hsieh, *et al.*, *Nature* **452**, 970 (2008).
11. L. Fu and C.L. Kane, *Phys. Rev. Lett.* **100**, 096407 (2008).
12. N. Read and D. Green, *Phys. Rev.* **B61**, 10267 (2000).
13. A. Kitaev, *Ann. Phys.* **303**, 2 (2003).

APPLICATION OF DIRECT AND INVERSE SPIN-HALL EFFECTS: ELECTRIC MANIPULATION OF SPIN RELAXATION AND ELECTRIC DETECTION OF SPIN CURRENTS

K. ANDO[1*] and E. SAITOH[1,2]

[1]*Department of Applied Physics and Physico-Informatics, Keio University, Yokohama 223-8522, Japan*

[2]*PRESTO, Japan Science and Technology Agency, Sanbancho, Tokyo, 102-0075, Japan*

[*]*E-mail: kazuya_ando@z8.keio.jp*

The spin-Hall effect (SHE) and the inverse spin-Hall effect (ISHE) coupled with magnetization dynamics were investigated using a simple $Ni_{81}Fe_{19}$/Pt film. A spin current generated by magnetization dynamics was detected electrically using ISHE. The observed magnetic field angle dependence of the ISHE signal is well reproduced by a model calculation based on the dc spin pumping and ISHE. In the same system, we found that spin relaxation in the $Ni_{81}Fe_{19}$ layer is manipulated electrically using SHE. An electric current applied to the Pt layer exerts the spin torque on the entire magnetization of the $Ni_{81}Fe_{19}$ layer via the macroscopic spin transfer induced by SHE, which modulates spin relaxation in the $Ni_{81}Fe_{19}$ layer. This spin-relaxation modulation enables quantitative measurements of spin currents without assuming any microscopic parameters.

Keywords: Spin-Hall effect; Spin pumping; Spin torque.

1. Introduction

A spin current, a flow of electron spins in a solid, plays a key role in spintronics.[1,2] A novel spintronic phenomenon, called the spin-Hall effect (SHE),[3–17] has been predicted. SHE refers to the generation of a spin current transverse to an applied electric current as shown in Fig. 1. Two different mechanisms have been proposed for SHE: one is due to asymmetry in electron scattering for up and down spins, which is the extrinsic SHE.[3,4,12] The other is due to topological band structures, which is the intrinsic SHE.[5,6] Both mechanisms ultimately originate from the relativistic effect in electron motion, or the spin-orbit interaction.

The spin-orbit interaction responsible for SHE is also expected to cause the inverse process of SHE: the inverse spin-Hall effect (ISHE),[9–11,13–15,17] a process that converts a spin current $\bar{\mathbf{J}}_s$ into an electric current $\mathbf{J}_c$. ISHE has the same symmetry as that of the relativistic transformation of magnetic moment into electric polarization, which is derived from the Lorentz transformation, as follows. Consider a magnet with the magnetic moment $\mathbf{M}$ moving at a constant velocity v along the z axis with respect to an observer. This motion of the magnet means an existence of a "spin current". In the observer's coordinate system (see Fig. 2), the Lorentz transformation converts a part of this magnetic moment into an electric dipole moment

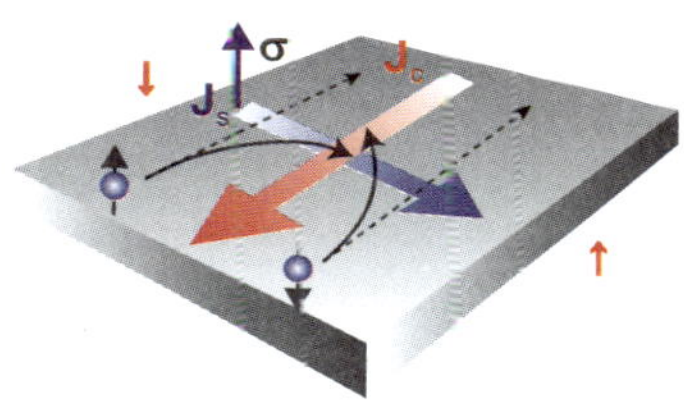

Fig. 1. A schematic illustration of the spin-Hall effect. $\mathbf{J}_c$, $\mathbf{J}_s$ and $\boldsymbol{\sigma}$ denote a charge current, the spatial direction of a spin current, and the spin-polarization vector of the spin current, respectively.

$$\mathbf{P}' = -\frac{1}{\sqrt{1-(v/c)^2}}(\epsilon_0 \mathbf{v} \times \mathbf{M}), \qquad (1)$$

where c and ϵ_0 are the light velocity and the electric constant, respectively. This means

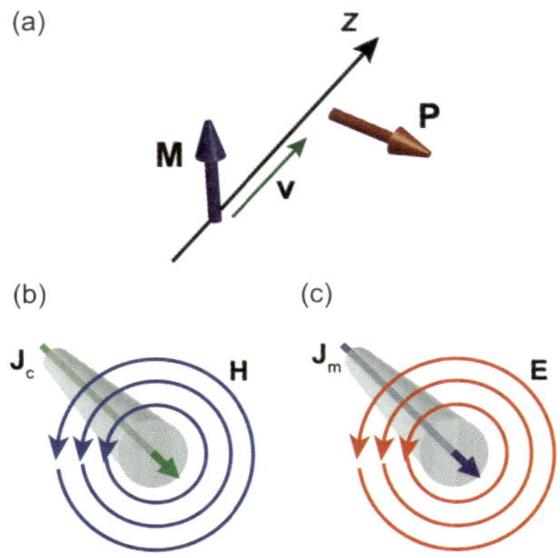

Fig. 2. (a) A schematic illustration of the Lorentz transformation of magnetic moment **M** into electric polarization **P**. (b) A schematic illustration of the magnetic-field generation from an electric current $\mathbf{J}_c$. (c) A schematic illustration of the electric-field generation from a hypothetical magnetic-monopole current $\mathbf{J}_m$.

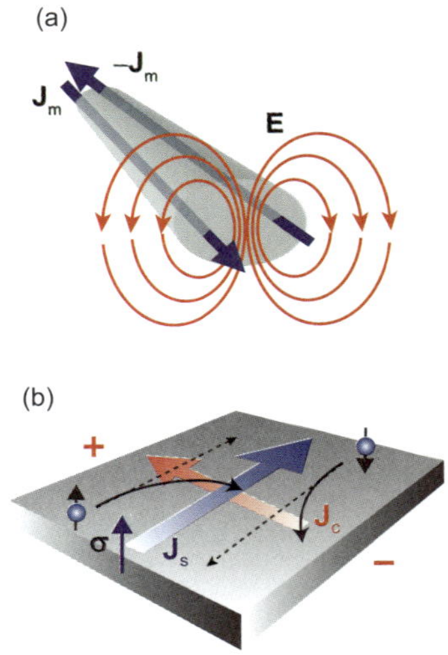

Fig. 3. (a) A schematic illustration of the electric-field generation from a pair of hypothetical magnetic-monopole currents, $\mathbf{J}_m$ and $-\mathbf{J}_m$, or a spin current. (b) A schematic illustration of the inverse spin-Hall effect induced by two electrons with opposite spins in a solid.

that electric polarization perpendicular to the direction of the magnetic-moment velocity is induced.

This electric-polarization generation can also be regarded as the spin-current version of Ampere's law as follows. As shown in Fig. 2(b), when an electric current flows, a circular magnetic field is induced around the electric current, according to Ampere's law. If hypothetical magnetic monopoles flow, a circular electric field **E** is expected to be induced around the monopole current[18] $\mathbf{J}_m$ according to rot$\mathbf{E} = \mathbf{J}_m$(see Fig. 2(c)), from the electromagnetic duality. Although this monopole has never been observed in reality, a spin current can be regarded as a pair of the hypothetical monopole currents flowing in the opposite directions along the spin current spatial direction (see Fig. 3(a)). Therefore, a spin current may generate an electric field and this field is the superposition of the two electric fields induced by this pair of the monopole current, as shown in Fig. 3(a). This spin-current-induced electric field is identical to the filed induced by the dipole moment described by Eq. 1.

In this way, electromagnetism and relativity predict that a spin current generates an electric field. According to Eq. 1, however, this electric field is too weak in a vacuum to be detected in reality. In a solid with strong spin-orbit interaction, in contrast, a similar but strong conversion between spin currents and electric fields appears. This is the inverse spin-Hall effect (ISHE).

In a solid, considering two electrons traveling in the opposite directions along $\mathbf{J}_s$, where $\mathbf{J}_s$ represents the spatial direction of the spin current $\bar{\mathbf{J}}_s$ (see Fig. 3(b)), these two electrons tend to have opposite spins: parallel and antiparallel to the spin-current polarization $\boldsymbol{\sigma}$. The spin-orbit interaction bends these two electrons in the same direction and induces a charge current transverse to $\mathbf{J}_s$, which is ISHE. The relation among $\mathbf{J}_s$, $\mathbf{J}_c$, and $\boldsymbol{\sigma}$ is therefore given by

$$\mathbf{J}_c \propto \mathbf{J}_s \times \boldsymbol{\sigma}. \tag{2}$$

This equation is similar to Eq. 1 but this effect may be enhanced by the strong spin-orbit interaction in solids. Harnessing both ISHE and SHE enables the reciprocal conversion between a spin current and a charge current, which will be essential in harmonizing spintronics with conventional electronics.

In this paper, we demonstrate the electric detection of spin currents using ISHE and the electric manipulation of spin relaxation using SHE in a $Ni_{81}Fe_{19}$/Pt film.

2. Electric detection of spin currents using the inverse spin-Hall effect

ISHE was recently observed using the spin-pumping operated by ferromagnetic resonance (FMR)[9,17,19] and by the non-local method.[10,11,14] ISHE enables the electric detection of spin currents, even in the absence of spin accumulation. In particular, the spin pumping method requires only a simple ferromagnet/paramagnet bilayer film, which allows us to explore spin currents in a wide range of materials.

The experimental method for detecting ISHE induced by the spin pumping in a ferromagnetic/paramagnetic metallic bilayer film is as follows.[9] In the FMR condition, the steady magnetization precession in a ferromagnetic layer is maintained by balancing the absorption of the applied microwave and the dissipation of the spin angular momentum of the magnetization. This dissipation is partly due to the transfer of spin angular momentum from the local spins to conduction electrons, which polarizes the conduction-electron spins. When a paramagnetic metal is connected to the ferromagnetic metal, this spin polarization propagates into the metal as a spin current, which is the spin pumping operated by FMR.[20,21] The injected spin current in the paramagnetic metal is converted into a charge current or an electric field by ISHE in the paramagnetic metal, which causes a difference of electric potential between the edges (see Fig. 4(b)). By measuring the electric potential difference, we can detect ISHE, or spin currents.

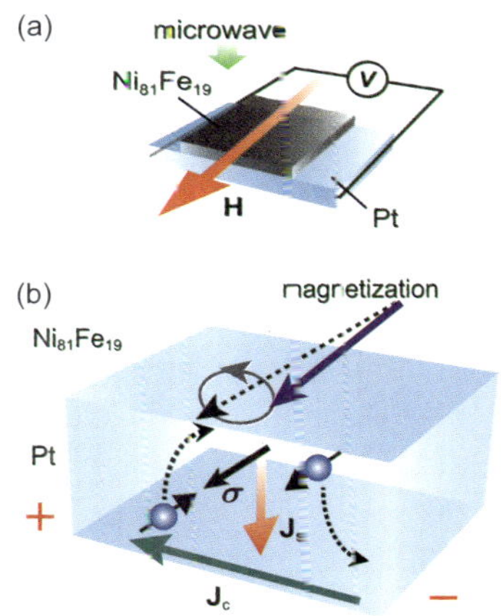

Fig. 4. (a) A schematic illustration of the $Ni_{81}Fe_{19}$/Pt system used in the present study. **H** is the external magnetic field. (b) A schematic illustration of the spin pumping and the inverse spin-Hall effect in the present system. **M** is the magnetization in the $Ni_{81}Fe_{19}$ layer.

2.1. *Experimental procedure*

Figure 4(a) shows a schematic illustration of the sample used in the present study. The sample is a $Ni_{81}Fe_{19}$/Pt bilayer film comprising a 10-nm-thick ferromagnetic $Ni_{81}Fe_{19}$ layer and a 10-nm-thick paramagnetic Pt layer. The Pt layer was sputtered on a thermally oxidized Si substrate and then the $Ni_{81}Fe_{19}$ layer was evaporated in a high vacuum. Two electrodes are attached to both ends of the Pt layer.

For the measurement, the sample system is placed near the center of a TE_{011} cavity at which the magnetic-field component of the microwave mode is maximized while the electric-field component is minimized. During the measurement, a microwave mode with frequency $f = 9.441$ GHz exists in the cavity and the external magnetic field **H** along the film plane is applied perpendicular to the direction across the electrodes, as illustrated in Fig. 4(a). When H and f fulfill the FMR condition, a pure dc spin current with a spin polarization $\boldsymbol{\sigma}$ is resonantly injected into the Pt layer by the spin pumping (see Fig. 4(b)). We measured the FMR signal and the electric potential difference V between the electrodes attached to the Pt layer. All the measurements were performed at room temperature.

2.2. *Results and discussion*

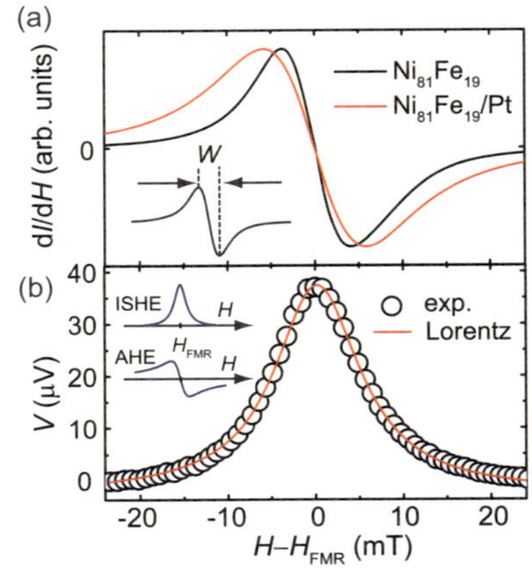

Fig. 5. (a) Field (H) dependence of the FMR signal $\mathrm{d}I/\mathrm{d}H$ for the $Ni_{81}Fe_{19}$/Pt film and the $Ni_{81}Fe_{19}$ film. Here, I denotes the microwave absorption intensity. H_{FMR} is the resonance field. The inset shows the definition of the spectral width W in the present study. (b) Field dependence of the electric-potential difference V for the $Ni_{81}Fe_{19}$/Pt film under the 200 mW microwave excitation. The open circles are the experimental data. The curve shows the fitting result using a Lorentz function for the V data. The inset shows the spectral shape of the electromotive force due to the inverse-spin Hall effect (ISHE) and the anomalous-Hall effect (AHE).

Figures 5(a) and 5(b) show the FMR spectrum $\mathrm{d}I(H)/\mathrm{d}H$ and the dc electromotive force signals V for the $Ni_{81}Fe_{19}$/Pt film under the 200 mW microwave excitation, respectively. Here, I denotes the microwave absorption intensity. For comparison, the FMR spectrum for a similar $Ni_{81}Fe_{19}$ film where the Pt layer is missing is shown in Fig. 5(a). The spectral width W for the $Ni_{81}Fe_{19}$ system is clearly enhanced by attaching the Pt layer. Since the spectral width W is proportional to the Gilbert damping constant[22] α, this enhancement of the spectral width shows that the magnetization-precession relaxation is enhanced by attaching Pt. This relaxation enhancement is attributed to the emission of a spin current from the magnetization-precession motion. Since a spin current carries spin angular momentum, this spin-current emission deprives the magnetization of its spin angular momentum and thus gives rise to additional magnetization-precession relaxation.

In the V spectra, an unconventional signal appears around the resonance field H_{FMR}. This unconventional signal is relevant to ISHE, as described in the literatures.[9,17,19] Notable is that this electromotive force signals are well reproduced using Lorentz functions as shown in Fig. 5. This indicates that the anomalous-Hall effect (AHE) contribution due to the in-plane component of a microwave-electric field is eliminated in this measurement; ISHE and AHE are distinguishable in terms of the spectral shape[19] as shown in the inset to Fig. 5.

The external magnetic field angle dependence of the electromotive force is well reproduced by the model calculation based on ISHE induced by the spin pumping. In Fig. 6, we show the out-of plane magnetic field-angle dependence of the normalized ISHE singal $V_{\mathrm{ISHE}}/V_{\mathrm{max}}$ measured for the $Ni_{81}Fe_{19}$/Pt film, where V_{ISHE} is estimated as the peak height of the resonance shape in the V spectrum. θ_H is the external magnetic field angle to the normal axis of the film plane (see the inset to Fig. 6). Notable is the drastic change observed in $V_{\mathrm{ISHE}}/V_{\mathrm{max}}$ across $\theta_H = 0$. This feature is well reproduced by a calculation of V_{ISHE} based on the LLG equation combined with the phenomenological model of the spin pumping and ISHE described in Eq. 2. Since the dc component of the spin current generated by the spin pumping is proportional to the time averaged damping torque, in our sample system, V_{ISHE} induced by ISHE and the spin pumping is the time average of the projection of $\mathbf{M} \times \mathrm{d}\mathbf{M}/\mathrm{d}t$ to the film plane, because of

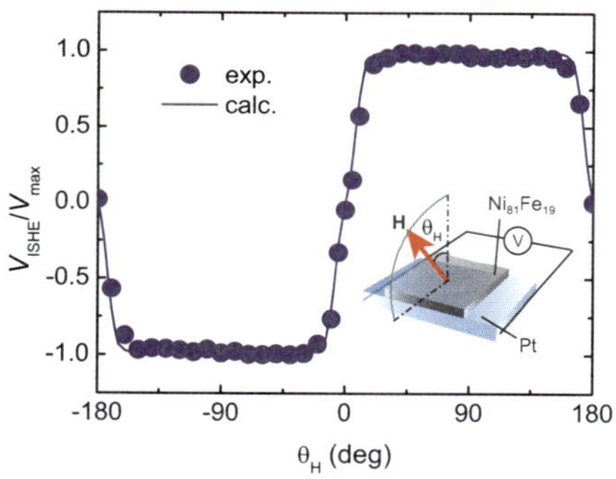

Fig. 6. The out-of-plane magnetic-field-angle dependence of the ISHE signal $V_{\rm ISHE}$ measured for the $Ni_{81}Fe_{19}$/Pt film. θ_H is the magnetic field angle.

the relation $\mathbf{J}_c \propto \mathbf{J}_s \times \boldsymbol{\sigma}$. This model leads the out-of plane magnetic field-angle dependence of $V_{\rm ISHE}$ as[17]

$$V_{\rm ISHE}(H_{\rm FMR}) \propto \frac{4\pi M_s \gamma h_{\rm ac}^2}{(4\pi\alpha)^2} \times \frac{\sin\theta_M \left[\hat{H}_{\rm FMR}\cos(\theta_M - \theta_H) - \cos 2\theta_M\right]}{\left[2\hat{H}_{\rm FMR}\cos(\theta_M - \theta_H) - \cos 2\theta_M - \cos^2\theta_M\right]^2}, \tag{3}$$

where $\hat{H}_{\rm FMR} = H_{\rm FMR}/4\pi M_s$. Here, $h_{\rm ac}$ and θ_M are the microwave magnetic field and the magnetization angle to the normal axis to the film plane, respectively. The magnetization angle θ_M is readily estimated at each θ_H value from the values of the ferromagnetic resonance fields $H_{\rm FMR}$.

We compared the experimental result with Eq. (3) and found that the experimentally measured field-angle dependence of $V_{\rm ISHE}$ is well reproduced by Eq. (3) as shown in Fig. 6 (solid line). This demonstrates again the validity of the ISHE model described in Eq. (2).

3. Electric manipulation of spin relaxation using the spin-Hall effect

In the above experiment, we found that, in a $Ni_{81}Fe_{19}$ film, the spectral width W or α in the FMR spectrum is enhanced by attaching the Pt layer as shown in Fig. 5(a). This enhancement is due to the emission and the absorption of a spin current induced by the spin pumping. Now, consider the inverse of the above electric current generation due to ISHE induced by the spin pumping; what happens when an electric current is injected in the metal with strong SO interaction attached to the magnet? One may expect from the reciprocity that spin relaxation α, namely, the width of the FMR spectra, may be modulated via SHE in the metal, enabling manipulation of the magnetization-precession relaxation of the film in an electric manner. In this section, we demonstrate that this is actually the fact in a metallic magnet attached to a Pt film. In a $Ni_{81}Fe_{19}$/Pt film, spin relaxation is manipulated by an electric current due to the macroscopic spin transfer induced by SHE. The model calculation based on the standard Valet-Fert model and the spin torque well reproduces the experimental results and indicates that this effect can also be used as a "spin meter".[16]

3.1. *Experimental procedure*

The sample is a $Ni_{81}Fe_{19}$/Pt bilayer film comprising a 10-nm-thick ferromagnetic $Ni_{81}Fe_{19}$ layer and a 10-nm-thick paramagnetic Pt layer, as illustrated in Fig. 7(a). The surface of the $Ni_{81}Fe_{19}$ layer is of a 0.48-mm^2 rectangular shape. The FMR spectra were measured with applying an electric current $\mathbf{J}_c$ through electrodes attached to both ends of the Pt layer.

In this Pt layer, SHE converts $\mathbf{J}_c$ into a pure spin current, $\mathbf{J}_s$, which propagates into the $Ni_{81}Fe_{19}$ layer through the $Ni_{81}Fe_{19}$/Pt interface. The spin polarization of this spin current is directed along $\mathbf{J}_c \times \mathbf{n}$ (see Fig. 7(b)),[4] where $\mathbf{n}$ represents the normal vector of the interface. During the measurement, a microwave mode with frequency of $f = 9.441$ GHz exists in the cavity, and an

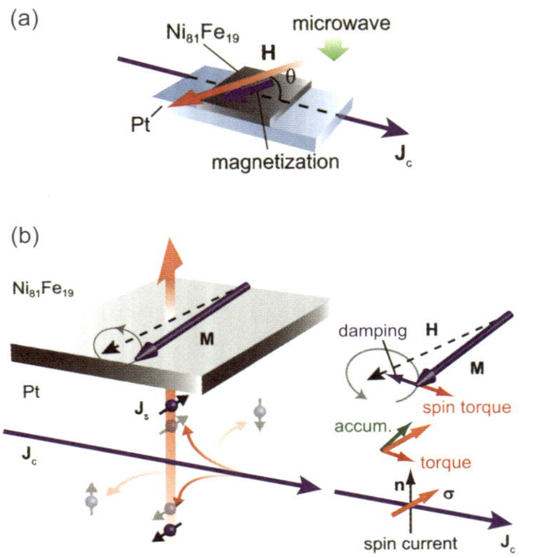

Fig. 7. (a) A schematic illustration of the $Ni_{81}Fe_{19}$/Pt system used in the present study. **H** and $\mathbf{J}_c$ represent the external magnetic field and the applied electric current, respectively. (b) A schematic illustration of the spin-Hall and the spin-torque effects in the present system. **M**, $\mathbf{J}_s$, and $\boldsymbol{\sigma}$ denote the magnetization in the $Ni_{81}Fe_{19}$ layer, the flow direction of the spin current, and the spin polarization vector of the spin current, respectively.

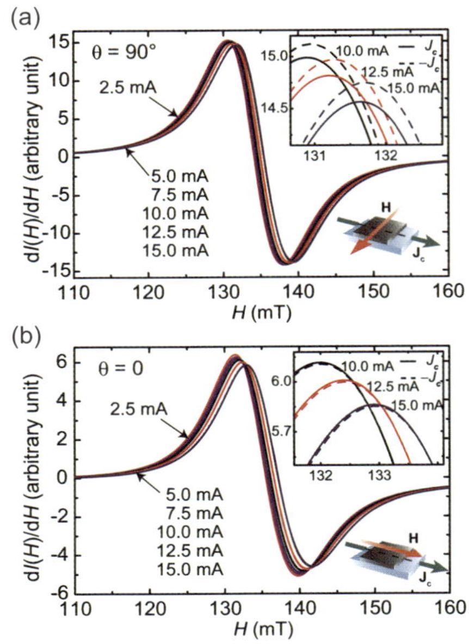

Fig. 8. The FMR spectra for the $Ni_{81}Fe_{19}$/Pt bilayer film measured at various electric current values J_c when the magnetic field direction is (a) $\theta = 90°$ and (b) $\theta = 0$. The inset shows magnified views around the peaks of the spectra, where the solid and dashed curves are the FMR spectra measured with electric currents J_c and $-J_c$, respectively.

external magnetic field **H** is applied along the film plane at the angle θ to the direction across the electrodes (see Fig. 7(a)). All the measurements were performed at room temperature.

3.2. *Results and discussion*

Figure 8(a) shows the FMR spectra measured at various electric current values J_c when $\theta = 90°$. Each spectrum is well reproduced by a simple Lorentz function. The rough current-dependent spectral features, that is, the resonance field increases and the spectral intensity decreases with $|\mathbf{J}_c|$, are attributed to joule heating due to the applied electric current.[23]

The most notable feature is found when exploring the current-polarity dependence of the FMR spectra. In the insets to Figs. 8(a) and 8(b), we show magnified views around the peaks of the spectra at $\theta = 90°$ and 0, respectively. At $\theta = 0$, the spectral shapes are little changed by reversing the electric current polarity (compare the solid and dashed curves), showing that the current-induced spectral change is due entirely to the heating effect when $\theta = 0$. In contrast, the spectral shapes are clearly modulated by reversing the current polarity when the external field direction is perpendicular to the applied electric currents ($\theta = 90°$, see the inset to Fig. 8(a)), demonstrating that the spectral shape is strongly dependent on the electric current at $\theta = 90°$.

In order to investigate in detail this unconventional current-induced modulation of the spectral shape, in Fig. 9, we plot $\Delta W \equiv W^*(J_c) - W^*(-J_c)$: the asymmetric component of spectral width W^* with respect to J_c, where $W^*(J_c) \equiv W(J_c)/W(0)$. ΔW at $\theta = 0$ is almost zero for all the J_c values as shown in the inset to Fig. 9. In contrast, when $\theta = 90°$, ΔW clearly increases with J_c in the $Ni_{81}Fe_{19}$/Pt film, demonstrating that the spin relaxation α is manipulated electrically when $\theta = 90°$.

The $\mathbf{J}_c$-induced modulation of the spin relaxation observed in the $Ni_{81}Fe_{19}$/Pt film cannot be attributed to magnetic-field effects. We measured the $\mathbf{J}_c$ dependence of ΔW for a $Ni_{81}Fe_{19}$/Cu film and a plain $Ni_{81}Fe_{19}$ film and found that these films show no ΔW modulation even when an electric current is applied at $\theta = 90°$ as shown in Fig. 9. This indicates that the observed $\mathbf{J}_c$-induced modulation of spin relaxation in the $Ni_{81}Fe_{19}$/Pt film is attributed not to a possible small flow of electric currents in the $Ni_{81}Fe_{19}$ layer or the inhomogeneity of the magnetic field (Oersted field) induced by the electric current.

The observed modulation of the spin relaxation is best interpreted in terms of the macroscopic spin transfer induced by the strong SHE in the Pt layer. When $\theta = 90°$, the spin torque acting on the magnetization in the $Ni_{81}Fe_{19}$ layer draws the magnetization toward ($J_c > 0$) or away from ($J_c < 0$) the external magnetic field direction. Since this torque is parallel or anti-parallel to the Gilbert-damping torque, it modulates the relaxation of the magnetization precession in the whole $Ni_{81}Fe_{19}$ layer.

To describe this effect quantitatively, we performed a model calculation based on the standard Valet-Fert model,[24] in which SHE parameterized by the spin-Hall angle θ_{SHE} (the ratio of the spin-Hall conductivity to the electrical conductivity) is considered. Spin relaxation coefficient α is calculated using the generalized LLG equation for local spins in the $Ni_{81}Fe_{19}$ layer:[25]

$$\frac{\mathrm{d}\mathbf{M}}{\mathrm{d}t} = -\gamma \mathbf{M} \times \boldsymbol{H}_{\mathrm{eff}} + \frac{\alpha}{M_{\mathrm{s}}} \mathbf{M} \times \frac{\mathrm{d}\mathbf{M}}{\mathrm{d}t} - \frac{\gamma J_{\mathrm{s}}}{M_{\mathrm{s}}^2 V_{\mathrm{F}}} \mathbf{M} \times (\mathbf{M} \times \boldsymbol{\sigma}), \quad (4)$$

where J_s and V_F are the spin current injected

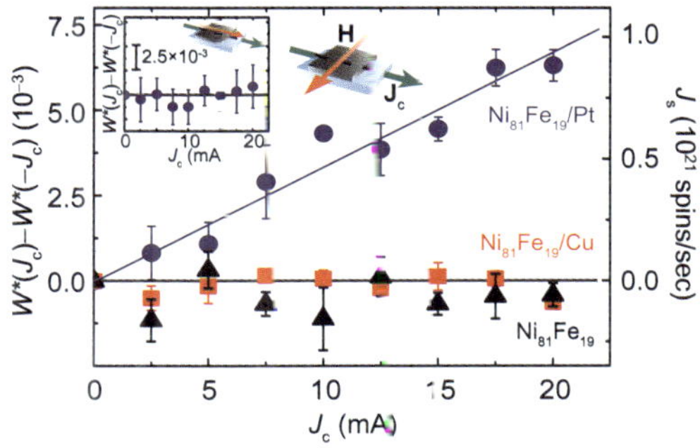

Fig. 9. The J_c dependence of $\Delta W \equiv W^*(J_c) - W^*(-J_c)$ for the $Ni_{81}Fe_{19}$/Pt bilayer film, a $Ni_{81}Fe_{19}$/Cu bilayer film, and a simple $Ni_{81}Fe_{19}$ film at $\theta = 90°$, where $W^*(J_c) \equiv W(J_c)/W(0)$. The inset shows ΔW for the $Ni_{81}Fe_{19}$/Pt bilayer film at $\theta = 0$. J_s is the spin current amplitude injected into the $Ni_{81}Fe_{19}$ layer estimated using Eq. 5.

into the ferromagnet and the volume of the ferromagnet, respectively. By ignoring the second order contribution of the precession amplitude, we obtained α as $\alpha = \alpha_0 + \Delta\alpha_{\mathrm{SHE}}$ at $\theta = 90°$, where α_0 is independent of J_c and

$$\Delta\alpha_{\mathrm{SHE}} = \left(\frac{\gamma}{2\pi f M_{\mathrm{s}} V_{\mathrm{F}}}\right) J_{\mathrm{s}} = \left(\frac{\hbar\gamma\eta\theta_{\mathrm{SHE}}}{2\pi f M_{\mathrm{s}} e A_{\mathrm{N}} d_{\mathrm{F}}}\right) J_{\mathrm{c}}. \quad (5)$$

Here, V_F, η, A_N, and d_F are the volume of the $Ni_{81}Fe_{19}$ layer, the spin injection efficiency, the cross-sectional area of the Pt layer, and the thickness of the $Ni_{81}Fe_{19}$ layer, respectively. The asymmetric component of the relaxation coefficient is given by $\Delta\alpha \equiv \alpha(J_c) - \alpha(-J_c) = 2\Delta\alpha_{\mathrm{SHE}}$ owing to the cancellation of α_0, and therefore $\Delta\alpha_{\mathrm{SHE}}$ is directly related to the W modulation, since $\alpha = (\sqrt{3}\gamma/4\pi f)W$.[22]

The calculated W modulation, i.e., $W(J_c) - W(-J_c) = (4\pi f/\sqrt{3}\gamma)2\Delta\alpha_{\mathrm{SHE}} = (4/\sqrt{3}M_s V_F)J_s$, is directly proportional to the amplitude of the spin current injected into the $Ni_{81}Fe_{19}$ layer and the proportionality coefficient comprises macroscopic parameters only. This situation allows us to measure the spin-current amplitude by monitoring W without assuming any microscopic material parameters. This modulation can thus be used as a spin-current meter. In Fig. 9, we show the amount of the spin current injected into the $Ni_{81}Fe_{19}$ layer estimated from the W modulation and Eq. 5.

4. Summary

The direct and the inverse spin-Hall effects have been investigated using a $Ni_{81}Fe_{19}/Pt$ film. A spin current generated by the magnetization dynamics was electrically detected using the inverse spin-Hall effect (ISHE). The magnetic field angle dependence of the ISHE signal is consistent with the model calculation based on the dc spin pumping and ISHE. In the same system, the spin relaxation in the $Ni_{81}Fe_{19}$ layer is electrically modulated. This modulation is attributed to the macroscopic spin transfer induced by the spin-Hall effect (SHE). The model calculation shows that this relaxation modulation due to spin injection enables the quantitative measurement of spin currents.

Acknowledgments

The authors thank S. Maekawa, S. Takahashi, J. Ieda, and G. Tatara for valuable discussions. This work was supported by a Strategic Information and Communications R&D Promotion Programme from The Ministry of Internal Affairs and Communications, by a Grant-in-Aid for Scientific research in Priority Area "Creation and control of spin current" from Ministry of Education, Culture, Sports, Science and Technology.

References

1. I. Zutic, J. Fabian and S. D. Sarma, *Rev. Mod. Phys.* **76**, 323 (2004).
2. S. Maekawa, *Concepts in Spin Electronics* (Oxford University Press, Oxford, 2006).
3. M. I. Dyakonov and V. I. Perel, *Phys. Lett.* **35A**, 459 (1971).
4. J. E. Hirsch, *Phys. Rev. Lett.* **83**, 1834 (1999).
5. S. Murakami, N. Nagaosa and S. C. Zhang, *Science* **301**, 1348 (2003).
6. J. Sinova, D. Culcer, Q. Niu, N. A. Sinitsyn, T. Jungwirth and A. H. MacDonald, *Phys. Rev. Lett.* **92**, 126603 (2004).
7. Y. K. Kato, R. C. Myers, A. C. Gossard and D. D. Awschalom, *Science* **306**, 1910 (2004).
8. J. Wunderlich, B. Kaestner, J. Sinova and T. Jungwirth, *Phys. Rev. Lett.* **94**, 047204 (2005).
9. E. Saitoh, M. Ueda, H. Miyajima and G. Tatara, *Appl. Phys. Lett.* **88**, 182509 (2006).
10. T. Kimura, Y. Otani, T. Sato, S. Takahashi and S. Maekawa, *Phys. Rev. Lett.* **98**, 156601 (2007).
11. S. O. Valenzuela and M. Tinkham, *Nature* **442**, 176 (2006).
12. S. Zhang, *Phys. Rev. Lett.* **85**, 393 (2000).
13. J. I. Ohe, A. Takeuchi and G. Tatara, *Phys. Rev. Lett.* **99**, p. 266603 (2007).
14. T. Seki, Y. Hasegawa, S. Mitani, S. Takahashi, H. Imamura, S. Maekawa, J. Nitta and K. Takanashi, *Nat. Mater.* **7**, 125 (2008).
15. S. Takahashi and S. Maekawa, *Phys. Rev. Lett.* **88**, 116601 (2002).
16. K. Ando, S. Takahashi, K. Harii, J. Ieda, K. Sasage, S. Maekawa and E. Saitoh, *Phys. Rev. Lett.* **101**, 036601 (2008).
17. K. Ando, Y. Kajiwara, S. Takahashi, S. Maekawa, K. Takemoto, M. Takatsu and E. Saitoh, *Phys. Rev. B* **78**, 014413 (2008).
18. P. A. M. Dirac, *Proc. R. Soc. London.* **A133**, 60 (1931).
19. H. Y. Inoue, K. Harii, K. Ando, K. Sasage and E. Saitoh, *Journal of Applied Physics* **102**, 083915 (2007).
20. Y. Tserkovnyak, A. Brataas and G. E. W. Bauer, *Phys. Rev. Lett.* **88**, 117601 (2002).
21. S. Mizukami, Y. Ando and T. Miyazaki, *Phys. Rev. B* **66**, 104413 (2002).
22. A. H. Morrish, *The Physical Principles of Magnetism* (Robert E. Krieger, New York, 1980).
23. S. M. Bhagat and E. P. Chicklis, *Phys. Rev.* **178**, 828 (1968).
24. T. Valet and A. Fert, *Phys. Rev. B* **48**, 7099 (1993).
25. J. C. Slonczewski, *J. Magn. Magn. Mater.* **159**, L1 (1996).

NOVEL CURRENT PUMPING MECHANISM BY SPIN DYNAMICS

A. TAKEUCHI*, K. HOSONO, and G. TATARA
Department of Physics, Tokyo Metropolitan University, Hachioji, Tokyo 192-0397, Japan
**E-mail: atake@phys.metro-u.ac.jp*

We show that a charge current and a spin current are pumped from the spin dynamics in the presence of the spin-orbit interaction. We are particularly interested the inverse spin Hall effect which converts spin currents into charge ones. This conversion mechanism is demonstrated based on the conservation laws of charge and spin. The result is compared with the microscopic calculation.

Keywords: current pumping; inverse spin Hall effect; spin current.

1. Introduction

Recent studies in the spintronics aim at manipulating both the charge and the spin degrees of freedom. Phenomena induced by the spin-orbit interaction have been investigated actively because the spin-orbit interaction works as a converter between the charge degrees of freedom and the spin one. The direct spin Hall effect [1,2,3,4,5] is one of the hottest recent topics, which enables to control magnetization by purely electrical means. In this effect, a spin current flows due to the spin-orbit interaction in the transverse direction to an applied electric voltage, i.e., the spin-orbit interaction converts a charge current into a spin current. Spin currents are not observable directly, so detection of the spin Hall effect was carried out first by measuring the spin accumulation at the edge of a sample. [4,5] Recently, the inverse spin Hall effect was proposed as a new technique for detecting spin currents by Saitoh *et al.*[6] A charge current is generated in the transverse direction to a pumped spin current via the inverse spin Hall effect as shown in Fig. 1, and therefore this effect enables electrical detection of spin currents. In fact in Ref. 6, a spin current was injected into Pt layer by use of the spin pumping, and it was observed as a electric voltage induced by the inverse spin Hall effect in the transverse direction to this pumped spin current. Experimental observations of the inverse spin Hall effect have recently been reported by several groups.[6,7,8,9,10]

Fig. 1. Schematic illustration of the inverse spin Hall effect by the spin pumping technique. A spin current (j_s) is generated by the local spin (S) dynamics in ferromagnets, and injected into a material with the spin-orbit interaction (s-o). This pumped spin current is converted into a charge current (j_c) transverse to it by the spin-orbit interaction.

Theoretical studies of the inverse spin Hall effect were done by discussing a charge current (or an electric voltage) in nonmagnetic materials when a spin current was injected from attached ferromagnets by use of the spin pumping. [11,12,13,14,15] Ohe *et al.* [12] considered a disordered two dimensional electron gas system with the spatially uniform Rashba spin-orbit interaction and the exchange interaction with dynamical magnetization. They calculated charge and spin currents driven by dynamical magnetization using the non-equilibrium Green function assuming a perturbative regime. They demonstrated that the dominant contribution to this charge current j_c could be written as

$$j_{c\mu} \simeq -\frac{3e\alpha_R\tau}{\hbar^2}\epsilon_{\mu\nu z}\nabla\cdot \boldsymbol{j}_s^\nu, \quad (1)$$

where j_s^ν represents a spin current polarized in the direction ν, α_R is the Rashba coupling constant, and τ denotes an elastic electron lifetime. The present authors extended the study of Ref. 12 to the case of a three-dimensional system with the inhomogeneous spin-orbit interaction. [15] They demonstrated that the contribution arising from inhomogeneity of the spin-orbit interaction would be essential in actual experiments.

The inverse spin Hall effect originated from the spin dynamics can be useful for a novel battery operated by the magnetism. Therefore understanding the role of spin currents in the inverse spin Hall effect and a relation between charge currents and spin currents is a very important task. In this paper, we investigate a linkage between these two currents based on the conservation laws of charge and spin.

2. Model

We consider an electron system in the presence of the spin-independent impurity scattering, the spin-orbit interaction, and interacting with the localized dynamical spins. The total Hamiltonian is $H(t) = H_0 + H_{so} + H_{ex}(t) + H_i$, where

$$H_0 = -\frac{\hbar^2}{2m}\sum_{\boldsymbol{x}}\psi_{\boldsymbol{x}}^\dagger \nabla^2\psi_{\boldsymbol{x}}, \quad (2a)$$

$$H_{so} = -\mathrm{i}\sum_{\boldsymbol{x}}\psi_{\boldsymbol{x}}^\dagger(\nabla U_{\boldsymbol{x}}\times\nabla)\cdot\boldsymbol{\sigma}\psi_{\boldsymbol{x}}, \quad (2b)$$

$$H_{ex}(t) = -J_{ex}\sum_{\boldsymbol{x}}\psi_{\boldsymbol{x}}^\dagger \boldsymbol{S}_{\boldsymbol{x},t}\cdot\boldsymbol{\sigma}\psi_{\boldsymbol{x}}, \quad (2c)$$

$$H_i = u\sum_{i=1}^{n_i}\psi_{\boldsymbol{r}_i}^\dagger\psi_{\boldsymbol{r}_i}. \quad (2d)$$

Here $\psi^{(\dagger)}$ is the annihilation (creation) operator of conducting electrons. The first term is the kinetic energy term of free electrons. The second term H_{so} describes the spin-orbit interaction, where $U_{\boldsymbol{x}}$ is a scalar potential causing the spin-orbit interaction and $\boldsymbol{\sigma}$ represent the Pauli matrices. The exchange interaction between conduction electrons and local spins is represented by $H_{ex}(t)$, where J_{ex} is an exchange coupling constant and $\boldsymbol{S}_{\boldsymbol{x},t}$ denotes a local spin. The last term denotes the spin-independent impurity scattering, where n_i is a number of impurities and u is the strength of impurity scattering, and this impurity scattering gives rise an elastic electron lifetime.

The charge density and the spin density are defined as $\rho_c \equiv -e\langle\psi_{\boldsymbol{x},t}^\dagger\psi_{\boldsymbol{x},t}\rangle$ and $\rho_s^\nu \equiv (\hbar/2)\langle\psi_{\boldsymbol{x},t}^\dagger\sigma^\nu\psi_{\boldsymbol{x},t}\rangle$, and continuity equations they satisfy are derived as

$$\dot\rho_c + \nabla\cdot(\boldsymbol{j}_c^n + \boldsymbol{j}_c^{so}) = 0, \quad (3)$$

$$\dot\rho_s^\nu + \nabla\cdot(\boldsymbol{j}_s^{n\nu} + \boldsymbol{j}_s^{so\nu}) = \frac{2m}{\hbar^2}\epsilon_{\mu\xi\alpha}\epsilon_{\nu\beta\alpha}\frac{\partial U_{\boldsymbol{x}}}{\partial x_\xi}j_{s\mu}^{n\beta} + \mathcal{T}_s^\nu. \quad (4)$$

Here each current density is defined as

$$j_{c\mu}^n = \frac{\mathrm{i}e\hbar}{2m}\langle\psi_{\boldsymbol{x},t}^\dagger\overleftrightarrow{\partial_\mu}\psi_{\boldsymbol{x},t}\rangle, \quad (5)$$

$$j_{c\mu}^{so} = -\frac{e}{\hbar}\epsilon_{\mu\nu\xi}\frac{\partial U_{\boldsymbol{x}}}{\partial x_\xi}\langle\psi_{\boldsymbol{x},t}^\dagger\sigma^\nu\psi_{\boldsymbol{x},t}\rangle, \quad (6)$$

$$j_{s\mu}^{n\nu} = -\frac{\mathrm{i}\hbar^2}{4m}\langle\psi_{\boldsymbol{x},t}^\dagger\sigma^\nu\overleftrightarrow{\partial_\mu}\psi_{\boldsymbol{x},t}\rangle, \quad (7)$$

$$j_{s\mu}^{so\nu} = \frac{1}{2}\epsilon_{\mu\nu\xi}\frac{\partial U_{\boldsymbol{x}}}{\partial x_\xi}\langle\psi_{\boldsymbol{x},t}^\dagger\psi_{\boldsymbol{x},t}\rangle, \quad (8)$$

where $\langle\cdots\rangle$ represents the expectation value estimated by the total Hamiltonian $H(t)$ and $\psi^\dagger\overleftrightarrow{\partial_\mu}\psi$ means $\psi^\dagger(\partial_\mu\psi) - (\partial_\mu\psi^\dagger)\psi$. Here superscript "n" of these currents describes normal parts and "so" means the correction due to the spin-orbit interaction. The spin source arising from dynamical spins is represented by $\mathcal{T}_s^\nu$:

$$\mathcal{T}_s^\nu = \frac{\mathrm{i}}{2}\langle\psi_{\boldsymbol{x},t}^\dagger[H_{ex}(t),\sigma^\nu]\psi_{\boldsymbol{x},t}\rangle. \quad (9)$$

3. Results

From definition of the normal part of the charge current density (Eq. (5)), the time derivative of $j_{\rm c}^{\rm n}$ is given by

$$\begin{aligned}\frac{\partial}{\partial t}j_{{\rm c}\mu}^{\rm n} =& \frac{{\rm i}e\hbar}{2m}\Big(\langle\dot{\psi}^\dagger\overleftrightarrow{\partial_\mu}\psi\rangle + \langle\psi^\dagger\overleftrightarrow{\partial_\mu}\dot{\psi}\rangle\Big)\\ =& -\frac{e}{2m}\bigg\{\frac{\hbar^2}{2m}\Big[-\langle(\boldsymbol{\nabla}^2\psi^\dagger)\overleftrightarrow{\partial_\mu}\psi\rangle\\ &+\langle\psi^\dagger\overleftrightarrow{\partial_\mu}(\boldsymbol{\nabla}^2\psi)\rangle\Big] + {\rm i}\epsilon_{\alpha\nu\xi}\frac{\partial U_{\boldsymbol{x}}}{\partial x_\xi}\\ &\times\Big[\langle(\partial_\alpha\psi^\dagger)\sigma^\nu\overleftrightarrow{\partial_\mu}\psi\rangle\\ &+\langle\psi^\dagger\sigma^\nu\overleftrightarrow{\partial_\mu}(\partial_\alpha\psi)\rangle\Big]\bigg\}\\ =&\frac{2e}{\hbar^2}\epsilon_{\alpha\nu\xi}\frac{\partial U_{\boldsymbol{x}}}{\partial x_\xi}\frac{\partial}{\partial x_\alpha}j_{{\rm s}\mu}^{{\rm n}\nu} + J_{{\rm c}\mu},\end{aligned} \quad (10)$$

where $J_{{\rm c}\mu} \equiv -(e\hbar^2/4m^2)\partial_\alpha\langle\psi_{\boldsymbol{x},t}^\dagger\overleftrightarrow{\partial_\mu}\overleftrightarrow{\partial_\alpha}\psi_{\boldsymbol{x},t}\rangle$. The correction term of the charge current density by the spin-orbit interaction can be written by using the spin density as $j_{{\rm c}\mu}^{\rm so} = -(2e/\hbar)\epsilon_{\mu\nu\xi}(\partial_\xi U_{\boldsymbol{x}})\rho_{\rm s}^\nu$. The time derivative of $j_{\rm c}^{\rm so}$ is given by use of Eq. 4 as

$$\begin{aligned}\frac{\partial}{\partial t}j_{{\rm c}\mu}^{\rm so} =& -\frac{2e}{\hbar^2}\epsilon_{\mu\nu\xi}\frac{\partial U_{\boldsymbol{x}}}{\partial x_\xi}\dot{\rho}_{\rm s}^\nu\\ =&\frac{2e}{\hbar^2}\epsilon_{\mu\nu\xi}\frac{\partial U_{\boldsymbol{x}}}{\partial x_\xi}\bigg[\boldsymbol{\nabla}\cdot(\boldsymbol{j}_{\rm s}^{{\rm n}\nu}+\boldsymbol{j}_{\rm s}^{{\rm so}\nu})\\ &-\frac{2m}{\hbar^2}\frac{\partial U_{\boldsymbol{x}}}{\partial x_\alpha}j_{{\rm s}\nu}^{{\rm n}\alpha} - \mathcal{T}_{\rm s}^\nu\bigg].\end{aligned} \quad (11)$$

By similar calculations, we get following identities for each spin current density:

$$\begin{aligned}\frac{\partial}{\partial t}j_{{\rm s}\mu}^{{\rm n}\nu} =& \frac{1}{2e}\epsilon_{\alpha\nu\xi}\frac{\partial U_{\boldsymbol{x}}}{\partial x_\xi}\frac{\partial}{\partial x_\alpha}j_{{\rm c}\mu}^{\rm n}\\ &+J_{{\rm s}\mu}^\nu + \tilde{\mathcal{T}}_{{\rm s}\mu}^\nu,\end{aligned} \quad (12)$$

$$\frac{\partial}{\partial t}j_{{\rm s}\mu}^{{\rm so}\nu} = \frac{1}{2e}\epsilon_{\mu\nu\xi}\frac{\partial U_{\boldsymbol{x}}}{\partial x_\xi}\boldsymbol{\nabla}\cdot(\boldsymbol{j}_{\rm c}^{\rm n}+\boldsymbol{j}_{\rm c}^{\rm so}), \quad (13)$$

where $J_{{\rm s}\mu}^\nu \equiv (\hbar/4m)\big[\delta_{\nu\gamma}(\hbar^2/2m)\partial_\alpha - \epsilon_{\alpha\xi\beta}\epsilon_{\nu\gamma\beta}(\partial_\xi U_{\boldsymbol{x}})\big]\langle\psi_{\boldsymbol{x},t}^\dagger\sigma^\gamma\overleftrightarrow{\partial_\mu}\overleftrightarrow{\partial_\alpha}\psi_{\boldsymbol{x},t}\rangle$, and $\tilde{\mathcal{T}}_{{\rm s}\mu}^\nu \equiv (\hbar/4m)\langle\psi_{\boldsymbol{x},t}^\dagger\big[H_{\rm ex}(t),\sigma^\nu\big]\overleftrightarrow{\partial_\mu}\psi_{\boldsymbol{x},t}\rangle$.

These results show that the conversion of charge (spin) currents into spin (charge) currents by the spin-orbit interaction indeed occurs. In Eq. (10), the charge current arises in the same direction (μ-direction) to the normal part of the spin current. In contrast, in Eq. (13), the spin-orbit part of the spin current arises from the divergence of the charge currents in the transverse direction (μ-direction) to the spin polarization (ν-direction), i.e., it is the direct spin Hall effect. Similarly Eq. (11) describes the inverse spin Hall effect. Noting $j_{\rm c}(t) = \sum_\omega j_{\rm c}(\omega){\rm e}^{{\rm i}(\omega+{\rm i}/\tau)t}$ and assuming $\omega\tau \ll 1$, we obtain $\partial_t j_{\rm c} \sim -(1/\tau)j_{\rm c}$ and the dominant part of Eq. (11) agrees with Eq. (1) derived microscopically (with $\alpha_{\rm R} = -\partial_z U_{\boldsymbol{x}}$).

4. Microscopic Calculation in Rashba System

In the above discussion, a relation between charge currents and spin currents in the spin-orbit interaction system was clarified. However explicit forms of these currents cannot be accessed by those general arguments. Here we show results of the microscopic calculation in the case of the uniform Rashba system ($\alpha_{\rm R} = -\partial_z U_{\boldsymbol{x}}$)[12] to see the phenomena in more detail. Defining the lesser component of the Green function $G_{x,x'}^< \equiv ({\rm i}/\hbar)\langle\psi_{\boldsymbol{x}',t'}^\dagger\psi_{\boldsymbol{x},t}\rangle$, we can write charge and spin current densities:

$$j_{{\rm c}\mu}^{\rm n} = \frac{e\hbar^2}{2m}(\partial_\mu - \partial'_\mu){\rm tr}\big[G_{x,x'}^<\big]_{x'=x}, \quad (14)$$

$$j_{{\rm c}\mu}^{\rm so} = -{\rm i}e\alpha_{\rm R}\epsilon_{\mu\nu z}{\rm tr}\big[\sigma^\nu G_{x,x}^<\big], \quad (15)$$

$$j_{{\rm s}\mu}^{{\rm n}\nu} = \frac{\hbar^3}{4m}(\partial'_\mu - \partial_\mu){\rm tr}\big[\sigma^\nu G_{x,x'}^<\big]_{x'=x}. \quad (16)$$

Here we neglect the correction of the spin current due to the spin-orbit interaction ($j_{\rm s}^{\rm so}$) because we are interested in the inverse spin Hall effect, i.e., the conversion of a spin current arising from dynamical spins into a charge current by the spin-orbit interaction.

We treat both the spin-orbit and the exchange interactions perturbatively. We show the Feynman diagrams of the dominant contribution to the inverse spin Hall effect in Fig. 2. (The other contributions are shown in Appendix A.) After straightforward calculations,[12,15] we obtain charge and spin current densities as

$$j_{{\rm c}\mu} \sim -\frac{3em\alpha_{\rm R}J_{\rm ex}^2\tau^2}{\pi\hbar^4}\epsilon_{\mu\nu z}$$

$$\times \sum_{\boldsymbol{x}_1} D_{\boldsymbol{x}-\boldsymbol{x}_1} \left(\boldsymbol{S}_{\boldsymbol{x},t} \times \dot{\boldsymbol{S}}_{\boldsymbol{x}_1,t}\right)^{\nu}, \quad (17)$$

$$j_{\mathrm{s}\mu}^{\nu} \sim -\frac{\varepsilon_{\mathrm{F}} J_{\mathrm{ex}}^2 \tau^3}{\pi \hbar^2} \frac{\partial}{\partial x_\mu} \sum_{\boldsymbol{x}_1,\boldsymbol{x}_2} D_{\boldsymbol{x}-\boldsymbol{x}_1}$$

$$\times D_{\boldsymbol{x}_1-\boldsymbol{x}_2} \left(\boldsymbol{S}_{\boldsymbol{x}_1,t} \times \dot{\boldsymbol{S}}_{\boldsymbol{x}_2,t}\right)^{\nu}, \quad (18)$$

where $D_{\boldsymbol{x}} \equiv (1/V) \sum_{\boldsymbol{q}} \mathrm{e}^{-\mathrm{i}\boldsymbol{q}\cdot\boldsymbol{x}} (m/\boldsymbol{q}^2 \varepsilon_{\mathrm{F}} \tau^2)$ describes diffusive motion of conducting electrons (V is the volume of the system and ε_{F} is the Fremi energy). From these results Eq. (1) is easily proved.

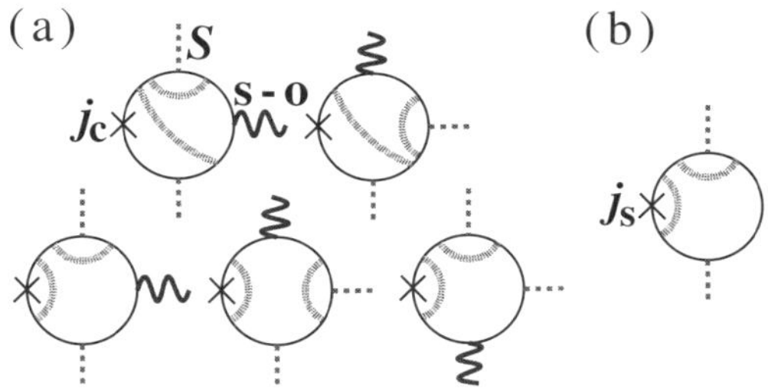

Fig. 2. Diagrammatic representations of currents contributing to the inverse spin Hall effect. Contribution from (a) is to the charge current (j_{c}) and (b) describes contribution to the spin current (j_{s}). (Correctly, we need to calculate diagrams overturned up and down.) Dotted, wavy and thick lines denote interacting with a local spin (S), the Rashba spin-orbit interaction (s-o) and the diffusion ladder due to the impurity scattering, respectively.

5. Discussions and Conclusions

We have demonstrated that the conversion of a spin current into a charge current indeed occurs due to the spin-orbit interaction based on two different theoretical formalisms. We first showed that the spin-charge and the charge-spin conservation mechanisms could be partially understood simply by looking into the continuity equation of charge and spin quite generally independent of the system details (Eqs. (10) - (13)). The results by the continuity equations contain two conversion processes. One is consistent phenomenologically with the inverse and the direct spin Hall effects shown in Eqs. (11) and (13), whereas another (Eqs. (10) and (12)) describes a different conversion mechanism which has not been observed in experiments. However, there are contributions which are not directly related to the spin-charge conversion, and the microscopic calculation is necessary to understand these contributions. (Contributions which are not interpreted as the inverse spin Hall effect are discussed in Appendix A.)

To summarize, we have shown that general argument based on the continuity equation without the microscopic calculation could be useful to understand the mutual conversion between a charge current and a spin current in the presence of the spin-orbit interaction.

Acknowledgments

The authors are grateful to J.-I. Ohe, E. Saitoh, H. Kohno, J. Shibata, and S. Murakami for helpful discussions and comments. This work was supported by a Grant-in-Aid for Scientific Research in Priority Area gCreation and control of spin current" (1948027) from the Ministry of Education, Culture, Sports, Science and Technology, Japan.

Appendix A. Other Terms

For appendix we comment the contribution by the microscopic calculations which is not interpreted as the inverse spin Hall effect in the uniform Rashba system. The charge current at the first order of the exchange interaction shown in Fig. 3 vanishes identically. This result suggests the peculiarity known in the direct spin Hall effect originating from the disordered Rashba spin-orbit interaction.[16] In the case of the inhomogeneous Rashba spin-orbit interaction, this cancellation does not occur and the charge current density is driven proportional to $(\partial \alpha_{\mathrm{R}\boldsymbol{x}})(\partial \boldsymbol{S}_{\boldsymbol{x},t})$.[15]

Other important contributions shown in

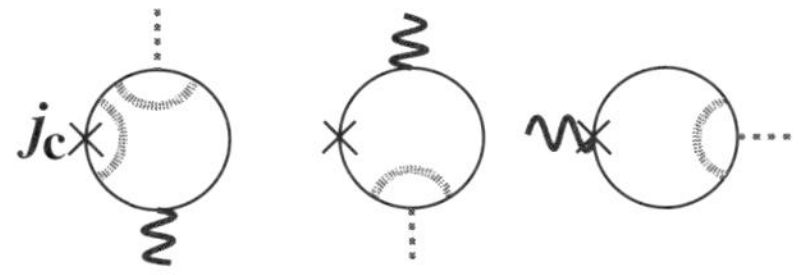

Fig. 3. Feynman diagrams of the charge current at the first order of the exchange interaction. These contributions turn out to vanish identically.

Fig. 4 are given as

$$j^{\phi}_{c\mu} \sim \frac{2e\alpha_R J^2_{ex}\tau^3}{\pi^2\hbar^3}\epsilon_{\xi\nu z}\frac{\partial}{\partial x_\mu}\sum_{\boldsymbol{x}_1,\boldsymbol{x}_2} D_{\boldsymbol{x}-\boldsymbol{x}_1} \times \frac{\partial D^{(2)}_{\boldsymbol{x}_1-\boldsymbol{x}_2}}{\partial x_{1\xi}}\left(\boldsymbol{S}_{\boldsymbol{x}_1,t}\times\dot{\boldsymbol{S}}_{\boldsymbol{x}_2,t}\right)^\nu, \quad (A.1)$$

$$j^{(1)\nu}_{s\mu} \sim \frac{\varepsilon_F J_{ex}\tau^2}{2\pi\hbar}\frac{\partial}{\partial x_\mu}\sum_{\boldsymbol{x}_1} D_{\boldsymbol{x}-\boldsymbol{x}_1}\dot{S}^\nu_{\boldsymbol{x}_1,t}, \quad (A.2)$$

where $D^{(2)}_{\boldsymbol{x}} \equiv (1/V)\sum_{\boldsymbol{q}} e^{-i\boldsymbol{q}\cdot\boldsymbol{x}}(m/\boldsymbol{q}^2\varepsilon_F\tau^2)^2$.

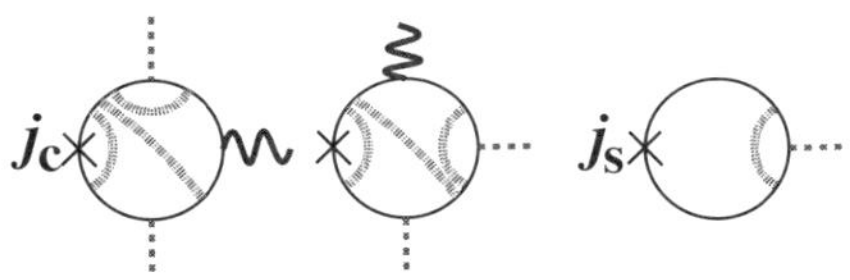

Fig. 4. Other important diagrams of the charge current and the spin current except contribution to the inverse spin Hall effect Contribution from (a) is written as j^{ϕ}_c in Eq. (A.1) and contribution from (b) is denoted by $j^{(1)}_s$ in Eq. (A.2).

In Eq. (A.1), $\boldsymbol{j}^{\phi}_c$ forms into $\boldsymbol{\nabla}\Phi$, where Φ is a scalar potential. This charge current is induced by the fictitious conservative potential Φ generated by the spin dynamics. This fictitious potential acts only on the charge having the spin degrees of freedom. The spin current at the first order of the exchange interaction (Eq. (A.2)) does not have its counter part in the charge current in the uniform Rashba case. Therefore, the Rashba spin-orbit interaction converts not all of spin currents into charge currents. The role of each spin current needs to be studied in more detail in the future works.

References

1. J.E. Hirsch, *Phys. Rev. Lett.* **83**, 1834 (1999).
2. S. Zhang, *Phys. Rev. Lett.* **85**, 393 (2000).
3. S. Murakami, N. Nagaosa, and S.C. Zhang, *Science* **301**, 1348 (2003).
4. Y.K. Kato, R.C. Myers, A.C. Gossard, and D.D. Awschalom, *Science* **306**, 1910 (2004).
5. J. Wunderlich, B. Kaestner, J. Sinova, and T. Jungwirth, *Phys. Rev. Lett.* **94**, 047204 (2005).
6. E. Saitoh, M. Ueda, H. Miyajima, and G. Tatara, *Appl. Phys. Lett.* **88**, 182509 (2006).
7. S.O. Valenzuela and M. Tinkham, *Nature* **442**, 176 (2006).
8. T. Kimura, Y. Otani, T. Sato, S. Takahashi, and S. Maekawa, *Phys. Rev. Lett.* **98**, 156601 (2007).
9. T. Seki, Y. Hasegawa, S. Mitani, S. Takahashi, H. Imamura, S. Maekawa, J. Nitta, and K. Takanashi, *Nat. Mater.* **7**, 125 (2008).
10. K. Ando, S. Takahashi, K. Harii, K, Sasage, J. Ieda, S. Maekawa, and E. Saitoh, *Phys. Rev. Lett.* **101**, 036601 (2008).
11. X. Wang, G.E W. Bauer, B.J. van Wees, A. Brataas, and Y. Tserkovnyak, *Phys. Rev. Lett.* **97**, 216602 (2006).
12. J.-I. Ohe, A. Takeuchi, and G. Tatara, *Phys. Rev. Lett.* **99**, 266603 (2007).
13. J.-I. Ohe, A. Takeuchi, G. Tatara, and B. Kramer, *Physica* **E40**, 1554 (2008).
14. J. Xiao, G.E.W. Bauer, and A. Brataas, *Phys. Rev.* **B77**, 180407(R) (2008).
15. A. Takeuchi and G. Tatara, *J. Phys. Soc. Jpn.* **77**, 074701 (2008).
16. J.-I. Inoue, G.E.W. Bauer, and L.W. Molenkamp, *Phys. Rev.* **B70**, 041303(R) (2004).

QUANTUM SPIN HALL PHASE IN BISMUTH ULTRATHIN FILM

S. MURAKAMI*

Department of Physics, Tokyo Institute of Technology, 2-12-1 Ookayama, Meguro-ku, Tokyo 152-8551, Japan

**E-mail: murakami@stat.phys.titech.ac.jp*

PRESTO, Japan Science and Technology Agency (JST), Saitama, Kawaguchi 332-0012, Japan

We theoretically study quantum phase transitions between the quantum spin Hall and insulator phases, both of which are insulating in the bulk. This gives us an alternative view to the topological order in the quantum spin Hall phase. We also explain our theoretical proposal of the ultrathin bismuth film as a candidate to the 2D quantum spin Hall system.

Keywords: Spin Hall effect; Quantum spin Hall effect; Bismuth

1. Introduction

From the recent theoretical and experimental studies of the spin Hall effect [1,2,3], a new state of matter, called the quantum spin Hall (QSH) phase in 2D [4,5,6], has been proposed theoretically. In the QSH phase, the bulk is gapped and insulating while there are gapless edge states. The edge states consist of two counter-propagating modes with opposite spins, thus carrying a spin current. This phase is a topological order characterized by the Z_2 topological number [4,5], which becomes manifest as topologically protected gapless edge states. This phase can be realized also in 3D[7,8,9], in which gapless surface states appear. These gapless edge states in the 2D QSH phase are robust against nonmagnetic disorder and interaction[10,11]. This robustness stems from the topological order.

In contrast to the quantum Hall (QH) phase, the QSH requires an absence of magnetic field, i.e. time-reversal invariance. Moreover, the QH phase is realized usually in 2D, but not easily realized in 3D because the motion along the magnetic field usually remains gapless. In contrast, the QSH phase is easily realized in 3D as well as in 2D, in materials with strong spin-orbit coupling.

The time-reversal symmetry relates the states with wavenumbers $\mathbf{k}$ and $-\mathbf{k}$. Therefore, the wavenumbers satisfying $\mathbf{k} \equiv -\mathbf{k}$ (modulo reciprocal lattice vector) play an important role. Such momenta are called time-reversal-invariant momenta (TRIM), and in 2D and 3D Brillouin zone there are 4 and 8 TRIMs, respectively [4,12,13]. The Z_2 topological number ν is defined as follows [12,13]. First we define a $(2N) \times (2N)$ matrix w, defined as $w_{nn'}(\mathbf{k}) = \langle u_{-\mathbf{k},n}|\Theta|u_{\mathbf{k},n'}\rangle$, where Θ is the time-reversal operator, $u_{\mathbf{k},n}$ $(n = 1, \cdots, 2N)$ is the periodic part of the Bloch wavefunction which lies below the Fermi energy, and N is the number of Kramers pairs below E_F. Then for each TRIM we define the index δ_i as $\delta_i \equiv \sqrt{\det w(\mathbf{\Gamma}_i)}/\mathrm{Pf}w(\mathbf{\Gamma}_i)$ for inversion-($\mathcal{I}$-)asymmetric systems[12], and $\delta_i \equiv \prod_{n=1}^{N} \xi_{2n}(\mathbf{\Gamma}_i)$, for $\mathcal{I}$-symmetric systems[13], where $\xi_{2n}(\mathbf{\Gamma}_i)$ $(= \pm 1)$ is the parity eigenvalue of the Kramers pairs $(u_{\mathbf{k},2n}, u_{\mathbf{k},2n-1})$ at $\mathbf{\Gamma}_i$. The Z_2 topological number ν is defined as $(-1)^\nu = \prod_{i=1}^{4} \delta_i$. Since δ_i is equal to ± 1, the Z_2 topological number ν takes only two different values, i.e. $\nu = 1$ and $\nu = 0$, corresponding to the QSH phase and the ordinary nonmagnetic insulator phase, respectively. In the QSH phase there exist robust gapless edge states, while not in the insulator phase.

2. Universal Phase Diagram

It is hard to calculate the Z_2 topological number and to capture its physical meaning intuitively. On the other hand, the change of the Z_2 topological number is easier to handle, because the change involves local information in $\mathbf{k}$ space. When the system transits between the QSH and insulator phases by changing an external parameter, the bulk gap must close, because otherwise the Z_2 topological number never changes.

In general, by changing system parameters, the bulk gap does not necessarily close due to level repulsion. The minimum number of tuned parameters to achieve degeneracy is called the codimension. The codimension is sensitive to the symmetry and the dimension of the system. Because the Z_2 topological number can change only when the gap closes, investigation of the codimension in various cases reveals the nature of the Z_2 topological number. By this motivation, we have investigated the criterion for the occurence of the phase transition in Refs. 14, 15, 16, and the result is summarized in Fig. 1. We emphasize that this gap-closing physics is equivalent to the physics of the Z_2 topological number.

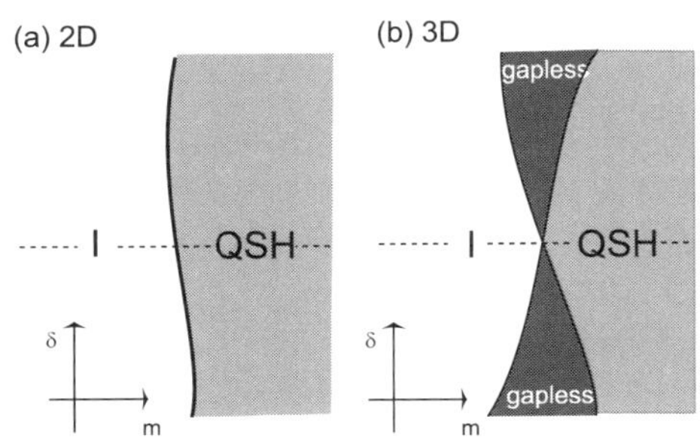

Fig. 1. Universal phase diagram between the QSH and the insulator (I) phases in (a) 2D and (b) 3D. δ is a parameter representing breaking of $\mathcal{I}$-symmetry. The dotted line corresponds to the $\mathcal{I}$-symmetric case. m is a parameter which drives the phase transition.

The feature of the phase transition is different whether the system is (i) $\mathcal{I}$-symmetric or (ii) $\mathcal{I}$-asymmetric[15,14]. The overall feature of gap closing is summarized in Fig. 2.

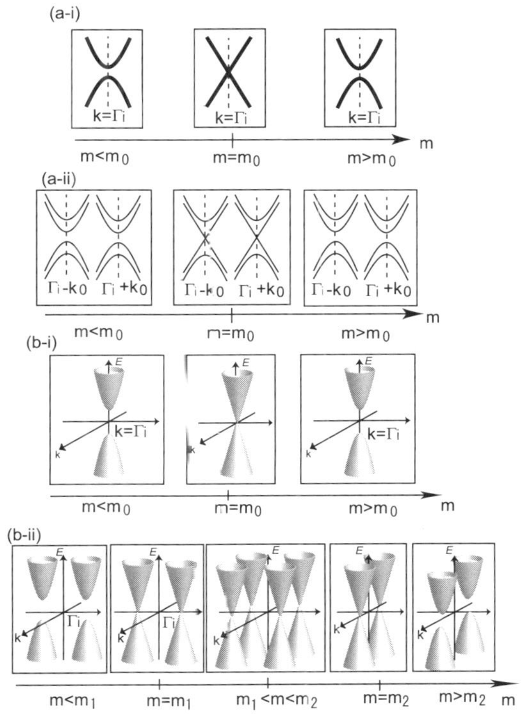

Fig. 2. Generic gap-closing for (a-i) 2D $\mathcal{I}$-symmetric, (a-ii) 2D $\mathcal{I}$-asymmetric, (b-i) 3D $\mathcal{I}$-symmetric and (b-ii) 3D $\mathcal{I}$-asymmetric cases, as we vary the parameter m, which drives the phase transition. In (a-i) and (b-i) all the states are doubly degenerate by Kramers theorem. In (a-i), (a-ii) and (b-i) the gap closes only at $m = m_0$, whereas in (b-ii) the gap closing points exist between $m = m_1$ and $m = m_2$, and they move as m is changed.

In (ii) $\mathcal{I}$-asymmetric systems, when $\mathbf{k} \neq \mathbf{\Gamma}_i$, each band is non-degenerate. At the gap-closing point, one valence band and one conduction band become degenerate. In this case it is shown that the codimension is three. In 2D system with an external parameter m, the codimension ($= 3$) is equal to the number of parameters. (k_x, k_y and m). Thus the gap can close at some $\mathbf{k}$ when the parameter m is tuned to a critical value[14]. In 3D, as is different from 2D, the codimension ($= 3$) is less than the number of parameters (m, k_x, k_y, k_z). This gives rise to a gapless phase between the QSH and the insulator phases, for $\mathcal{I}$-asymmetric 3D systems [15,16].

In $\mathcal{I}$-symmetric systems, all the states are doubly degenerate for any $\mathbf{k}$ by the Kramers theorem. From a general counting argument, the codimension is five[14], which exceeds the number of tunable parameters, i.e. (m, k_x, k_y) in 2D or (m, k_x, k_y, k_z) in 3D. Therefore, the gap does not close in general. However, this counting holds true only when $\mathbf{k} \neq \mathbf{\Gamma}_i$. At the $\mathbf{k} = \mathbf{\Gamma}_i$, on the other hand, the codimension changes to 1 when the parities of the conduction and valence bands are opposite. Thus, only in such case, the gap closes by fine-tuning a single parameter.

A detailed analysis shows that these gap-closing cases by tuning m involves a change of the Z_2 topological number, and the resulting universal phase diagram is as shown in Fig. 1[14,15,16]. As a by-product, we get low-energy effective Hamiltonian describing the phase transition for the respective cases. From the effective model thus obtained, we can calculate the edge states appearing at the boundary between the QSH and the insulator phase[14]. Thus we have reached the physics of the Z_2 topological number, through the study of gap closing by single parameter tuning. Several known models fall into one of the above-mentioned cases of gap-closing. For example, the Kane-Mele model on the honeycomb lattice[5] belongs to the class (a-ii) in Fig. 2 while the HgTe quantum well model[17,18] belongs to the class (a-i). The 3D Fu-Kane-Mele model[7] is (b-i) $\mathcal{I}$-symmetric. By introducing an $\mathcal{I}$-symmetry-breaking term into the 3D Fu-Kane-Mele model. we have verified that when the $\mathcal{I}$-symmetry is broken, the gapless region appears in the phase diagram, in accordance with our universal phase diagram[16].

3. Bismuth Ultrathin Films

The QSH phase requires an absence of the magnetic field. Hence some materials might realize the QSH phase in nature. The only necessary conditions for the QSH systems are (a) nonmagnetic insulator, and (b) odd Z_2 topological number. The latter condition means that the spin-orbit coupling should be strong enough. Suppose the system is an ordinary insulator in the absence of the spin-orbit coupling. When the spin-orbit coupling is made stronger, some systems can undergo a phase transition into the QSH phase. In this sense, the gap in the QSH phase should be opened by the spin-orbit coupling. In Ref. 19 we clarified that systems with large susceptibility are good candidates for the QSH phase, by showing that the formula for the susceptibility contains one term proportional to the spin Hall conductivity. Bismuth is one of the promising candidate for the 2D QSH phase, because it is a strong diamagnet due to interband matrix elements by the spin-orbit coupling. However, bismuth is a nonmagnetic semimetal, not an insulator. A gap should be opened to be the QSH phase. One possibility is to make it into thin film, as the (111) 1-bilayer bismuth is proposed to have a gap, from first-principle calculations[20]. In Ref. 19 we studied a simple tight-binding model of the (111) 1-bilayer bismuth. By calculating the Z_2 topological number, the band structure in the ribbon geometry (Fig. 3), and the parity eigenvalues, we predicted the (111) 1-bilayer bismuth to be the 2D QSH phase.

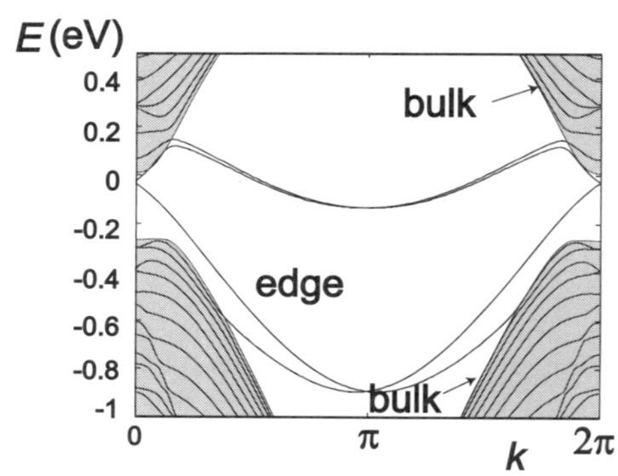

Fig. 3. Band structure of (111) ultrathin bismuth film in a narrow ribbon geometry, calculated from a simple tight-binding model.

Another way to make it gapped is by alloying with Sb. This leads to the theoretical proposal of 3D QSH in Ref. 7. In the ARPES measurement[21] of the surface states of $Bi_{0.9}Sb_{0.1}$, the Fermi curve crosses the $\bar{\Gamma}$-$\bar{M}$ line odd times. This shows that the surface state of $Bi_{0.9}Sb_{0.1}$ agrees with that expected from the 3D QSH system.

4. Concluding Remarks

In the present paper we have studied the quantum phase transition between the QSH and insulator phases. We study the condition of gap-closing in generic time-reversal-invariant systems with a bulk gap, by tuning a single external parameter. The phase transition is different between $\mathcal{I}$-symmetric and asymmetric systems. In particular, in 3D $\mathcal{I}$-asymmetric systems, there appears a gapless region in the phase diagram between the QSH and the insulator phases.

In the QH and the QSH phases, robust edge states exist, due to nontrivial topological numbers defined only for insulators. On the other hand, graphene, being a zero-gap semiconductor, cannot have such topological numbers; it means that its edge states are not robust against perturbations. In each case with robust gapless edge states, edge states carry some kind of a current. In the 2D QH and QSH systems, it is the charge current and the spin current, respectively. It is interesting to note other classes of systems with similar kinds of stable edge states. One is the superconductor with broken time-reversal symmetry, for example with gap function $p_x + ip_y$. Its edge states carry a current of Majorana fermion, and the edge states are the surface Andreev bound states. Another interesting case is the superconductor/superfluid with time-reversal symmetry; in this case, the edge carries a spin current of the Majorana fermion[22].

Acknowledgements

We are grateful to N. Nagaosa and S.-C. Zhang for fruitful discussions. This research is supported in part by Grant-in-Aids from the Ministry of Education, Culture, Sports, Science and Technology of Japan.

References

1. S. Murakami, N. Nagaosa and S.-C. Zhang, *Science* **301**, 1348 (2003).
2. J. Sinova *et al.*, *Phys. Rev. Lett.* **92**, 126603 (2004).
3. S. Murakami, N. Nagaosa, and S.-C. Zhang, *Phys. Rev. Lett.* **93**, 156804 (2004).
4. C. L. Kane and E. J. Mele, *Phys. Rev. Lett.* **95**, 226801 (2005).
5. C. L. Kane and E. J. Mele, *Phys. Rev. Lett.* **95**, 146802 (2005).
6. B. A. Bernevig and S.-C. Zhang, *Phys. Rev. Lett.* **96**, 106802 (2006).
7. L. Fu, C. L. Kane and E. J. Mele, *Phys. Rev. Lett.* **98**, 106803 (2007).
8. J. E. Moore and L. Balents, *Phys. Rev. B* **75**, 121306(R) (2007).
9. R. Roy, cond-mat/0607531.
10. C. Wu, B. A. Bernevig and S.-C. Zhang, *Phys. Rev. Lett.* **96**, 106401 (2006).
11. C. Xu and J. E. Moore, *Phys. Rev. B* **73**, 045322 (2006).
12. L. Fu and C. L. Kane, *Phys. Rev. B* **74**, 195312 (2006).
13. L. Fu and C. L. Kane, *Phys. Rev. B* **76**, 045312 (2007).
14. S. Murakami *et al.*, *Phys. Rev. B* **76**, 205304 (2007).
15. S. Murakami, *New J. Phys.* **9**, 356 (2007); (Corrigendum) *ibid.* **10**, 029802 (2008).
16. S. Murakami and S. Kuga, arXiv:0806.3309.
17. B. A. Bernevig, T. L. Hughes and S.-C. Zhang, *Science* **314**, 1757 (2006).
18. M. König *et al.*, *Science* **318**, 766 (2007).
19. S. Murakami, *Phys. Rev. Lett.* **97**, 236805 (2006).
20. Yu. M. Koroteev *et al.*, *Phys. Rev. B* **77**, 045428 (2008).
21. D. Hsieh *et al.*, *Nature* **452** 970 (2008).
22. X.-L. Qi, T. L. Hughes, S. Raghu and S.-C. Zhang, arXiv:0803.3614.

ANOMALOUS HALL EFFECT DUE TO THE VECTOR CHIRALITY

K.TAGUCHI and G.TATARA

Department of Physics, Tokyo Metropolitan University,Hachioji, Tokyo 192-0397, Japan
**E-mail: katsuhisa-taguchi@ed.tmu.ac.jp*

We theoretically study the local anomalous Hall effect due to vector spin chirality carried by the local spins in the *s*-*d* model. We will show that the vector spin chirality contribution appears locally in the present of spin polarization, The global Hall effect vanishes if electron transport is homogeneous.

Keywords: Anomalous Hall effect; Spin chirality; Keldysh formula.

1. Introduction

The anomalous Hall effect is the Hall effect due to the magnetization in ferromagnets. The anomalous contribution is usually attributed to the spin-orbit interaction, and is usually proportional to the magnetization[1]. Recently, it was found that some manganites and ferromagnetic pyrochlores can exhibit an abnormal anomalous Hall effect, which is different from the conventional anomalous Hall effect[2,3]. The behavior was explained by the spin Berry phase associated with non-trivial spin configuration (spin chirality) driven thermally or by the geometrical frustration[2,3]. Such the anomalous Hall effect was first pointed out in the strong Hund-coupling limit[2], considering a half-metallic nature of the experimental systems. A weak coupling limit was studied by applying the linear response theory and the perturbation expansion of *s*-*d* interaction[4]. In the perturbation regime, anomalous Hall effect was found to be induced by the scalar spin chirality, $\boldsymbol{S}_i \cdot (\boldsymbol{S}_j \times \boldsymbol{S}_k)$, which is made from three localized spin $\boldsymbol{S}_i$, $\boldsymbol{S}_j$, and $\boldsymbol{S}_k$(i, j, and k represents the position). The scalar spin chirality becomes a nonzero when the spin configuration is noncoplanar. The scalar chirality reduces to the spin Berry phase in the slowly varying spin structures[5].

The spin Berry phase and scalar chirality has been shown to induced spontaneous charge current[6,7].In fact, this spin circulating current is the origin of the Hall effect by the spin scalar chirality[5].

In the context of spintronics, another spin chirality associated with the spin current is of interest. This is the vector chirality, $\boldsymbol{S}_i \times \boldsymbol{S}_j$. The vector spin chirality has been found to be related to the spin current caused by the static local spins. The vector spin chirality has been shown to drive electric polarization in insulators and multi-ferroic of manganese oxide[8,9].

In this paper, we will investigate the role of vector spin chirality in the anomalous Hall effect in metals. We calculate the Hall conductivity in treating the exchange interaction perturbatively in the *s*-*d* model, and the spin-orbit interaction is not taken account of.

2. *s*-*d* Hamiltonian

The Hamiltonian we consider is written as $\mathcal{H}_0 + \mathcal{H}_{\rm sd}$, where

$$\mathcal{H}_0 = \sum_{k,\omega\sigma=\pm} \epsilon_{k\sigma} c^\dagger_{k,\omega\sigma} c_{k,\omega\sigma} \qquad (1)$$

$$\mathcal{H}_{\rm sd} = -J_{\rm sd} \sum_{kk',\omega} \boldsymbol{S}_{k'-k} \cdot c^\dagger_{k',\omega} \boldsymbol{\sigma} c_{k,\omega}. \qquad (2)$$

Here $\sigma = \pm$ denotes the electron spin, and $c_{k\sigma}$ and $c^\dagger_{k\sigma}$ are electron creation and annihilation operators respectively. The conduction electron energy is $\epsilon_{k\sigma} (\equiv \frac{\hbar^2 k^2}{2m} - \epsilon_F - \sigma M)$, where m is the electron mass, ϵ_F is the Fermi energy. M is a uniform spin polarization of conduction electron, which is caused by the uniform magnetization or the external magnetic field. Uniform polarization is chosen as along z axis. Frequency of the electron is ω and $\sigma^{\alpha}(\alpha = x, y, z)$ are the 2×2 Pauli matrices. The locally localized spin at position $\boldsymbol{x}$ is uniform as, $S(\boldsymbol{x}) = \sum_k e^{-i\boldsymbol{k}\cdot\boldsymbol{x}}\boldsymbol{S}_k$.

The effect of the impurity scattering is taken account by time lifetime, τ, of the electrons. The electric field can be written as the Hamiltonian, $\mathcal{H}_{em} = -\int d\boldsymbol{x}^3 \boldsymbol{j}(\boldsymbol{x})\cdot\boldsymbol{a}(\boldsymbol{x})$, where $\boldsymbol{a}(\boldsymbol{x})$ is vector potential. Assuming the case of uniform electric field, E, this Hamiltonian is given as

$$\mathcal{H}_{em} = \sum_{k,\omega} \frac{ie^{i\Omega t}}{\Omega}\frac{e\hbar}{m} k^{\nu} E^{\nu} c^{\dagger}_{k,\omega\sigma} c_{k,\omega+\Omega\sigma}, \quad (3)$$

the Ω is the frequency external electric field. The charge current is defined as

$$\boldsymbol{j}(\boldsymbol{x},t) = -i\frac{e\hbar}{2mV}\sum_{k,k',\omega}(\boldsymbol{k}+\boldsymbol{k'})\, e^{i(\boldsymbol{k'}-\boldsymbol{k})\cdot\boldsymbol{x}} \times \mathrm{Tr}\left[G^{<}_{k,k',\omega}\right], \quad (4)$$

where Tr represents trace over spin indices and $G^{<}_{k,k',\omega}$ is a lesser Green's function defined as $G^{<}_{k,k',\omega} \equiv i\langle c^{\dagger}_{\omega,w} c_{k',\omega}\rangle$. We include the exchange interaction to the second order, considering the weak exchange interaction, $J_{\mathrm{sd}} \ll \hbar/\tau$.

3. Global Hall conductivity

We calculate the global Hall conductivity due to the vector spin chirality in the s-d model by use of the conventional liner response theory, $\sigma_{xy} = \lim_{\omega,\delta\to 0}\langle Q(\hbar\omega + i\delta) - Q(i\delta)\rangle/(-i\omega)$. Here $Q(i\omega_\ell) \equiv \frac{1}{\beta V}\langle J_x(i\omega_\ell) J_y(-i\omega_\ell)\rangle$ is current-current correlation function in the imaginary time, the brackets denote an average, and the thermal frequency is $\omega_\ell = \frac{2\pi}{k_B T}\ell$, where ℓ is integer, and k_B is Boltzman and T is temperature.

The Hall conductivity is obtained perturbatively by expanding in $\mathcal{H}_{ex}$ to the second order as

$$\sigma^{(\mathrm{A})}_{xy} = \frac{1}{2V}\left(\frac{e\hbar}{m}\right)^2 \sum_{k,k'} k_x k_y S^{\alpha}_{k-k'} S^{\beta}_{k'-k} \mathrm{Tr}[G^r_k \sigma_\alpha G^r_{k'} \sigma_\beta G^r_k G^a_k], \quad (5)$$

where G^R_k, $\mathrm{G}^A_k \left(= (\mathrm{G}^R_k)^*\right)$ are retarded and advance Green's functions in the limit of zero-frequency, $G^R_{k\sigma} = (-\epsilon_{k\sigma} + \frac{i\hbar}{2\tau})^{-1}$. The Green's functions include the effect of the uniform magnetization and finite lifetime, the G^R has only diagonal components. The contribution of $\sigma^{(\mathrm{A})}_{xy}$ turns out to be zero, because of the identity of Hall conductivity, $\sigma_{xy} = -\sigma_{yx}$. Similarly, the contribution of process A* vanishes. The other term of the Figure.1 can be estimated as

$$\sigma^{(\mathrm{B})}_{xy} = -\frac{1}{2V}\left(\frac{e\hbar}{m}\right)^2 \sum_{k,k'} k_x k'_y S^{\alpha}_{k-k'} S^{\beta}_{k'-k} \mathrm{Tr}[G^r_k \sigma_\alpha G^r_{k'} G^a_{k'} \sigma_\beta G^a_k], \quad (6)$$

Equation.(6) can be evaluated by noting $\sigma_{xy} = -\sigma_{yx}$ and by using the identity($\alpha \neq \beta$),

$$\mathrm{Tr}[G^R_k \sigma_\alpha G^R_{k'} G^A_{k'} \sigma_\beta G^A_k] = i\epsilon_{\alpha\beta z}\sum_{\sigma=\pm}\sigma |G^R_{k'-\sigma}|^2 |G^R_{k'\sigma}|^2, \quad (7)$$

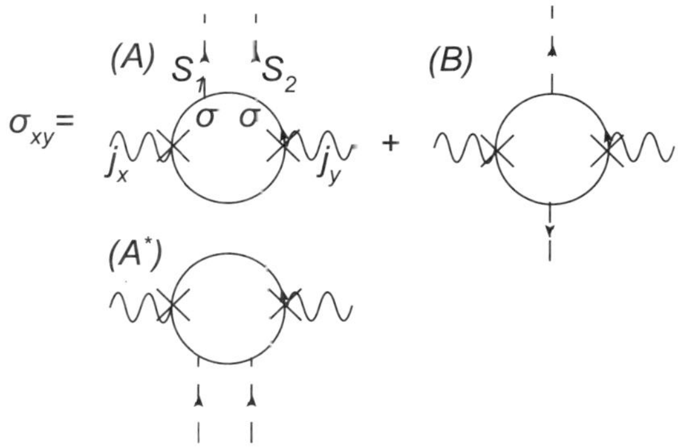

Fig. 1. The contributions to the global Hall conductivity, which are the second order with respect to the exchange interactions. The wave lines refer to charge current. The dots line refer to the exchange interaction. The curved line is electron motion including the life time τ.

as

$$\sigma_{xy}^{(\mathrm{B})} = \frac{1}{2iV^3}\left(\frac{e\hbar}{m}\right)^2 \iint d\boldsymbol{x}_1 d\boldsymbol{x}_2 (\boldsymbol{S}_1 \times \boldsymbol{S}_2)^{\mathrm{z}}$$
$$\sum_{kk'\sigma} \sigma k_x k'_y |G^r_{k\sigma}|^2 |G^r_{k',-\sigma}|^2 e^{i(\boldsymbol{k}-\boldsymbol{k'})\cdot \boldsymbol{x}_{12}}, \quad (8)$$

where $\boldsymbol{S}_1 \equiv \boldsymbol{S}(\boldsymbol{x}_1)$, $\boldsymbol{x}_{ij} = \boldsymbol{x}_i - \boldsymbol{x}_j$ and $\boldsymbol{x}_i$, $\boldsymbol{x}_j$ as the positions of local spins. This contribution $\sigma_{xy}^{(\mathrm{B})}$ is again zero. Therefore, the global Hall conductivity caused by the vector chirality is zero.

4. Local Hall conductivity

Although we have seen that the Hall effect due to the vector chirality vanishes if averaged over the system, the local Hall effect can be finite. We will have investigated this possibility by using Keldysh Green functions. We calculate the local($\boldsymbol{x}$-dependent) current density. The first contribution in Figure.2 now becomes Equation.(9)

$$j_y^{(\mathrm{A})}(\boldsymbol{x}, t) = \frac{-e^{i\Omega t}}{2V\Omega} J_{\mathrm{sd}}^2 \left(\frac{e\hbar}{m}\right)^2 \sum_{kk'k''\sigma} \sum_{\omega}$$
$$\times (k+k')_y k'_x E_x S^{\alpha}_{k-k''} S^{\beta}_{k''-k'} e^{i(\boldsymbol{k'}-\boldsymbol{k})\cdot \boldsymbol{x}}$$
$$\times \mathrm{Tr}[G_{k,\omega}\sigma_\alpha G_{k'',\omega}\sigma_\beta G_{k',\omega} G_{k',\omega+\Omega}]^{<}. \quad (9)$$

The lesser component can be expanded by using these Green's function,

$$[G_k \sigma_\alpha G_{k''} \sigma_\beta G_{k'} G_{k',\Omega}]^{<}$$
$$= f(\omega+\Omega) G^R_{k,\omega}\sigma_\alpha G^R_{k'',\omega}\sigma_\beta G^R_{k',\omega} G^A_{k',\omega+\Omega}$$
$$- f(\omega) G^R_{k,\omega}\sigma_\alpha G^R_{k'',\omega}\sigma_\beta G^R_{k',\omega} G^A_{k',\omega+\Omega}$$
$$+ f(\omega+\Omega) G^R_{k,\omega}\sigma_\alpha G^R_{k'',\omega}\sigma_\beta G^R_{k',\omega} G^R_{k',\omega+\Omega}$$
$$- f(\omega) G^A_{k,\omega}\sigma_\alpha G^A_{k'',\omega}\sigma_\beta G^A_{k',\omega} G^A_{k',\omega+\Omega}, \quad (10)$$

where $f(w)$ is the Fermi distribution function which is given as $f(w) = -\theta(w)$ at zero temperature($\theta(w)$ is a step function). We neglect the terms only G^R's or G^A's which are higher order of $\frac{\hbar}{\epsilon_F \tau} \ll 1$ compared with the contribution of the G^R and G^A mixed. Then Equation (10) can be written as,

$$[G_{k,\omega}\sigma_\alpha G_{k'',\omega}\sigma_\beta G_{k',\omega} G_{k',\omega+\Omega}]^{<}$$
$$\simeq -\Omega\delta(\Omega) G^R_{k,\omega}\sigma_\alpha G^R_{k'',\omega}\sigma_\beta G^R_{k',\omega} G^A_{k',,\omega+\Omega}. \quad (11)$$

From the Equation (9) and (10), the local Hall conductivity can be written as,

$$\sigma_{xy}^{(\mathrm{A})}(\boldsymbol{x}) = \iint d\boldsymbol{x}_1 d\boldsymbol{x}_2$$
$$A^{(A)}(\boldsymbol{x}, \boldsymbol{x}_1, \boldsymbol{x}_2)\ (\boldsymbol{S}_1 \times \boldsymbol{S}_2)^{\mathrm{z}}, \quad (12)$$

and

$$A^{(A)}(\boldsymbol{x}, \boldsymbol{x}_1, \boldsymbol{x}_2) = \frac{-J_{\mathrm{sd}}^2 \tau}{\hbar V^3}\left(\frac{e\hbar}{m}\right)^2 \sum_\sigma \sigma$$
$$\times I_{-\sigma}(r_{21}) \partial_{x_{10}} I_\sigma(r_{10}) \partial_{y_{20}} \mathrm{Re}(I_\sigma(r_{20})) \quad (13)$$

, where the correlation function is $I_\sigma(r_{ij}) = \sum_k e^{i\boldsymbol{k}\cdot(\boldsymbol{x}_i - \boldsymbol{x}_j)} G^R_{k,\sigma} \simeq \frac{1}{2\pi}\frac{mV}{\hbar^2 r_{ij}} e^{-\frac{\hbar}{4\epsilon_F \tau} k_F r_{ij}} e^{ik_{F_\sigma} r_{ij}}$, where $k_{F_\sigma} \equiv (1 - \sigma\frac{M}{2\epsilon_F})k_F$. The $I(r_{ij})$ depends only on the distance, $r_{ij} \equiv |\boldsymbol{x_{ij}}|$. It oscillats and decreases within the distance of the mean free path, ℓ.

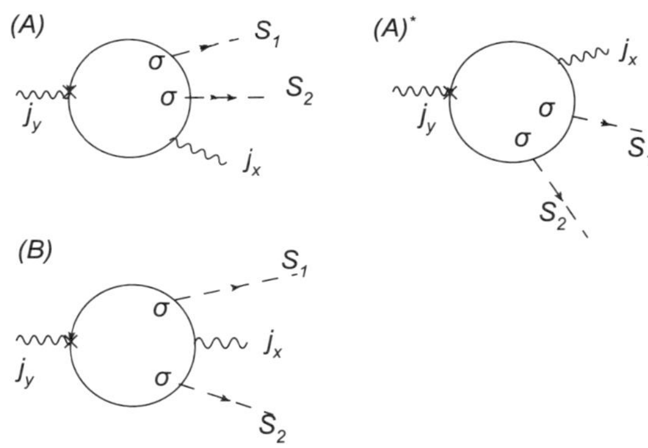

Fig. 2. Diagrams representing the contributions to the local Hall current density, which are the second order with respect to the exchange interactions.

Similarly, we obtain the contribution (B) in Figure.2 as

$$\sigma_{xy}^{(\mathrm{B})}(\boldsymbol{x}) = \iint d\boldsymbol{x}_1 d\boldsymbol{x}_2$$
$$\mathrm{Re}(A^{(B)}(\boldsymbol{x}, \boldsymbol{x}_1, \boldsymbol{x}_2))\ (\boldsymbol{S}_1 \times \boldsymbol{S}_2)^{\mathrm{z}}, \quad (14)$$

where

$$A^{(B)}(\boldsymbol{x},\boldsymbol{x}_1,\boldsymbol{x}_2) = \sum_\sigma \sigma \frac{-J_{\rm sd}^2\tau}{\hbar V^3}\left(\frac{e\hbar}{m}\right)^2 \times I_\sigma^*(r_{20})\partial_{x_{10}} I_\sigma(r_{10})\partial_{y_{21}}{\rm Re}(I_{-\sigma}(r_{21})). \quad (15)$$

Therefore, the total local Hall conductivity, $\sigma_{xy}(\boldsymbol{x}) = \sigma_{xy}^{(A)}(\boldsymbol{x}) + \sigma_{xy}^{(A)^*}(\boldsymbol{x}) + \sigma_{xy}^{(B)}(\boldsymbol{x})$, can be written as

$$\sigma_{xy}(\boldsymbol{x}) = \iint d\boldsymbol{x}_1 d\boldsymbol{x}_2 \; {\rm Re}\left(A(\boldsymbol{x},\boldsymbol{x}_1,\boldsymbol{x}_2)\right)\ (\boldsymbol{S}_1\times\boldsymbol{S}_2)^{\rm z}, \quad (16)$$

where

$$A(\boldsymbol{x},\boldsymbol{x}_1,\boldsymbol{x}_2) = (\boldsymbol{x}_{10}\times\boldsymbol{x}_{20})_{\rm z}\sum_\sigma \sigma \left(\frac{I_{-\sigma}(r_{21}){\rm Re}I'_\sigma(r_{20})}{r_{20}} + \frac{{\rm Re}(I'_{-\sigma}(r_{21}))I_\sigma^*(r_{20})}{r_{21}}\right) \times \frac{-J_{\rm sd}^2\tau}{\hbar V^3}\left(\frac{e\hbar}{m}\right)^2 \frac{I'_\sigma(r_{10})}{r_{10}}. \quad (17)$$

5. Discussion

From the above result, we notice that the Hall effect is induced by the vector chirality at least locally. The local Hall effect appears only in the presence of the spin polarization. The Hall effect is zero in the absent of the spin polarization. In fact, if the electron polarization is zero, the coefficient A identically vanishes.

The local Hall effect due to the vector spin chirality would be observed, for example, in small systems contain one magnetic vortex, as one can easily see that coefficient of A is finite in this case. The Hall effect could arise more generally if electron transport has asymmetry with respect to the spin structure.

In the care of insulator, manganese oxides, the vector spin chirality has been pointed out to induce the electric polarization as $\boldsymbol{P} = \alpha(\boldsymbol{x}_2 - \boldsymbol{x}_1)\times(\boldsymbol{S}_1\times\boldsymbol{S}_2)$, where α is a constant proportional to the spin-orbit interaction[8,9]. The relation between the Hall effect in metals and electric polarization in insulator, both induced by the vector spin chirality, would be an interesting future subject.

6. Conclusion

We calculated the Hall conductivity caused by the vector spin chirality in the *s-d* model. The result indicates that the Hall effect is generated indeed by a vector spin chirality if the present of the electron spin polarization. We note that the vector chirality effect is local Hall effect. The global vecter-chirality effect is canceled.

This work was supported by a Gran-in-Aid for Scientific Research in Priority Area "Creation and control of spin current" (1948027) from the Ministry of Education, Culture, Sports, Science and Technology, Japan.

References

1. R.Karplus and J.M.Luttinger, *Phys. Rev.* **95**, 1154 (1954).
2. J.Ye, *et al.*, *Phys. Rev. Lett.* **83**, 3737 (1999).
3. Y.Taguchi, Y.Oohara, H.Yoshizawa, N.Nagaosa and Y.Tokura, *Science.* **281**, 2573 (2001).
4. G.Tatara and H.Kawamura, *J. Phys. Soc. Jpn.* **71**, 2613 (2002).
5. M.Onoda, G.Tatara, and N.Nagaosa, *J. Phys. Soc. Jpn.* **73**, 2624 (2004).
6. G.Tatara, H.Kohno, *Phys.Rev.B* **67**, 113316 (2003).
7. D.Loss, P.Goldbart, *Phys. Rev.B* **43**, 13762 (1991).
8. H.Katsura, N.Nagaosa, *Phys. Rev. Lett* **95**, 057205 (2005).
9. M.Mostovoy, *Phys. Rev. Lett* **96**, 067601 (2006).

SPIN CURRENT DISTRIBUTIONS AND SPIN HALL EFFECT IN NONLOCAL MAGNETIC NANOSTRUCTURES

R. SUGANO* and M. ICHIMURA

Advanced Research Laboratory, Hitachi, 1-280, Higashi-Koigakubo, Kokubunji-shi, Tokyo 185-8601, Japan

*E-mail: ryoko.sugano.qq@hitachi.com

S. TAKAHASHI and S. MAEKAWA

Institute for Materials Research, Tohoku University, Sendai, 980-8577, Japan

We study the spatial distribution of spin/charge current in the nonlocal geometry of the Py/Cu/Pt nanostructure, where the Py/Cu and Cu/Pt structures are employed as a spin injection/accumulation and a spin/charge current converter, respectively, by applying a finite element method in three-dimensions. Calculating two types of the device operations, we have found that the inhomogeneous current distribution appears in Cu/Pt area, suggesting the possibility of the spin-signal enhancement by controlling the shape of the Cu/Pt structure

Keywords: spin injection; spin-orbit interaction; finite element method.

1. Introduction

Spin-dependent transport phenomena in magnetic nanostructures, spin-orbit (SO) interactions play the role of a converter between charge and spin currents, that is, spin/charge-current induced spin Hall effects (SHE) [1]. Since a nonmagnetic materials with strong SO interaction act as a spin injector/detecor instead of the ferromagnetic materials, the spin injection Hall device is expected to be a sensor device. A spin injection Hall device has been theoretically discussed and the nonlocal Hall voltage was evaluated [2]. In the recent experiments in the geometry of nonlocal measurement, the reversible spin Hall effect has been observed in one device [3,4]. This suggests the possibility of the spin Hall device for the MRAM application. By using a pure spin current due to nonlocal spin injection, it is also reported that the spin accumulation signal is enhanced by reducing contact sizes [5].

We treat a two operation device with the nonlocal geometry as shown in Fig.1(a). A nonmagnetic Cu wire is laterally connected to a ferromagnetic Py electrode and a nonmagnetic Pt electrode. The applied current arrangements alter the operation of the two lateral contacts Py/R_H/Cu (R_H: high resistivity layer) and Pt/Cu. The charge current I from one end on Pt to the other, as shown in Fig.1(b), SO interaction converts the charge current to the spin current in Pt, and provokes a pure spin-current injection from Pt into Py through Cu. On the other hand, I from Py to Cu along the negative direction of x, as shown in Fig (c), gives a pure spin current from Py into Pt and SO interaction convert the spin current to the charge current in Pt. Thus the charge accumulation is induced on Pt along y. By controlling only a pure spin current in both operations, we can catch the electric signals. These signals depend on the contact size and the distance L between electrodes [2].

In this work, in order to study the contact size dependence on the induced accumulation signal in the nonlocal geometry of the mag-

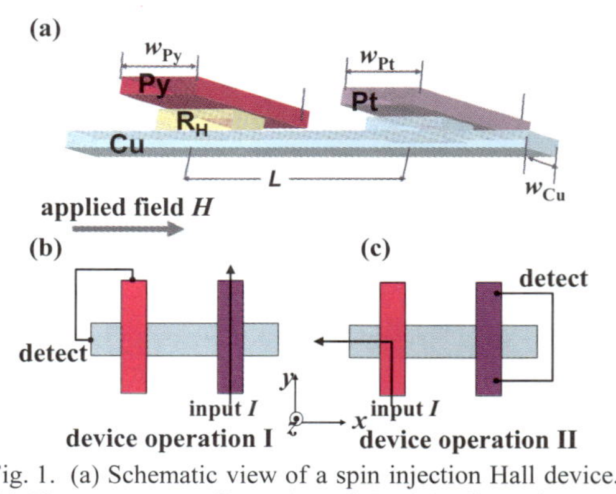

Fig. 1. (a) Schematic view of a spin injection Hall device. (b) Charge current I flows through Pt. (c) I flows from Py to the left edge of Cu.

netic nanostructure, we calculate the spatial distribution of spin/charge current by performing a finite element method (FEM) in three-dimensions under the open circuit condition [6].

2. Model and Simulation

We start with drift-diffusion of the spin-dependent electrochemical potential $\mu_\nu(\nu=\downarrow\uparrow)$ (ECP) [7]. In each region of the device, the spin-dependent current $\mathbf{j}_\nu$ is related to the spin-dependent conductivity σ_ν by $\mathbf{j}_\nu=-(\sigma_\nu/e)\nabla\mu_\nu$. In addition to longitudinal charge and spin current $\mathbf{j}q=\mathbf{j}_\uparrow+\mathbf{j}_\downarrow$ and $\mathbf{j}s=\mathbf{j}_\uparrow-\mathbf{j}_\downarrow$, transverse charge and spin current $\mathbf{j}q\mathrm{H}=\alpha\mathrm{H}[\mathbf{s}\ \mathrm{x}\ \mathbf{j}s]$ and $\mathbf{j}s\mathrm{H}=\alpha\mathrm{H}[\mathbf{s}\mathrm{x}\ \mathbf{j}q]$ are induced. Here αH is the ratio of the spin Hall conductivity to the electrical one and **s** is the magnetization direction. Total currents are given by $\mathbf{j}_\nu=-(\sigma_\nu/e)[(\mathbf{I}+\alpha_H\mathbf{w_s})\nabla\mu_\nu]$. **I** is unit matrix. Here we introduce the matrix $\mathbf{w_s}$ satisfying $\mathbf{s}\mathrm{x}\mathbf{u}=\mathbf{w_s}\mathbf{u}$. In the presence of spin diffusion, SHE is also known to obey the spin diffusion equation derived from semiclassical Boltzmann equation [2,8]. Thus, in a whole device, ECPs satisfy the following equations:

$$\begin{pmatrix} \mathbf{J}_q \\ \mathbf{J}_s \end{pmatrix} = \begin{pmatrix} \mathbf{I} & \alpha_H \mathbf{w}_s \\ \alpha_H \mathbf{w}_s & \mathbf{I} \end{pmatrix} \begin{pmatrix} \mathbf{j}_q \\ \mathbf{j}_s \end{pmatrix} \tag{1}$$

$$\begin{pmatrix} \nabla\left(\sigma_\uparrow^{-1} + \sigma_\downarrow^{-1}\right)\mathbf{J}_q \\ \nabla\left(\sigma_\uparrow^{-1} + \sigma_\downarrow^{-1}\right)\mathbf{J}_s \end{pmatrix} = -\frac{1}{e\lambda^2} \begin{pmatrix} 0 \\ \mu_\uparrow - \mu_\downarrow \end{pmatrix} \tag{2}$$

Here λ^2 is the spin-diffusion length [2,7]. By solving the combined equations of Eqs. (1) and (2), we obtain the spin/charge current distribution and the spin Hall voltage.

We use the experimental values for λ (nm), conductivity $\rho(\mu\Omega^{-1}\mathrm{cm}^{-1})$, the tunneling spin polarization P, and αH. They are λPy=30, λCu= 500, λPt=14, σPy=0.065, σCu=0.48, σPt=0.064, P=0.2, and αH=0.37 [3,9]. We note here the value of has been corrected to 0.0037 [3]. Magnetization of Py is due to the applied magnetic field H (H // x). The thicknesses of Py, Pt and Cu are fixed to 20 nm. The distance between Py and Pt is also fixed by L= 0.8λCu. The thickness of Py/RH/Cu contact is 30 nm. Boundary conditions are set by two alternative conditions as follows. In Fig. 1(b), I= 100 μA is applied to Pt and the voltage difference between Py and Cu due to magnetization configurations is measured for detecting the spin accumulation generated by charge-current induced SHE (direct SHE) in Pt. In Fig.1 (c), I =100 μA is sent from Py into the left end of Cu, and the voltage difference between both ends of Pt indicates charge accumulation generated by spin-current induced SHE (inverse SHE). We set the charge current normal to the boundary surface of the device to be zero. In these arrangements, we study the spin/charge accumulation signal by changing wPy, wPt and wCu., where wPy, wPt and wCu are width of Py, Pt and Cu.

3. Result and Discussion

Firstly, we examine direct SHE in Fig.1(b). Figure 2(a) shows ECP profile of each spin channel $\mu^\uparrow$ and $\mu^\downarrow$ for the top of the whole Cu wire along x for wCu=100nm, wPy=100nm, and wPt=100nm. The spin polarized current injected from the Pt/Cu interface into the Cu wire diffuses in both $\mp$x

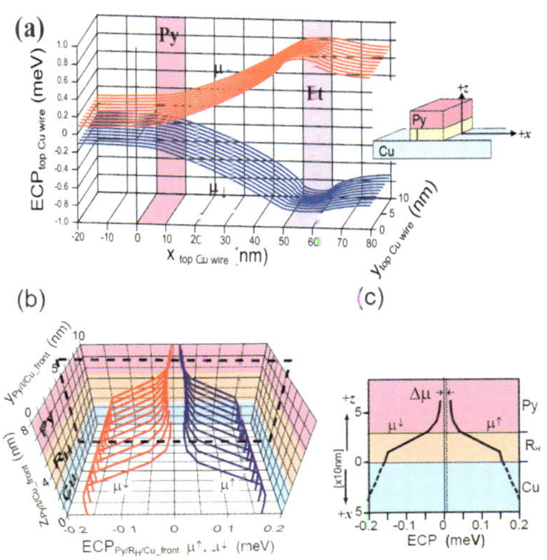

Fig. 2. ECP profiles: (a) Up and down spin channels ($\mu^\uparrow$, $\mu^\downarrow$) on the top of a whole Cu wire along the x. (b) $\mu^\uparrow$ and $\mu^\downarrow$ at x=10 through the Py/$\mathrm{R_H}$/Cu interface area along the z-direction. (c) $\mu^\uparrow$ and $\mu^\downarrow$ from the top center of the Cu (the right region of 0) to the Py (the left region of 0) along the arrow of the right illustration in Fig.2(a)

directions and decays exponentially. Figure 2(b) displays at x=10 through the Py/RH/Cu interface along z. We can see sharp drop in $\mu^{\uparrow}$ and $\mu^{\downarrow}$. The ECP behavior seems homogeneous along y. Figure 2(c) indicates $\mu^{\uparrow}$ and $\mu^{\downarrow}$at the center on the top of the Cu wire (y=5, x=10) along x (the lower region of 0) indicated by dotted curves and penetrated into Py along z (the upper region of 0) from the Py/Cu interface by solid curves, as shown by the arrow in the right illustration of Fig. 2(a). The voltage drop between Cu and Py: $E=-\nabla z(\mu^{\uparrow}+\mu^{\downarrow})/2=\mu^{\uparrow}+\mu^{\downarrow}$ detected by the spin accumulation signal $\Delta\mu$. These ECP profiles are consistent with theoretical prediction[2]. Figure 3 displays $\Delta\mu$ as a function of the Py/Cu interface area SPy=wPywCu for various values of wPy, wCu, and wPt. The $\Delta\mu$ in Fig. 2 gives the $\Delta\mu$ by 0.65μeV. The corrected $\Delta\mu$ by αH= 0.0037[3] matches up to the experimentally observed SHE signal[10] by considering that the injected spin current through the Pt/Cu interface becomes proportional to αH. Almost all signals are collapsed into the reciprocal plot of SPy except the extremely narrow wire that has the oblong shaped contact. Although we found the contact condition of the deviation from the reciprocal plot as max (wCu/wPy,wCu/w_{Pt})<0.5 suggesting a possibility of artifact in calculation. However, this suggests that a proper choice of the Py/Cu contact area may bring an enhancement of spin signal.

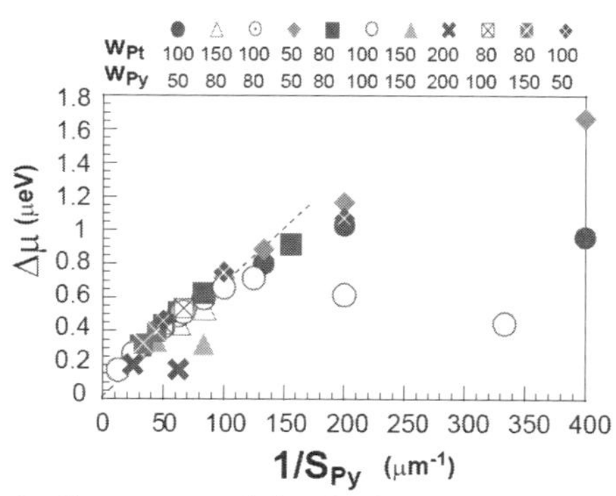

Fig. 3. Charge-current induced spin accumulation signal $\Delta\mu$ between Py and Cu as a function of Py/Cu contact area $S_{Py}=w_{Py}w_{Cu}$ for various values of w_{Py}, w_{Cu}, and w_{Pt}

We next discuss spin-current induced SHE (inverse SHE) in the device of Fig. 1(c). As distinct from direct SHE, we found the wCu-independent behavior accompanied by peculiar inhomogeneous distributions of the up and down spin currents reflecting the contact shape in inverse SHE. Figure 4(b) displays the voltage drop, which corresponds to the spin current inflow, through the Py/Cu interface. At the interface, the injected spin current is inversely proportional to the contact area SPy. Figure 4(b) shows the calculated SHE signal $\Delta\mu$ as a function of a cross sectional area wPtdPt of Pt for various values of wPy, wCu, and wPt. Each type of the symbol corresponds to the combination of two electrode widths (wPt, wPy) for various values of wCu ranging from 50 to 200nm. Same symbols overlap each other, clearly showing that the signal does not depend on the wire width. The spin current absorption into Pt from Cu through the Py/RH/Cu interface, which is proportional to 1/SPy(=1/wCuwPy), is proportional to the Pt/Cu interface area SPt=wCuwPt. This shows that the spin current inflow into Pt is propor-

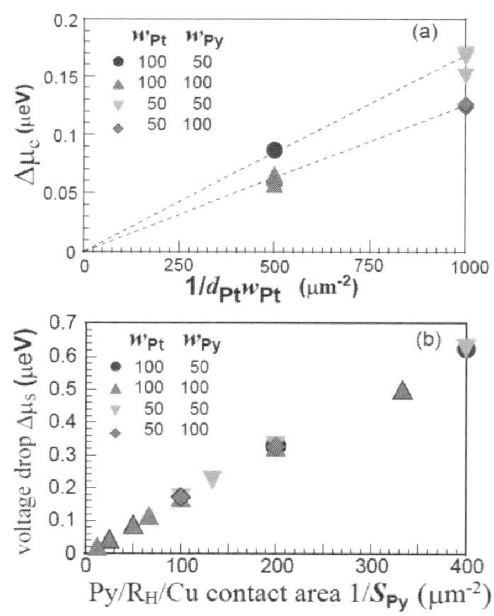

Fig. 4. (a) SHE signal $\Delta\mu$ for various values of w_{Py}, w_{Cu}, and w_P as a function of $1/d_{Pt}w_{Pt}$. Each symbol corresponds to the combination of the width of two electrodes (w_{Pt}, w_{Py}) for various values of w_{Cu}. (b) Spin current inflow $\Delta\mu_s$ through the Py/R_H/Cu interface as a function of $1/S_{Py}$.

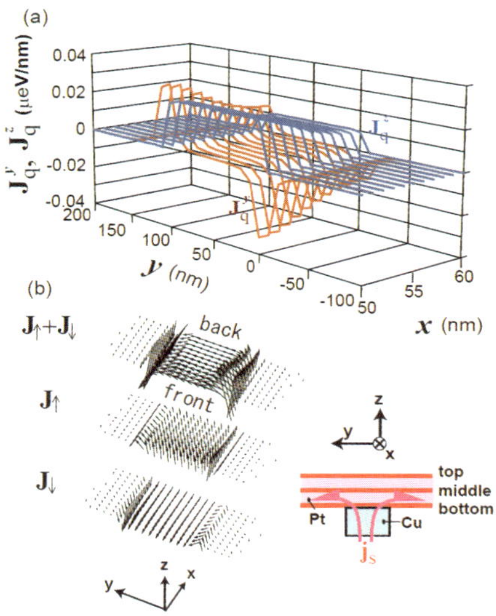

Fig. 5. Spatial distribution: (a) *y*- and *z*- components J_q^y and J_q^z of absorbed spin current $\mathbf{J}_q = \mathbf{J}_\uparrow + \mathbf{J}_\downarrow$ into Pt through Pt/Cu at the front bottom of Pt as a function of *y*. (b) Charge current $\mathbf{J}_q = \mathbf{J}_\uparrow + \mathbf{J}_\downarrow$ and up and down spin cur-rents $\mathbf{J}_\uparrow$, $\mathbf{J}_\downarrow$. Inset: Side view of spin current absorption.

tional to only *w*Pt/*w*Pt, indicating the SHE signal $\Delta\mu$ independent of w_{Cu}. This allows us room to choice the convenient wire width in the device design.

Then we also obtain how the current is absorbed into Pt through the Pt/Cu interface. As shown in Fig. 5(b), the Pt/Cu cross area the inhomogeneous current distribution is observed. Almost all current flows along both edges of the Cu wire and two separated flowing channels are formed. While the current flow parallel to the magnetization direction (//x) has no contribution to the Hall voltage, the absorbed spin current by Pt changes to the motion along y due to the SH matrix term in Eq. (2). In Fig. 5(a), the absorbed current at the front bottom of the Pt electrode are shown as a function of y and x. In the x-direction, the current monotonically decays. Since the spin current tends to diffuse in the opposite direction (//y) in Pt, the effects of the SH matrix works in reverse. At one side of the Pt contact, the absorption of the spin current is enhanced, and at the same time, the spin current on the other side is strongly modified by the SO interaction. As a result, the up spins flow into Pt at the one edge and the down spins flow out along the opposite edge as shown in Fig. 5 (b). This two channel situation may be an important factor for the charge accumulation and indicates that the current distribution is sensitive to the contact shape. It also suggests that the distribution of two species of spins (up and down) can be spatially localized.

In conclusion, we study the spatial distribution of spin/charge current in the nonlocal geometry of the magnetic nanostructure Py/R_H/Cu/Pt by applying a finite element method in three-dimensions. We have found that the inhomogeneous current distribution appears in the Cu/Pt contact, suggesting the possibility of the spin-signal optimization by proper choice of the shape and width of the electrodes.

Acknowledgments

This work is supported by Next Generation Supercomputing Project Nanoscience Program, MEXT, Japan.

References

1. J. E. Hirsch, Phys. Rev. Lett. 83, 1834 (1999).
2. S. Takahashi, et al., in Concepts in Spin Electronics, edited by S. Maekawa (Oxford University, Oxford, 2006).
3. T. Kimura, et al., Phys. Rev. Lett. 98, 156601(2007); Phys. Rev. Lett. 98, 249901 (2007)
4. T. Seki, et al., nature materials vol. 7, 125(2008).
5. T. Kimura, et al., Phys. Rev. B73, 132405(2006).
6. M. Ichimura, et al., J. Appl. Phys. 95, 7255 (2004)
7. S. Takahashi and S. Maekawa, Phys. Rev. B 67, 052409 (2003).
8. S. Zhang, Phys. Rev. Lett. 85, 393 (2000).
9. H. Kurt, et al., J. Appl. Phys. 81, 4787 (2002)
10. L. Vila, et al., Phys. Rev. Lett. 99, 226604 (2007).

NEW BOUNDARY CRITICAL PHENOMENON AT THE METAL-QUANTUM SPIN HALL INSULATOR TRANSITION

HIDEAKI OBUSE

Department of Physics, Kyoto University, Kyoto 606-8502, Japan
**E-mail: obuse@scphys.kyoto-u.ac.jp*

We investigate bulk and boundary multifractality at the metal-quantum spin Hall (QSH) insulator transition driven by disorder in two-dimensions. Recently, we have shown that boundary multifractality at this transition is different from that of the symplectic class by using the network model with random spin-orbit interactions. To give the another evidence for this result, we investigate multifractality at the metal-QSH insulator transition by using a different model, namely, the network model with a constant spin-orbit interaction. It is found that bulk multifractality at the metal-QSH insulator transition is that of the symplectic class while boundary multifractality is different, confirming identical with the previous result.

Keywords: Quantum spin Hall effect; multifractality; Anderson transition.

1. Introduction

The Anderson transition is a metal-insulator transition driven by disorder. Since the localization length diverges at the critical point of this transition, the wave function at criticality shows a scale-invariant multifractal nature reflecting no characteristic length in the system. While multifractality in the bulk region has been investigated for many disordered systems so far[1,2], boundary multifractality which characterizes a boundary critical phenomenon at the Anderson transition had not been investigated. Recently, boundary multifractality is investigated for the system with strong spin-orbit interaction (SOI), namely, symplectic class in two-dimensions[3] and the quantum Hall plateau transition.[4,5]

Furthermore, we have investigated multifractality at the metal-quantum spin Hall (QSH) insulator transition[6], where the QSH insulator is a new kind of topological insulator which possesses time-reversal symmetry but broken spin-rotational symmetry.[7,8] We have shown that boundary multifractality at this transition is different from that of the ordinary symplectic class while bulk multifractality is the same.

In this work, we investigate multifractality at the Anderson transition in the QSH system by using a different model than that used in the previous work in order to give the another evidence of that finding. To this end, we treat the network model with a constant SOI in the present work, while the network model with the spatially random SOI is used in Ref. 6. Through the detailed studies, we have confirmed that bulk multifractality at the metal-QSH insulator transition is identical with that of the ordinary symplectic class while boundary multifractality at this transition is different. These results give the strong evidence for the validity of the previous work.

2. Model

Here we briefly recall the network model for the QSH system introduced in Ref. 9. In this network model, the system is composed from nodes connected to two in-coming and two out-going links of a Kramers pair. Each node is represented by a scattering matrix S that preserves time-reversal symmetry but breaks spin-rotational symmetry. This 4×4 S matrix is described by

$$S = \begin{pmatrix} r\boldsymbol{\sigma}_0 & tQ \\ -tQ^\dagger & r\boldsymbol{\sigma}_0 \end{pmatrix}, \qquad (1)$$

where

$$r = \tanh X, \qquad t = (\cosh X)^{-1},$$
$$Q = i\boldsymbol{\sigma_0} \cos\theta \sin\varphi_1 + \boldsymbol{\sigma_1} \sin\theta \cos\varphi_2 \quad (2)$$
$$- \boldsymbol{\sigma_2} \sin\theta \sin\varphi_2 + \boldsymbol{\sigma_3} \cos\theta \cos\varphi_1.$$

Here $\boldsymbol{\sigma_0}$ is the 2×2 unit matrix and $\boldsymbol{\sigma_{1,2,3}}$ are the Pauli matrices. There are two important parameters X and θ which determines the phase diagram. The parameter X corresponding to the chemical potential characterizes the transmission and reflection amplitudes t and r to neighboring plaquettes. The parameter θ characterizes the strength of the SOI. There are two ways to introduce the SOI; introducing different θ's for different nodes or the same θ for all nodes. $\varphi_{1,2}$ are chosen as random between 0 and 2π.

Since it is known that the metallic phase is surrounded with the ordinary and topological insulating phases in this network model,[9] there are two critical points X_s and X_l which satisfies $X_s < X_l$. The important property of QSH insulator is the existence of edge states of the Kramers doublet, which are realized by imposing the reflecting boundary conditions (RBC) to the network. For the node on the reflecting boundary, the in-coming and outgoing links are connected by the probability unity without any spin-flips.

By tuning the chemical potential X to the critical point, the critical properties for the metal-QSH insulator transition are investigated. In Ref. 6, the network model with random SOI is investigated. In this model, the contribution of the irrelevant length due to the spin-relaxation length can be negligibly small at the critical point X_l. Unfortunately, the finite size effect originating from the symplectic to unitary class cross-over becomes significant at X_s, because X_s locates close to $X = 0$ where the system belongs to the unitary class.[9] To obtain the accurate value of two kinds of boundary multifractality by focusing only on X_l in the random SOI model, the position imposed RBC should be changed alternatively, which permutes the

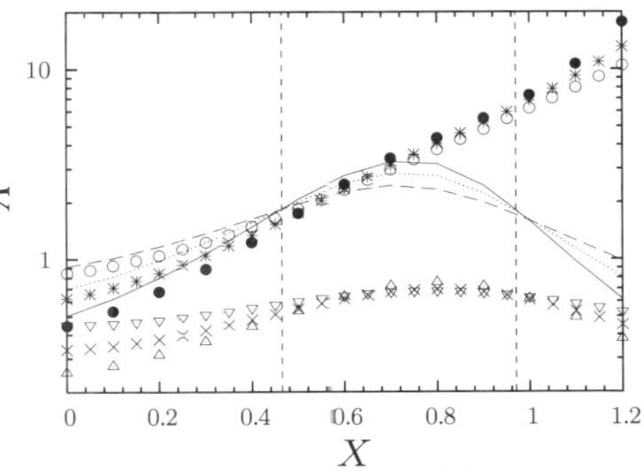

Fig. 1. The localization length $\Lambda^{(1)}(X, M)$ in the Q1D systems with RBC as a function of X for the width $M = 4$ (○), 8 (∗), and 16 (•). The localization length $\Lambda^{(2)}(X, M)$ are shown by $M = 4$ (▽), 8 (×), and 16 (△). The localization length $\Lambda^{(1)}_{\text{PBC}}(X, M)$ are shown by the curves for $M = 4$ (dashed), 8 (dotted), and 16 (solid). The dashed vertical lines indicate critical point $X_s = 0.467$ and $X_l = 0.97$.

ordinary insulating phase and the QSH insulating phase.[6] This trick is, however, far from physical realizations. In the present work, we treat the network model with fixed $\theta = 3\pi/16$ which is free from the above trick because X_s is not close to $X = 0$, while the spin relaxation length becomes larger.

3. Results

At first, we consider effects of RBC for localization length in the quasi-one dimensional (Q1D) system. Figure 1 shows the X dependence of the normalized localization length $\Lambda^{(i)}(X, M) \equiv \xi^{(i)}(X, M)/M$, where $\xi^{(i)}(X, M)$ is calculated from the inverse of the ith smallest Lyapunov exponent in the Q1D system with the RBC imposed for the transverse direction. The width M of the system is counted by the number of nodes for the transverse direction and the Lyapunov exponents are calculated by the standard transfer matrix method.[10] The normalized localization lengths for the system with the periodic boundary condition (PBC) for the transverse direction $\Lambda^{(1)}_{\text{PBC}}(X, M)$ are shown by curves for comparison. The dashed vertical lines represent critical points at $X_s = 0.465$ and $X_l = 0.97$ which are obtained from the finites-size scaling analysis for $\Lambda^{(1)}_{\mathrm{PBC}}$.

Figure 1 clarifies that changing boundary conditions significantly affect behaviors of normalized localization lengths, especially for larger X. Indeed, the phase for $X > X_l$ corresponds to the QSH insulating phase with edge states in the present setup of RBC. This is understood from the increasing behavior of $\Lambda^{(1)}(X > X_l, M)$ but decreasing $\Lambda^{(2)}(X > X_l, M)$ with increasing M. $\Lambda^{(1,2)}(X, M)$ in the range $X < X_s$ and $X_s < X < X_l$ correspond to the ordinary insulating and metallic behaviors, respectively, which are qualitatively consistent to $\Lambda^{(1)}_{\mathrm{PBC}}$. The fact that the crossing points of $\Lambda^{(1,2)}(X, M)$, except $\Lambda^{(1)}(X \approx X_l, M)$, do not coincide with X_s or X_l implies the effect of the irrelevant length is enhanced by imposing RBC. Since $\Lambda^{(1)}(X > X_l, M)$ reflects metallic properties of the edge states, $\Lambda^{(1)}(X \approx X_l, M)$ increases for both phases around X_l with increasing M. Although $\Lambda^{(1)}(X \approx X_l, M)$ would lose the M dependence at X_l for $M \to \infty$, the present system sizes are not enough to observe such a behavior.

To investigate details of critical properties we calculate multifractality of wave functions $\psi_\sigma(x, y)$ at X_s and X_l. Multifractality is characterized by anomalous dimensions $\Delta_q^{(\kappa,\gamma)}$ defined as the power law exponent for the system size L dependence of the qth moment of wave function amplitudes,

$$\overline{|\psi_\sigma(x,y)|^{2q}}/(\overline{|\psi_\sigma(x,y)|^2})^q \propto L^{-\Delta_q^{(\kappa,\gamma)}}. \quad (3)$$

Here $\kappa(= 2, 1)$ indicates bulk or boundary anomalous dimensions and $\gamma(= Z_2, \mathrm{O})$ represents the critical point between the metallic and QSH insulating phases or the metallic and ordinary insulating phases. The overline means the ensemble averaging. The multifractal spectrum $f^{(\kappa,\gamma)}(\alpha^{(\kappa,\gamma)})$ is related to $\Delta_q^{(\kappa,\gamma)}$ by the Legendre transform,

$$f^{(\kappa,\gamma)}(\alpha_q^{(\kappa,\gamma)}) = (\alpha_q^{(\kappa,\gamma)} - 2)q - \Delta_q^{(\kappa,\gamma)} + d_\kappa,$$
$$\alpha_q^{(\kappa,\gamma)} - 2 = d\Delta_q^{(\kappa,\gamma)}/dq, \quad (4)$$

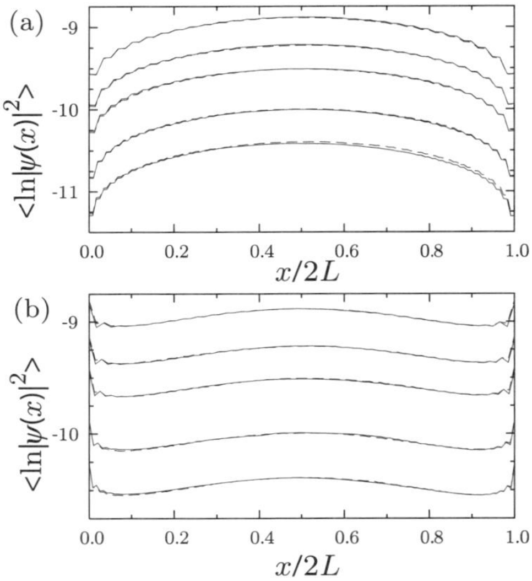

Fig. 2. Dependence on the position x along the cylinder axis of $\langle \ln |\psi(x)|^2 \rangle_L$ for $L = 30, 35, 40, 50$, and 60 from top to bottom at critical point X_s (a) and X_l (b). The dashed curves in (a) and (b) represent $\langle \ln |\psi(x)|^2 \rangle_L$ for the network model with random SOI.

where $d_2(d_1) = 2(1)$.

To investigate bulk and boundary multifractality, we calculate wave functions of the two dimensional QSH network model with a constant SOI. Here, we impose RBC for the x direction and PBC for the y direction then the system becomes the cylindrical geometry. The dynamics of wave functions on links of the network is governed by a unitary evolution operator U. For each disorder realization, we numerically diagonalize U and obtain a critical wave function whose eigenvalue is closest to 1. The system size is changed as $L = 30, 35, 40, 50$, and 60. The number of ensemble is 5×10^4 for each system size.

To clarify the effect of boundaries, we compute the logarithm of squared wave function amplitudes averaged over samples and the circumference direction defined by

$$\langle \ln |\psi(x)|^2 \rangle_L \equiv \frac{1}{2L} \sum_{y=1}^{2L} \overline{\ln(\sum_{\sigma=\uparrow,\downarrow} |\psi_\sigma(x,y)|^2)},$$

here x and y give the position along the axis and circumference directions, respectively.

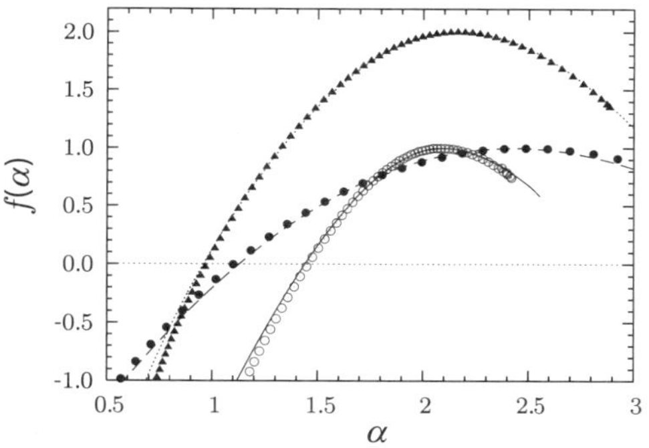

Fig. 3. The multifractal spectra $f^{(\kappa)}(\alpha)$ calculated for the network model with a constant SOI. The triangles, open circles, and filled circles represent $f^{(\kappa,\gamma)}(\alpha)$ in the bulk at metal-QSH insulator transition X_l $[(\kappa,\gamma)=(2,Z_2)]$, on the boundary at metal-QSH insulator transition X_l $[(\kappa,\gamma)=(1,Z_2)]$, and on the boundary at the metal-insulator transition X_s $[(\kappa,\gamma)=(1,\mathrm{O})]$. The dotted, solid, and dashed curves corresponds to $f^{(2)}(\alpha)$,$f^{(1,Z_2)}(\alpha)$, and $f^{(1,\mathrm{O})}(\alpha)$ obtained from the network with the random SOI.

Figures 2 (a) and (b) show the x dependence of $\langle\ln|\psi(x)|^2\rangle_L$ at the metal-ordinary insulator transition X_s and the metal-QSH insulator transition X_l, respectively. As a comparison, $\langle\ln|\psi(x)|^2\rangle_L$ calculated for the network model with random SOI are shown by the dashed curves. In Fig. 2 (a), wave function amplitudes near boundaries is smaller than that in the bulk. Remarkably, at the metal-QSH insulator transition, the wave function amplitude near the boundaries becomes larger than that in the bulk as shown in Fig. 2 (b).

By using Eqs. (3) and (4), the bulk and boundary multifractal spectra are numerically calculated. Figure 3 shows multifractal spectra $f^{(2,Z_2)}(\alpha)$, $f^{(1,Z_2)}(\alpha)$, and $f^{(1,\mathrm{O})}(\alpha)$ calculated from the network model with a constant SOI. The multifractal spectra $f^{(2)}(\alpha)$, $f^{(1,Z_2)}(\alpha)$, and $f^{(1,\mathrm{O})}(\alpha)$ calculated from the network model with the random SOI which is obtained in Ref. 6 are also shown by the dotted, solid, and dashed curves, respectively. By comparing the results, it is found that the multifractal spectra corresponding to the same κ and γ are almost overlapped each other. These results, therefore, provide the strong evidence that bulk multifractality at the metal-QSH insulator transition is same with those of the ordinary symplectic class while boundary multifractality at the metal-QSH insulator transition is different from that of the symplectic class. Taking into account of the nature of the topological insulator, this result implies that edge states existing on the adjacent QSH insulating phase significantly affect the boundary critical phenomenon at the metal-QSH insulator transition.

Acknowledgments

H.O. was supported by JSPS Research Fellowship for Young Scientists. Numerical calculations have been mainly done using the facilities of the Supercomputer Center, ISSP, University of Tokyo.

References

1. M. Janssen, *Phys. Rep.* **295**, 1 (1998).
2. F. Evers and A. D. Mirlin, *Rev. Mod. Phys.* **80**, 1355 (2008).
3. H. Obuse, A. R. Subramaniam, A. Furusaki, I. A. Gruzberg, and A. W. W. Ludwig, *Phys. Rev. Lett.* **98**, 156802 (2007).
4. H. Obuse, A. R. Subramaniam, A. Furusaki, I. A. Gruzberg, and A. W. W. Ludwig, *Phys. Rev. Lett.* **101**, 116802 (2008).
5. F. Evers, A. Mildenberger, and A. D. Mirlin, *Phys. Rev. Lett.* **101**, 116803 (2008).
6. H. Obuse, A. Furusaki, S. Ryu, and C. Mudry, *Phys. Rev.* **B78**, 115301 (2008).
7. C. L. Kane and E. J. Mele, *Phys. Rev. Lett.* **95**, 226801 (2005).
8. M. König, S. Wiedmann, C. Brüne, A. Roth, H. Buhmann, L. W. Molenkamp, X. -L. Qi, and S. -C. Zhang, *Science* **318**, 766 (2007).
9. H. Obuse, A. Furusaki, S. Ryu, and C. Mudry, *Phys. Rev.* **B76**, 075301 (2007).
10. A. MacKinnon and B. Kramer, *Z. Phys. B: Condens. Matter* **53**, 1 (1983).

ON SCALING BEHAVIORS OF ANOMALOUS HALL CONDUCTIVITY IN DISORDERED FERROMAGNETS STUDIED WITH THE COHERENT POTENTIAL APPROXIMATION

SHIGEKI ONODA*

Condensed Matter Theory Laboratory, RIKEN, 2-1, Hirosawa, Wako, Saitama 351-0198, Japan
**E-mail: s.onoda@riken.jp*

Roles of scattering events on the anomalous Hall effect in disordered ferromagnets at zero temperature are studied theoretically with a particular emphasis on scaling relations between the longitudinal conductivity σ_{xx} and the anomalous Hall conductivity σ_{xy}. The coherent potential approximation, which remains efficient even for the poorly conducting regime, to the two-dimensional ferromagnetic Rashba model has revealed three regimes and the associated scaling behaviors. In particular, it establishes the robustness of the moderately dirty intrinsic regime with σ_{xy} being insensitive to σ_{xx} as well as the relation $\sigma_{xy} \propto \sigma_{xx}^{\varphi}$ with $\varphi \approx 1.6$ in the poorly conducting regime which has been experimentally observed in a wide class of disordered ferromagnets.

Keywords: Anomalous Hall effect; Berry phase; Disorder.

1. Introduction

The anomalous Hall effect (AHE) in ferromagnets has been one of the most fundamental and intriguing issues among various quantum transport phenomena. In spite of intensive and extensive studies from both experimental and theoretical viewpoints, the theoretical understanding of the AHE has been controversial. In particular, central in the controversy is the issue of the intrinsic[1] versus extrinsic[2,3] mechanisms[4]. The intrinsic mechanism proposed by Karplus and Luttinger[1] gives the anomalous Hall conductivity σ_{xy} being insensitive to the scattering rate or the longitudinal conductivity σ_{xx} since it is mediated by the interband processes. On the other hand, the extrinsic skew-scattering mechanism[2,3] yields $\sigma_{xy} \propto \sigma_{xx}$ through the intraband processes. Recently, Karplus-Luttinger's expression of σ_{xy} has been reformulated in the form of the Th ouless-Kohmoto-Nightingale-Nijs formula[5] for the quantum Hall effect[6,7];

$$\sigma_{xy}^{\mathrm{TKNN}} = -\frac{e^2}{h} \sum_n \int \frac{d^d \boldsymbol{k}}{(2\pi)^d} b_{n,\boldsymbol{k}}^z f(\varepsilon_{n,\boldsymbol{k}}), \quad (1)$$

with the Fermi distribution function f, and the Berry-phase curvature $\boldsymbol{b}_{n,\boldsymbol{k}} = \boldsymbol{\nabla}_k \times \boldsymbol{a}_{n,\boldsymbol{k}}$ and connection $\boldsymbol{a}_{n,\boldsymbol{k}} = i\langle n, \boldsymbol{k}|\boldsymbol{\nabla}_k|n, \boldsymbol{k}\rangle$ of the Bloch wavefunction $|n, \boldsymbol{k}\rangle$ with the band index n and the wavevector $\boldsymbol{k}$.

A key to theoretically resolving the controversy has been given by taking into accout both the intrinsic and extrinsic mechanisms on an equal footing in a fully quantum-mechanical transport theory[8]. It has revealed three regimes depending on what kind of approximate scaling behaviors appear between σ_{xy} and σ_{xx}[8,9,10].

(i) In the superclean case with $\sigma_{xx} \gtrsim 10^3 3e^2/h$ in two dimensions, which is translated into $\sigma_{xx} \sim 10^3 w^2/ha \gtrsim 10^6\ \Omega^{-1}\ \mathrm{cm}^{-1}$ with the lattice constant $a \sim 4\text{Å}$, the skew-scattering contribution to σ_{xy} dominates over the others, leading to an asymptotic scaling relation $\sigma_{xy} \propto \sigma_{xx}$.

(ii) In the moderately dirty regime, the skew-scattering due to the vertex correction rapidly decays and the intrinsic AHE can be dominant, making σ_{xy} insensitive to a change in σ_{xx}.

(iii) In the dirty but metallic regime satisfying $k_F\ell \sim 1$ with k_F and ℓ being the Fermi velocity and the mean-free path, respectively, there occurs another crossover to an approximate scaling behavior of $\sigma_{xy} \propto \sigma_{xx}^{\varphi}$ with $\varphi \approx 1.6$.

Indeed, recent intensive and extensive experimental works[11,12,13,14,15,16,17,18] have evidenced these three different regimes. However, the previous calculation was based on the assumption of metallic conduction, whereas some experimental results in nearly localized regime also obey the scaling behavior with $\varphi \approx 1.6$-1.7 in a wide range of conductivity. The similar behavior has also been reported for magnetic semiconductors, Mn-doped GaAs[14]. A theory that accouts for this scaling behavior even in nearly localized regime is called for.

Here, I will present a minimal theory valid for a wide range of conductivity, for instance, $\sigma_{xx} \gtrsim e^2/h$ in two dimensions, to establish the existence of the previously obtained regimes (i), (ii), and (iii). For simplicity, I take a simple minimal model for disordered ferromagnets, the two-dimensional Rashba model[19] with spin polarization,

$$\hat{H}_0 = \frac{\hbar^2 \boldsymbol{k}^2}{2m} + v\boldsymbol{k} \times \hat{\boldsymbol{\sigma}} \cdot \boldsymbol{e}^z - \Delta_0 \hat{\sigma}^z, \qquad (2)$$

with δ-functional short-range scatters which have a different local energy level by v_{imp} and a concentration n_{imp}. In the previous paper, the intirnsic and extrinsic contributions to the AHE in this model have been calculated in the self-consistent T-matrix approximation[8,9]. Here, instead, I employ the coherent potential approximation (CPA)[20] to study the disordered system where the self-consistent T-matrix approximation and the semiclassical approaches are not applicable. It is found that the CPA also gives a similar scaling relation for the poor conductors and that the intrinsic regime appears more robust than in the self-consistent T-matrix approximation.

2. Quantum transport equations

Fully quantum-mechanical linear transport equations equipped with the integral equations for the self-energy parts and the vertex parts have been developed on the basis of the Keldysh Green function formalism in the gauge-covariant Wigner space[21]. Here, the conductivity tensor is expressed as

$$\sigma_{ij} = \frac{e^2}{2} \int \frac{d\varepsilon}{2\pi i} \int \frac{d^d \boldsymbol{k}}{(2\pi)^d} \mathrm{Tr}\left[\hat{v}_i(\boldsymbol{k}) \hat{G}^<_{E_j}(\varepsilon, \boldsymbol{k})\right], \qquad (3)$$

where $\hat{v}_i(\boldsymbol{k}) \equiv \hbar^{-1}\partial_{k_i}\hat{H}_0(\boldsymbol{k}) = \hbar k_i/m + \epsilon_{ijz}\hat{\sigma}^j$ is the velocity for the Hamiltonian $\hat{H}_0(\boldsymbol{k})$ with the 3-rank fully antisymmetric tensor $\epsilon_{ij\ell}$, and $-e$ is the electronic charge. Henceforth, $\hat{G}^{\mathrm{A/R/<}}_{E_y}$ represents the $(e\hbar E_y)$-linear coefficient of the advanced/retarded/lesser Green function where E_y represents the electric field applied in the y direction. Note that the lesser ($<$) component corresponds to the quantum distribution function multiplied by $2\pi i$. The electric field also modifies the self-energy part. The $(e\hbar E_y)$-linear coefficient is denoted as $\hat{\Sigma}^{\mathrm{A/R/<}}_{E_y}$, which plays the role of the vertex correction[21].

Now I decompose $\hat{G}^<_{E_y}$ and $\hat{\Sigma}^<_{E_y}$ into two:

$$\hat{G}^<_{E_y} = \hat{G}^<_{E_y,I}\partial_\varepsilon f(\varepsilon) + \left(\hat{G}^A_{E_y} - \hat{G}^R_{E_y}\right) f(\varepsilon),$$
$$\hat{\Sigma}^<_{E_y} = \hat{\Sigma}^<_{E_y,I}\partial_\varepsilon f(\varepsilon) + \left(\hat{\Sigma}^A_{E_y} - \hat{\Sigma}^R_{E_y}\right) f(\varepsilon), \qquad (4)$$

with the Fermi distribution function f. Below, the arguments $(\varepsilon, \boldsymbol{k})$ are suppressed. Then, the Dyson equation linearized in E_y gives the self-consistent equations

$$\hat{G}^<_{E_y,I'} = \hat{G}^R_0 \left[\hat{\Sigma}^<_{E_y,I'} - i\hat{v}_y\right] \hat{G}^A_0, \qquad (5)$$

$$\hat{G}^\alpha_{E_y} = \hat{G}^\alpha_0 \hat{\Sigma}^\alpha_{E_y} \hat{G}^\alpha_0 - \frac{i}{2}\left(\hat{G}^\alpha_0 \hat{v}_y (\partial_\varepsilon \hat{G}^\alpha_0) - (\partial_\varepsilon \hat{G}^\alpha_0)\hat{v}_y \hat{G}^\alpha_0\right). \qquad (6)$$

with $\alpha = \mathrm{R}, \mathrm{A}$, where I have introduced

$$\hat{G}^<_{E_y,I'} \equiv \hat{G}^<_{E_y,I} - \frac{i}{2\hbar}\partial_{k_y}\left[\hat{G}^R_0 + \hat{G}^A_0\right] \qquad (7)$$

$$\hat{\Sigma}^<_{E_y,I'} \equiv \hat{\Sigma}^<_{E_y,I} - \frac{i}{2\hbar}\partial_{k_y}\left[\hat{\Sigma}^R_0 + \hat{\Sigma}^A_0\right]. \qquad (8)$$

The quantities $\hat{G}_0^{A/R}$ and $\hat{\Sigma}_0^{A/R}$ represent the equilibrium advanced/retarded Green function and self-energy part in the absence of the electric field, respectively.

3. Coherent potential approximation

In the Keldysh formalism, the CPA gives[22]

$$\hat{\Sigma}_0^{\alpha}(\varepsilon) = n_{\rm imp} u_{\rm imp} \hat{C}_0^{\alpha}(\varepsilon), \tag{9}$$

$$\hat{C}_0^{\alpha} = \left[1 - \hat{g}_0^{\alpha}\left(u_{\rm imp} - \hat{\Sigma}_0^{\alpha}\right)\right]^{-1}, \tag{10}$$

$$\hat{\Sigma}_{E_y,I}^{<} = n_{\rm imp} u_{\rm imp} \hat{C}_0^{R}\left[\hat{g}_{E_y,I}^{<}\left(u_{\rm imp} - \hat{\Sigma}_0^{A}\right) - \hat{g}_0^{R}\hat{\Sigma}_{E_y,I}^{<}\right]\hat{C}_0^{A}, \tag{11}$$

$$\hat{\Sigma}_{E_y}^{\alpha}(\varepsilon) = n_{\rm imp} u_{\rm imp} \hat{C}_0^{\alpha}\left[\hat{g}_{E_y}^{\alpha}\left(u_{\rm imp} - \hat{\Sigma}_0^{\alpha}\right) - \hat{g}_0^{\alpha}\hat{\Sigma}_{E_y}^{\alpha}\right]\hat{C}_0^{\alpha}, \tag{12}$$

$$\hat{g}_{0,E_y}^{\alpha}(\varepsilon) = \int \frac{d^d\boldsymbol{k}}{(2\pi)^d}\hat{G}_{0,E_y}^{\alpha}(\varepsilon,\boldsymbol{k}), \tag{13}$$

for $\alpha = {\rm A}, {\rm R}$, and

$$\hat{g}_{E_y,I}^{<}(\varepsilon) = \int \frac{d^d\boldsymbol{k}}{(2\pi)^d}\hat{G}_{E_y,I}^{<}(\varepsilon,\boldsymbol{k}). \tag{14}$$

I solve the above set of self-consistent equations and then calculate σ_{xx} and σ_{xy} through Eq. (3).

4. Scaling behaviors

The numerical results on the relation between σ_{xx} and σ_{xy} with an internal parameter of $n_{\rm imp}$ are shown in Fig.1, for a particular choice of parameters, $\Delta_0 = 0.1$, $2mv = 3.59$, $2mu_{\rm imp}/\hbar^2 = -10$ in a energy unit of the energy level of the bottom of the minority band measured from that of the majority band. The chemical potential is also chosen as $\mu = 1$. The energy cutoff has been introduced in the band dispersion at 3 in the same energy unit, which also yields a wavevector cutoff k_c with $\hbar^2 k_c^2/2m \sim 7.4$. This result supports the presence of the three

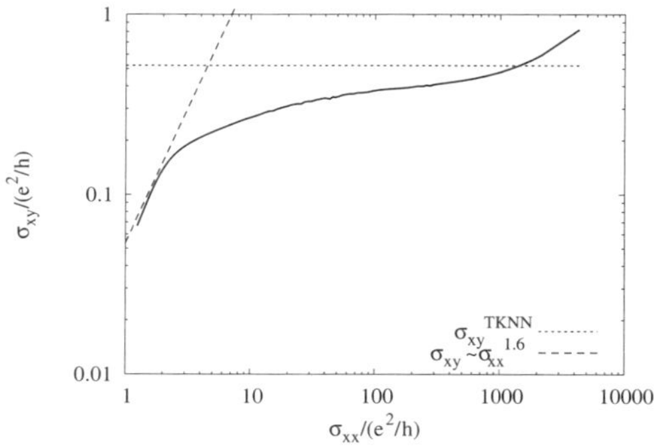

Fig. 1. Scaling plot of σ_{xy} versus σ_{xx} obtained with the CPA (solid curve). $\sigma_{xy}^{\rm TKNN}$ is obtained from Eq. (1).

regimes and the associated crossovers, as previously obtained by assuming the metallic conduction[8,9,10]:

Clean limit – With the relaxation rate $\hbar/\tau$ being much smaller than the spin-orbit interaction energy scale given by Δ_0, the AHE is mostly dominated by the skew-scattering contribution leading to an asymptotic scaling behavior $\sigma_{xy} \propto \sigma_{xx}$. Nevertheless, comparing with the previous results obtained in the self-consistent T-matrix approximation[8,9,10], it is found that this skew-scattering contribution is appreciably suppressed.

Moderately dirty regime – With increasing the concentration $n_{\rm imp}$ of impurity scatters, the skew-scattering contribution rapidly decays. Then, the AHE is almost determined by the subdominant terms, the intrinsic contribution and the side-jump contribution. Since the strength of the skew-scattering contribution is much smaller, this intrinsic regime with a slowly changing σ_{xy} is clearly observed in a wider range of σ_{xx} as it decreases by two orders of magnitude. Then, the agreement with experimental findings[11] appears to be much better in this case.

Dirty regime — When $\sigma_{xx} \sim e^2/h$, the relation $\sigma_{xy} \propto \sigma_{xx}^{\varphi}$ appear with the exponent $\varphi \approx 1.6$. In the context of the AHE, this exponent was first obtained for the dirty metallic regime in the self-consistent T-matrix approximation. Since the CPA gives a good approximation even for such highly disordered regime, the results reasonably well explain recent experimental finding on the scaling behavior with $\varphi \approx 1.5$-1.7 at least for the Fe epitaxial thin film [18].

Here, I stress that the above bahaviors are not universal. If σ_{xy} suffers from a sign change, for instance, because of the interplay of intrinsic and skew-scattering contributions with the opposite sign[10], it is needless to say that the scaling analysis becomes useless. Systematic CPA study on the AHE in this model will be presented elsewhere[22].

5. Summary

Various regimes of scaling behaviors in the anomalous Hall conductivity versus the longitudinal conductivity have been studied for a minimal model by means of the coherent potential approximation. In particular, it explains the robustness of the scattering-insensitive regime, or intrinsic regime, which has been improved from the previous results justified in the dilute impurity limit[8,9,10]. Furthermore, it has been recognized that the scaling behavior $\sigma_{xy} \propto \sigma_{xx}^{\varphi}$ with $\varphi \approx 1.6$, which has been previously obtained in the self-consistent T-matrix approximation[8,9,10], indeed appear as $\sigma_{xx} \sim e^2/h$ is approached. This may be relevant to recent experimental findings on Fe epitaxial thin films. Detailed theoretical investigations on the effects of the weak localization, the inelastic scattering in both the metallic and localized regimes, and the criticality of the metal-insu lator transition are left for future studies.

Acknowledgments

The author thanks N. Nagaosa, K. Nomura, J. Inoue, and J. Sinova for stimulating discussions. The work was supported by Grants-in-Aid from Scientific Research on Priority Area under the grant numbers 20029006 and 20046016 from the Ministry of Education, Culture, Sports, Science, and Technology of Japan. The calculation was performed using RIKEN Super Combined Cluster.

References

1. R. Karplus and J. M. Luttinger, *Phys. Rev.* **95**, 1154 (1954).
2. J. Smit, *Physica* **21**, 877 (1955).
3. J. M. Luttinger, *Phys Rev.* **112**, 739 (1958).
4. F. Maranzana. *Phys. Rev.* **160**, 421 (1967).
5. D. J. Thouless, M. Kohmoto, M. P. Nightingale, and M. den Nijs, *Phys. Rev. Lett.* **49**, 405 (1982).
6. M. Onoda and N. Nagaosa, *J. Phys. Soc. Jpn.* **71**, 19 (2002).
7. T. Jungwirth, Q. Niu, and A. H. MacDonald, *Phys. Rev. Lett.* **88**, 207208 (2002).
8. S. Onoda, N. Sugimoto, and N. Nagaosa, *Phys. Rev. Lett.* **97**, 126602 (2006).
9. S. Onoda, N. Sugimoto, and N. Nagaosa, *Phys. Rev.* **B 77**, 165103 (2008), and references therein.
10. A. A. Kovalev, Y. Tserkovnyak, K. Vyborny, and J. Sinova, unpublished.
11. T. Miyasato, *Phys. Rev. Lett.* **99**, 086602 (2007).
12. T. Fukumura *et al.*, *Jap. J. Appl. Phys.* **46**, L642 (2007).
13. S. Iguchi, N. Hanasaki, and Y. Tokura, *Phys. Rev. Lett.* **99**, 077202 (2007).
14. S. Shen *et al.*, *J. Appl. Phys.* **103**, 07D134 (2008).
15. J. G. Checkelsky *et al.*, *Phys. Rev.* **B 77**, 014433 (2008)
16. A. Fernández-Pacheco *et al.*, *Phys. Rev.* **B 77**, 100403 (2008).
17. D. Venkateshvaran *et al.*, *Phys. Rev.* **B 78**, 092405 (2008).
18. S. Sangiao *et al.*, *Phys. Rev.* **B 79**, 014431 (2009).
19. Y. Bychkov and E. Rashba, *J. Phys.* **C 17**, 6039 (1984).
20. T. Yonezawa and K. Morigaki, *Prog. Theor. Phys. Suppl.* **53**, 1 (1973).
21. S. Onoda, N. Sugimoto, and N. Nagaosa, *Prog. Theor. Phys.* **116**, 61 (2006).
22. S. Onoda, unpublished.

DYNAMICAL MAGNETOELECTRIC EFFECTS IN MULTIFERROICS

Y. TOKURA
Department of Applied Physics, University of Tokyo, Tokyo 113-8656, Japan
ERATO Multferroics Porject, JST, c/o University of Tokyo, Tokyo 113-8656, Japan
Cross-Correalted Materials Research Group(CMRG), ASI, RIKEN, Wako 351-0198, Japan
**E-mail: tokura@ap.t.u-tokyo.ac.jp*

Multiferroics, the materials in which both (anti)ferromagnetism and ferroelectricity can coexist, are the prospective host of the gigantic magnetoelectric (ME) effect. The multiferroics based on the spin-current (or inverse Dzyaloshinskiy-Moriya interaction) mechanism have recently been proved to realize in many cycloidal and conical spin states of frustrated magnets, in which the clamping between the magnetic and ferroelectric domains can show up. Recent advance in the research on the dynamical ME effects toward the electrical control of magnetism is presented.

Keywords: multiferroics; magnetoelectric effect; spin current

1. Introduction

Highly efficient control of magnetism in terms of electric field or current in a solid may widen the bottle-neck of the contemporary spin-electronics technology. Since the mageneoelectric (ME) effect, meaning magnetic (electric) induction of polarization P (magnetization M), was first confirmed in 1959-60, many magnetic materials have been demonstrated to show this effect [1]. Nevertheless, the magnitude of the observed ME effect has been too small to apply to any practical devices.

Multiferroics, the materials in which both ferromagnetism and ferroelectricity can coexist (Fig.1), are the prospective candidate which can potentially host the gigantic ME effect.

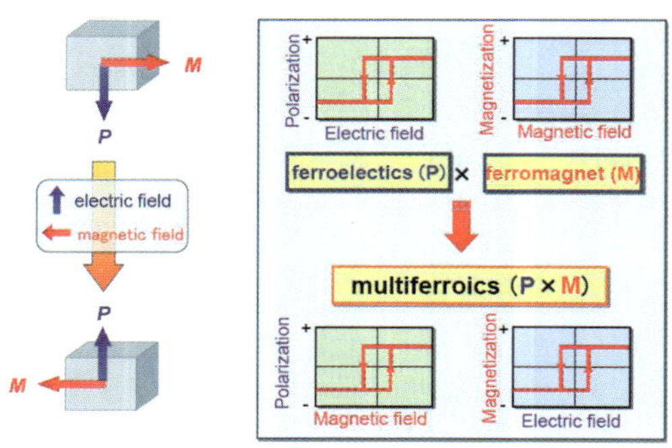

Fig. 1 Multiferroics and magneto-electric control.

Thus, a naïve expectation is that the large ME effect may be driven by weak electric or magnetic field, if the M and P shows the close coupling. There can be several possible ways to materialize the multiferroics. However, the coexistence alone of the P and M does not necessarily lead to the effective mutual coupling. In this context, the magnetically driven ferroelectricity is the most plausible candidate in which the enhanced M-P coupling is anticipated.

2. Cycloidal spin ferroelectrics

One of the promising strategies for exploring such multiferroics as showing the strong M-P coupling has recently been clarified experimentally [2] and theoriteically [3] with the focus of pervskite rare-earth manganites $RMnO_3$ (R=Tb and Dy). When the spins on the adjacent atomic sites are canted from each other (Fig. 2(a)), the horizontal mirror-plane symmetry is lost, meaning the possible generation of polarization along the vertical direction. In other words, the spontaneous spin current flows between the mutally conted spin; in analogy to the charge current produing the magnetic field, such a spin current can produce the electric dipole [3]. Recently, it has been theoretically shown

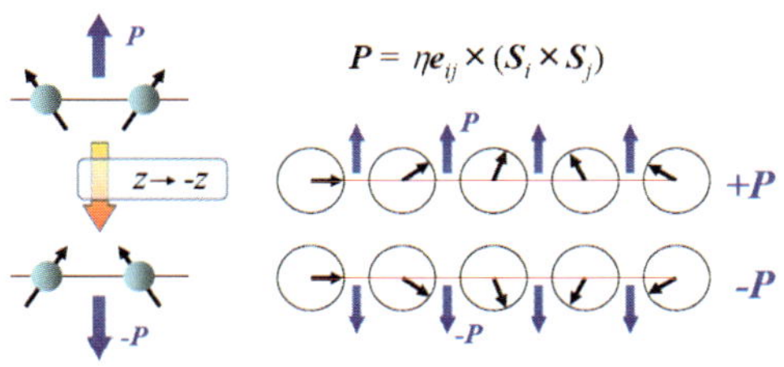

Fig. 2 (a) Interacting canted spins can generate the electric polarization under the presence of spin-orbit interaction. (b) When the spins show the transverse-spiral or cycloidal configuration, each neighboring spin pair shows the local polarization, and as the sum leads to the macroscopic polarization. The polarity depends on the helicity of the spiral mode, either clockwise or counter-clockwise.

[3-5] that the overlap of the electronic wave function between the two atomic sites with canted spins can generate the observable electronic polarization via the spin-orbit interaction. When the spins form the cycloidal modulation along the specific crystallographic direction (Fig. 2(b)), then every nearest-neighbor pair produces the local unidirectional polarization and hence the macroscopic P of the magnetic origin should be generated. The direction of the polarization can be totally determined by the clockwise or counter-clockwise rotation of the spin in proceeding along the spiral propagation axis, that is the spin helicity.

The spiral spin states are widely seen in complex transition-metal compounds like spinels and perovskites, where competing exchange interactions of neighboring spins can produce such a periodically modulated spin structure. In those compounds, the spin structure being responsible for the spontaneous polarization can often be controlled by an external magnetic field of specific direction. A prototypical example is shown in Fig.3 for the case of $TbMnO_3$[2]. Successive anomalies in the temperature variation of the specific heat and the magnetization represent the magnetic phase transitions to the collinear sinusoidal spin modulation with the moment direction and

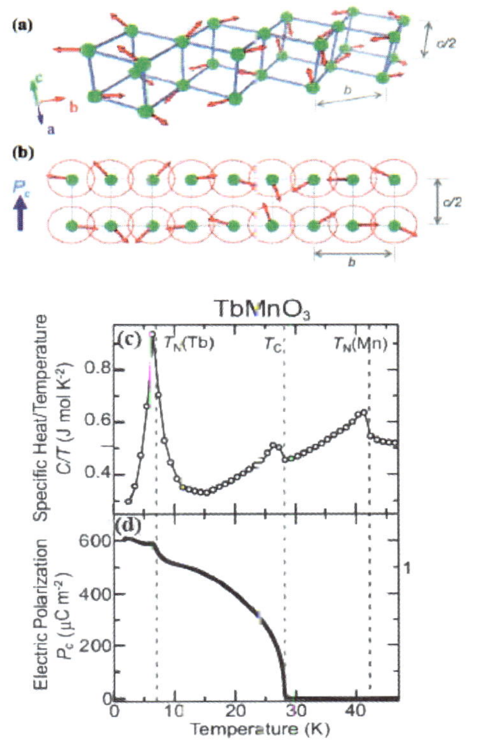

Fig.3 (a) Cycloidal spin structure of TbMnO3 in the ferroelectric phase below T_C(=28K). (b) The relation between the spin helicity and the polarization (P) vector as determined by a polarized neutron scattering study [6]. (c) and (d): Temperature dependence of the specific heat and spontaneous P[2].

the magnetic modulation vector (q_m~$0.27b^*$) along the b (orthorhombic) axis, and then to the transverse-spiral (cycloidal) structure on the bc plane and along the b-axis (see Fig.3(a)). Only the latter phase shows the ferroelectric polarization along the c axis, in accord with the above expectation from the spin current [3] or inverse Dzyaloshinskii–Moriya (D-M) [4] model.

By a recent study on the spin-polarized neutron scattering [6], the relation between the P direction and the spin helicity is deduced as shown in Fig. 3(b). It was confirmed that the reversal of the P direction in terms of the external electric field leads to the reversal of the spin helicity, i.e. from clockwise to counter-clockwise. This is a sort of chirality control of "spin enantiomer" with electric field. Furthermore, the spin ellipticity as defined as the ratio of the shorter (//*c*) to longer (//*b*) spin component in the *bc*-plane spin cycloid evolves toward to unity with lowering temperature. The temperature dependence of *P* can be calculated from these data in terms of the spin current model, which precisely agrees with the observed temperature evolution of the *P* (Fig.2 (d)).

3. Multiferroics with conical spin structures

To realize the real multiferroics, namely the concurrently ferromagnetic and ferroelectric state, with use of spiral (cycloidal) spin configuration, one may consider the conical spin state (Fig.4(a)), that can be viewed as the sum of the ferromagnetic component and transverse-spiral one. A remarkable characteristic expected for this kind of multiferroics is the electric (magnetic) reversal of the *M* (*P*) vector, as schematically shown in Fig. 1 [7]. Namely, the multiferroics of all spin origin may show the clamping of the ferromagnetic and ferroelectric domain walls.

Recently, such a multiferroic state has been observed for the spinel oxide, $CoCr_2O_4$. It has long been known that the Cr-oxide spinels

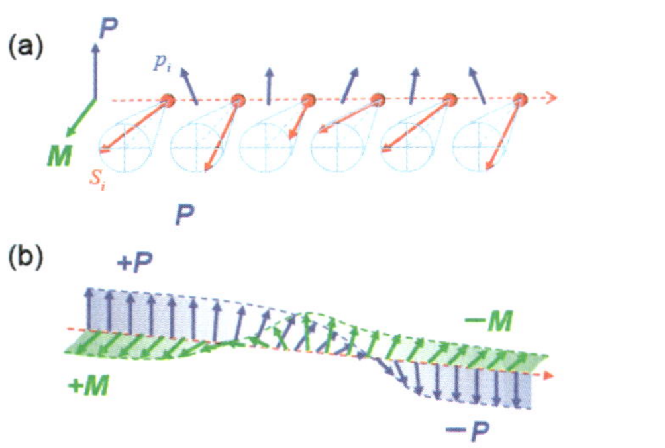

Fig. 4 (a) Spin conical structure as the candidate of multiferroics of all magnetic origin. p_i represents the local polarizations, the sum of which generates the macroscopic polarization *P*. Thus, the magnetization *M* of the conical spin is always orthogonal to *P*. (b) Possible domain wall clamping of the multiferroics of the magnetic origin.

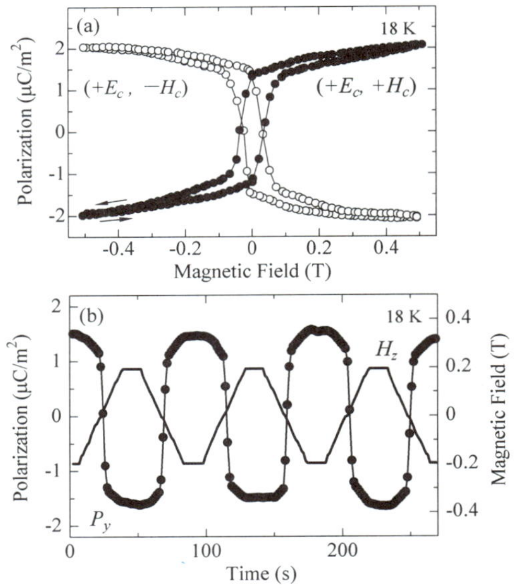

Fig. 5 (a) Magnetic field dependence of electric polarization at the multiferroic state (18K) in $CoCr_2O_4$ with conical spin structure for the respective cases of electric (E_c) and magnetic (H_c) cooling procedures. (E_c=400kV/m and H_c=0.5T) (b) The synchronized reversal of the polarization (P_y) with reversal of the external magnetic field (H_z). The reversal of H means that of the magnetization, which is orthogonal (//*z*) to the direction of polarization (//*y*). Reproduced from ref.[9].

MCr_2O_4 (M=Mn, Co) show the conical spin structure due to the combined effect of the geometrical spin frustration on the Cr sites and the exchange interaction between the Cr and M spins [8]. According to the scenario shown in Figs. 2 and 4, these should be ferroelectrics, which was in fact confirmed experimentally [9].

Figure 5 as reproduced from ref.[9] shows the magnetic field induced reversal of the ferroelectric polarization in $CoCr_2O_4$. As shown in Fig. 5(a), the crystal, which was once cooled under low electric (E_c=400kV/m) and magnetic (H_c=0.5T) fields to prepare a single multiferroic-domain, shows the reversal of the polarization with reversal of the magnetization, irrespective of starting from $+H_c$ or $-H_c$ side. When the magnetic field scan enabling the reversal of the magnetization is repeated, the polarization is also reversed in a synchronized manner, as shown in Fig. 5(b). These experimental observations clearly indicate the multiferroic domain wall clamping, as shown in Fig. 4(b).

Specifically to the multiferroics with conical spin structure, the spin helicity (say, clockwise) is expected to be simply reversed (say, to counter-clockwise) by the reversal of the magnetization along the cone axis (see Fig. 7). This is quite reasonable in case the spin spiral habit is not violated across the magnetic domain wall.

Such a synchronized reversal of the ferroelectric domain with that of the ferromagnetic domain strongly indicates the possibility of electric-field reversal of the magnetic domain. (In $CoCr_2O_4$, unfortunately, the spontaneous polarization is too small to reverse the ferroelectric domain by a manageable electric field, in spite of the easy and unusual reversal in terms of an external magnetic field.)

4. Toward electric control of magnetism

The multiferroic state of magnetic origin may in most cases compete with another ME phase, say a paraelectric-antiferromagnetic phase. In the phase-competing region, therefore, the application of magnetic (or electric) field may induce the first-order phase transition between those ME phases, resulting in the magnetic (or electric) control of the ME phases and hence the P (or M). In fact, this is the microscopic origin of the magnetic P-control as observed for the spiral-spin ferroelectrics[2]. The critical tuning of the ME phase competition may realize the colossal ME phenomena, such as the electric-field induction of the ferromagnetic phase, i.e. the function of quantum "electromagnet".

Even a more remarkable characteristic expected for the multiferroics is the electric reversal of the M vector. The multiferroics of all spin origin may show the clamping of the ferromagnetic and ferroelectric domain walls, that fixes the relative direction of M and spin helicity (or equivalently P) across the domain wall. The clamping of the ME domain walls are clearly manifested by the magnetic-field control of the P direction also not only for $CoCr_2O_4$ [9] but for $ZnCr_2Se_4$ [10] with transverse-conical spin structure, as partly described in the preceding section. As the inverse effect, such multiferroic conical magnets are expected to show the electric field reversal of the M vector. This may find application in the future low-energy dissipation electronic devices.

Another important issue related to the potential for future spintronics is the electrically driven magnetic excitation, or the magnetic resonance with ac electric field. In fact, the strong light (THz) absorption is observed for RMnO3 (R=Tb, Dy, etc) with unique light polarization characteristics both for the *bc* spiral ($P//c$) and *ab* spiral ($P//a$) phases [11,12]. Nevertheless, the microscopic origin of the electric-dipole active magnetic excitations such as Goldstone magnon and two-magnon has not been fully clarified as yet. Solving the problem will lead to the unique electric generation/detection of the spin excitations: a new fundamental function of spintronics.

Acknowledgments

The author would like to thank the collaborators for the work presented here, in particular, T. Kimura, T. Arima, Y. Onose, Y. Yamasaki, and Y. Kaneko.

References

[1] M. Fiebig, *J. Phys. D : Appl. Phys.* **38**, R123 (2005).
[2] T. Kimura et al., *Nature*, **426**, 55 (2003).
[3] H. Katsura, N. Nagaosa, and A.V. Balatsky, *Phys. Rev. Lett.* **95**, 057205 (2005).
[4] I.A. Sergienko and E. Dagotto, *Phys. Rev. B,* **73**, 094434 (2006).

[5] M. Mostovoy, *Phys. Rev. Lett.* **96**, 067601 (2006).
[6] Y. Yamasaki et al., *Phys. Rev. Lett.* **98**, 147204 (2007).
[7] Y. Tokura, *Science*, **312**, 1481 (2006).
[8] D.H. Lyons, T.A. Kaplan, K. Dwight, adn N. Menyuk, P*hys. Rev. B*, **126**, 540 (1964).
[9] Y. Yamasaki, S. Miyasaka, Y. Kaneko, J.P. He, and Y. Tokura, *Phys. Rev. Lett.* **96**, 20724 (2006).
[10] H. Murakawa, Y. Onose, K. Ohgushi, S. Ishiwata, and Y. Tokura, *J. Phys. Soc.Jpn.* 77, 043709(2008).
[11] A. Pimenov, V. Y. Ivanov, V.D. Travkin, A.M. Balbashov, and A. Loidl, *Nature Phys.* **2,** 97 (2006).
[12] N. Kida et al., *Phys. Rev.B*, **78**, 104414 (2008).

EXCHANGE-STABILIZATION OF SPIN ACCUMULATION IN THE TWO-DIMENSIONAL ELECTRON GAS WITH RASHBA-TYPE OF SPIN-ORBIT INTERACTION

HENRI M. SAARIKOSKI and GERRIT E.W. BAUER

Kavli Institute of NanoScience, Delft University of Technology,
Lorentzweg 1, 2628 CJ Delft, The Netherlands

Spin accumulation is a crucial but imprecise concept in spintronics, the science that strives to harness the electron spin for the benefit of information and communication technology. In metal-based spintronics it is characterized in terms of semiclassical distribution functions. In semiconductors with a strong spin-orbit coupling the spin accumulation is instead described as superpositions of coherent eigenstates. Here we present theoretical arguments that both views of the spin accumulation can be reconciled by taking into account the electron-electron interaction: a sufficiently strong self-consistent exchange field reduces a spin accumulation to a chemical potential difference between the two spin bands even in the presence of spin-orbit coupling. We demonstrate the idea on a clean two-dimensional electron gas (2DEG) by showing how the exchange field protects a spin accumulation from dephasing.

Keywords: Spin accumulation; Spin-orbit interaction; Rashba.

Metal-based spintronics[1] has evolved into a mature field in which spin phenomena are routinely exploited in versatile applications. [2] However, integration of spin-based functionalities into semiconductor circuits is still a pressing challenge. Much of recent research in this area has been motivated by device concepts, such as the seminal Datta-Das transistor,[3] which requires injection and detection of spins by ferromagnetic contacts to a narrow channel of a two-dimensional electron gas with gate-controlled spin-orbit interaction. In spite of progress to inject, modulate, transport, and detect spin polarization all electrically[4,5,6,7,8] as well as evidence that the spin-orbit interaction can indeed be tuned by external gates,[9] the route to a working spin transistor appears to be still full of obstacles. In the meantime, many insights have been obtained on the spin accumulation and its dynamics by optical methods, especially time-dependent Kerr and Faraday rotation spectroscopy.[10,15,11,12,14,13]

A spin accumulation can be defined as a non-equilibrium spin polarized state injected optically or electrically into a non-magnetic material. In metal-based spintronics it is a semiclassical concept involving chemical potential difference between spin up and down bands.[16] However, in semiconductors, in which the spin-orbit interaction (SOI) prominently affects the electronic structure and transport properties,[17] the spin accumulation is viewed as an intrinsically time-dependent quantum superposition of coherent eigenstates. Whether spin accumulation is a semiclassical or quantum concept is not just a semantic question since it is essential for the functionality of spintronic devices. Here we offer a mean-field theory for the spin accumulation in both metals and semiconductors that give a unified view on the spin accumulation in both metals and semiconductors. The crucial parameter that turns out to decide whether the spin accumulation is a robust chemical potential difference or a time-dependent beating pattern is the ratio between spin-orbit interaction and the electron exchange interaction. The general idea is illustrated at the hand of weakly disordered 2DEGs with Rashba SOI[18], in which the disorder-scattering lifetime broadening is much smaller than the spin-orbit splitting at the Fermi-level. We first describe long-lived excited states in the parameter space of electron density and spin-orbit coupling strength that support a given spin accumulation, followed by results for the spin accumulation dynamics and finally a discussion of the effects of impurity scattering.

Let us consider an infinitely extended homogeneous 2DEG. To leading order in the electron momentum $\mathbf{k} = -i\nabla$ the (non-interacting) Rashba Hamiltonian reads[17] $H_0 = \frac{\hbar^2 k^2}{2m^*} + \alpha(\sigma_x k_y - \sigma_y k_x)$, where $m^* = m_r m_e$ is the effective electron mass, m_e is the bare electron mass, and σ_i the Pauli matrices. α is the Rashba SOI strength parameter that originates from structure inversion asymmetry of the confinement potential.[18] We include the electron-electron interaction within the density-matrix functional theory (DMFT)[19] which is a generalization of the Hohenberg-Kohn-Sham density-functional theory[20] that can handle excited states such like a spin accumulation. Compared to the Hartree-Fock (HF) method, exchange and correlation effects in the DFMT can be treated within local approximations. The reduced density matrix is $\Gamma(\mathbf{z},\mathbf{z}') = \sum_i^\infty n_i \chi_i(\mathbf{z})\chi_i^\dagger(\mathbf{z}')$, in terms of its eigenstates χ_i are its eigenstates (natural orbitals) and $0 < n_i < 1$ are eigenvalues (natural occupation numbers) and $\mathbf{z} = (\mathbf{r},\sigma)$ denotes the space-spin coordinates. We define $\Gamma_\mathbf{s}$ as the subset of all density matrices which correspond to a given electron density $\rho(\mathbf{r}) = \sum_{i\sigma} n_i \chi_i^\dagger(\mathbf{z})\chi_i(\mathbf{z})$ and spin-polarization $\mathbf{s}(\mathbf{r}) = \sum_i n_i \chi_i^\dagger(\mathbf{z})\boldsymbol{\sigma}\chi_i(\mathbf{z})$. The density-matrix functional is defined via minimization of the total energy $E[\Gamma_\mathbf{s}] = \min_{\Psi[\Gamma_\mathbf{s}]}\langle\Psi[\Gamma_\mathbf{s}]|H|\Psi[\Gamma_\mathbf{s}]\rangle$ in the space of all many-body wave functions that correspond to a given density matrix $\Gamma_\mathbf{s}$.

We now assume that the exact density matrix can be generated by a non-interacting system of pseudo particles

$$[H_0 + V_{ext} + V_H + V_{\rm xc}]\phi_i = \epsilon_i \phi_i, \quad (1)$$

where V_{ext}, V_H, and $V_{\rm xc}$ are the external, the Hartree, and the exchange-correlation potential, respectively, such that with $f_i = \{0,1\}$

$$\Gamma_\mathbf{s}(\mathbf{z},\mathbf{z}') \cong \sum_i^N f_i \phi_i(\mathbf{z})\phi_i^\dagger(\mathbf{z}'). \quad (2)$$

The exchange-correlation potential

$$V_{\rm xc}(\mathbf{z},\mathbf{z}',[\Gamma_\mathbf{s}]) = \frac{\delta E_{\rm xc}[\boldsymbol{\Gamma}_\mathbf{s}]}{\delta\Gamma_\mathbf{s}(\mathbf{z},\mathbf{z}')} \quad (3)$$

in terms of the exchange-correlation energy

$$E_{\rm xc}[\boldsymbol{\Gamma}_\mathbf{s}] = E[\Gamma_\mathbf{s}] - \sum_i^N f_i\langle\phi_i|H_0 + \frac{1}{2}V_H|\phi_i\rangle. \quad (4)$$

For the spatially homogeneous 2DEG $V_{\rm xc}(\mathbf{z}',\mathbf{z}) \to V_{\rm xc}(\mathbf{r}-\mathbf{r}',\sigma,\sigma')$. HF-calculations for the Rashba Hamiltonian, following Ref. [22], confirm that the effects of the spin-orbit interaction on the exchange potential is negligible for small spin polarizations. Assuming that this holds for the correlation part as well, the potential only depends on the electron density ρ and the spin polarization : $[\Gamma_\mathbf{s}] \to V_{\rm xc}(\rho,\mathbf{p})$,where $\mathbf{p} = \mathbf{s}/\rho = p\hat{\mathbf{p}}$ is the polarization with $0 \le p \le 1$. Making the local approximation, we finally arrive at

$$V_{\rm xc}(\mathbf{z},\mathbf{z}',[\Gamma_\mathbf{s}]) \approx \delta(\mathbf{r}-\mathbf{r}')(V_0(\rho,\mathbf{p})+J_{\rm xc}(\rho,s)\hat{\mathbf{p}}\cdot\sigma),$$

where $J_{\rm xc}(\rho,s)$ ($\simeq J(\rho)p$ for small p) is the exchange-correlation field vector. The scalar V_0 can be dropped in homogeneous systems.

We may approximate J by the Hartree-Fock exchange energy of the strictly 2DEG.[21] $V_{\rm xc} \approx J_{\rm x}(\rho,s)\hat{\mathbf{p}}\cdot\sigma \approx J(\rho)\mathbf{p}\cdot\sigma = -\sqrt{2}m_r/(\pi r_s\epsilon^2){\rm Ry}\,\mathbf{p}\cdot\sigma$, where $r_s = m_r/(a_B\epsilon\sqrt{\pi\rho})$ is the dimensionless density parameter, ϵ is the relative dielectric constant of the medium, and Ry = 13.6 eV. Published parameterization of the correlation energy for a non-SO coupled 2DEG can be used as well (see e.g. [23]). The effective

Hamiltonian for our system is therefore of the form

$$H(\mathbf{s}) = H_0 + J\mathbf{p}\cdot\boldsymbol{\sigma}, \tag{5}$$

where $J < 0$ the effective exchange(-correlation) potential. For typical electron densities ρ the exchange energies (a few meV) of the same order of magnitude as SOI energies at the Fermi-level in III–V[9] and II–VI[24] semiconductor-based 2DEG's.

The eigenstates of the Hamiltonian in the non-polarized ground state are split into two bands with a chiral spin pattern (see Fig. 1 (a-b)). When a spin accumulation

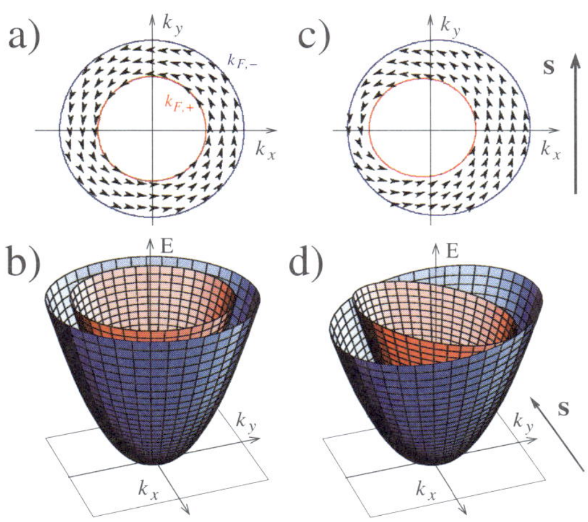

Fig. 1. (Left panel) The symmetric pattern of the eigenstate spins (a) and energies (b) of the ground-state spin-split bands in a non-interacting Rashba 2DEG. The inner and outer circles correspond to the Fermi energies of the spin-split bands. The doubly occupied states (inside the inner circle) are spin-compensated. (Right panel) The band structure for an in-plane spin accumulation **s** is modified by the self-consistent exchange field. Eigenstates with spins parallel (antiparallel) to the field are lowered (raised) in energy.

is injected, $\mathbf{p} \neq 0$, the exchange field deforms the spin-split bands as shown in Fig. 1 (c-d) for in-plane and in Fig. 2(a-b) for perpendicular fields. Our task is to find the self-consistent single-determinant eigenfunction of $H(\mathbf{p})$ with polarization $\mathbf{p}$, which according to DMFT is unique. Introducing the

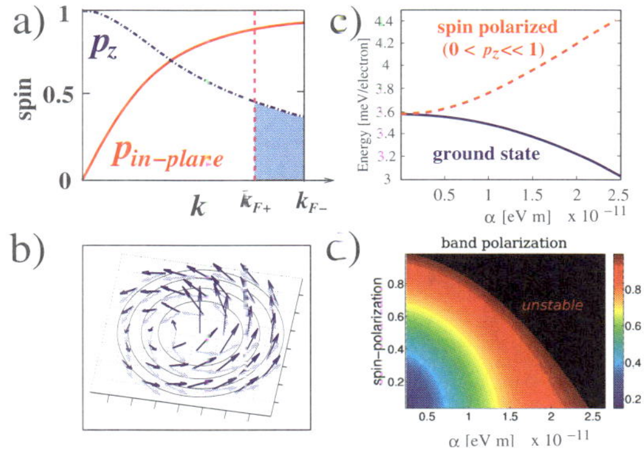

Fig. 2. (a-b) Spin direction of the lower spin-split ($-$) states in the presence of a spin polarization normal to the 2DEG plane. The spins of the higher band ($+$) point in the opposite direction. The total spin polarization is determined by the shaded area between $k_{F,-}$ and $k_{F,+}$. c) Energies of the ground and excited states with (small) spin accumulation normal to the 2DEG. d) Band polarization of spin-accumulation states at fixed $J = -2$ meV. Above $\alpha = 2.4 \times 10^{-11}$ eV m a self-consistent solution cannot be found. Material parameters are $\epsilon = 12.7$, $m^* = 0.067$, $\rho = 2 \times 10^{15}/\mathrm{m}^2$.

occupation numbers $f_{\mathbf{k}\lambda} = \{0, 1\}$ of the spin-split states $\phi_{\mathbf{k}\lambda}$, where $\mathbf{k}$ is a wave vector and $\lambda = \{+, -\}$ the spin polarization reads

$$\mathbf{p} = \sum_{\mathbf{k}\lambda} f_{\mathbf{k}\lambda}\langle\phi_{\mathbf{k}\lambda}|\boldsymbol{\sigma}|\phi_{\mathbf{k}\lambda}\rangle/\rho. \tag{6}$$

The wave function we are looking for minimizes the energy under the constraint (6), with single particle wave functions and distribution of the occupation number as variational variables. We solve the problem either analytically in limiting cases or numerically by a stochastic minimization method, which uses penalty functions to fix the spin polarization and the Metropolis sampling method to find the global energy minimum.

Since the system is not spontaneously magnetic, a spin polarization requires a redistribution of the occupation numbers in essentially two different ways. An unequal occupation of spin bands, $N_- = \sum_{\mathbf{k}-} f_{\mathbf{k}-} > N_+$ can be parameterized by a chemical potential difference or a band polarization $p_b = (N_- - N_+)/N > 0$. Occupations can also shift in momentum space, creating spin-galvanic currents. We now discuss separately spin accumulations in the plane of the 2DEG and normal to it. The large differences in the results are manifestations of the SO-coupling induced magnetic anisotropy in the 2DEG.

A spin polarization in the plane of the 2DEG can be generated at low energy cost by shifting the Fermi circles as shown in Fig. 1(c-d). This leads to currents[25] via a spin-galvanic Hall effect: the minimum energy state for a fixed p_x is associated with a charge current in the perpendicular y-direction

$$j_{c,y} = -e\frac{\alpha}{\hbar}\left(1 + \frac{Jm^*}{\pi\rho\hbar^2}\right)p_x. \quad (7)$$

Since electrons move with constant drift velocity, there are no intrinsic spin-Hall currents which arise only under acceleration.[26] Non-equilibrium spin currents are induced also, but vanish to first order in s_x.

A spin accumulation perpendicular to the 2DEG surface cannot cause charge currents for symmetry reasons. A spin accumulation can then be generated by a combination of a band polarization p_b (in excess of that of the ground state) [a] and the exchange field that pops the in-plane spin textures out of the plane (Fig. 2(b)). Excited eigenstates corresponding to a finite spin accumulation normal to the plane are therefore, in contrast to the in-plane case, separated from the non-polarized ground state by a finite energy gap (Fig. 2(c))

$$E_g = \left(\frac{\alpha m^*}{\hbar J}\right)^2 \left(J + \frac{\hbar^2\pi\rho}{m^*}\right) + \mathcal{O}(\alpha^4). \quad (8)$$

Spin accumulation vectors that are tilted into finite angles therefore feel a torque corresponding to a 2DEG "easy" plane, analogous to the coercivity in ferromagnets induced by magnetic crystalline anisotropies.[27]

[a]Since states at the Brillouin zone center are easiest to polarize, we find that at large SOI it becomes energetically favorable to instead depopulate a large fraction of states in the majority spin band around $k = 0$.

A sufficiently small exchange-stabilized spin accumulation perpendicular to the 2DEG always exists. The maximum spin accumulation that can be accommodated is determined by the total spin polarization of a single occupied band ($p_b = 1$). In the exchange-only approximation the self-consistency criterion (6) in this limit can be fulfilled only when

$$\alpha < 2|J|/k_F = 2|J|/\sqrt{4\pi\rho} = 0.32 \text{ eV nm}/\epsilon, \quad (9)$$

Figure 2(d) shows the band polarization of spin accumulation eigenstates as a function of α and s_z.

In spin valves with a semiconductor spacer, such as in the Datta-Das transistor, spins are injected and detected by ferromagnetic electrodes at energies close to the Fermi energy.[3] Even when the magnetizations of the two electrodes are parallel to each other, transport depends on the magnetization direction. When magnetizations are oriented in the current direction, the injected spins can be easily accommodated by the semiconductor by shifted distribution functions. The spins does not precess in the channel and, hence, not react to a gate voltage that modulates the SOI. For magnetizations perpendicular to the plane, the transport properties depend on the applied bias. Spins that are injected at low energies, do not find corresponding eigenstates, and therefore precess in the channel with a gate-voltage dependent wave length, as originally envisioned by Datta and Das. However, when the spins are injected above the exchange-induced gap, they end up in stable eigenstates again, meaning that the transistor will not work anymore.

In real samples, additional spin relaxation processes exists, due to electron-electron interactions and impurity scattering. A large phase space available for scattering processes which makes a large spin accumulation susceptible to a fast decay, so the stability limit in Figure 2(d) should not be interpreted as a sharp phase boundary. For not too highly excited systems, the Dyakonov-Perel[29] mechanism by random scattering at defects is believed to be the dominant source of finite T_1 spin life times, especially in clean systems in which spin precession is effectively randomized when the distance between impurities is larger than the precession length. Since in an exchange stabilized 2DEG the precession in the SO field is suppressed, the efficiency of the Dyakonov-Perel mechanism should be strongly diminished for systems in the clean limit considered here. The opposite (dirty) regime can be handled by spin-coherent kinetic[31,30] and diffusion[32,33,34] equations or numerical simulations.[13,35] Also in the dirty regime exchange has been found to increase T_1[31,30], but the effect on T_2 and possible anisotropies have not yet been studied.

The results here provide a general theoretical framework to treat spin accumulation in the presence of spin-orbit coupling. In the space of the parameters provided by material and excitation condition the spin accumulation features both semiclassical and quantum properties, anisotropies, and spin-galvanic effects. Our method can be extended to treat three-dimensional, inhomogeneous and/or finite systems as well as the Dresselhaus SOI due to bulk inversion asymmetry.[36] The importance of the exchange field and thus the spin accumulation properties can be engineered by electron density, excitation intensity, spin direction and electric currents, and should therefore be considered in advanced spintronic device concepts.[4,14]

This work has been supported by Stichting FOM and NWO. One of us (H.S.) acknowledges support from the Academy of Finland. We acknowledge useful discussions with İnanç Adagideli and Klaus Capelle.

References

1. I. Žutić, J. Fabian, and S. Das Sarma, Rev. Mod. Phys. **76**, 323 (2004).
2. D. D. Awschalom and M. E. Flatté, Nature Phys. **3**(3), 153 (2007).
3. S. Datta and B. Das, Appl. Phys. Lett. **56**, 665 (1990).
4. Xiaohua Lou, Christoph Adelmann, Scott A. Crooker, Eric S. Garlid, Jianjie Zhang, K. S. Madhukar Reddy, Soren D. Flexner, Chris J. Palmstrøm and Paul A. Crowell, Nature Phys. **3**, 197 (2007).
5. P. Chen, J. Moser, P. Kotissek, J. Sadowski, M. Zenger, D. Weiss, and W. Wegscheider, Phys. Rev. B **74**, 241302 (2006).
6. E. J. Koop, B. J. van Wees, D. Reuter, A. D. Wieck, and C. H. van der Wal, arXiv:0801.2699.
7. S. M. Frolov, A. Venkatesan, W. Yu, S. Luescher, W. Wegscheider, and J. A. Folk, arXiv:0801.4021.
8. S. A. Crooker, M. Furis, X. Lou, C. Adelmann, D. L. Smith, C. J. Palmstrom, and P. A. Crowell, Science **309**, 2191 (2005).
9. J. Nitta, T. Akazaki, H. Takayanagi, and T. Enoki, Phys. Rev. Lett. **78**, 1335 (1997).
10. J. M. Kikkawa, I. P. Smorchkova, N. Samarth, and D. D. Awschalom, Science **277**, 1284 (1997).
11. S. A. Crooker, D. D. Awschalom, and N. Samarth, IEEE J. Sel. Top. Quantum Electron. **1**, 1082 (1995).
12. A. Pugžlys, P. J. Rizo, K. Ivanin, A. Slachter, D. Reuter, A. D. Wieck, C. H. van der Wal, and P. H. M. Loosdrecht, J. Phys.: Condens. Matter **19**, 295206 (2007).
13. W. J. H. Leyland, R. T. Harley, M. Henini, D. Taylor, A.J. Shields, I. Farrer, and D. A. Ritchie, arXiv:0707.4180 (2007).
14. L. Meier, G. Salis, I. Shorubalko, E. Gini, S. Schön, and K. Ensslin, Nature Phys. **3**, 651 (2007).
15. Y. K. Kato, R. C. Myers, A. C. Gossard, and D. D. Awschalom, Science **306**, 1910 (2004).
16. A. G. Aronov, Pis'ma Zh. Eksp. Teor. Fiz. **24**, 37 (1976) [JETP Lett. **24**, 32 (1976)]; M. Johnson and R. H. Silsbee, Phys. Rev. Lett. **55**, 1790 (1985); F. J. Jedema, A. T. Filip, and B. J. van Wees, Nature **410**, 345 (2001).

17. Roland Winkler, Spin-orbit coupling effects in two-dimensional electron and hole systems, Springer Tracts in Modern Physics **191**, Springer-Verlag (2003).
18. Y. A. Bychkov and E. I. Rashba, J. Phys. C: Solid State Phys. **17**, 6039 (1984) [Y. A. Bychkov, E. I. Rashba, JETP Lett. **39**(2), 78 (1984)].
19. G. Zumbach and K. Maschke, J. Chem. Phys. **82**, 5604 (1985).
20. W. Kohn and L. J. Sham, Phys. Rev. **140**, A1133 (1965).
21. U. von Barth and L. Hedin, J. Phys. C: Solid State Phys. **5**, 1629 (1972).
22. S. Chesi, G. Simion, and G. F. Giuliani, cond-mat/0702060 (2007).
23. C. Attaccalite, S. Moroni, P. Gori-Giorgi, and G. B. Bachelet, Phys. Rev. Lett. **88** 256601 (2002).
24. Y. S. Gui, J. Liu, V. Daumer, C. R. Becker, H. Buhmann, and L. W. Molenkamp, Physica E **12**, 416 (2002).
25. S. D. Ganichev, E. L. Ivchenko, V. V. Bel'kov, S. A. Tarasenko, M. Sollinger, D. Weiss, W. Wegscheider, and W. Prettl, Nature **417**, 153-156 (2002).
26. J. Wunderlich, B. Kaestner, J. Sinova, and T. Jungwirth, Phys. Rev. Lett. **94**, 047204 (2005).
27. S. Chikazumi, Physics of Ferromagnetism, 2nd edition (Oxford University Press, Oxford, 1997).
28. D. Stich, J. Zhou, T. Korn, R. Schulz, D. Schuh, W. Wegscheider, M. W. Wu, and C. Schüller, Phys. Rev. Lett. **98**, 176401 (2007).
29. M. I. Dyakonov and V. I. Perel, Fiz. Tverd. Tela **13**, 3581 (1971) [Sov. Phys. - Solid State **13**, 3023 (1971)].
30. M. Q. Weng and M. W. Wu, Phys. Rev. B **68**, 075312 (2003).
31. M. M. Glazov and E. L. Ivchenko, JETP Lett. **75**, 403 (2002)
32. A. A. Burkov, Alvaro S. Núñez, and A. H. MacDonald, Phys. Rev. B **70**, 155308 (2004).
33. E. G. Mishchenko, A. V. Shytov, and B. I. Halperin, Phys. Rev. Lett. **93**, 226602 (2004).
34. İ. Adagideli and G. E. W. Bauer, Phys. Rev. Lett. **95**, 256602 (2005).
35. E. J. Koop, B. J. van Wees, and C. H. van der Wal, arXiv:0804.2968.
36. G. Dresselhaus, Phys. Rev. **100**, 580 (1955).

ELECTRONIC AHARONOV-CASHER EFFECT IN INGAAS RING ARRAYS

JUNSAKU NITTA* and MOKOTO KOHDA
Department of Materials Science, Tohoku University, Sendai, 980-8579, Japan
**E-mail: nitta@material.tohoku.ac.jp*

TOBIAS BERGSTEN
Cavendish Laboratory, Cambridge University, Cambridge CB3 0HE, United Kingdom
E-mail: betb2@cam.ac.uk

We demonstrate a spin interference effect in a small array of InGaAs mesoscopic rings The spin interference is based on the time reversal Aharonov-Casher (AC) effect. The AC interference oscillations are controlled over several periods. This result shows evidence for electrical manipulation of the spin precession angle in an InGaAs two-dimensional gas channel. We observe a reproducible half-oscillation, which may be attributed to the competition between the Rashba SOI and the Dresselhaus SOI.

Keywords: Rashba spin-orbit interaction; Spin interference; Aharonov-Casher effect.

1. Introduction

An electron acquires a phase around magnetic flux due to the vector potential leading to the Aharonov-Bohm (AB) effect in an interference loop [1]. From the view point of inherent symmetries between magnetic field and electric field in the Maxwell equations, Aharonov and Casher have predicted that a magnetic moment acquires a phase around a charge flux line [2]. It should be noted that the original Aharonov-Casher (AC) effect was proposed for charge neutral particle since the electric field modifies the trajectory of charged particle in the same sense as the original AB effect was predicted in the situation where magnetic flux should not exist in an electron path. It is pointed out that the AC phase shift can be derived from spin-orbit interaction (SOI) [3]. The major difference of the AC effect from the AB effect is that the AC effect is not observable if the electric field is not in the paths although the AB effect can take place even if there is no magnetic field in the electron paths. Cimmino *et al.* [4] managed to perform the AC interference experiment in a neutron (having spin 1/2 but no charge) beam loop using a voltage of 45 kV to create the electric field. However, modified precession angle of neutron spin was only 2.2 mrad since the SOI is not strong in vacuum.

Mathur and Stone have theoretically shown that the effects of SOI in disordered conductors are manifestations of the AC effect in the same sense as the effects of weak magnetic fields are manifestations of AB effect [5]. They have proposed the electronic AC effect in a mesoscopic interference loop made of GaAs 2DEG with the Dresselhaus SOI [6]. It is emphasized that a thousand-fold improvement in its experiment can be expected in the electronic AC effect since the SOI in semiconductors is much enhanced compared with that in vacuum.

We have proposed a spin interferometer in a mesoscopic ring [7] based on the *gate controllable* Rashba SOI [8]. However, the spin interference effect is not easily observable since the gate electric field tunes not only SOI but also carrier density, namely electron wave length which affects the interference. In this paper, we have measured time reversal symmetric Al'tshuler-Aronov-Spivak (AAS) oscillations in an array of mesoscopic loops [9] in order to distinguish spin interference from other spurious effects. The AAS oscillation amplitude which does not depend on the electron wavelength but depends on spin precession shows an oscillatory behavior as a function of gate voltage. The observed spin interference period is consistent with gate voltage dependence of SOI. This result shows that the spin precession can be controlled by the gate voltage and is an experimental demonstration of Aharonov-Casher effect in time reversal paths [10].

2. Spin interferometer

We have proposed a spin-interference device based on the Rashba SOI [7]. The AC phases acquired by the spin wave functions during a cyclic evolution are calculated in a mesoscopic ring in the presence of the Rashba SOI. The origin of the AC spin interference is the spin precession angle difference between the left and right branches of the ring. The phases acquired in the left and right branches are not the same: they have an opposite sign because the precession orientation is opposite. The origin of the AC phase difference in both cases is related to the spin precession. By using the spin interference device, we can investigate how much the spin precession angle can be controlled by a gate voltage. Furthermore, an observation of spin interference induced by SOI in a mesoscopic ring is an experimental demonstration of the electronic AC effect.

The precession angle θ of an electron moving along a straight narrow channel is given by [11]

$$\theta = \frac{2\alpha m^*}{\hbar^2} L \ , \qquad (1)$$

with m^* being the effective mass and L being the travel distance. The modulation of the h/2e oscillation amplitude can be expressed as a function of the SOI parameter α ,

$$\frac{\delta R_\alpha}{\delta R_{\alpha=0}} = \cos\left\{ 2\pi\sqrt{1+\left(\frac{2m^*\alpha}{\hbar^2} r\right)^2} \right\} \ , \qquad (2)$$

where δR_α and $\delta R_{\alpha=0}$ is the h/2e amplitude with and without SOI, respectively. In relating the result of the spin interference experiment to the spin precession angle, the argument of cosine in Eq. (2) reduces to the spin precession angle θ in the limit of strong SOI or large ring radius because the distance traveled around the ring is $2\pi r$.

This equation shows that the conductance of the ring oscillates as a function of the strength α of Rashba SOI. This SOI dependence is very similar to the conductance of the Spin-FET proposed by Datta and Das, in which they need ferromagnetic electrodes for spin injection and detection [11]. This proposed spin interferometer works without ferromagnetic electrodes.

3. Electronic Aharonov-Casher experiments

The resistance of a mesoscopic ring is affected by several quantum interference effects. The well known AB effect results in a resistance oscillation with a magnetic flux period of h/e. The AB effect is sample specific and very sensitive to the Fermi wave length, therefore, the interference pattern is rapidly changed by the gate voltage. In order to detect the AC effect we used another quantum interference phenomenon, the Al'tshuler-Aronov-Spivak (AAS) effect [12]. The AAS effect is an AB effect of time reversal symmetric paths, where the two wave function parts go all around back to the origin on identical paths, but in opposite directions. In this situation any phase which is due to path geometry will be identical and will not affect the interference. This also means that it is independent of the Fermi energy E_F (and consequently the carrier density *Ns*). However, the AAS effect is sensitive to the AC spin phase when the SOI plays a role. If there is magnetic flux inside the paths the resistance will oscillate with the period of h/2e. When the flux is increased the resistance oscillates with the period h/2e, but the AAS oscillation amplitude decays after a several periods because of averaging between different paths in the ring, with different areas. If there is SOI in the ring, the electron spin will start precessing around the effective magnetic field and change the interference at the entry point. Note that the effective magnetic field due to the SOI is much stronger than the external magnetic field to pick up AAS oscillations. The precession axes for the two parts of the wave function are opposite and therefore the relative precession angle is twice the angle of each part. If the relative precession angle is π the spins of the two parts are opposite and can not interfere, and the AAS oscillations disappear. If the relative angle is 2π the two parts will have the same spin but opposite signs because of the 1/2 spin quantum laws (a 4π rotation is required to return to the original wave function), effectively changing the phase of the AAS oscillations by π, which we interpret as a negative amplitude.

By using arrays rather than single rings we get a stronger spin signal and we average out some of the universal conductance fluctuations (UCF) and sample specific AB oscillations.

The ring arrays were etched out in an electron cyclotron resonance (ECR) dry-etching process from an InP/InGaAs/InAlAs based 2DEG, the same as used for the Shubnikov- de Haas (SdH) measurements. The electron mobility was 7-11 m^2/Vs depending on the carrier density and the effective electron mass m^* was $0.050m_0$ as determined from the temperature dependence of SdH oscillation amplitudes. Figure 1 shows an example of the ring array which consists of 4 × 4 rings of 1.0 μm radius. Note that the actually measured sample was a 5 × 5 ring array. The rings were covered with a 50 nm thick SiO_2 insulator layer, and an Au gate electrode, used to control the carrier density and the SOI parameter α. In the present sample, we design the array with a small number of rings in order to escape a gate tunneling leakage problem. The advantage of using a small number of rings rather than a large array is that the gate tunneling leakage is much smaller and we can use a relatively high gate voltage. This makes it possible to see several oscillations of AC interference [10]. Earlier experiments on square loop arrays with very large number of loops showed convincing spin interference results, but only up to one interference period [9].

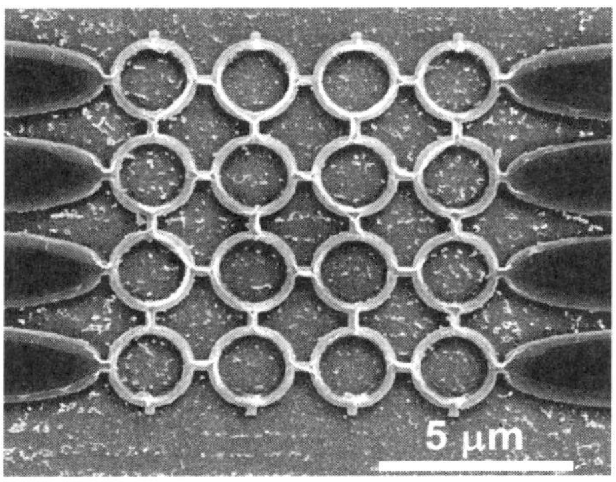

Fig. 1. An SEM image of an array of 4x 4 rings made of an InP/ InGaAs/InAlAs heterostructure. The whole area is covered with a gate electrode to control SOI.

The experiment was carried out in a ^{3}He cryostat at the base temperature which varied between 220 mK and 270 mK. We measured the resistance R of the ring array simultaneously with the Hall resistance R_H of the Hall bar close to the rings, while stepping the magnetic field and the gate voltage V_g. Close to the arrays and in the same current path and under the same gate was a Hall bar, 5 μm wide and 20 μm long, used to measure the carrier density. We calculated the carrier density N_s from the slope of the R_H vs B ($N_s^{-1} = e\,dR_H / dB$) and the carrier concentration is linearly increased with the gate voltage V_g.

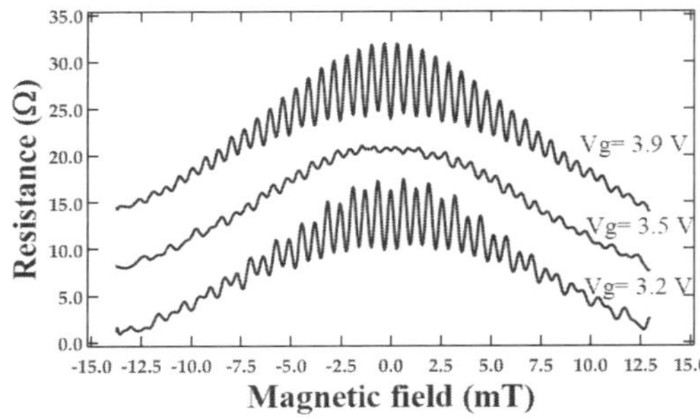

Fig. 2. Magnetoresistance oscillations with period of h/2e due to the AAS effect at three different gate voltages. The oscillations in the top and bottom curves are reversed because of the AC effect. The curves are shifted for clarity.

In order to reduce noise and UCF effects we averaged ten resistance versus magnetic field (R vs B) curves with slightly different gate voltages. This averaging preserves the AAS oscillations but the averaging of M curves reduces the AB amplitude roughly as $M^{-1/2}$. In Fig. 2, we display the h/2e magnetoresistance oscillations due to the AAS effect at three different gate voltages. The oscillations in the top and bottom curves are reversed compared to the middle one because of the AC effect. The second curves have almost no oscillations the spin precession rotates the spins of the two wave-function parts to opposite directions. We can clearly see the oscillations switching phase as we increase the gate voltage. We then plotted the amplitude against the gate voltage as shown in Fig. 3. The AAS amplitude oscillates as a function of the gate voltage which changes the SOI parameter α . As we discuss below using Eqs. (1) and (2) the amplitude crosses zero, inverting the AAS oscillations. Each period represents one extra 2π spin precession of an electron moving around a ring.

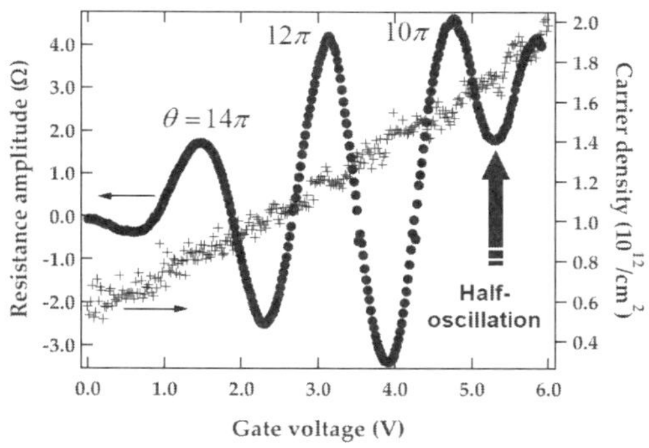

Fig. 3. Aharonov-Casher oscillations as a function gate voltage. The precession angles for electron round trip of the ring are indicated.

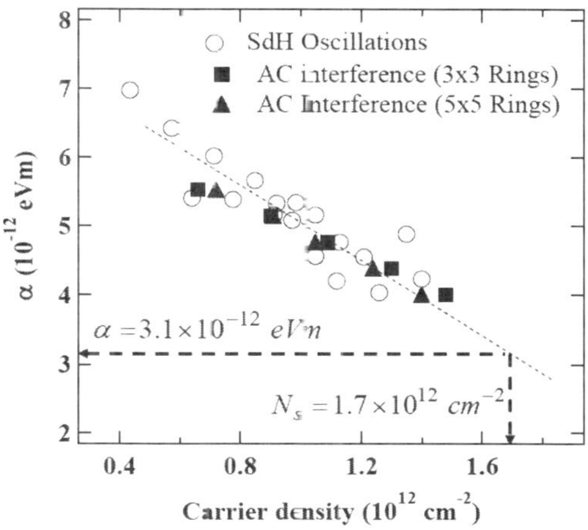

Fig. 4. Carrier density dependence of the Rashba SOI parameters measured by SdH oscillations and AC interference.

We also measured SdH oscillations at different carrier densities in a separate Hall bar. The SOI strength α obtained from the beating pattern of the SdH oscillations shows carrier density dependence because the Rashba SOI depends on the structural inversion asymmetry (SIA) therefore the shape of the quantum well (QW) which is modified by applying the gate voltage. The gate voltage sensitivity $\Delta\alpha / \Delta V_g$ is about 0.51 x 10^{-12} eVm/V in the present heterostructure. From the AC oscillation periods, we can also deduce the gate voltage sensitivity of α. We plot the estimated α from the AC oscillations period into the carrier density dependence of α obtained from the SdH experiment in order to make the best fit. Then, we estimate the spin precession angles at several different gate voltages (carrier densities).

The estimated spin precession angles at the peaks of the AC oscillations are shown in Fig. 3. It is found that the spin precession angle is controlled over the range of 4π by the gate electric field.

In Fig. 3, a mysterious half-oscillation is found at Ns = 1.7 x 10^{12} cm^{-2}, which does not cross over zero. This half-oscillation is reproducible and, is also observed in 3 x 3 ring array sample at the same carrier density. It is predicted that the non-zero-crossing oscillation occurs when the Rashba SOI parameter α is equal to the Dresselhaus SOI parameter β. From Fig. 4, the corresponding α value at Ns = 1.7 x 10^{12} cm^{-2} is about 3.1 x 10^{-12} eVm. The Dresselhaus SOI parameter depends on the material parameter $\langle b_{41}^{6c6c} \rangle$ and the confinement of the QW thickness $\langle k_z^2 \rangle \approx (\pi / d_{QW})^2$ [13]. The estimated β is about 2.7 x 10^{-12} eVm with $d_{QW} = 10\,nm$ InGaAs channel. When $\alpha = \beta$, the total effective magnetic field due to the Rashba and to the Dresselhaus SOIs becomes uni-axial. In this case, the spin rotational operator is commutable, and the spin precession effect is cancelled out [14]. One possible explanation for the half-oscillation is that the Rashba SOI α approaches to the Dresselhaus SOI β. However, higher harmonic AAS oscillations with period of h/4e shows zero cross oscillation at Ns = 1.7 x 10^{12} cm^{-2} (Not shown in this paper). Further study is necessary for the origin of half-oscillation.

4. Conclusion

The AC effect is of fundamental importance for quantum interference phenomena. We have discussed an AC spin interferometer based on the Rashba SOI. We have demonstrated the electronic AC effect dual to the magnetic AAS effect in a small array of rings. This clear demonstration of the AC interference controlled by the gate electric field can be attributed to the fact that the SOI is much enhanced in semiconductors compared to the SOI in vacuum. A reproducible half-oscillation is observed, which may be attributed to the competition between the Rashba SOI and the Dresselhaus SOI.

Acknowledgments

This work is partly supported by Grant-in Aids from CREST-JST and MEXT.

References

1. A. Tonomura, N. Osakabe, T. Matsuda, T. Kawasaki, and J. Endo, Phys. Rev. Lett. **56**, 792 (1986).
2. Y. Aharonov and A. Casher, Phys. Rev. Lett. **53**, 2964 (1984).
3. A. Balatsky and B. Al'tshuler, Phys. Rev. Lett. **70**, 1678 (1993).
4. A. Cimmino, G. I. Opat, A. G. Klein, H. Kaiser, S. A. Werner, M. Arif, and R. Clothier, Phys. Rev. Lett. **63**, 380 (1989).
5. H. Mathur, A. D. Stone, Phys. Rev. Lett. **68**, 2964 (1992).
6. G. Dresselhaus, Phys. Rev. **100**, 580 (1955).
7. J. Nitta, F. E. Meijer, and H. Takayanagi, Appl. Phys. Lett. **75**, 695 (1999).
8. J. Nitta, T. Akazaki, H. Takayanagi, and T. Enoki, Phys. Rev. Lett. **78**, 1335 (1997); T. Koga, J. Nitta, T. Akazaki, and H. Takayanagi, *ibid.* **89**, 046801 (2002).
9. T. Koga, Y. Sekine, and J. Nitta, Phys. Rev. **B74**, 041302 (R) (2006).
10. T. Bergsten, T. Kobayashi, Y. Sekine, and J. Nitta, Phys. Rev. Lett. **97**, 196803 (2006).
11. S. Datta and B. Das, *Appl. Phys. Lett*. **56**, 665 (1990).
12. B. L. Al'tshuler, A. G. Aronov, and B. Z. Spivak, JETP Lett. **33**, 94 (1981).
13. R. Winkler, "Spin-Orbit Coupling Effects in Two-Dimensional Electron and Hole Systems", Springer Tracts in Modern Physics **191** (2003).
14. M. J. van Veenhuizen, T. Koga, and J. Nitta, Phys. Rev. **B73**, 235315 (2006).

MICROSCOPIC THEORY OF CURRENT-SPIN INTERACTION IN FERROMAGNETS

H. KOHNO*, S. KAWABATA, T. NOGUCHI, S. UETA

Department of Materials Engineering Science, Graduate School of Engineering Science, Osaka University, Toyonaka, Osaka 560-8531, Japan
**E-mail: kohno@mp.es.osaka-u.ac.jp*

J. SHIBATA

Kanagawa Institute of Technology, Atsugi, Kanagawa, 243-0292, Japan

G. TATARA

Graduate School of Science, Tokyo Metropolitan University, Hachioji, Tokyo 192-0397, Japan

Interplay between magnetization dynamics and electric current in a conducting ferromagnet is theoretically studied based on a microscopic model calculation. First, the effects of the current on magnetization dynamics (spin torques) are studied with special attention to the "dissipative" torques arising from spin-relaxation processes of conduction electrons. Next, an analysis is given of the "spin motive force", namely, a spin-dependent 'voltage' generation due to magnetization dynamics, which is the reaction to spin torques. Finally, an attempt is presented of a unified description of these effects.

Keywords: Current-driven magnetization dynamics; domain wall motion; spin torque; spin-transfer torque; spin relaxation; Gilbert damping; spin motive force; gauge field; effective action

1. Introduction

The fact that electrons have spin degree of freedom as well as electric charge enables us to control, in principle, magnetism by electrical means, and vice versa, without recourse to the relativistic effect of spin-orbit coupling. This type of magnetoelectric coupling has been actively studied over these two decades based on nanostructured ferromagnets, where the interplay of electric current and magnetization leads to giant/tunnel magnetoresistance, current-induced magnetization reversal, and so on.[1,2]

Microscopic origin of such phenomena is the s-d exchange interaction

$$H_{sd} = -M \int d^3x\, \boldsymbol{n}(x)\cdot\hat{\boldsymbol{\sigma}}(x), \qquad (1)$$

between the spin $\hat{\boldsymbol{\sigma}}(x)$ of conduction electrons and magnetization $\boldsymbol{n}(x)$. For example, if an electron moves through a magnetization texture $\boldsymbol{n}(x)$, its spin feels a time-dependent 'field' $M\boldsymbol{n}$ and is affected. The electron, in turn, exerts a reaction torque[3,4] (spin torque)

$$\boldsymbol{t}_{sd} = M\boldsymbol{n}(x) \times \langle\hat{\boldsymbol{\sigma}}(x)\rangle, \qquad (2)$$

on the magnetization, which enables us to control the magnetization by current.

In this paper, we present our microsopic study on the spin torque and its reciprocal effect (spin motive force). The magnetization is treated as a classical object, whereas electrons are treated quantum-mechanically.

2. Spin torques

2.1. *Case of domain wall*

To illustrate how an electric current flowing in a ferromagnet affects the magnetization dynamics, let us first consider a magnetic domain wall (DW) as an example.[3,4,5,6]

For a rigid DW, there are two distinct effects of the current. If a conduction electron

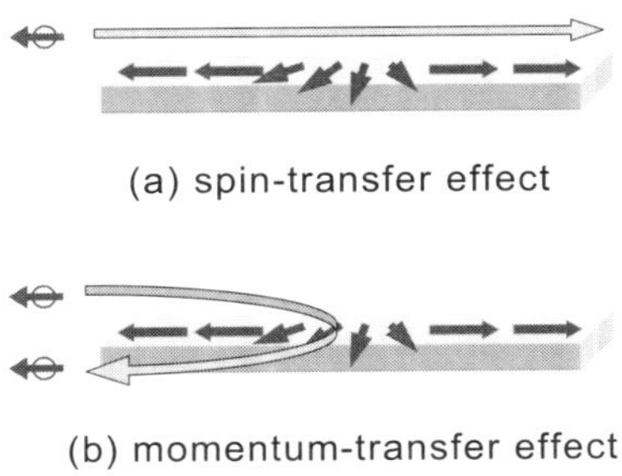

Fig. 1. Two effects of electric current on a domain wall (DW) via the s-d exchange interaction. (a) Adiabatically transmitted electron transfers spin angular momentum to the DW, and exerts a torque (in a narrow sense) on the DW. (b) Reflected electron transfers linear momentum to the DW, and exerts a force on the DW.

passes through the DW adiabatically and its spin is flipped after the passage (Fig.1(a)), this change of electron spin should be compensated by the change of magnetization owing to total angular momentum conservation, thereby driving the DW. This is the celebrated spin-transfer effect. If, instead, an electron is reflected by the DW, a linear momentum is transferred to the DW and the electron exerts a force on it (Fig.1(b)).

The latter process is nonadiabatic, and will be negligible for a 'thick' DW as realized in typical metallic magnets. However, if the electron system admits spin-relaxation processes, a new *adiabatic* torque (called β-term, see below) arises which has the same effect (*i.e.*, force) on a DW and crucially affects the dynamics of the DW.[7,8,9,10]

2.2. *Landau-Lifshitz-Gilbert equation under current*

For general but slowly-varying (in space and time) magnetization configurations, the dynamics is described by the Landau-Lifshitz-Gilbert (LLG) equation,

$$\dot{\boldsymbol{n}} = \gamma_0 \boldsymbol{H}_{\rm eff} \times \boldsymbol{n} + \alpha_0 \dot{\boldsymbol{n}} \times \boldsymbol{n} + \boldsymbol{t}'_{sd}. \tag{3}$$

Here, $\boldsymbol{n} = \boldsymbol{n}(\boldsymbol{r}, t)$ is a unit-vector field representing the d-spin direction, and the dot represents time derivative. The first term, a precessional torque around the effective field $\gamma_0 \boldsymbol{H}_{\rm eff}$, and the second term (Gilbert damping) come from processes without conduction electrons. The effects of conduction electrons are contained in the third term, $\boldsymbol{t}'_{sd} \equiv \boldsymbol{t}_{sd} \times (a^3/\hbar S)$ (S: magnitude of d spin, a^3: volume per d-spin), called spin torque.

In this paper, we focus on *adiabatic spin torques*,[11] which are first order in space/time derivative and are expressed as

$$\begin{aligned}\boldsymbol{t}'_{sd} = &-(\boldsymbol{v}^0_{\rm s} \cdot \boldsymbol{\nabla})\, \boldsymbol{n} - \beta_{\rm sr}\, \boldsymbol{n} \times (\boldsymbol{v}^0_{\rm s} \cdot \boldsymbol{\nabla})\, \boldsymbol{n} \\ &-\alpha_{\rm sr}\, (\boldsymbol{n} \times \dot{\boldsymbol{n}}) - \frac{\delta S}{S}\, \dot{\boldsymbol{n}}.\end{aligned} \tag{4}$$

The first term on the right-hand side is the celebrated spin-transfer torque,[12] where

$$\boldsymbol{v}^0_{\rm s} = -\frac{a^3}{2eS}\, \boldsymbol{j}_{\rm s} \tag{5}$$

is the (unrenormalized) "spin-transfer velocity", with $\boldsymbol{j}_{\rm s} = \boldsymbol{j}_\uparrow - \boldsymbol{j}_\uparrow$ being the spin-current density. The second term, called 'β-term',[8] comes from spin-relaxation processes of electrons,[7] and acts as a force on a rigid DW. Here $\beta_{\rm sr}$ is a dimensionless constant. The third term is the Gilbert damping, also resulting from spin relaxation of electrons.

The fourth term contributes as a "renormalization" of spin;[7] it can be combined with the term on the left-hand side of eq.(3) to form $(1 + \delta S/S)\, \dot{\boldsymbol{n}} = (S_{\rm tot}/S)\, \dot{\boldsymbol{n}}$, where

$$S_{\rm tot} = S + \delta S, \tag{6}$$

is the total ("renormalized") spin with δS being the contribution from conduction electrons. Then, dividing both sides of the equation by $S_{\rm tot}/S$, we arrive at

$$\begin{aligned}\dot{\boldsymbol{n}} = &\gamma \boldsymbol{H}_{\rm eff} \times \boldsymbol{n} - \alpha\, (\boldsymbol{n} \times \dot{\boldsymbol{n}}) \\ &-(\boldsymbol{v}_{\rm s} \cdot \boldsymbol{\nabla})\, \boldsymbol{n} - \beta\, \boldsymbol{n} \times (\boldsymbol{v}_{\rm s} \cdot \boldsymbol{\nabla})\, \boldsymbol{n},\end{aligned} \tag{7}$$

where $\gamma = (S/S_{\rm tot})\, \gamma_0$, $\alpha = (S/S_{\rm tot})(\alpha_0 + \alpha_{\rm sr})$, $\beta = \beta_{\rm sr}$, and

$$\boldsymbol{v}_{\rm s} = \frac{S}{S_{\rm tot}}\, \boldsymbol{v}^0_{\rm s} = -\frac{a^3}{2eS_{\rm tot}} \boldsymbol{j}_{\rm s}, \tag{8}$$

is the "renormalized" spin-transfer velocity. Note that β is not renormalized by this procedure.

In the parameter space of the LLG equation (7), the manifold of $\alpha = \beta$ provides a very special case for the dynamics. For example, any static solution $\boldsymbol{n}(\boldsymbol{r})$ in the absence of spin current can be used to construct a solution $\boldsymbol{n}(\boldsymbol{r} - \boldsymbol{v}_s t)$ in the presence of spin current $\boldsymbol{v}_s$ if $\alpha = \beta$. Since the controversy on the current-driven domain-wall motion,[6,9] whether the relation $\alpha = \beta$ holds generally or not has been a theoretical issue.

The relation $\alpha = \beta$ was originally suggested in ref.[9] based on the assumption of Galilean invariance of the system. Although one may argue that the Galilean invariance should be valid for long-wavelength and low-frequency dynamics in which the underlying lattice structure is irrelevant, the α and β come from spin-relaxation processes,[7] which are usually intimately related to the lattice, *e.g.*, through the spin-orbit coupling. Also, for a many-electron system having Fermi surfaces, Galilean invariance is not an obvious property. Therefore, it is desired to carry out a fully microscopic calculation without introducing any phenomenological assumptions once a microscopic model is fixed.

2.3. *Microscopic model*

For conceptual simplicity, we take a localized picture for ferromagnetism, and consider the so-called *s*-*d* model consisting of localized *d* spins, $\boldsymbol{S} = S\boldsymbol{n}$, and conducting *s* electrons (as we already used in the previous sections). They are mutually coupled *via* the *s*-*d* exchange interaction H_{sd} [eq.(1)] and obey, respectively, the LLG equation (3) and the Schrödinger equation

$$i\hbar\dot{c} = \left[-\frac{\hbar^2}{2m}\nabla^2 - M\boldsymbol{n}\cdot\boldsymbol{\sigma} + V_{\rm imp}\right]c. \quad (9)$$

The impurity potential $V_{\rm imp}$ includes potential scattering as well as spin scattering

$$V^{\rm s}_{\rm imp} = u_{\rm s}\sum_j \boldsymbol{S}_j\cdot\boldsymbol{\sigma}\,\delta(\boldsymbol{r} - \boldsymbol{R}'_j) \quad (10)$$

due to quenched magnetic impurities $\boldsymbol{S}_j$. The latter has been introduced as a microscopic modeling of spin-relaxation processes. The averaging over the impurity spin direction is taken as $\overline{S^\alpha_i} = 0$ and

$$\overline{S^\alpha_i S^\beta_j} = \frac{1}{3}S^2_{\rm imp}\delta_{ij}\delta^{\alpha\beta}. \quad (11)$$

To obtain the torque $\boldsymbol{t}_{sd}$, we calculate the *s*-electron spin density $\langle\hat{\boldsymbol{\sigma}}_\perp\rangle_{\rm ne}$ [see eq.(2)]. ($\perp$ means perpendicular component to $\boldsymbol{n}$.) Here the average $\langle\cdots\rangle_{\rm ne}$ is taken in the following nonequilibrium states for electrons depending on the type of the torque.

(a) Nonequilibrium states under the influence of uniform but *time-dependent magnetization*. This leads to torques with time derivative of $\boldsymbol{n}$, namely, Gilbert damping and spin renormalization.

(b) Nonequilibrium states with *current flow* under static but *spatially-varying magnetization*. This leads to current-induced torques, namely, spin-transfer torque and the β-term.

2.4. *Small-amplitude method*

In the presence of spin rotational symmetry in the electron system (except for H_{sd}), adiabatic spin torques are expressed as

$$\boldsymbol{t}_{sd} = a_\mu\partial_\mu\boldsymbol{n} + b_\mu\,(\boldsymbol{n}\times\partial_\mu\boldsymbol{n}), \quad (12)$$

where a_μ and b_μ are the coefficients, and summing over $\mu = 1, 2, 3$ (space components) and 0 (time) is understood. The corresponding *s*-electron spin polarization is given by

$$\langle\boldsymbol{\sigma}_\perp\rangle_{\rm ne} = \frac{1}{M}\left[\,b_\mu\partial_\mu\boldsymbol{n} - a_\mu\,(\boldsymbol{n}\times\partial_\mu\boldsymbol{n})\,\right]. \quad (13)$$

The coefficients, a_μ and b_μ, can be determined by considering small transverse fluctuations, $\boldsymbol{u} = (u^x, u^y, 0)$, $|\boldsymbol{u}| \ll 1$, around a uniformly magnetized state, $\boldsymbol{n} = \hat{z} \equiv (0, 0, 1)$, such that $\boldsymbol{n} = \hat{z} + \boldsymbol{u} + \mathcal{O}(u^2)$, and retain the terms first order in $\boldsymbol{u}$ as [13,14,15]

$$\langle\boldsymbol{\sigma}_\perp\rangle_{\rm ne} = \frac{1}{M}\left[\,b_\mu\partial_\mu\boldsymbol{u} - a_\mu(\hat{z}\times\partial_\mu\boldsymbol{u})\,\right]. \quad (14)$$

Then a_μ and b_μ are given as linear-response coefficients, which are evaluated in the uniformly magnetized state, $\boldsymbol{u} = \boldsymbol{0}$.

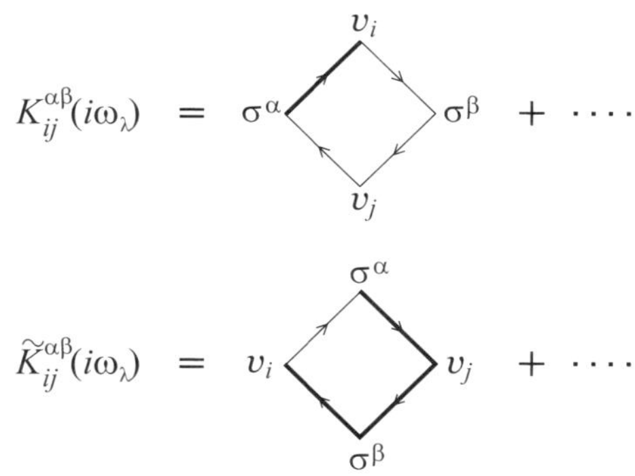

Fig. 2. Diagrammatic expression for spin torque (upper panel) and spin motive force (lower panel). The thick lines carry the external frequency $i\omega_\lambda$.

To calculate current-induced torques for example, we assume a static configuration, $\boldsymbol{n}(\boldsymbol{r}) = \hat{z} + \boldsymbol{u}(\boldsymbol{r})$, and introduce a d.c. electric field $\boldsymbol{E}$ to produce a current-carrying state. We calculate $\boldsymbol{\sigma}_\perp$ by first applying the linear-response theory to extract $\boldsymbol{E}$ as

$$\langle \hat{\sigma}_\perp^\alpha(\boldsymbol{q}) \rangle_{\rm ne} = \lim_{\omega \to 0} \frac{K_i^\alpha(\boldsymbol{q}, \omega + i0)}{i\omega} E_i. \quad (15)$$

The linear-response coefficient

$$K_i^\alpha(\boldsymbol{q}, i\omega_\lambda) = \int_0^\beta d\tau \, {\rm e}^{i\omega_\lambda \tau} \langle {\rm T}_\tau \, \hat{\sigma}_\perp^\alpha(\boldsymbol{q}, \tau) \, J_i \rangle \quad (16)$$

is the correlation function of spin $\hat{\sigma}$ and electric current $\boldsymbol{J}$, which can be non-vanishing in the presence of non-uniform spin texture $\boldsymbol{u}(\boldsymbol{r}) = u_{\boldsymbol{q}} {\rm e}^{i\boldsymbol{q}\cdot\boldsymbol{r}}$. Extracting u^β and q_j as

$$K_i^\alpha(\boldsymbol{q}, i\omega_\lambda) = -eMK_{ij}^{\alpha\beta}(i\omega_\lambda) q_j u_{\boldsymbol{q}}^\beta, \quad (17)$$

we have calculated the coefficient $K_{ij}^{\alpha\beta}$, which is expressed by the upper diagram in Fig. 2.

The results are given, in the lowest non-trivial order in the electron damping, by [14,15]

$$\delta S = \frac{1}{2} \rho_{\rm s} a^3, \quad (18)$$

$$\boldsymbol{v}_{\rm s} = -\frac{a^3}{2e\,(S + \delta S)} \boldsymbol{j}_{\rm s}, \quad (19)$$

$$\alpha = \frac{a^3 \nu_+}{4(S + \delta S)} \cdot \frac{\hbar}{\tau_{\rm s}} + \frac{S}{S + \delta S} \alpha_0, \quad (20)$$

$$\beta = \frac{\hbar}{2M\tau_{\rm s}}. \quad (21)$$

Here $\rho_{\rm s} = n_\uparrow - n_\downarrow$ is the s-electron spin density, $\nu_\pm = \nu_\uparrow \pm \nu_\downarrow$ is the density of states, and $\boldsymbol{j}_{\rm s} = \sigma_{\rm s} \boldsymbol{E} = \boldsymbol{j}_\uparrow - \boldsymbol{j}_\downarrow$ is the spin current, with $\sigma_{\rm s} = \sigma_\uparrow - \sigma_\downarrow$ being the "spin conductivity". ($\sigma_{\uparrow(\downarrow)}$ is the conductivity of majority-(minority-) spin electrons.) We have defined the spin-relaxation time $\tau_{\rm s}$ by

$$\frac{\hbar}{\tau_{\rm s}} = \frac{4\pi}{3} n_{\rm s} u_{\rm s}^2 \, S_{\rm imp}^2 \, \nu_+. \quad (22)$$

As expected, only the spin scattering ($\sim \tau_{\rm s}^{-1}$) contributes to α and β, and the potential scattering does not.

The ratio β/α cannot be unity in general for the two-component s-d model, since it contains mutually independent quantities, *e.g.*, S of d electrons and δS of s electrons. For a single-band itinerant ferromagnet, where δS gives the total moment, the results are obtained by simply putting $S = 0$ and $\alpha_0 = 0$ in eqs.(18)-(21). We still see that $\alpha \neq \beta$, but it was pointed out that the ratio

$$\frac{\beta}{\alpha} = \frac{\rho_{\rm s}}{M\nu_+} \simeq 1 + \frac{1}{12} \left(\frac{M}{\varepsilon_{\rm F}} \right)^2 \quad (23)$$

is very close to unity.[13] Even so, if we generalize eq.(11) to the anisotropic one,

$$\overline{S_i^\alpha S_j^\beta} = \delta_{ij} \delta_{\alpha\beta} \times \begin{cases} \overline{S_\perp^2} & (\alpha, \beta = x, y) \\ \overline{S_z^2} & (\alpha, \beta = z) \end{cases} \quad (24)$$

we have

$$\frac{\beta}{\alpha} = \frac{3\overline{S_\perp^2} + \overline{S_z^2}}{2\,(\overline{S_\perp^2} + \overline{S_z^2})}, \quad (25)$$

which ranges from 1/2 (for $\overline{S_\perp^2} \ll \overline{S_z^2}$) to 3/2 (for $\overline{S_\perp^2} \gg \overline{S_z^2}$). Therefore, we conclude that $\alpha \neq \beta$ in general, and that the value β/α is very sensitive to the details of the spin-relaxation mechanism.

The "β-term" due to spin relaxation was first derived by Zhang and Li based on a phenomenological spin-diffusion equation.[7] Their results can be written as

$$\alpha_{\rm ZL} = \frac{\delta S}{S + \delta S} \cdot \frac{\hbar}{2M\tau_{\rm s}}, \quad (26)$$

and $\beta_{\rm ZL} = \hbar/2M\tau_{\rm s}$, thus predict "$\alpha = \beta$" for a single-band itinerant ferromagnet, $S = 0$.

So far, all phenomenologial theories predict $\alpha = \beta$, in contrast to the present microscopic results[14] showing $\alpha \neq \beta$ in general.

2.5. *Gauge-field method*

The treatment in the previous subsection is based on the assumption of rotational symmetry in spin space of electrons; otherwise it is limited to small-amplitude magnetization dynamics around a uniformly magnetized state. To treat finite-amplitude dynamics directly, we introduce in this section a local/instantaneous spin frame ("adiabatic frame") for s electrons.[16,17] In this frame, the spin quantization axis of s electrons is taken to be the local/instantaneous d-spin direction, $\boldsymbol{n}$. The electron spinor $a(x)$ in the new frame is related to the original spinor $c(x)$ as $c(x) = U(x)a(x)$, where U is a 2×2 unitary matrix satisfying $c^\dagger(\boldsymbol{n}\cdot\boldsymbol{\sigma})c = a^\dagger\sigma^z a$. The a-electrons then obey the equation,

$$i\hbar\left(\frac{\partial}{\partial t} + iA_0\right)a(x) \qquad (27)$$
$$= \left[-\frac{\hbar^2}{2m}\left(\nabla_i + iA_i\right)^2 - M\sigma_z + \tilde{V}_{\rm imp}\right]a(x),$$

which is characterized by a constant magnetization $M\sigma_z$ and an SU(2) gauge field

$$A_\mu = -iU^\dagger(\partial_\mu U) = A_\mu^\alpha\sigma^\alpha \equiv \boldsymbol{A}_\mu\cdot\boldsymbol{\sigma}. \quad (28)$$

This gauge field expresses the influence of temporal ($\mu = 0$) or spatial ($\mu = 1,2,3$) variation of $\boldsymbol{n}$.

The adiabatic torques in eq.(12) follow from the following expression[18]

$$\langle\tilde{\boldsymbol{\sigma}}_\perp\rangle_{\rm ne} = \frac{2}{M}\left[a_\mu\boldsymbol{A}_\mu^\perp + b_\mu(\hat{z}\times\boldsymbol{A}_\mu^\perp)\right], \quad (29)$$

obtained in the first order in $\boldsymbol{A}_\mu$. Here $\langle\tilde{\boldsymbol{\sigma}}\rangle \equiv \langle a^\dagger\boldsymbol{\sigma}a\rangle$ is the electron spin density in the adiabatic frame, and $\tilde{\boldsymbol{\sigma}}_\perp$ and $\boldsymbol{A}_\mu^\perp$ are those projected onto the xy-plane. The coefficients a_μ and b_μ can be calculated as linear-response coefficients. The results for $\delta S, \boldsymbol{v}_{\rm s}$ and β thus obtained coincide with eqs.(18), (19), and (21). However, it leads to $\alpha_{\rm sr} = 0$ and fails to produce the Gilbert damping.

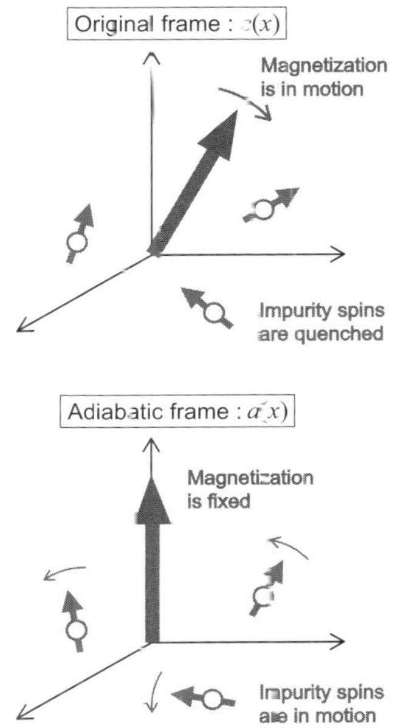

Fig. 3. Upper panel (lower panel) shows magnetization vector $\boldsymbol{n}(t)$ ($\hat{z}$) and impurity spins $\boldsymbol{S}_j$ ($\tilde{\boldsymbol{S}}_j(t)$) in the original frame (adiabatic frame).

This difficulty has been resolved[18] by noting that the impurity spins, which are static (quenched) in the original frame, become time-dependent in the adiabatic frame:

$$\tilde{\boldsymbol{S}}_j(t) = {}^t\mathcal{R}(t)\boldsymbol{S}_j. \quad (30)$$

(See Fig. 3.) Here $\mathcal{R}$ is a 3×3 orthogonal matrix representing the same rotation as U but acting on three-component vectors. From the time dependence of $\tilde{\boldsymbol{S}}_j(t)$ or $\mathcal{R}(t)$, the SU(2) gauge field can arise as

$$[\mathcal{R}(t)\,{}^t\dot{\mathcal{R}}(t)]^{\alpha\beta} = 2\varepsilon^{\alpha\beta\gamma}A_0^\gamma(t). \quad (31)$$

In fact, explicit evaluation of $\langle\tilde{\boldsymbol{\sigma}}_\perp\rangle_{\rm ne}$ in second order in $\tilde{\boldsymbol{S}}_j(t)$ (nonlinear response) gives

$$\langle\tilde{\boldsymbol{\sigma}}_\perp\rangle_{\rm ne} = -\frac{2\pi\hbar}{3M}n_{\rm s}u_{\rm s}^2S_{\rm imp}^2\nu_+^2\,(\hat{z}\times\boldsymbol{A}_0^\perp), \quad (32)$$

leading to the Gilbert damping which coincides with the first term of eq.(20).

The above calculation provides us a new picture of Gilbert damping; while the spins of s electrons tend to follow $\boldsymbol{n}(t)$, it is at the same time pinned by the quenched impurity spins, and this frustration gives rise to the Gilbert damping. This picture also applies to the case where spin relaxation originates from spin-orbit coupling.[19]

3. Spin motive force

As a reaction to spin torques, magnetization dynamics in turn exerts a spin-dependent force, called spin motive force, on electrons.[20,21,22,23,24,25,26,27,28,29] According to Stern,[22] this effect arises from the time-dependent spin Berry phase, which we interpret in our context as arising as a combined effect of temporal variation and spatial variation of magnetization. Here we present a simple argument using the results obtained in the previous section.

We apply the small-amplitude method, and consider a small fluctuation of the form,

$$\boldsymbol{u}(\boldsymbol{r},t) = \boldsymbol{u}_1\,\mathrm{e}^{-i\omega t} + \boldsymbol{u}_2\,\mathrm{e}^{i\boldsymbol{q}\cdot\boldsymbol{r}}, \qquad (33)$$

to calculate the current density in the first order in $\dot{\boldsymbol{u}}$ and $\nabla\boldsymbol{u}$, *i.e.*, in $\omega\boldsymbol{u}_1$ and $q_j\boldsymbol{u}_2$:

$$\langle j_i(\boldsymbol{q})\rangle_{\rm ne} = -eM\frac{\tilde{K}^{\alpha\beta}_{ij}(\omega)}{i\omega}\,\omega\, u_1^{\alpha}\cdot q_j u_2^{\beta}. \quad (34)$$

The coefficient $\tilde{K}^{\alpha\beta}_{ij}$ (see Fig.2) can be shown to be related to $K^{\alpha\beta}_{ij}$ of the spin torque as

$$K^{\alpha\beta}_{ij}(i\omega_\lambda) = \tilde{K}^{\alpha\beta}_{ij}(-i\omega_\lambda). \qquad (35)$$

Therefore, using the results of §2-4, we readily obtain $\boldsymbol{j} = \sigma_{\rm s}\boldsymbol{E}_{\rm s}$, where

$$E_{{\rm s},i} = \frac{\hbar}{2e}\left[\boldsymbol{n}\cdot(\partial_i\boldsymbol{n}\times\dot{\boldsymbol{n}}) + \beta\,(\dot{\boldsymbol{n}}\cdot\partial_i\boldsymbol{n})\right]. (36)$$

From $\boldsymbol{j} = \sigma_\uparrow\boldsymbol{E}_{\rm s} + \sigma_\downarrow(-\boldsymbol{E}_{\rm s})$, we may identify $\boldsymbol{E}_{\rm s}$ to be a spin-dependent 'electric' field, or $-e\boldsymbol{E}_{\rm s}$ to be the spin motive force, in the sense that majority- (minority-) spin electrons feel an effective 'electric' field of $\boldsymbol{E}_{\rm s}$ ($-\boldsymbol{E}_{\rm s}$). The second term, containing the same β parameter as the spin torque, is due to spin relaxation, and was first reported by Duine.[24]

More general calculation without assuming the form of eq.(33),[30] as well as the one based on the gauge-field method[31] will be reported elsewhere.

4. Effective gauge-field action

Spin torque and spin motive force are action and reaction to each other, and should be derived from the same term in the effective action. This kind of study has been done by Duine *et al.*[15] based on the real-time, small-amplitude formalism. Here we present a treatment based on the imaginary-time, gauge-field formalism. It should be noted that dynamical/dissipative processes can also be treated with imaginary time.

We introduce an electromagnetic vector potential $\boldsymbol{A}^{\rm em}$ to drive the non-equilibrium Ohmic current in a ferromagnet, and eliminate the a-electrons. Up to the second order in $\boldsymbol{A}^{\rm em}$ and the SU(2) gauge field $\boldsymbol{A}^{\alpha}$, the effective action $\mathcal{S}$ is obtained as[32]

$$\begin{aligned}\mathcal{S} = &\int_0^{\beta'} d\tau \int \frac{d\boldsymbol{r}}{a^3}\left[2i\hbar S_{\rm tot}A_0^z + \frac{J_{\rm eff}}{2}(\partial_i\boldsymbol{n})^2\right] \\ &+ \int_0^{\beta'} d\tau \int_0^{\beta'} d\tau' \int d\boldsymbol{r} I(\tau-\tau') \\ &\times\left\{\left[\sigma_{\rm s}\frac{\hbar}{e}\boldsymbol{A}^z(\tau) + \frac{\sigma_{\rm c}}{2}\boldsymbol{A}^{\rm em}(\tau)\right]\cdot\boldsymbol{A}^{\rm em}(\tau') \right.\\ &\left. + c_{\alpha\beta}\left[\mathcal{R}(\tau)^t\mathcal{R}(\tau')\right]^{\alpha\beta}\right\}. \end{aligned} \qquad (37)$$

Here $\beta' \equiv (k_{\rm B}T)^{-1} = \infty$ is the inverse temperature, $J_{\rm eff} = J_{dd}S^2 + J_{ss}(\delta S)^2$,[17] and

$$c_{\alpha\beta} = \frac{\pi}{6}n_{\rm s}u_{\rm s}^2 S_{\rm imp}^2\left[2\nu_\uparrow\nu_\downarrow\delta^{\alpha\beta} + \nu_-^2\delta^{\alpha z}\delta^{\beta z}\right]. \qquad (38)$$

The kernel $I(\tau-\tau') = -[\pi(\tau-\tau')^2]^{-1}$ describes dissipative processes characterized by Ohmic damping, as is familiar since the work by Caldeira and Leggett[33] on macroscopic quantum tunneling. The coupling $\boldsymbol{A}^z\cdot\boldsymbol{A}^{\rm em}$ describes the spin-transfer torque and spin motive force. The term containing $\mathcal{R}(\tau)^t\mathcal{R}(\tau')$ describes Gilbert damping.

In fact, by taking the variation of $\mathcal{S}$ with respect to $\boldsymbol{n}$, and perform an analytic continuation, $\tau\to it$, we obtain the LLG equation consistent with eqs.(7), (18)-(20) but with $\beta = 0$. Similarly, the electric current density is obtained from $\boldsymbol{j} = -\delta\mathcal{S}/\delta\boldsymbol{A}^{\rm em}$ as

$$\boldsymbol{j} = -\frac{\hbar}{e}\sigma_{\rm s}\dot{\boldsymbol{A}}^{z} - \sigma_{\rm c}\dot{\boldsymbol{A}}^{\rm em}, \qquad (39)$$

from which we can read the existence of the spin motive force as $-e\boldsymbol{E}_{\rm s} = \hbar\dot{\boldsymbol{A}}^{z} = (\hbar/2)\,\boldsymbol{n}\cdot(\dot{\boldsymbol{n}} \times \partial_i\boldsymbol{n})$. The effective coupling describing the β-term remains to be derived.

5. Summary and remarks

We have developed a microscopic theory of spin torques and spin motive force, and their unified description. Although the present magnetic impurity model may not be quite realistic as the origin of spin relaxation, we expect the present calculation already captures the essential features of the current-spin interaction including spin-relaxation effects. For quantitative information such as the value of β/α, calculations with realistic spin-relaxation mechanisms are necessary.

Acknowledgments

We would like to thank G. Bauer, A. Brataas, R. Duine, H. Fukuyama, A. H. MacDonald, S. Maekawa, Y. Nakatani, Q. Niu, H. Ohno, T. Ono, E. Saitoh, J. Sinova, M. Stiles, Y. Suzuki, A. Thiaville and Y. Tserkovnyak for valuable discussions. H. K. is indebted to K. Miyake for his continual encouragement.

References

1. *Concepts in Spin Electronics*, Ed. S. Maekawa (Oxford University Press, Oxford, 2006).
2. *Spintronic Materials and Technology*, Eds. Y.B. Xu and S.M. Thompson, (Taylor & Francis, 2007).
3. L. Berger, *J. Appl. Phys.* **55**, 1954 (1984).
4. L. Berger, *J. Appl. Phys.* **71**, 2721 (1992).
5. A. Yamaguchi *et al.*, *Phys. Rev. Lett.* **92**, 077205 (2004); E. Saitoh *et al.*, *Nature* **432**, 203 (2004); M. Yamanouchi *et al.*, *Nature* **428**, 539 (2004); M. Kläui *et al.*, *Phys. Rev. Lett.* **94**, 106601 (2005); M. Hayashi *et al.*, *Nature Phys.* **3**, 21 (2007).
6. G. Tatara and H. Kohno, *Phys. Rev. Lett.* **92**, 086601 (2004); **96** 189702 (2006).
7. S. Zhang and Z. Li, *Phys. Rev. Lett.* **93**, 127204 (2004).
8. A. Thiaville, Y. Nakatani, J. Miltat and Y. Suzuki, *Europhys. Lett.* **69**, 990 (2005).
9. S.E. Barnes and S. Maekawa, *Phys. Rev. Lett.* **95**, 107204 (2005); **96**, 189701 (2006).
10. G. Tatara, H. Kohno and J. Shibata, *Phys. Rep.* **468**, 213 (2008).
11. There are *non-adiabatic torques* in addition if the magnetization varies rapidly, which we do not discuss in this report. See G. Tatara *et al.*, *J. Phys. Soc. Jpn.* **76**, 054707 (2007).
12. Ya. B. Bazaliy, B. A. Jones, and S.-C. Zhang, *Phys. Rev.* **B 57**, R3213 (1998).
13. Y. Tserkovnyak, H.J. Skadsem, A. Brataas and G.E.W. Bauer, *Phys. Rev.* **B74**, 144405 (2006).
14. H. Kohno, G. Tatara and J. Shibata, *J. Phys. Soc. Jpn.* **75**, 113706 (2006).
15. R.A. Duine, A.S. Núñez, J. Sinova and A.H. MacDonald, *Phys. Rev.* **B75**, 214420 (2007).
16. V. Korenman *et al.*, *Phys. Rev.* **B16**, 4032 (1977).
17. G. Tatara and H. Fukuyama, *J. Phys. Soc. Jpn.* **63**, 2538 (1994).
18. H. Kohno and J. Shibata, *J. Phys. Soc. Jpn.* **76**, 063710 (2007).
19. S. Kawabata, Master thesis (Osaka Univ., 2008).
20. L. Berger, *Phys. Rev.* **B33**, 1572 (1986).
21. G.E. Volovik, *J. Phys.* **C 20**, L83 (1987).
22. A. Stern, *Phys. Rev. Lett.* **68**, 1022 (1992).
23. S. E. Barnes and S. Maekawa, Chap.7 of ref.1; *Phys. Rev. Lett.* **98**, 246601 (2007).
24. R.A. Duine, *Phys. Rev.* **B77**, 014409 (2008).
25. M. Stamenova, T.N. Todorov and S. Sanvito, cond-mat/0708.1167.
26. S.A. Yang, D. Xiao and Q. Niu, cond-mat/0709.1117.
27. Y. Tserkovnyak and M. Mecklenburg, *Phys. Rev.* **B77**, 134407 (2007).
28. S.A. Yang, G. Beach, C. Knutson, D. Xiao, Q. Niu, M. Tsoi and J.L. Erskine, preprint.
29. A similar phenomenon in a FN junction system was studied in, A. Brataas, Y. Tserkovnyak, G. Bauer and B.I. Halperin, *Phys. Rev.* **B66**, 060404 (2002).
30. T. Noguchi, Master thesis (Osaka Univ., 2008).
31. J. Shibata and H. Kohno, in preparation.
32. S. Ueta, Master thesis (Osaka Univ., 2008).
33. A. O. Caldeira and A. J. Leggett, *Phys. Rev. Lett.* **46**, 211 (1981); *Ann. Phys.* **149**, 374 (1983).

SPIN-POLARIZED CARRIER INJECTION EFFECT IN FERROMAGNETIC SEMICONDUCTOR/DIFFUSIVE SEMICONDUCTOR/SUPERCONDUCTOR JUNCTIONS

H. TAKAYANAGI*

Research Institute for Science and Technology, Tokyo University of Science, 1-3 Kagurazaka, Shinjuku Tokyo, 162-8601, Japan
International Center for Materials Nanoarchitectonics, National Institute for Materials Science, 3-13 Sakura, Tsukuba, Ibaraki 305-0003, Japan
**E-mail: h-taka@rs.kagu.tus.ac.jp*

T. AKAZAKI

NTT Basic Research Laboratories, NTT Corporation, 3-1 morinosato-Wakamiya, Atsugi, Kanagawa 243-0198, Japan

N. NISHIZAWA

International Center for Materials Nanoarchitectonics, National Institute for Materials Science, 3-13 Sakura, Tsukuba, Ibaraki 305-0003, Japan

Y. SAWA, T. YOKOYAMA and Y. TANAKA

Department of Applied Physics, Nagoya University, Nagoya, Aichi 464-8603, Japan

A. A. GOLUBOV

Faculty of Science and Technology, University of Twente, Enschede, The Netherlands

H. MUNEKATA

Image Science and Engineering Lab., Tokyo Institute of Technology, Yokohama, Kanagawa 226-8503, Japan

We study the transport properties of a *p*-InMnAs/*n*-InAs/Nb junction, where a *p*-InMnAs can be regarded as a spin injector. We fabricate the junctions at different distances between InMnAs and Nb electrodes and measure the differential conductance of the *n*-InAs channel. The conductance varies significantly depending on the direction of the injection current and the distance between the spin injector and the superconductor. We also calculate the conductance in the *n*-InAs channel by taking into account the exchange field in the InAs channel that is induced by the ferromagnetic semiconductor InMnAs. The difference between the conductance behaviors for different injection current directions can be explained by the inverse proximity effect in which the exchange field is also induced in the superconductor. From the experimental and theoretical results, it can be said that we successively observed the phenomena due to the inverse proximity effect that depends on a spin-polarized carrier injection.

Keywords: Spin injection; Inverse proximity effect; Ferromagnetic semiconductor.

1. Introduction

Today, there is a growing interest in the realization of electronic devices with new functionalities using spin degree of freedom. This new frontier of electronics is called 'Semiconductor Spintronics'. Typically, there are two approaches for injecting spin-polarized electrons in semiconductor devices. The first approach involves the development of semiconductors that exhibit properties of ferromagnetism; this property enables all conducting electrons to be intrinsically aligned in semiconductor devices. Since the discovery of the ferromagnetism in Mn-doped III-V semiconductors [1], intensive research has been carried out for developing high transition temperature (T_C) ferromagnetic semiconductors; results of such discovery will help us to better understand the mechanism of ferromagnetism. The second method involves the injection of spin-polarized electrons, their control and detection from a ferromagnetic material to a semiconductor with the help of optics or by applying an external electrical and magnetic field. Numerous studies have been carried out on the process of spin injection using junctions with tunnel barriers [2,3], because there is a large difference between the conductivities of a ferromagnetic metal and a semiconductor, and spin polarized carriers cannot be injected directly into semiconductors [4]. However, the spin injection efficiency can be improved by using ferromagnetic semiconductors having a small difference in conductivity with a semiconductor than compared to ferromagnetic metals.

Interesting phenomena have been reported in case of ferromagnetic material (F)/ superconductor (S) junctions [5–7]. When there is a clean interface between S and normal metal (N) including F, Cooper pairs can easily penetrate into N from S and maintain their coherence from the interface at considerable distances. This phenomenon, called the proximity effect, has been studied for many years. The spin-polarized carriers penetrate into S that is in contact with F, and the exchange field can be induced in S. This effect is called the inverse proximity effect [8]. However, to the best of our knowledge, there exists neither an elaborate theoretical description nor a detailed experimental study of this effect.

In the present study, we have investigated the transport properties of a p-InMnAs/n-InAs/Nb junction in order to clarify the correlation between the inverse proximity effect and the spin injection effect. The ferromagnetic semiconductor p-InMnAs, whose conductivity is slightly different from that of n-InAs, acts as a spin injector for both the normal semiconductor n-InAs and superconductor Nb.

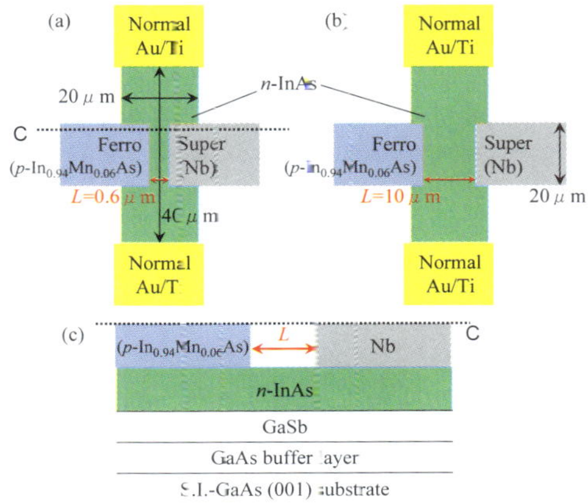

Fig. 1. Schematic diagrams of p-InMnAs / n-InAs / Nb junctions: Top view of the junctions with the distance L is (a) 0.6 μm, (b) 10 μm. (c) Cross-sectional view of diagram.

2. Sample Fabrication

We fabricated p-$In_{0.94}Mn_{0.06}As$/n-InAs/Nb junctions as shown in Fig. 1. The p-InMnAs/n-InAs heterostructure was grown by the molecular-beam epitaxy (MBE) method. GaAs and GaSb buffer layers were grown on a GaAs (100) substrate, and then, an n-InAs layer with a thickness of 70 nm was grown on these buffer layers. Subsequently, a p-InMnAs layer of around 15 nm in thickness was grown on the n-InAs layer. After the growth of the heterostructure, electrode patterns were fabricated by using the conventional process. First, the p-InMnAs layer except for the electrode parts was removed by chemical etching. Second, the oxidized layer on the surface of the n-InAs layer was removed by Ar sputtering in a Nb evaporation chamber. Third, Nb was deposited by electron beam deposition and a Nb electrode pattern was formed by the lift-off process. Finally, two Au/Ti electrodes were deposited on the n-InAs layer. It was designed such that the distance L between p-InMnAs and Nb was (a) 0.6 μm and (b) 10 μm.

The magnetization of the p-$In_{0.94}Mn_{0.06}As$ layer was measured using a superconducting quantum interference device (SQUID) magnetometer. The ferromagnetic transition temperature T_C of the InMnAs layer was estimated to be around 60 K. When the temperature is sufficiently low below T_C, the fabricated p-InMnAs electrode can be considered as a ferromagnetic material that acts as a spin source. Because the p-n junction was formed at the p-InMnAs/n-InAs interface and the carrier concentration in the p-InMnAs layer was two orders of magnitude higher than the concentration in n-InAs, the depletion layer between them was very thin (several nanometers in width). Therefore, the spin-polarized carriers in p-InMnAs were diffusively injected into the nonmagnetic InAs channel by Zener tunneling. It is thought that spin accumulation occurs in the n-InAs channel. Moreover, when the temperature was below the transition temperature of the Nb electrode (~ 8.8 K), the p-InMnAs/n-InAs/Nb junction contained both a ferromagnet/semiconductor (p-InMnAs/n-InAs) interface and a superconductor/diffusive-semiconductor (Nb/n-InAs) interface.

3. Experimental Results

Fig. 2 shows the bias voltage V dependence of the differential conductance between Au/Ti electrodes on the junction as a function of the

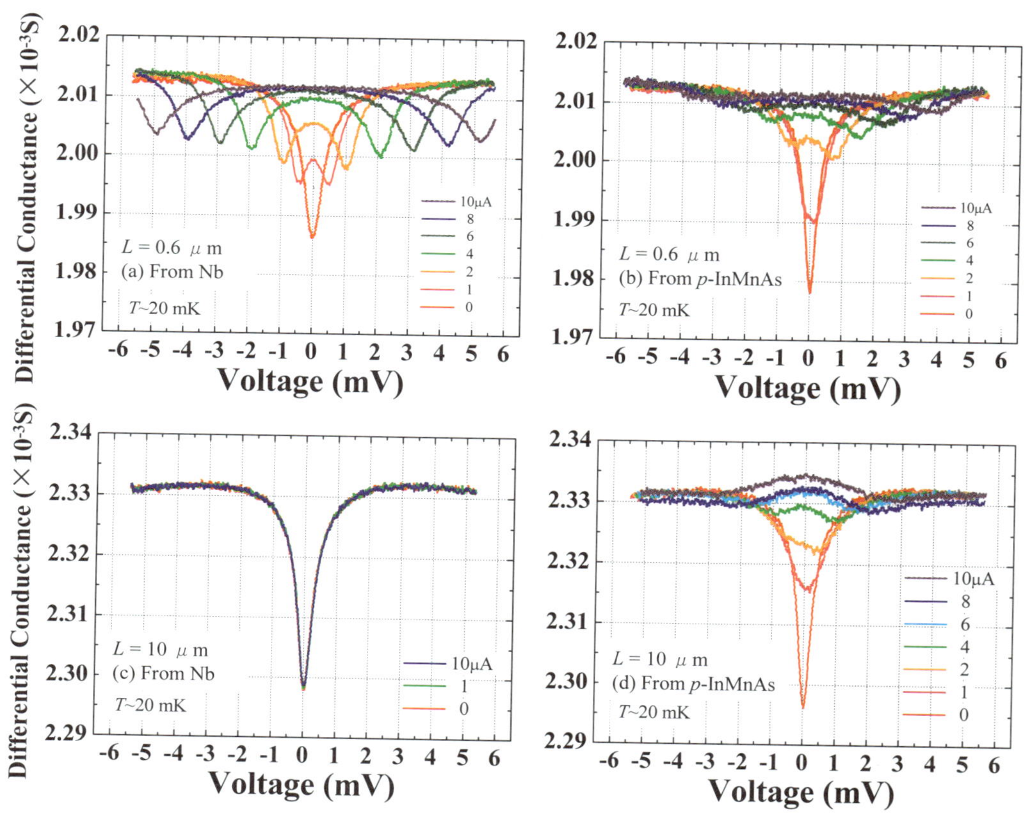

Fig. 2. Bias voltage V dependence of the differential conductance of the n-InAs channel between two Au/Ti electrodes on the junction as a function of injection current (a)(c) from Nb and (b)(d) from p-InMnAs at about 20 mK. The distances between p-InMnAs and Nb electrodes are 0.6 μm ((a), (b)) and 10 μm ((c),(d)).

injection current from InMnAs or from Nb at about 20mK. Figs. 2 (a) and (b) show the results for the sample with L = 0.6 μm. The conductance minimum is observed at zero bias voltage without current injection. With an increase in the injection current from Nb, the conductance minimum splits into two dips that shift toward higher bias, as shown in Fig. 2(a). In contrast, with an increase in the injection current from InMnAs, the conductance minimum gradually decreases and disappears (Fig. 2(b)) by splitting into two dips. These splits are not observed in Figs. 2 (c) and (d), which show the results for L = 10 μm. Moreover, in the case of current injection from the Nb electrode (Fig. 2(c)), the conductance behavior shows no change.

4. Theory

For theoretical calculation, we used a model of a one-dimensional diffusive ferromagnet/super-conductor–superconductor/diffusive ferro-magnet (DF/S-S/DF) junction, as shown in Fig. 3, where we assume that the diffusive semiconductor n-InAs becomes a DF due to the exchange field in the ferromagnetic semiconductor InMnAs. R_d, R_b', and R_b denote the resistance of the DF, resistance at the electrode/DF interface, and the resistance at the DF/S interface, respectively.

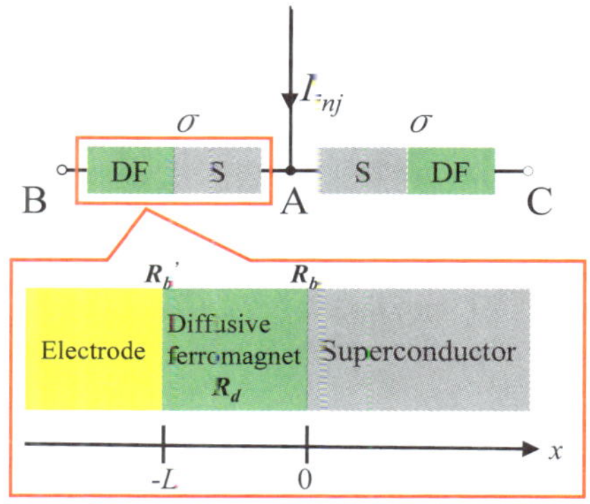

Fig. 3. Theoretical Model

We employ the quasi-classical Green's function in DF (S) that are expressed as

$$G_{F(S)}(x,\varepsilon) = \cos\theta_{F(S)}(x,\varepsilon)$$
$$\text{and} \tag{1}$$
$$F_{F(S)}(x,\varepsilon) = \sin\theta_{F(S)}(x,\varepsilon),$$

where $\theta_{F(S)}$ for majority (minority) spin in the DF (S) obeys the Usadel equations [7] with the diffusion constant $D_{F(S)}$ in the DF (S) and the exchange field $h(x)$. Here, ε is the energy of the quasi-particle. The exchange field is assumed to be given by

$$h(x) = \begin{cases} h_F & -L < x < 0 \\ h_S \exp(-x/\xi_S) & x > 0 \end{cases} \tag{2}$$

where $\xi_{F(S)} = \sqrt{D_{F(S)}/2\pi T_C}$ is the thermal coherence length in the DF (S). This

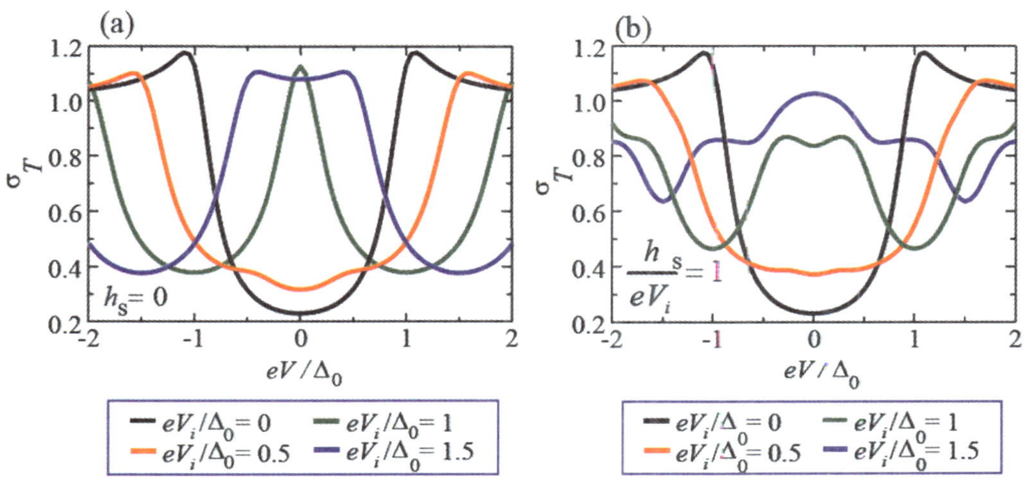

Fig. 4. Calculated V dependence of σ_T for various values of V_i in the case of (a) h_S=0 and (b) $h_S=eV_i$.

expression indicates that the injected spin-polarized current is considered to penetrate S and induce an exchange field in it. Solving the Usadel equation and the gap equation self-consistently [9], we calculate both $\Delta(x)$ and $\theta_{F(S)}$, where $\Delta(x)$ denotes the pair potential in S. For $x \rightarrow \infty$, $\Delta(x)$ corresponds to the bulk value of the pair potential $\Delta(x)$. In the following calculation for experiment, we fix the parameters as $T/T_C = 0.1$, $R_d/R_b' = R_d/R_b = 1$, $(D_F/L^2)/\Delta_0 = 0.01$, $h_F/\Delta_0 = 10$, and we focus on V dependences of the normalized conductance $\sigma_T = \sigma_S/\sigma_N$, where σ_S is the total conductance of the junction and σ_N is the conductance in the normal state.

Fig. 4 (a) shows the V dependence of σ_T for various values of V_i and $h_S = 0$. Peaks at $V = \pm\Delta_0/e$ move to $V = \pm(\Delta_0/e + V_i)$, and the zero-bias minimum splits into two dips due to the injection current I_{inj}. As a result, the zero bias conductance peak appears for $V_i \geq \Delta_0/e$. On the other hand, Fig. 4 (b) shows the V dependence of σ_T for various values of V_i and $h_S = eV_i$; these values are in agreement with the experimental results for the case in which spin-polarized current injection induces the exchange field in S. It is noted that the equation $h_S = eV_i$ indicates that the exchange field in S is proportional to the injection current. The exchange field in S should destroy the superconductivity of S near the DF interface and the zero bias conductance minimum.

5. Discussion

To understand the split of conductance, as shown in Fig. 2, we considered the model shown in Fig. 3. When additional current is injected from InMnAs or Nb, the injection current I_{inj} shifts the chemical potential to a point A (shown in Fig. 3) by $\sigma^{-1}I_{inj}$, where σ is assumed to be the conductance of the InAs/Nb single junction. Therefore, the difference in potential between B and C (shown in Fig. 3) becomes $\sigma^{-1}I_{inj}$. The conductance of each junction shifts by $\pm 1/2\sigma^{-1}I_{inj}$, and the bias voltage dependence of the total conductance shows two dips. However, we cannot quantitatively explain the width of splitting because the n-InAs channel that has a resistance of approximately 300 Ω is connected in series between Au and Nb electrode.

To investigate the origin of the conductance minimum, we also measured the transport properties of the path through a single Nb/InAs interface. The conductance minimum at zero bias voltage was observed even if the distance between the Nb and InMnAs electrodes was 10 μm. If the proximity effect works well, the conductance of the Nb/InAs interface should increase at zero bias. However, we did not observe any influence of the proximity effect on the conductance behavior at the Nb/InAs interface. From this result, we infer that there are two possible origin of the conductance minimum. One is the destruction of Cooper pairs in Nb by ferromagnetic carriers because of spin accumulation in the n-InAs channel. In this case, when the spin source is far from Nb, the spin polarization of InAs channel near Nb should become weak and the dip structure of conductance should change. However, the experimental results do not indicate any such behaviors. Another possibility is that the barrier at the InAs / Nb interface is considerably high. The conductance minimum reflects the energy gap of Nb, which does not depend on L. The suppression of the conductance minimum due to current injection from InMnAs, as shown in Figs. 2 (b) and (d), can be explained by the reflection of the energy gap. The injection of spin-polarized carriers from InMnAs enhances the spin accumulation in InAs. When spin accumulation occurs in the n-InAs channel near the interface with Nb, the superconductivity of Nb is suppressed. Therefore, conductance minimum decreases and disappears. The behaviors of conductance can be understood by this mechanism. The behaviors of conductance are observed clearly in the sample with L = 10 μm. The effect of the path through the Nb electrode on the transport properties is relatively small in both directions of injection. Hence, the split of the conductance minimum cannot be observed. When the current is injected from InMnAs, the conductance minimum of the sample with L = 10 μm is suppressed. Current injection from Nb, in contrast, does not influence the conductance behavior. With the assumption that the distance L between InMnAs and Nb is shorter than the spin diffusion (relaxation) length in *n*-InAs, we can conclude that the current injection from Nb does not change the spin distribution in the *n*-InAs layer; further in the case of current injection from the InMnAs layer, the spin-polarized carriers diffuse up to InAs near Nb; as a result, the conductance minima reflecting the energy gap of Nb are suppressed. This conclusion is also supported by theoretical calculations. Fig. 4(a) shows that the conductance minimum is hardly suppressed, but split into two dips. This result is qualitatively consistent with the experimental result in the case of current injection from Nb (Fig. 2 (a)). Fig. 4 (b) shows the suppression of conductance minima due to the inverse proximity effect, which is consistent with the result obtained when current is injected from *p*-InMnAs (Figs. 2(b), (d)).

6. Conclusions

We measured the transport properties of *p*-InMnAs/*n*-InAs/Nb junctions with different value of L. The differential conductance of the *n*-InAs channel showed a conductance minimum at zero bias voltage. With an increase in spin-polarized carrier injection from InMnAs, the conductance minimum was suppressed, while current injection from Nb split into two dips in the case of a sample with short L. We also self-consistently calculated the bias voltage dependence of the conductance of the *n*-InAs channel by using the Usadel equation and the gap equation. We found that spin-polarized current injection from InMnAs induced an exchange field in the Nb electrode, and this so-called inverse proximity effect could be used to explain the different conductance behaviors measured in the experiments. From these experimental and theoretical results, we can conclude that the exchange field penetrated into Nb destroyed the superconductivity. This results in the suppression of the conductance minimum. Thus, we successfully observed the phenomena due to the inverse proximity effect that depends on the spin injection effect.

References

1. H. Ohno, H. Munekata, T. Penney, S. von Molnár and L. L. Chang, *Phys. Rev. Lett.* **68**, 2664 (1992).
2. H. J. Zhu, M. Ramsteiner, H. Kostial, M. Wassermeier, H. P. Schonherr and K. H. Ploog, *Phys. Rev. Lett.* **87**, 016601 (2001).
3. C. Adelmann, X. Lou, J. Strand, C. J. Palmstrøm and P. A. Crowell, *Phy. Rev. B* **71**, 121301R (2005).
4. E. I. Rashba, *Phys. Rev. B* **62**, R16267 (2000).
5. V. V. Ryazanov, V. A. Oboznov, A. Yu. Veretennikov, A. A. Golubov and J. Aarts, *Phys. Rev. Lett.* **86**, 2427 (2001).
6. D. Beckmann, H. B. Weber and H. v. Löneysen, *Phys. Rev. Lett.* **93,** 197003 (2004)
7. K. D. Usadel, *Phys. Rev. Lett.* **25**, 507 (1970).
8. F. S. Bergeret, A.F.Volkov and K.B. Efetov, *Review of Modern Physics* **77,** 1321 (2005).
9. A. A. Golubov, E. F. Houwman, J.G. Gijsbertsen, V. M. Krasov, J. Flokstra, H. Rogall and M. Yu. Kupriyanov *Phys. Rev. Lett. B* **51**, 1073 (1995)

LOW VOLTAGE CONTROL OF FERROMAGNETISM IN A SEMICONDUCTOR P-N JUNCTION

J. WUNDERLICH*(1), M. H. S. OWEN(1, 2), A. C. IRVINE(2) and S. OGAWA(1)

(1) Hitachi Cambridge Laboratory, Cambridge CB3 0HE, United Kingdom

**E-mail: jw526@cam.ac.uk*

(2) Microelectronics Research Centre, Cavendish Laboratory, University of Cambridge, CB3 0HE, UK

K. VYBORNY(3), K. OLEJNIK(3), and T. JUNGWIRTH(3,4)

(3) Institute of Physics ASCR, v.v.i., Cukrovarnicka 10, 162 53 Praha 6, Czech Republic

(4) School of Physics and Astronomy, University of Nottingham, Nottingham NG72 RD, UK

Controlling the magnetization by low-voltage charge depletion in field effect transistors has been a formidable challenge due to the typically large carrier concentrations in ferromagnets compared to semiconductors. Here we demonstrate that this concept is viable in an all-semiconductor, p-n junction transistor utilizing a thin-film ferromagnetic (Ga,Mn)As channel. We report gate-dependent Curie temperature and magnetoresitance, and persistent magnetization switchings induced by short electrical pulses of a few volts.

Keywords: Electric field controlled magnetization; Magnetic semiconductors.

1. Introduction

One of the major driving forces in spintronics research is to allow the field to progress from sensors and memories to logic devices which requires development of three-terminal spintronic transistors. Ferromagnetic semiconductors such as (Ga,Mn)As are particularly favorable materials in this research area for their rich phenomenology of magnetoresistive characteristics and the potential to combine and integrate spintronics with conventional semiconductor microelectronics [1, 2]. Only a few works have so far reported electrical gating, via charge depletion, of magnetic properties of (III,Mn)V ferromagnetic semiconductor field-effect transistors (FETs). Curie temperature and coercive field variations have been demonstrated by applying tens of Volts on a top-gate FET with an AlOx dielectric separating a metal gate from an (In,Mn)As or (Ga,Mn)As channel [3, 4, 5]. The difficulty to achieve efficient depletion in the highly doped ferromagnetic semiconductor films has motivated research in several alternative routes to the field-control of magnetism in (Ga,Mn)As, including the electro-mechanical gating by piezo-stressors [6, 7, 8] or fabrication of small island Coulomb blockade transistors with magnetization dependent single-electron charging energy [9]. In this paper we discuss an all-semiconductor, epitaxial p-n junction FET allowing for a large depletion of (Ga,Mn)As thin films at a few Volts. We demonstrate various gatable magnetic characteristics ranging from the anisotropic magnetoresistance and the Curie temperature to coercivity and magnetization switchings induced by short electrical pulses.

2. Device Structure and Simulations

The schematic cross-section of the III-V heterostructure used in our study is shown in Fig.1(a). It is a semiconductor p-n junction FET specially designed to accommodate ferromagnetism in the p-type region and its efficient depletion by low voltages. From the top, the structure comprises a 5 nm thick approximately 2.5% Mn-doped GaAs capped by 2 nm of undoped GaAs to prevent oxidation of the underlying transition metal doped semiconductor film. These two top layers were

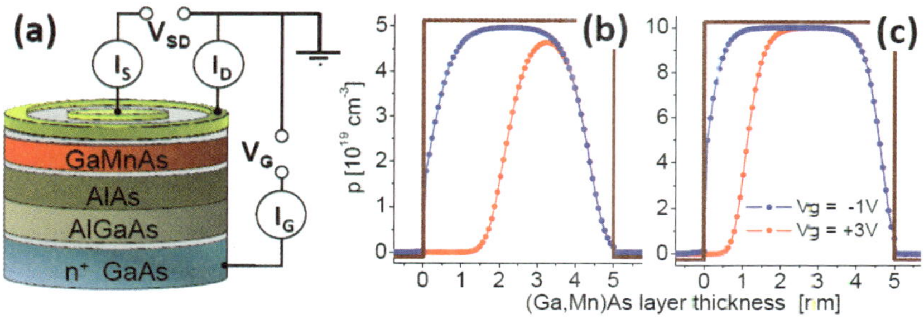

Fig. 1. (a) Schematic of the ferromagnetic p-n junction FET structure and the Corbino disk geometry of the source and drain contacts. (b) and (c) Numerical simulations of the hole density profile at -1 V (accumulation) and +3 V (depletion),considering a 2×10^{19} cm^{-3} electron doping in the n-GaAs and 5×10^{19} cm^{-3} and 10^{20} cm^{-3} hole doping in the p-(Ga,Mn)As.

grown by low-temperature molecular-beam-epitaxy (MBE) to avoid Mn precipitation. The 2.5% doping was chosen to pass the insulator-to-metal transition threshold which for the moderately deep MnGa acceptor is between 1-2% and to achieve robust ferromagnetic state with Curie temperature Tc ≈ 30 K, while still minimizing the number of unintentional interstitial-Mn impurities [1, 2, 10]. (The interstitial Mn is highly mobile at the growth temperature and its diffusion into the p-n junction would result in detrimental leakage currents.) The Curie temperature measured by SQUID in an unpatterned piece of the wafer is comparable to maximum Tc's achieved at the same Mn-doping in thicker films, indicating a very good quality of our ultra-thin ferromagnetic semiconductor epilayer. The n-type gate electrode is formed by a highly Si-doped (2×10^{19} cm^{-3}) GaAs grown by high-temperature MBE. The large electron doping is required in order to achieve appreciable and voltage dependent depletion of the ferromagnetic p-region with hole doping ~10^{20} cm^{-3}. The built-in electrostatic barrier due to the depletion effect at the p-n junction is further supported by inserting a 10 nm $Al_{0.3}Ga_{0.7}As$ spacer layer with a large conduction band off-set to the neighboring n-GaAs and a 10 nm AlAs spacer with a large valence band off-set to the neighboring p-(Ga,Mn)As.

Self-consistent numerical simulations, shown in Fig. 1(b), confirm that sizable depletions are achievable by gating our heterostructure with less than 4 Volts. Measurements discussed below were done at voltages between −1 V (forward bias) to +3 V (reverse bias) for which the leakage currents between the n-GaAs gate and p-(Ga,Mn)As channel were more than two orders of magnitude smaller than the channel currents. The (Ga,Mn)As channel was lithographically patterned in a low-resistance Corbino disk geometry with the inner contact diameter of 500 μm and the outer diameter of 600 μm.

3. Voltage control of Curie temperature and magnetoresistance

In Fig. 2(a) we plot the measured channel resistances vs. gate-voltage at temperature 4-40 K. Both at low temperatures and near Tc we observe a marked increase of the channel resistance at positive voltages. It is consistent with the depletion of the (Ga,Mn)As channel as predicted by the simulations in Fig. 1(b) and with the vicinity of the metal-insulator transition which causes the superlinear increase of R with Vg. At 4 K, the increase of R between -1 and +3 V is by more than 100%.

In Fig. 2(b) we show the voltage-dependence of the Curie temperature in the ferromagnetic p-n junction. Our measurement technique is distinct from previous studies which relied on approximate extrapolation schemes based on Arrot plot measurements at finite magnetic fields [3, 5, 11]. Recent observation and interpretation by the authors [12] of the peak in the zero-field temperature derivative of the resistance at the Curie point in good quality (Ga,Mn)As materials has provided the tool for direct transport measurements of Tc in microdevices without relying on any extrapolation schemes. In Fig. 2(b) we plot differentiated resistivity curves obtained in our device at 0 and +3 V. The data show a clear shift of the Curie temperature, i.e., the magnetization can be turned on and off in parallel with accumulating and depleting holes in the ferromagnetic semiconductor channel by biasing the p-n junctions with a few Volts.

Curie temperature variations provide the key physical demonstration of the low-voltage control of magnetization. Nevertheless, for most spintronic functionalities it is not required to destroy the ordered state of spins but only to

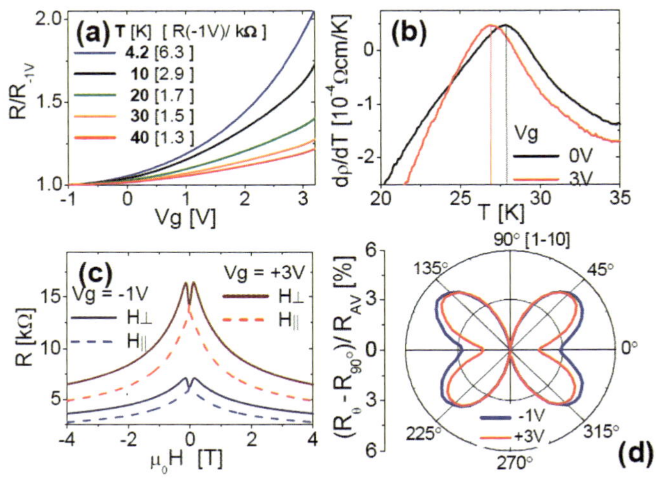

Fig. 2. (a) Gate voltage dependence of the p-(Ga,Mn)As channel resistance at temperatures 4-40 K. (b) Temperature derivative of the measured channel resistivity at 0 and +3 V. The maximum corresponds to Tc. (c) In-plane (H||) and perpendicular-to-plane (H⊥) magnetic field sweep measurements of the channel resistance at -1 and +3 V. The difference between the Hk and H⊥ sweeps above the reorientation field of ≈150 mT corresponds to the out-of-plane AMR. (d) In-plane AMR measured at saturation in a rotating in-plane field at -1 and +3 V.

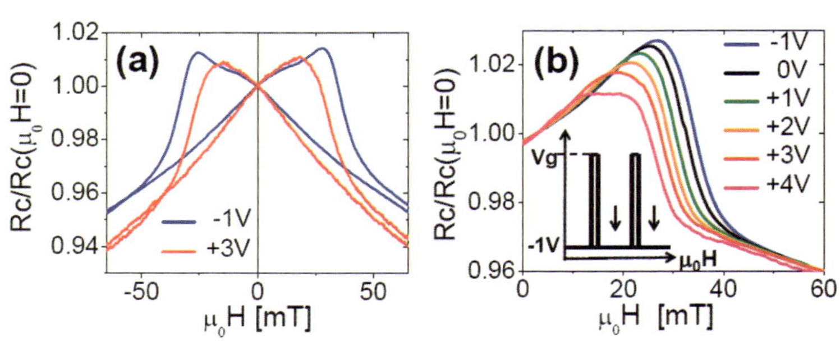

Fig. 3. (a) Hysteretic field sweep measurements at field angle $\theta = 90°$ for constant gate voltages of -1 and +3 V. (b) Up-sweeps of the $\theta = 90°$ in-plane field at constant -1V gate voltage, and for measurements with the gate voltage set to a base value of -1 V and with short additional voltage pulses corresponding to a total peak voltage of 0, +1 , +2, +3, and +4 V, respectively.

change their collective orientation. We therefore focus on effects related to reorientations of the unit vector of the macroscopic moment. To avoid thermal fluctuations of the magnetization, all measurements are done far from the Curie point at 4 K. In Fig. 2(c) we show magnetoresistance traces recorded during in-plane and perpendicular-to-plane sweeps of an external magnetic field, at gate voltages of -1 and +3 V. Apart from the negative isotropic magnetoresistance (IMR), the data indicate a remarkably large anisotropic magnetoresistance (AMR) effect which at saturation reaches ~30%. (Note that AMR sensors fabricated in transition metal ferromagnets with AMR ratios of a few per cent [13] marked the dawn of spintronics in the early 1990's.) The resistance is larger for the perpendicular-to-plane magnetization orientation and the size of the effect is enhanced by depletion. The electrical response of our system to magnetization rotations is both large and tuneable by low gate voltages. The magnetoresistance traces in Fig. 2(c) indicate that the film has a magnetic anisotropy favoring in-plane magnetization, which is overcome by an external field of approximately 150 mT. At weaker magnetic fields, magnetization switching effects are confined to the plane of the ferromagnetic film. The qualitative nature of the in-plane magnetic anisotropy landscape which determines the switching processes can be scanned in our Corbino microdevice by recording the AMR at a rotating in-plane saturation field. Unlike in the out-of-plane rotation AMR, contributions depending on the relative angle between the in-plane magnetization and current average out over the radial current lines. The in-plane AMR then depends purely on the angle between magnetization and crystallographic axes [14]. It reflects therefore the same underlying symmetry breaking crystal fields as the magnetic anisotropy. The measurements, shown in Fig. 2(d), unveil a cubic anisotropy along the [110]/[1-10] crystal axes and an additional uniaxial term breaking the symmetry between the [110] and [1-10] directions. Although the specific responses to these symmetries can be very different for the AMR and for the magnetic anisotropy, the presence of the cubic and uniaxial AMR terms and their sensitivity to the gate voltage observed in Fig. 2(d) suggest that the in-plane magnetization orientation itself can be switched at weak magnetic fields by the low voltage charge accumulation or depletion.

4. Persistent magnetization switching with short voltage pulses

A variable width of hysteretic magnetization loops measured at different constant gate voltages, shown in Fig.3(a), is the prerequisite for observing electrically assisted magnetization switchings. Note that electrical measurements of magnetization reorientations utilized in Figs. 3 and 4 are facilitated in our system by the IMR which responds to abrupt changes of the total magnetic induction upon a 180° reversal, and by a combined effect of the IMR and of the AMR for intermediate switchings by less than 180°. The amplitude of the AMR and the IMR contributions are similar in our experiments. The switchings by short low-voltage pulses are demonstrated in Fig. 3(b) and analyzed in detail in Fig. 4. The experiments were performed at constant field-sweep rate of 0.1 mT per second starting from negative saturation field of 1 T. The gate voltage was set to a base value of -1 V and then after each measurement step spanning 1 second we applied a 10 ms voltage pulse of a fixed magnitude and then returned to the base voltage. The technique allows us to demonstrate magnetic response to short electric pulses and the persistence of induced reorientations of the magnetization vector. It also removes potentially obscuring variations among the resistance traces in regions away from magnetization switchings which are caused by different slopes of the negative IMR at different gate voltages.

In Fig. 3(b) we compare measurement with no pulses (constant -1 V gate voltage) and data acquired at 0 to +4 V peak voltages. The field was swept along the [1-10] crystal direction (θ = 90°, where θ is the in-plane field angle measured from the [110] direction). As argued in detail in Fig. 4 and confirmed by SQUID magnetization measurements on an unpatterned part of the wafer, [1-10] is the main magnetic easy axis. The negative IMR then allows us to observe the drop in R corresponding to a 180° reversal from an antiparallel to a parallel configuration of field and magnetization and a corresponding increase of the magnetic induction. As the applied peak voltage increases, the magnetization reversals consistently shift to lower magnetic fields and the magnetization remains switched when the peak voltage pulse is turned off.

5. Theoretical discussion

To discuss the detail phenomenology of these persistent low-voltage induced magnetization switchings we present in Figs. 4(a) and (b) field-sweep measurements at fixed field angles spanning the whole interval in 5° steps. In panels (a) and (b) we show color-maps of the resistance as a function of the field magnitude and angle for -1 V constant voltage and for the +3 V peak-voltage measurements, respectively. The main effect observed in these plots is the overall suppression of the magnitude of the switching fields by depletion. Additionally, the relative suppression is stronger at $\theta = 0$ than at 90°, as highlighted in Fig. 4(c). This indicates that both the magnitude and ratio between the uniaxial and cubic anisotropy fields is modified by the gate voltage. To quantify the depletion induced modification of the magnetic anisotropy we extracted the anisotropy constants from fitting the measured θ = 0 and 90° switching fields to a single domain

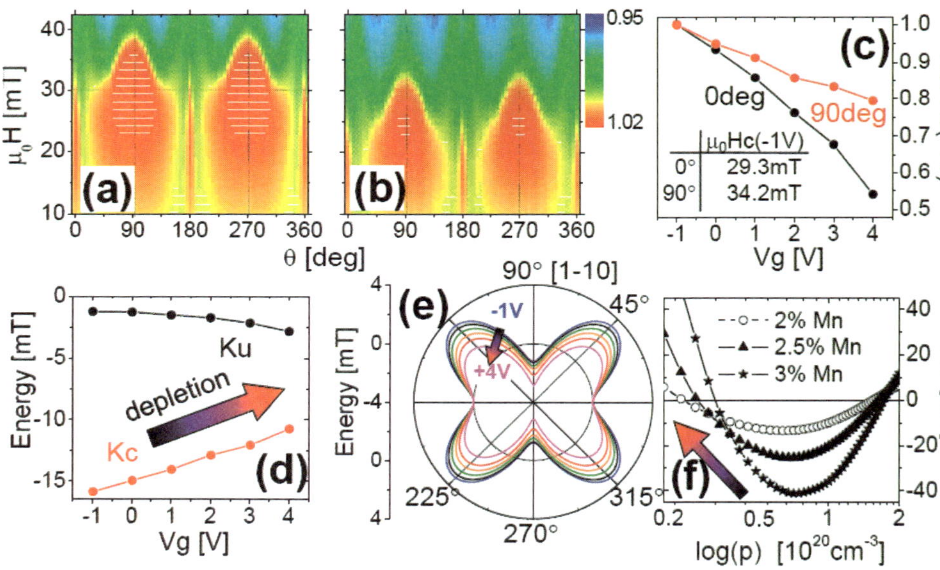

Fig. 4. (a) and (b) Color maps of channel resistance as a function of the inplane field angle and magnitude (normalized to H = 0 resistance) for -1V constant voltage and +3V peak voltage measurements, respectively. (c) Switching fields at field angles θ = 0 has a stronger suppression than at 90° as a function of the gate voltage. (d) Uniaxial and cubic anisotropy constants and (e) corresponding anisotropy energy profiles derived from the measured θ = 0 and 90° switching fields. (f) Microscopic calculations of the cubic anisotropy constant. Arrows in (d) and (f) highlight the common trend with depletion.

anisotropy energy model, $E(\theta,\phi) = K_u \sin^2\phi - K_c \sin^2 2\phi/4 - M H \cos(\theta - \phi)$, where H and M are the magnitudes of the external field and magnetization, respectively, and ϕ is the magnetization angle. The uniaxial constant K_u is relatively weak compared to the cubic constant K_c, as shown in Fig. 4(d). They both have a negative sign corresponding to the magnetic easy directions along the [1-10] and [110] axes and the most easy direction along [1-10]. As shown also apparent from Fig. 4(d), the dominant effect of depletion is in reducing the magnitude of K_c. Fig. 4(e) shows how the corresponding anisotropy energy profiles at $H = 0$ evolve with depletion.

We now discuss the key experimental observations by employing the $k{\cdot}p$ semiconductor theory approach combined with the mean-field kinetic-exchange model of hole mediated ferromagnetism in (Ga,Mn)As [1, 2]. Calculations for 2.5% local moment doping and hole density p ~ 1×10^{20} cm^{-3}, for which the simulations in Fig.1(b) predict hole depletions consistent with the measured variations of the channel resistance at temperatures near *Tc*, yield *Tc* ~ 20 K and $dTc/dp \approx 1\times10^{-19}$ Kcm^3. Both the absolute value of the Curie temperature and the few Kelvin suppression of *Tc* at a ~ 20% hole depletion predicted by the theory are consistent with our p-n junction simulations and the measured gate-dependent *Tc* values. The semiconductor theory modelling which includes strong spin-orbit coupling effects in the host semiconductor valence band captures also the sensitivity of magnetocrystalline anisotropies in (Ga,Mn)As to hole density variations. The cubic anisotropy is included by accounting in the $k{\cdot}p$ model for the zincblende crystal structure of GaAs. The additional weak uniaxial anisotropy is often present in (Ga,Mn)As epilayers but its microscopic origin is not known and we will therefore focus only on the stronger cubic anisotropy term. As shown in Fig. 4(f), the microscopically calculated K_c constant changes sign at hole density of approximately 1.5×10^{20} cm^{-3}.

Below this density it favors the [110]/[1-10] magnetization directions consistent with the experimental data. The typical magnitudes of Kc of ~ 10 mT are also consistent with experiment and considering the large gate action seen at low temperatures we can also associate, semiquantitatively, the decreasing magnitude of the experimental K_c at depletion with the behavior of the theoretical Kc at low hole densities.

To conclude we have reported low-voltage control of magnetic properties of a p-n junction FET via depletion effect in the ferromagnetic semiconductor channel. We have shown variable *Tc* and AMR, and demonstrated magnetization switchings induced by short electric field pulses of a few volts. Our concept of the spintronic transistor is distinct from previously demonstrated high-voltage metal-oxide-semiconductor ferromagnetic FETs [3, 5, 11] or electro-mechanically gated ferromagnets by piezo- stressors [15, 16 17, 18, 6, 7, 8]. It is realized in an all-semiconductor epitaxial structure and offers a principally much faster operation. In basic physics research, we expect broad utility of our results in studies of carrier-mediated ferromagnetism and in interdisciplinary fields combining ferromagnetism and spin-orbit coupling effects with localization and quantum-coherent transport phenomena [19] controlled by carrier depletion.

Acknowledgments

We acknowledge helpful discussions with R. P. Campion, M. Cukr, B. L. Gallagher, M. Marysko, J. Sinova, and J. Zemek, and from EU Grant IST-015728, FP7 Project 214499 NAMASTE, from Czech Republic Grants FON/06/E001, FON/06/E002, AV0Z1010052, KAN400100652, and LC510, and from U.S. Grant SWAN-NRI.

References

[1] F. Matsukura, et al., Ferromagnetic semiconductors. In K. H. J. Buschow, editor, *Handbook of Magnetic Materials*, volume 14, page 1. Elsevier, Amsterdam, 2002.

[2] T. Jungwirth, et al., Rev. Mod. Phys., 78:809, 2006.

[3] H. Ohno, et al., Nature, 408:944, 2000.

[4] D. Chiba, et al., Science, 301:943, 2003.

[5] D. Chiba, et al., Appl. Phys. Lett., 89:162505, 2006.

[6] A. W. Rushforth, et al., Phys. Rev. B78, 085314, 2008.

[7] M. Overby, et al., Appl. Phys. Lett., 92:192501, 2008.

[8] S. T. B. Goennenwein, et al., phys. stat. sol., (RRL) 2:96, 2008.

[9] J. Wunderlich, et al., Phys. Rev. Lett., 97:077201, 2006.

[10] T. Jungwirth, et al., Phys. Rev., B 76:125206, 2007.

[11] I. Stolichnov, et al., Nature Mater., 7:464, 2008.

[12] V. Nov´ak, et al., Phys. Rev. Lett., 101:077201, 2008.

[13] T. McGuire et al., IEEE Trans. Magn., 11:1018, 1975.

[14] A. W. Rushforth, et al., Phys. Rev. Lett., 99:147207, 2007.

[15] Sang-Koog Kim, et al., J. Magn. Magn. Mater., 267:127, 2003.

[16] Jeong-Won Lee, et al., Appl. Phys. Lett., 82:2458, 2003.

[17] Bernhard Botters, et al., Appl. Phys. Lett., 89:242505, 2006.

[18] H. Boukari, et al., J. Appl. Phys., 101:054903, 2007.

[19] D. Neumaier, et al., Phys. Rev. Lett., 99:116803, 2007.

MEASUREMENT OF NANOSECOND-SCALE SPIN-TRANSFER TORQUE MAGNETIZATION SWITCHING

K. ITO*, , K. MIURA, J. HAYAKAWA, and H. TAKAHASHI
ARL, Hitachi, Ltd., 1-280, Higashi-Koigakubo, Kokubunji-shi, Tokyo 185-8601, Japan
**E-mail: kenchi.ito.mb@hitachi.com*

T. DEVOLDER, P. CROZAT, J. V. KIM, C. CHAPPERT,
IEF, CNRS UMR 8622, Batiment 220, Universite Paris Sud, 91405 Orsay, France

S. IKEDA and H. OHNO
RIEC, Tohoku University, 2-1-1, Katahira, Aoba-ku, Sendai 980-8577, Japan

We study the single-shot response of magnetic tunnel junctions induced by spin transfer torque, using ultra-short pulse measurement. After the pulse is applied, the switching proceeded by ns-scale incubation delay, followed by 400 ps transition, terminated by a ringing with the frequency of 1.4 GHz. While, the incubation delay has large distribution, the other resistance traces are reproducible.

1. Introduction

Spin-transfer torque (STT) magnetization switching is one of the promising candidates for the new recording scheme to magnetic random access memory (MRAM). However, the entanglement between magnetization dependent transport properties and current dependent magnetization dynamics in STT experiments makes the deep understanding of the dynamics a difficult task. Stroboscopic techniques have already given us several understandings to follow the ensemble-averaged magnetization response.[1-4] However, any variance of STT switching was suppressed in the previous experiments because of their stroboscopic nature.

In this paper, we report the time-resolved measurements of ns-scale STT magnetization reversal, using high bandwidth resistance measurement.

2. Experimental

A magnetic tunnel junction (MTJ) pillar of CoFeB(2.5)/MgO(0.9)/CoFeB(6)/Ru(0.8)/CoFe(5)/MnIr(8), with thickness in nm, is fabricated in the center part of the high frequency coplanar circuit. The switched layer is the CoFeB(2.5) free layer. The MTJ is patterned in a 100 x 300nm rectangle, using EB lithography and ion milling technique. The tunnel magnetoresistance ratio of the MTJ is 50% with resistance-area products of 11 $\Omega\mu m^2$.

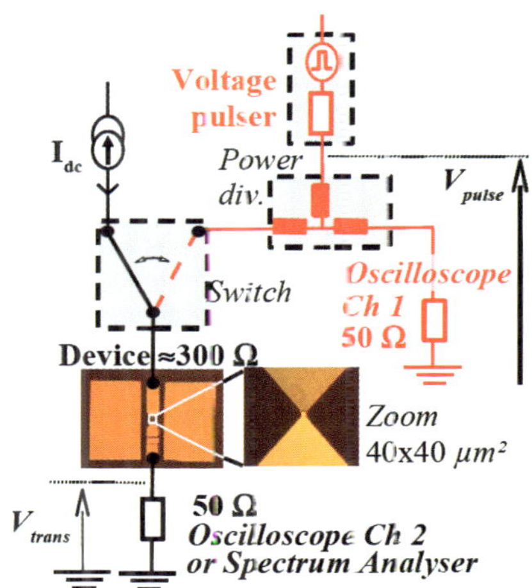

Figure 1 Experimental setup.

Figure 1 shows the schematic diagram of our experimental setup. A switch exchanges the low (static) and high frequency setup. The inset optical micrograph shows the high frequency coplanar waveguide and a zoom of the central part around MTJ. The measurement details are as follows: a magnetic field enough to saturate the free layer to the parallel (P) state is applied. A negative voltage step V_{pulse} is applied with a rise time 55 ps. A power divider routes half of the pulse to the MTJ. The transmitted voltage V_t is measured and stored with a single-shot 13 GHz oscilloscope. In this setup, the voltage drop across the MTJ is very close to V_{pulse}. V_t abruptly decreases at the switching instance [inset of Fig. 2]. No magnetic field was applied during the time-resolved measurement.

3. Results and Discussion

In Fig. 2, 20 responses V_t are displayed when the voltage step is -1 V. To obtain better evidence of V_t, we calculated resistance traces, assuming that when the switching fails, the device resistance stays constant in time.

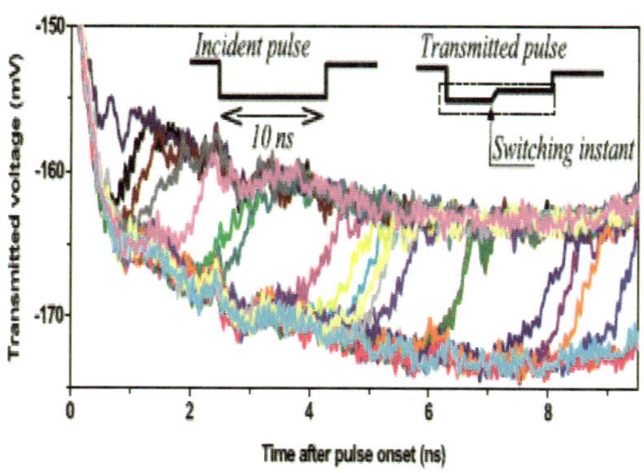

Fig. 2 The transmitted voltage pulse for -1 V incident voltage step.

The traces in Fig. 3 show the resistance records of single-shot events. The noisy trace of the bottom curve in Fig. 3 is due to the large bandwidth of 13 GHz. To reduce the noise, we have chosen a given switching instant and average the full bandwidth traces corresponding to events that have specific switching instant. There is an incubation delay before a switching event starts. It is followed by a rounded departure from the parallel resistance R_p to $R_p + 0.2\Delta R$, where ΔR is the resistamce change from the parallel state to anti-parallel state. Next, the transition from $R_p + 0.2\Delta R$ to $R_{ap} - 0.2\Delta R$ takes 400 ps. Finally, there is a pronounced ringing.

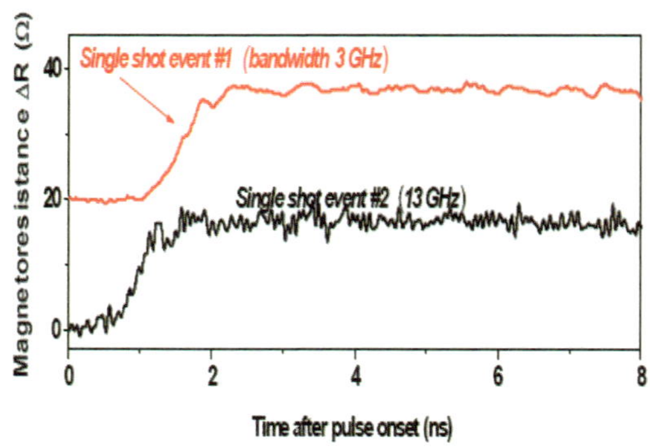

Fig. 3 Single-shot switching traces with band width of 3 GHz and 13 GHz

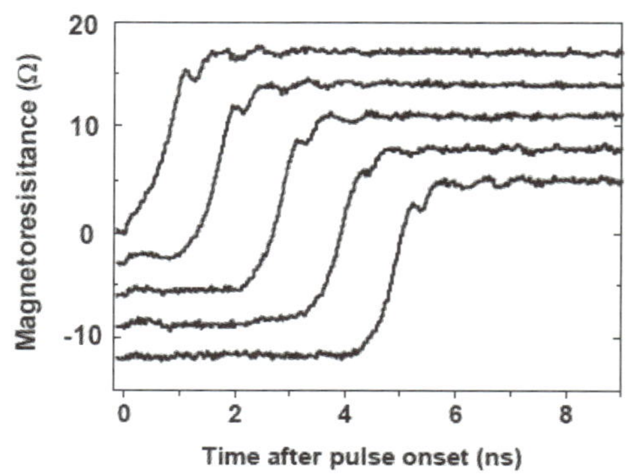

Fig. 4 Averaged single-shot switching traces with different incubation delay.

The curves in Fig. 4 show the averaged single-shot switching events measured with 13 GHz bandwidth, those having the same switching instant. From top to bottom, the switching instants are 0.7, 1.6, 2.8, 3.9 and 4.9 ns, respectively. The subsequent resistive signature of the switching, such as ringing amplitude, its frequency and damping rate, are completely reproducible. The only incubation delay is stochastic.

The absence of any electrical signature before switching indicates that the switching is not triggered by the pumping of the FMR mode with a macrospin-like reversal.[5] Also, the switching does not be driven by the amplification and beating of two mirror symmetric C-bending states induced by Oersted field,[6] because this scenario requires the resistance oscillation before switching, which is not observed in Fig. 3 and 4. It is also unlikely that spin transfer torque excites a spin wave mode with an odd nodal lines,[7] due to the same reason. The only possible scenario may be as follows: a tip of the nano-magnet may react first, until its magnetization is sufficiently tilted. Then, it propagates the disturbance to the nano-magnet center and trigger the fast reversal. Since the initial micromagnetic configurations are thermally incubation time fluctuates from pulse to pulse.

After switching, the time-resolved traces have an oscillation with a frequency of 1.4GHz, which indicates the motion is rather spatially coherent. At switching instance, the system can be viewed as undergoing a strong excitation. Among the oscillation eigenmodes, there is a quasi-uniform precession mode, whose frequency recalls that of ringing. In the anti-parallel state after switching, negative current provides an additional source of damping approximately equal to the natural dumping.[8] This expected decay rate is in pretty good agreement with the measured decay time of the ringing, indicating a high coherence of the sample behavior after ringing.

4. Conclusion

We have studied the ns-scale switching dynamics induced by spin transfer torque for rectangular MTJs. The switching starts after a random ns-scale incubation delay. Afterwards, the switching proceeds in a transition lasting 400 ps, and is always followed by a pronounced ringing. The frequency and decay rate of this ringing suggest a high spatial coherence after switching, recalling the macospin feature.

Acknowledgements

This work was supported in part by Research and Development for Next-Generation Information Technology by MEXT.

References

[1] S. I. Kiselev, J. C. Sankey, I. N. Krivorotov, N. C. Emley, R. J. Schoelkopf, R. A. Buhrman and D. C. Ralph, *Nature*, **425**, 380 (2003)

[2] N. C. Emley, I. N. Krivorotov, O. Ozatay, A. G. F. Garcia, J. C. Sankey, D. C. Ralph and R. A. Buhrman, *Phys. Rev. Lett.* **96**, 247202 (2006).

[3] Y. Acremann, J. P. Strachan, V. Chembrolu, S. D. Andrews, T. Tyliszczak, J. A. Katine, M. J. Carey, B. M. Clemens. H. C. Siegmann and J. Stohr, *Phys. Rev. Lett*, **96**, 217202 (2006).

[4] T. Devolder, C. Chappert, J. A. Katine, M. J. Carey and K. Ito, *Phys. Rev.* **B75**, 064402 (2007).

[5] J. Z. Sun, *Phys. Rev.* **B62**, 570 (2000).

[6] K. Ito, T. Devolder, C. Chappert, M. J. Carey and J. A. Katine, *J. Phys.* **D40**, 1261 (2007).

[7] R. D. McMichael and M. D. Stiles, *J. Appl. Phys.* **97**, 10J901 (2005).

[8] I. N. Krivorotov, N. C. Emley, J. C. Sankey, S. I. Kiselev, D. C. Ralph and R. A. Buhrman, *Science*, **307**, 228 (2005).

CURRENT-INDUCED DOMAIN WALL CREEP IN MAGNETIC WIRES

J. IEDA* and S. MAEKAWA

Institute for Materials Research, Tohoku University, Sendai 980-8577, Japan;
CREST, Japan Science and Technology Agency, Kawaguchi 332-0012, Japan
**E-mail: ieda@imr.tohoku.ac.jp*

S. E. BARNES

Physics Department, University of Miami, Coral Gables, FL 33124, USA.

Magnetic domain wall motion induced by magnetic fields and spin-polarized electrical currents is experimentally well established. A full understanding of the underlying mechanisms, however, remains elusive. We reported a detail comparison of such motions in the thermally activated subthreshold, or "creep," regime, where the wall velocity obeys an Arrhenius law. Experiment shows the velocity scales in an evidently different way with a driving current and field, proving the nonequivalence of two drives. Scaling theory explains the important features of experiment.

Keywords: Magnetic domain wall; Creep motion; Spin transfer; Magnetic semiconductor

1. Introduction

A magnetic domain wall (DW) interacts with electrical currents[1] as well as magnetic fields, showing the control of the magnetization direction. In particular, for both technological and fundamental reasons, the effect of electrical current in such non-uniform spin texture is attracting much attention[2,3]. This physics is reflected not only in possible spintronics devices[4,5] but also in new concepts such as spin-motive forces[6].

Systematic experimental data on the dependence of the DW velocity spanning five decades in current density have been obtained in a microstructure made from a ferromagnetic semiconductor (Ga,Mn)As[7]. This data provides a deep insight into the physics of DW dynamics. The quantitative analyses showed that the current-driven motion in (Ga,Mn)As can be explained by spin-transfer mechanism for currents beyond a threshold value. The linear velocity for a given current tempts ones to expect that there is always an equivalent magnetic field which has the same effect upon DWs.

In order to clarify whether a "field-like" effect due to currents on DW motion exists or not, we made a detailed comparison between these two sources of drive, but in the subthreshold, "creep," regime, where the velocity obeys an Arrhenius law with a scaling function of the drives[7]. The observed scaling exponent for the two drives was incompatibly different from each other, i.e., the effect of a driving current and field are not equivalent in the (Ga,Mn)As. In this report we review theory which explains the important features of experiment.

2. Formulation

The effect of pinning by randomly distributed weak point defects can be well described by the collective pinning model[8]. In this framework the response of the disordered elastic system to an external force f is studied. At zero temperature $T = 0$, the systems exhibit a deppining transition at a threshold driving force f_c. In the case of $T > 0$, the creep motion is possible for the subthreshold regime $f < f_c$. For a drive well below the threshold value, the motion is extremely slow and the dynamical problem can be addressed by considering thermal activation over barriers which would arise in a static treatment. Scaling arguments based on the static properties illustrate that upon decreasing f to zero, the barrier height U_b increases to infinity in the form $U_b \sim U_0(f_c/f)^\mu$ where $\mu \leq 1$. This implies a velocity v versus force characteristic $v \sim \exp[-U_0(f_c/f)^\mu/k_BT]$ where k_B is the Boltzmann constant. The inverse dependence of U_b on f manifests the glassy behavior, a clear distinction from the case of a constant barrier height for $f \ll f_c$, as might arise, when instead of the collective week pinning, the system is strongly pinned at a few points (Fig.1a). Then, $U_b \sim U_0[1-(f/f_c)]^\alpha$, and, as $f \to 0$, the velocity remains finite.

The creep exponent μ measures strength of the nonlinear response to an external force and depends on the disorder. Hence, measurements of creep motion in the same sample, i.e., the identical impurity distribution with different driving forces provide us with fundamental information about these drives.

2.1. *Field induced creep*

We consider a single DW lying in a ferromagnetic wire (the thickness d, width W) containing random disorder. We choose the coordinate system as y to be along the wire (the hard axis) while z to be normal to the film (the easy axis). When the very small field H, less than the threshold value H_c, is applied along the z axis, due to the localized pinning site that collectively prevents a uniform DW displacement, only a segment of the DW with length ℓ can be displaced by the same amount for x direction. The total free energy related to this displacement is estimated as

$$F(\ell) = e_0\ell - e_1\ell^2, \qquad (1)$$

where $e_0 = 4(A_sK_u)^{1/2}d$, with the spin exchange stiffness A_s, and the uniaxial anisotropy energy density K_u, denotes the energy cost due to the partial wall extension of a unit length and $e_1 = MHd$ represents the areal Zeeman energy gain of the magnetization M, relative to the original (flat) configuration. As usual for creep problems, we assume the characteristic energy scale is *unique* and in consequence the barrier height U_b scales in the same manner as the total free energy. Maximizing $F(\ell)$ with respect to a given field H leads to an optimal length $\ell_0 = e_0/2e_1 \propto H^{-1}$ so that the DW spends the largest time in such a configuration. An Arrhenius law: $v \sim \exp(-U_b/k_BT)$ implies that the thermally activated system against the barrier $U_b \sim F(\ell_0)$ obeys the formula: $\ln(v/v_c) \sim a - bH^{-\mu}$, with $\mu = 1$.

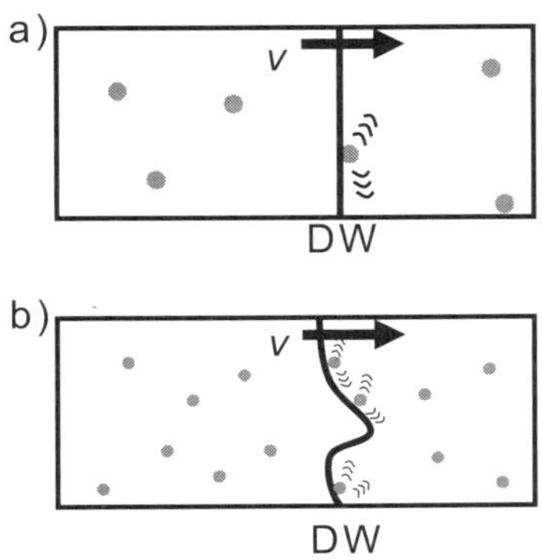

Fig. 1. Domain wall (DW) motion subject to disorder (dots) residing in ferromagnetic thin films, a) in a rigid wall limit, and b) in a "glassy" state.

2.2. *Current induced creep*

The creep formula for the current j can be derived by following the same steps on the above treatment. However, a difference arises in the manner the characteristic length scale is identified because there is no contribution to the free energy reflecting the applied current. Due to the disorder potential V_d, the angle ϕ has to advance by π in order for the wall to move forward. In this case, creep occurs through phase slips, i.e., the angle ("phase" in analogy with a Josephson junction[9]) ϕ advances by π within a finite segment of the DW as depicted in Fig.2b).

In addition to V_d, we have to consider a second barrier $\propto K\sin^2\phi$ with $K \simeq M^2/2\mu_0$ the hard axis anisotropy. Let us start with the free energy density without disorder,

$$\mathcal{F}[\theta,\phi] = -\frac{1}{2}K_\mathrm{u}\cos^2\theta + \frac{1}{2}K\sin^2\theta\sin^2\phi + \frac{1}{2}A_\mathrm{s}[(\nabla\theta)^2 + \sin^2\theta(\nabla\phi)^2], \quad (2)$$

where two Euler angles θ and ϕ are introduced to specify the direction of magnetization as usual. Substituting the spin transfer term[2] into the Landau-Lifshitz equation leads to $\partial_t\vec{M} + u\,\partial_y\vec{M} = -\gamma\vec{M}\times\vec{H}_\mathrm{eff} + \mathcal{R}$, where $\partial_t = \partial/\partial t$, $u = (g\mu_\mathrm{B}Pj)/(2eM)$ with μ_B the Bohr magneton and P the spin polarization, $\gamma = g\mu_\mathrm{B}/\hbar$, $\vec{H}_\mathrm{eff} = -\delta\mathcal{F}/\delta\vec{M}$, and $\mathcal{R}$ denotes damping. Using Eq. (2), we obtain

$$D_t\cos\theta = \frac{\gamma}{M}\Big[A_\mathrm{s}\vec{\nabla}\cdot\sin^2\theta(\vec{\nabla}\phi) - K\sin^2\theta\sin\phi\cos\phi\Big], \quad (3)$$

$$\sin\theta D_t\phi = \frac{\gamma}{M}\Big[A_\mathrm{s}\nabla^2\theta - K_\mathrm{u}\cos\theta\sin\theta - K\cos\theta\sin\theta\sin^2\phi\Big]. \quad (4)$$

where $D_t \equiv \partial_t + u\,\partial_y$. We have discarded damping terms because this does not make any significant changes to our arguments below. Under equilibrium condition $\partial_t = 0$, $u = 0$ and a suitable boundary condition it follows from Eqs. (3), (4) that $\theta = \theta_0 \equiv 2\tan^{-1}\exp(y/\Delta)$, where $\Delta = (A_\mathrm{s}/K_\mathrm{u})^{1/2}$ is the DW width parameter. Introducing the wall center coordinate $q(t)$, by replacing $y \to y - q(t)$ in θ_0, and substituting this into Eq. (3) leads to

$$\dot{q}(t) = u + \frac{\gamma K\Delta}{2M}\sin 2\phi - \frac{\gamma A_\mathrm{s}\Delta}{M}\frac{\partial^2\phi}{\partial x^2}, \quad (5)$$

where $\dot{q}(t) = dq(t)/dt$. The last term derives from the spatial dependence of ϕ along the x axis as shown in Fig.2b, which plays a crucial role. Now we introduce the disorder potential V_d which, as a function of position, exerts a force on the wall: $\dot{\phi}(t) = -(\gamma V_\mathrm{d}')/(2MWd)$ where $V_\mathrm{d}' = \partial V_\mathrm{d}/\partial y$. The coupled equations for q and ϕ give $\ddot{\phi} = -(\Omega^2/2)\sin 2\phi - f_j + f_A$, with $\Omega^2 = C\langle V_\mathrm{d}''\rangle/m_\mathrm{D}$ where $\langle\cdot\rangle$ represents the disorder average, m_D denotes the Döring mass[3], and C is a numerical factor. There are two inhomogeneous terms due to current: $f_j \propto j$, and to the phase slip: $f_A \propto A_\mathrm{s}\partial^2\phi/\partial x^2$, i.e., the local winding of ϕ by π along the x axis. To realize a quasistatic situation of creep motion, the inhomogeneous terms in the equation of motion must vanish. The scaling of a characteristic length for the phase slip is thus determined by the requirement, $f_j \sim f_A$. Replacing ∂_x^2 by ℓ_0^{-2} within this relation we obtain $\ell_0 \propto j^{-1/2}$. The barrier for the phase slip is estimated as $U_\mathrm{b} \sim K \times \ell_0\Delta d$. The Arrhenius law dictates for $\ln(v/v_\mathrm{c}) = a - bj^{-\mu}$ with $\mu = 1/2$.

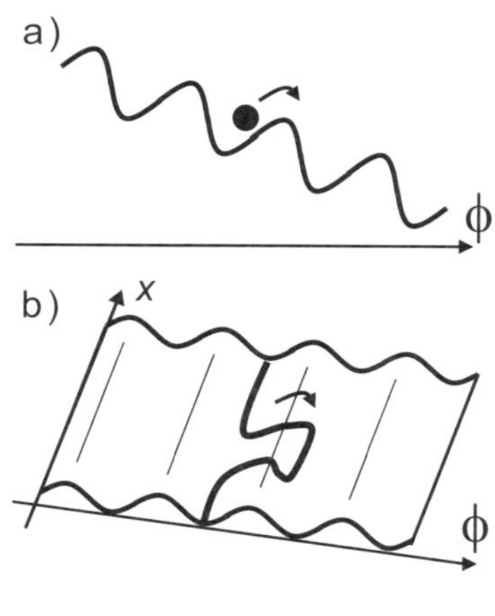

Fig. 2. a, b) Potential landscape of the DW's chirality, ϕ, and its fundamental excitation corresponding to the case of Fig. 1a), and b), respectively.

3. Discussion

The creep exponent measured in the field induced DW motion $\mu^\mathrm{exp} = 1.2 \pm 0.1$ is close to that expected from the scaling theory $\mu = 1$. In the above scaling arguments we have intentionally assumed in Eq. (1) the dimensions of the wall displacement have an equal parallel and perpendicular scale in the DW plane. This occurs when the disordered system belongs to the random field universality class[10]. It may be the case for the (Ga,Mn)As because the impurity potential

Table 1. Creep exponents for different drives in a semiconductor and a metal.

	semiconductor	metal
field-dirve	$\mu = 1$ random field	$\mu = 1/4$ random bond
current-drive	$\mu = 1/2$ new class	-

mainly comes from the internal field modulation due to the randomly distributed Mn moments. A different exponent $\mu = 1/4$ corresponding to the random bond university class was discussed for Co films where the surface roughness is the main source of disorder[10].

The creep exponent measured in the current driven case $\mu^{\text{exp}} = 0.33 \pm 0.06$ is slightly less than that expected from the scaling theory $\mu = 0.5$. In view of the simplicity of the scaling arguments developed above, agreement of the theory with experiment is fairly good. Needless to say, since the current induced wall creep corresponds to a completely new class of driven disordered systems, further theoretical investigation is required in order to obtain deeper understanding of experiment. If one imposes the rigid wall approximation $\partial_x \phi = 0$ the system reduces to the so-called tilted washboard model[11,12]. However, as mentioned above, the structural deformation of the DW is essential for the inverse dependence of the barrier on the driving force and hence for the correct description of the creep motion observed in experiment.

4. Summary

We observed the scaling behavior for the field driven case can be interpreted in the framework of creep in a random field while the case of current induced creep does not fit to any known such model. To account for the current-induced creep, we studied the DW's chirality which couples to DW motion. Its response to the current dramatically changes depending on whether the elasticity is infinite or not. The finite elasticity is a key to understand the observed scaling relation.

A proper understanding of creep clarifies a number of issues to do with spin-torque-transfer in magnetic semiconductors and metals. Particularly exciting in the understanding of creep itself, are the fact that there are two distinct drives possible and that the current driven case seems to correspond to a new class of creep motion. This merits further detailed research.

Acknowledgments

We acknowledge M. Yamanouchi, F. Matsukura, and H. Ohno for valuable discussions. This work was supported by MEXT, Grant-in-Aid for Scientific Research on Priority Areas and Grant-in-Aid for Young Scientists (B).

References

1. S. E. Barnes and S. Maekawa, *Phys. Rev. Lett.* **95**, 107204 (2005).
2. *Concepts in Spin Electronics*, ed. S. Maekawa, (Oxford University Press, Oxford, 2006).
3. G. Tatara, H. Kohno, and J. Shibata, *J. Phys. Soc. Jpn.* **77**, 031003 (2008).
4. S. E. Barnes, J. Ieda, and S. Maekawa, *Appl. Phys. Lett.* **89**, 122507 (2006).
5. S. S. P. Parkin, M. Hayashi, and L. Thomas, *Science* **320**, 190 (2008).
6. S. E. Barnes and S. Maekawa, *Phys. Rev. Lett.* **98**, 246601 (2007).
7. M. Yamanouchi, J. Ieda, F. Matsukura, S. E. Barnes, S. Maekawa, and H. Ohno, *Science* **317**, 1726 (2007).
8. A. I. Larkin and Yu. V. Ovchinnikov, *J. Low Temp. Phys.* **34**, 409 (1979).
9. M. Tinkham, *Introduction to Superconductivity*, 2nd ed., (Dover Publications Inc., New York, 1996).
10. P. Chauve, T. Giamarchi, and P. Le Doussal, *Phys. Rev.* **B62**, 6241 (2000).
11. G. Tatara, N. Vernier, and J. Ferré, *Appl. Phys. Lett.* **86**, 252509 (2005).
12. R. A. Duine, A. S. Núñez, and A. H. MacDonald, *Phys. Rev. Lett.* , (2006).

PURE SPIN CURRENT INJECTION INTO SUPERCONDUCTING NIOBIUM WIRE

K. OHNISHI[1], T. KIMURA[1,2] and Y. OTANI[1,2]

[1]*Institute for Solid State Physics, University of Tokyo, 5-1-5 Kashiwanoha, Kashiwa, Chiba 277-8581, Japan*

[2]*ASI, RIKEN, 2-1 Hirosawa, Wako, Saitama 351-0198, Japan*

We have investigated the spin transport in superconducting Nb wire by using a lateral multi-terminal superconductive/normal/ferromagnetic metal junction. The spin polarized current exhibits a longer Cooper-pair recombination time than the unpolarized one because of the spin-flipping. A pure spin current injection is also found to induce nonlocally the quasi-particle and supercurrents in the superconductor.

Keywords: spintronics; superconductor; quasi-particle

1. Introduction

In past few years, spin-dependent transport in nano-structured systems has attracted considerable attention owing to potential applications in spintronics.[1,2] Spin current, a flow of spin-angular momentum, plays an important role in the operation of the spintronic device. Therefore, manipulating spin currents is an important issue for developing the future spintronics. So far, spin currents in normal- or semi-conductors have been investigated intensively.[1,2,3] However, spin transports in superconductors have not been studied so much. Since superconductor itself has great potentialities in applications such as Josephson circuit and quantum computing[4], the study on spintronic devices combined with superconductors may open the way for newly emerging technology in spintronics. Indeed, several functional superconductive spintronic devices have been proposed theoretically.[5,6]

In the superconductor, up- and down-spin electrons are bound together to form a Cooper pair. One thus naively expects that the spin current is not allowed to flow in it. However, the spin current can enter into S as quasi-particles. Quasi-particles in S show nontrivial properties such as charge imbalance and non-linear voltage drop.[7,8] Therefore, the spin current in superconductors may exhibit novel spin dependent transport as a result of interaction between the spin current and superconductivity. Moreover, such phenomena may be useful for demonstrating new functionalities in spintronic devices. Here, we study the properties of the spin current in a superconductor by means of electrical spin injection technique.

2. Experimental

We employed a multi-terminal lateral structure to perform the detailed study of

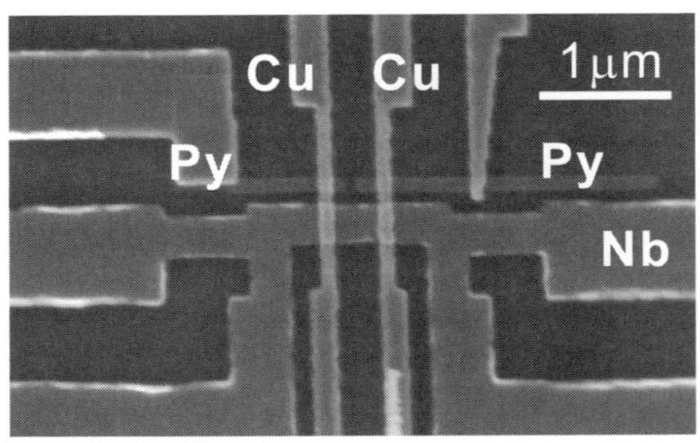

Fig. 1. SEM image of a measured multi-terminal superconductive/normal/ferromagnetic metal device. Ni-Fe (Py:Permalloy), Cu, Nb are a ferromagnetic metal, a normal metal, and a superconductor, respectively.

spin transport properties because of variable probe configurations in the same device. A measured lateral superconductive(S)/normal(N)/ferromagnetic(F) metal hybrid structure consists of Nb, Cu and Ni-Fe (Py:Permalloy) wires as shown in Fig. 1, where the Py wire is connected to the Nb wire via Cu to prevent ferromagnetic Py to directly interfere with superconductive Nb. First, Py nanowires 100 nm in width and 30 nm in thickness were fabricated by the electron-beam evaporation with a conventional lift-off process. Then, a Nb wire 300 nm in width and 40 nm in thickness was fabricated by a specially developed lift-off process with a magnetron sputtering.[9] Finally, Cu wires 100 nm in width and 100 nm in thickness were deposited by Joule evaporator. The interfaces of Nb/Cu and Py/Cu were well cleaned by low-voltage Ar ion milling prior to the Cu evaporation. Here, the center-center distance between the Py and Nb wires was 300 nm, shorter than the spin diffusion length of the Cu wire.[10] The efficiency of the spin current injecting into the Nb wire was evaluated from the nonlocal spin valve measurements. We found that the transition temperature T_c of Nb is 2.8 K.

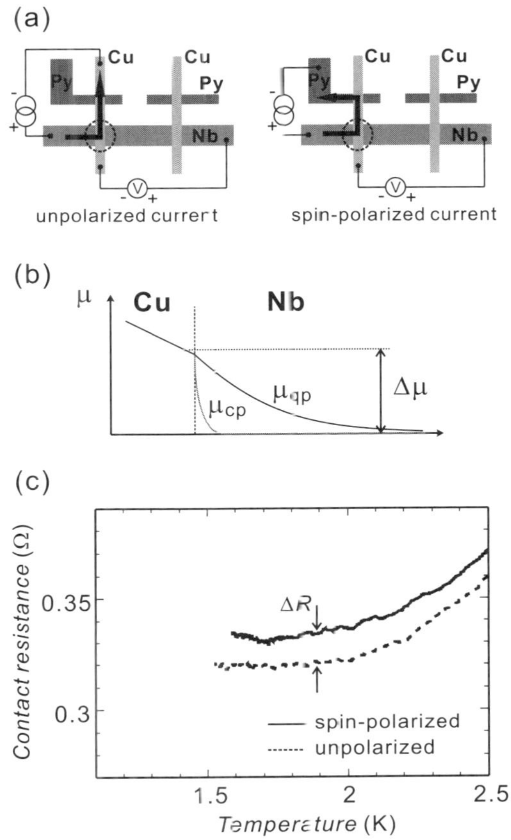

Fig. 2. (a) Measurement configurations of contact resistances with unpolarized and spin-polarized currents. The dashed circles indicate the measured contacts. (b) Electrochemical potentials at the Cu/Nb interface. $\mu_{\rm cp}$ and $\mu_{\rm qp}$ are electrochemical potentials of Cooper pairs and quasi-particles, respectively. The measured contact resistances are proportional to $\Delta\mu$. (c) Temperature dependence of contact resistances. Solid line and broken line are measured under spin-polarized and unpolarized current injection, respectively.

3. Results & Discussions

In the S/N interface, the electrons from N are mainly injected into S as quasi-particles especially at $T \approx T_c$.[7] The injected quasi-particles are converted to the Cooper pairs while traveling in S. Since a Cooper pair is formed by the combination of up- and down-spin electrons, the excess recombination time of quasi-particles is expected under spin-polarized current injection because of different population between up- and down-spins. In the device shown in Fig. 1, we can switch the spin polarization of the injecting current into the Nb wire by choosing either the Py or Cu current probes. Since the quasi-particle current induces the voltage drop in the Nb wire, the recombination time can be detected by the contact resistance of the Cu/Nb interface[8] as schematically shown in Fig. 2(b).

Figure 2(c) shows the temperature dependence of contact resistance under spin-polarized current injection together with that under unpolarized current injection. Contact resistances were measured using the probe configuration shown in Fig. 2(a) with the excitation current of 1 μA. The contact resistance with spin injection was larger than that without spin injection. This is consistent with our prediction that the spin-polarized quasi-particle current shows a longer recombination time than unpolarized one.

The resistance difference between the above two currents ΔR is 14 mΩ below 2 K. From this value, we can roughly estimate the spin-flip length λ_S of the quasi-particle as follows: In the spin-polarized current, the currents with up and down spins are given by $\frac{(1+P)}{2}I$ and $\frac{(1-P)}{2}I$, respectively. Here, P and I are the spin-polarization at the Nb/Cu interface and the total charge current. In this case, $(1-P)I$ can form the Cooper pair with the same process as the unpolarized one. However, the excess current of PI with up spins still remains. Assuming that the excess resistance is caused by the excess current flow during the spin-flipping, we can roughly calculate the resistance as $PI\rho\lambda_S/S$. Here, ρ and S are the resistivity of Nb and the cross section for quasi-particles current. Since P, ρ and S were 0.02, 64 $\mu\Omega$cm and 1.2×10^{-14} m^2, we obtain λ_S as 15 nm.

The value is 3 times longer than the spin diffusion length for Nb at 10 K.[11] The reason is not clear at present. Since the contact resistance includes the resistance due to reflection at the interface, the difference in reflection between two currents may contribute to the resistance difference.[12] Moreover, a local current flowing in the S/N interface induces spurious effects such as self heating and heat transport.[13] This may induce the additional resistance difference.

To investigate the influence of a spin current in superconductor more precisely, we measured the resistance of Nb wire under pure spin injection. When the charge current flows as in the inset of Fig. 3(a), a nonlocal spin current is induced in the bottom Cu strip. The induced spin current is absorbed as quasi-particles into the Nb wire without any charge flow. In this way, a pure spin current can be injected into the superconductive Nb wire.

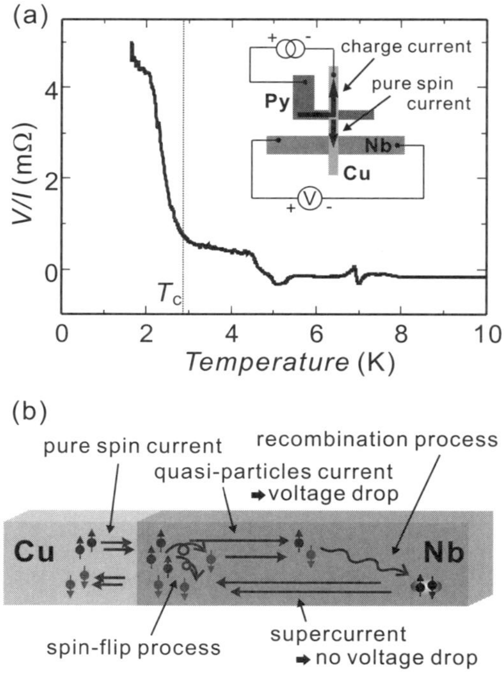

Fig. 3. (a) Temperature dependence of nonlocal resistance under the pure spin current injection together with the probe configuration. (b) Schematic processes and currents in the superconducting Nb under pure spin current injection.

Figure 3(a) shows the temperature dependence of nonlocal resistance in the Nb wire under pure spin current injection. A finite resistance appeared when the Nb wire becomes superconductive. Moreover, the resistance increased with decreasing the temperature. The origin of the resistance is due to the voltage drop of quasi-particles current along the Nb wire. Since the Nb wire is under a open circuit condition, the total current flowing in the Nb wire should be zero. However, an inversely flowing supercurrent satisfies this condition, as schematically shown in Fig. 3(b). In the ideal case, the induced current is symmetrically divided to the left and right part of Nb wire. This leads to no voltage drop along the Nb wire because of the cancellation of the voltage. However, in realistic case, there is a slight difference between the left and right part of the resistance. The observed voltage drop can be understood by the asymmetric diffusion of the quasi-particles.

4. Summary

We investigate spin relaxation in the superconductor by using a lateral multi-terminal S/N/F device. We found that the contact resistance under spin-polarized current was larger than that under unpolarized one because of the excess recombination time with the spin-flipping. A pure spin current injection is found to induce the quasi-particles and supercurrent in the superconductor nonlocally.

References

1. I. Žutić, J. Fabian and S. Das Sarma, *Rev. Mod. Phys.* **76**, 323 (2004)
2. S. A. Wolf, D. D. Awschalom, R. A. Buhrman, J. M. Daughton, S. von Molnár, M. L. Roukes, A. Y. Chtchelkanova and D. M. Treger, *Science* **294**, 1488 (2001)
3. S. A. Crooker, M. Furis, X. Lou, C. Adelmann, D. L. Smith, C. J. Palmstrom and P. A. Crowell, *Science* **309**, 2191 (2005)
4. Y. Nakamura, Yu. A. Pashkin and J. S. Tsai *Nature* **398**, 786 (1999)
5. T. Yamashita, K. Tanikawa, S. Talahashi and S. Maekawa, *Phys. Rev. Lett.* **95**, 097001 (2005)
6. F. Giazotto and F. Taddei, *Phys. Rev. B* **77**, 132501 (2008)
7. J. Clarke, *Phys. Rev. Lett.* **28**, 1363 (1972)
8. P. Cadden-Zimansky and V. Chandrasekhar, *Phys. Rev. Lett.* **97**, 237003 (2006)
9. K. Ohnishi, T. Kimura and Y Otani, *Appl. Phys. Express* **1** 021701 (2008)
10. T. Kimura, J. Hamrle and Y. Otani, *Phys. Rev. B* **72**, 014461 (2005)
11. K. Ohnishi *et al.*, unpublished
12. G. E. Blonder M. Tinkham and T. M. Klapwijk, *Phys. Rev. B* **25**, 4515 (1982)
13. K. Yu. Arutyunov, D. A. Presnov, S. V. Lotkhov, A. B. Pavolotski and L. Rinderer, *Phys. Rev. B* **59**, 6487 (1999)

SWITCHING OF A SINGLE ATOMIC SPIN INDUCED BY SPIN INJECTION: A MODEL CALCULATION

S. KOKADO
Faculty of Engineering, Shizuoka University, Hamamatsu 432-8561, Japan
**E-mail: tskokad@ipc.shizuoka.ac.jp*

K. HARIGAYA
Nanotechnology Research Institute, AIST, Tsukuba 305-8568, Japan

A. SAKUMA
Graduate School of Engineering, Tohoku University, Sendai 980-8579, Japan

Toward the manipulation of a single atomic spin, we theoretically study the switching of a localized quantum spin $\boldsymbol{S}=(S_x, S_y, S_z)$ induced by spin injection in electrode/$\boldsymbol{S}$/electrode junctions. This $\boldsymbol{S}$ has a uniaxial anisotropy energy, $-|D|S_z^2$, which shows the bistable potential between $S_z=-S$ and S, with D being an anisotropy constant. Furthermore, $\boldsymbol{S}$ interacts with the atomic vibration. For the initial state of $S_z=-S$, we consider a situation in which up-spin electrons exhibit the spin-flip tunneling from the left electrode to the right one through an exchange interaction between the electron spin and $\boldsymbol{S}$. Using the master equation approach, we investigate the time t dependence of the current I, the expectation value of S_z, $\langle S_z \rangle$, and that of the vibration quantum number, $\langle n \rangle$, of an $S=2$ system, which corresponds to an Fe atom on CuN surface. The systems exhibit the switching or nonswitching depending on the transition probability due to spin-atomic vibration interaction within a period of 10 ns. In addition, the t dependence of I and $\langle n \rangle$ is explained on the basis of $\langle S_z \rangle$ and the probability distribution.

Keywords: atomic memory; single atomic spin; spin switching.

1. Introduction

The single-atom memory is expected to be an ultimate microscopic element of the data storage device. Toward development of such a memory, physical properties of atomic scale materials have been extensively studied.[1,2,3] Recently, it has been shown from a scanning tunneling microscope experiment and its theoretical analysis that a single atomic spin ($S=2$) of an Fe atom on CuN surface has a uniaxial anisotropy energy, $-|D|S_z^2$, which shows a bistable potential between $S_z=-S$ and S states, with S_z being z component of the spin.[1] Here, D represents an anisotropy constant, and it is evaluated to be $|D|=1.55 \pm 0.01$ meV.[1]

Regarding the single atomic spin, we consider the manipulation of spin states. As a method, we bear in mind a theoretical study on "a localized spin reversal by the sequential injection of spins in a spin quantum dot", in which the transition from $S_z=-S$ to S with $1 \leq S < \infty$ is induced by the spin injection.[2] In this theory, however, the atomic vibration and the spin relaxation due to the vibration have not been taken into account. The vibration may give strong influence to the spin switching of single atomic spin systems, because the vibration energy is usually in the range of 0.4 meV - 40 meV (100 GHz - 10 THz) which is comparable to energy-level spacing of spin systems, and hence the spin relaxation may be enhanced.

In this paper, we study the switching of a spin $\boldsymbol{S}=(S_x, S_y, S_z)$ induced by the spin injection. This $\boldsymbol{S}$ has the uniaxial anisotropy energy, and it is coupled to the atomic vibration. We investigate the time dependence of the current, the expectation values of S_z, and that of the vibration quantum number of S=2 systems. The switching properties strongly depend on the transition probability due to spin-atomic vibration interaction.

2. Model and method

In Fig. 1, we show a model consisting of electrode/spin quantum dot/electrode junctions, where the dot has the localized quantum spin $\boldsymbol{S} = (S_x, S_y, S_z)$ with $S \geq 1$. The spin $\boldsymbol{S}$ has $-|D|S_z^2$ showing the bistable potential, and also $\boldsymbol{S}$ is coupled to the atomic vibration. Regarding conduction electrons, we choose the most efficient way to give rise to the spin switching; that is, the left electrode (L) has conduction electrons with up-spin, while the right electrode (R) has those with down-spin. When the voltage $-V$ is applied to the L, electrons are injected from the L into the dot and they move into the R via the spin-flip tunneling.

The Hamiltonian is given explicitly by,

$$\hat{\mathcal{H}} = \hat{\mathcal{H}}_{\rm e} + \hat{\mathcal{H}}_S + \hat{\mathcal{H}}_{\rm v} + \hat{V}_{\sigma-S} + \hat{V}_{S-{\rm v}}, \quad (1)$$

with

$$\hat{\mathcal{H}}_{\rm e} = \sum_{\boldsymbol{k}_L} (\epsilon_{\boldsymbol{k}_L,\uparrow} + eV) c^\dagger_{\boldsymbol{k}_L,\uparrow} c_{\boldsymbol{k}_L,\uparrow} + \sum_{\boldsymbol{k}_R} \epsilon_{\boldsymbol{k}_R,\downarrow} c^\dagger_{\boldsymbol{k}_R,\downarrow} c_{\boldsymbol{k}_R,\downarrow}, \quad (2)$$

$$\hat{\mathcal{H}}_S = -|D|S_z^2, \quad (3)$$

$$\hat{\mathcal{H}}_{\rm v} = \sum_i \hbar\omega_i a_i^\dagger a_i, \quad (4)$$

$$\hat{V}_{\sigma-S} = T_J \sum_{\boldsymbol{k}_L,\boldsymbol{k}_R} \left(c^\dagger_{\boldsymbol{k}_L,\uparrow} c_{\boldsymbol{k}_R,\downarrow} S_+ + c^\dagger_{\boldsymbol{k}_R,\downarrow} c_{\boldsymbol{k}_L,\uparrow} S_- \right), \quad (5)$$

$$\hat{V}_{S-{\rm v}} = \sum_i \Lambda_i (S_x S_z + S_z S_x + S_y S_z + S_z S_y)(a_i + a_i^\dagger), \quad (6)$$

where $\hat{\mathcal{H}}_{\rm e}$, $\hat{\mathcal{H}}_S$, and $\hat{\mathcal{H}}_{\rm v}$ are Hamiltonians for the electron, $\boldsymbol{S}$, and the atomic vibration, respectively. Furthermore, $\hat{V}_{\sigma-S}$ is the tunnel Hamiltonian with the exchange interaction between the conduction electron spin and $\boldsymbol{S}$, and T_J is the coefficient for the tunneling, while $\hat{V}_{S-{\rm v}}$ is the $\boldsymbol{S}$-vibration interaction.[3,4] The operator $c_{\boldsymbol{k}_L,\uparrow}$ ($c^\dagger_{\boldsymbol{k}_R,\downarrow}$) denotes the annihilation (creation) operator of an electron with the wave vector $\boldsymbol{k}$ of the L (R), and $\epsilon_{\boldsymbol{k}_L,\uparrow}$ ($\epsilon_{\boldsymbol{k}_R,\downarrow}$) represents energy levels with $\boldsymbol{k}$ of the L (R), where e is the magnitude of electric charge. The operator a_i ($a_i^\dagger$) is the annihilation (creation) operator of the vibration of the mode i ($i \leq 3$), and $\hbar\omega_i$ is the vibration energy of the mode i, where $i \leq 3$ means that the vibration has three degree-of-freedom at a maximum. The coefficient Λ_i is a coupling constant between $\boldsymbol{S}$ and the atomic vibration of the mode i.[3]

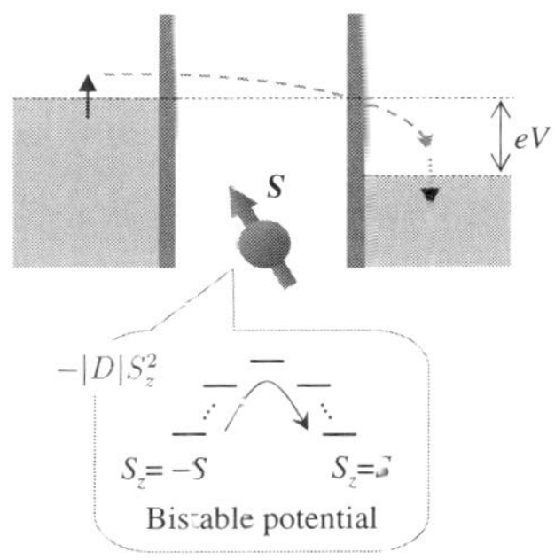

Fig. 1. Schematic illustration of electrode/spin quantum dot/electrode junctions. The dot has a localized quantum spin $\boldsymbol{S}$. This $\boldsymbol{S}$ has a uniaxial anisotropy energy, $-D|S_z^2$, and it is connected to the left electrode (L) and the right one (R). The L has conduction electrons with up-spin, while the R has conduction electrons with down-spin. Furthermore, $\boldsymbol{S}$ is coupled to the atomic vibration. An arrow drawn in $-|D|S_z^2$ represents the transition from S_z=$-S$ to S.

The state of the spin quantum dot is represented by $|S_z, \{n_j\}\rangle = |S_z\rangle|\{n_j\}\rangle$, with $S_z = -S, -S+1, \cdots, S-1, S$, and $\{n_j\} = (n_1, n_2, n_3)$, where n_j is the eigenvalue of the number operator of the vibration of the mode j, i.e., the vibration quantum number.

The probability of the state of dot, $|S_z, \{n_j\}\rangle$, is written by $P_{S_z,\{n_j\}}$, where

$P_{S_z,\{n_j\}}$ satisfies the condition of $\sum_{S_z,\{n_j\}} P_{S_z,\{n_j\}}=1$. An initial state of $\boldsymbol{S}$ is now assumed to be $S_z=-S$, i.e., $P_{-S,\{n_j\}} = \exp(-\beta\sum_\ell \hbar\omega_\ell n_\ell)/\sum_{\{n_\ell\}}\exp(-\beta\sum_\ell \hbar\omega_\ell n_\ell)$, and $P_{S',\{n_j\}}=0$ for $S' \neq -S$, with $\beta=(k_\mathrm{B}T)^{-1}$, where T is temperature and k_B is the Boltzmann constant. Here, $P_{-S,\{n_j\}}$ represents that the atomic vibration is in the thermal equilibrium state. Under the initial condition, we obtain the time t dependence of $P_{S_z,\{n_j\}}$ by solving the master equation,[5] which is described by the transition probabilities per unit time due to $\hat{V}_{\sigma-S}$ and $\hat{V}_{S-\mathrm{v}}$.[4] The transition probability due to $\hat{V}_{S-\mathrm{v}}$, $W_{S-\mathrm{v}}$, is the Lorentzian function[3,5] and its maximum value is proportional to $1/\tau$, where τ is a parameter related to the spin relaxation time.[3]

The current I, the expectation value of S_z, $\langle S_z\rangle$, and that of the vibration quantum number of the mode i, $\langle n_i\rangle$, are written by using $P_{S_z,\{n_j\}}$; that is,

$$I = e \sum_{S_z,\{n_j\},S'_z,\{n'_j\}} \left(P_{S_z,\{n_j\}} W^{\sigma-S,\ L\to R}_{S_z,\{n_j\}\to S'_z,\{n'_j\}} - P_{S'_z,\{n'_j\}} W^{\sigma-S,\ R\to L}_{S_z,\{n_j\}\to S'_z,\{n'_j\}}\right), \quad (7)$$

$$\langle S_z\rangle = \sum_{S_z,\{n_j\}} P_{S_z,\{n_j\}} \langle S_z,\{n_j\}|S_z|S_z,\{n_j\}\rangle, \quad (8)$$

$$\langle n_i\rangle = \sum_{S_z,\{n_j\}} P_{S_z,\{n_j\}} \langle S_z,\{n_j\}|a_i^\dagger a_i|S_z,\{n_j\}\rangle, \quad (9)$$

where $W^{\sigma-S,\ L\to R}_{S_z,\{n_j\}\to S'_z,\{n'_j\}}$ denotes the transition probability of the electron transfer from the L to the R, while $W^{\sigma-S,\ R\to L}_{S_z,\{n_j\}\to S'_z,\{n'_j\}}$ is the transition probability from the R to the L.[5,6,7]

3. Application to S=2 systems

On the basis of the single Fe atom on CuN surface,[1] we apply this theory to S=2 systems. The energy levels of S=2 are schematically shown in inset in upper panel of Fig. 2.

The parameter set is $k_\mathrm{B}T$=1.0 meV, $|T_J|$=10.0 meV, $|D|$=1.0 meV, $\hbar\omega_1=3.5|D|$, $\hbar\omega_2=1.5|D|$, $\hbar\omega_3=0$, V=5.0 mV, $|\Lambda_1|$=0.2 meV, $|\Lambda_2|$=0.2 meV, and $|\Lambda_3|=0$, where $|\Lambda_1|$ and $|\Lambda_2|$ are estimated on the basis of the spin-orbit interaction with a small displacement of nucleus.[3] To simplify the calculation, n_j of $|S_z,\{n_j\}\rangle$ is limited to $n_j \leq 10$.

In addition, we define the achievement of the spin switching as $\langle S_z\rangle/S \geq 0.9$ to make a judgment on whether the switching takes place. We here focus on a period of 0 ns $\leq t \leq$ 10 ns.

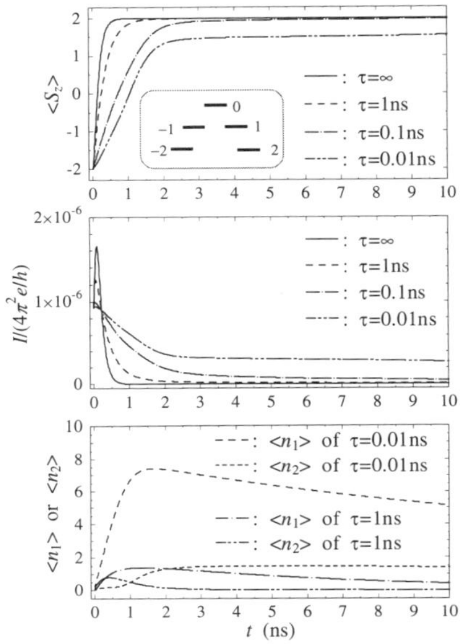

Fig. 2. The time t dependence of $\langle S_z\rangle$, I, $\langle n_1\rangle$, and $\langle n_2\rangle$ of S=2 systems. Upper panel: $\langle S_z\rangle$. Middle panel: I. Lower panel: $\langle n_1\rangle$ and $\langle n_2\rangle$. Inset in upper panel: Schematic illustration of energy levels of $-|D|S_z^2$ with S=2.

3.1. $\langle S_z\rangle$

In upper panel of Fig. 2, we show t dependence of $\langle S_z\rangle$ for $\tau=\infty$, 1 ns, 0.1 ns, 0.01 ns. The systems of $\tau=\infty$ bring about the spin switching. In the case of finite τ, the switching is not realized for τ=0.01 ns within a period of 10 ns, while it is done for τ=1 ns,

0.1 ns. On the basis of the achievement of the switching of $\langle S_z \rangle / S \geq 0.9$, the switching times for τ=∞, 1 ns, and 0.1 ns are evaluated to be 0.44 ns, 0.94 ns, and 2.1 ns, respectively.

On the other hand, the constant $\langle S_z \rangle$ with $\langle S_z \rangle \sim S$ indicates that the probability of the spin state is almost 1 for S_z=S. The constant $\langle S_z \rangle$ with $\langle S_z \rangle < 0.9S$ represents that the probability of the spin state is almost distributed to S_z=$-S$ and S.

3.2. I

In middle panel of Fig. 2, we show t dependence of I. When $\langle S_z \rangle$ for each τ becomes constant, each I takes a constant value, too. The constant I of the finite τ is larger than that of τ=∞, where I of τ=∞ takes almost 0. This behavior is explained on the basis of $\langle S_z \rangle$. When $\boldsymbol{S}$ has, for example, $\langle S_z \rangle \sim$ 1.5 for τ=0.01 ns, the probability is almost distributed to S_z=$-S$ and S. The electrons then bring about the spin-flip tunneling along with the transition from S_z=$-S$ to $-S+1$. In contrast, when $\boldsymbol{S}$ has $\langle S_z \rangle \sim S$ for τ=∞, the probability is almost 1 for S_z=S and hence the electrons are scarcely able to exhibit the spin-flip tunneling.

3.3. $\langle n_1 \rangle$, $\langle n_2 \rangle$

In lower panel of Fig. 2, we show t dependence of $\langle n_i \rangle$ (i=1, 2) for τ=1 ns, 0.01 ns. Roughly speaking, $\langle n_i \rangle$ increases with decreasing τ. The reason is that $W_{S-\mathrm{v}}$ increases with decreasing τ, and then the energy transfer from $\boldsymbol{S}$ to the atomic vibration increases with $W_{S-\mathrm{v}}$; that is, $\langle n_i \rangle$ becomes large.

The t dependence of $\langle n_i \rangle$ is explained by considering that of $\langle S_z \rangle$. In the region of $\langle S_z \rangle \sim 0$, in which the probability is almost in the excited states in $-|D|S_z^2$, $\langle n_i \rangle$ increases with increasing t, because the energy can transfer from $\boldsymbol{S}$ to the atomic vibration. When $\langle S_z \rangle$ is constant reflecting that the probability is almost in the ground states in $-|D|S_z^2$, $\langle n_i \rangle$ tends to become nonincreasing because the energy is scarcely able to move from $\boldsymbol{S}$ to the atomic vibration. The decrease of $\langle n_i \rangle$ means that the energy is returned from the vibration to $\boldsymbol{S}$, where the energy is subsequently moved to the electron systems.

4. Conclusion

We studied the effect of spin relaxation on the spin switching induced by spin injection using the master equation approach. The systems of S=2 exhibited the switching or nonswitching depending on τ ($\neq \infty$) within a period of 10 ns, although the systems of τ=∞ brought about the switching. Also, the t dependence of I and $\langle n_i \rangle$ was explained on the basis of $\langle S_z \rangle$ and the probability distribution.

Acknowledgments

This work has been supported by a Grant-in-Aid for Young Scientists (B) (No. 20710076) from the Japan Society for the Promotion of Science.

References

1. C. F. Hirjibehedin, C.-Y. Lin, A. F. Otte, M. Ternes, C. P. Lutz, B. A. Jones, and A. J. Heinrich, *Science* **317**, 1197 (2007).
2. S. Kokado, K. Ueda, K. Harigaya, and A. Sakuma, *Phys. Rev.* **B76**, 054451 (2007), *phys. stat. sol. (a)* **204**, 1937 (2007).
3. S. Kokado, K. Harigaya, and A. Sakuma, preprint.
4. For the transition probability due to the spin-phonon interaction, see F. Hartman-Boutron, P. Politi, and J. Villain, *Int. J. Mod. Phys.* **B10**, 2577 (1996).
5. C. P. Slichter, *Principles of Magnetic Resonance, Third Enlarged and Updated edition* (Springer, New York, 1996), chap. 5.
6. M. Misiorny and J. Barna *Phys. Rev.* **B77**, 172414 (2008).
7. J. Appelbaum, *Phys. Rev.* **154**, 633 (1967), *Phys. Rev. Lett.* **17**, 91 (1966).

SPIN TRANSFER TORQUE IN MAGNETIC TUNNEL JUNCTIONS WITH SYNTHETIC FERRIMAGNETIC LAYERS

M. ICHIMURA*, T. HAMADA
Advanced Research Laboratory, Hitachi Ltd., Kokubunji, Tokyo, 185-8601, Japan
**E-mail: masahiko.ichimura.yp@hitachi.com*

H. IMAMURA
Nanotechnology Institute, AIST, Tsukuba, 305-8568, Japan

S. TAKAHASHI, and S. MAEKAWA
Institute for Material Research, Tohoku University, Sendai ,980-8577, Japan
JST, CREST, Sanbancho, Chiyoda-ku,, Tokyo ,102-0075, Japan

Based on the spin-polarized free-electron model, we derive the spin transfer torque in the magnetic tunnel junctions with the synthetic ferrimagnetic layers by analyzing the spin and charge transports in the ballistic regime. In the realistic junctions, the spin torque exerted on the magnetizations of two ferromagnetic layers in the synthetic ferrimagnetic layer shows the unidirectional rotation. It is suggested that, through the antiferromagnetic interlayer coupling in the synthetic ferrimagnetic layer, this unidirectional rotation induces the cooperative reversal of magnetizations in two ferromagnetic layers, and expected that this cooperative rotation reduces the critical current for the magnetization reversal in the synthetic ferrimagnetic layer.

Keywords: Spintronics; Magnetic Tunnel Junction; Spin Transfer Torque.

1. Introduction

When the current density is adequately high, the spin-polarized current can induce a transfer of spin angular momentum that exerts a torque on a magnetization [1,2]. Such a spin-transfer torque (STT) in magnetic nano-structures is an indispensable technology for modern magnetic random access memories (MRAMs). With use of the tunnel magnetic junctions (MTJ), the quite different behavior of the STT is reported in the theoretical calculation and experiments, compared with the metallic current perpendicular plane giant magnetoresistance (CPP-GMR) devices; i.e., the anomalous bias dependence of the in-plane STT (parallel to the interface) and appearance of the out-of-plane STT (normal to the interface) [3,4]. For the development of MRAM, understanding of the STT in MTJ is important.

The synthetic ferrimagnetic (SyF) layers have been introduced to the free layers of MTJ in order to stabilize the thermal motion of magnetization in MRAM [5]. In addition, compared with the case of MTJ with the single free layer, the reduction of the critical current (J_c) for magnetization reversal has been reported in the MTJ with the SyF free layers [5]. Since the SyF free layers are considered to take advantage of both the reduction of J_c and the sustaining the thermal stability, the detailed analysis of the STT in SyF free layers is desired. In this work, we present the STT on SyF free layers in MTJ in the ballistic transport regime, which is the similar approach to the Slonczewski's torque [1].

2. Formulation

We analyze the wave function in the MTJ shown in Fig.1, where the plane waves with unit flux are incident from the left side of Fig.1. Based on the spin-polarized free-electron-model, the wave function $\psi_{k,s}$ with the wave number k and the spin s is calculated by numerical recursion method [6] at each coordinates. The spin and particle currents, $j_{i,k}$ (i=x, y, z) and $j_{p,k}$, respectively, are defined as follows,

$$j_{i,k} = \mathrm{Im} \sum_{ss'} \psi^*_{k,s} \sigma_i \partial_y \psi_{k,s'}$$

$$j_{p,k} = \mathrm{Im} \sum_{s} \psi^*_{k,s} \partial_y \psi_{k,s},$$

where σ_i are Pauli matrices. The total spin and particle currents $J_\mu(y) = \sum_k j_{\mu k}(y)$ $(\mu = x, y, z, p)$ are obtained by the similar manner to Ref.7. The STT is defined as the change of the spin current passing through the ferromagnetic layers in a SyF free layer, normalized by the particle current, $T_i = (J_i(y_{in}) - J_i(y_{out}))/J_p(y_{in})$, where y_{in} and y_{out} are the left and right coordinates close to the corresponding ferromagnetic layer.

3. Results and Discussion

We show the schematic picture of MTJ with the pinned layer, the insulating barrier, and the SyF free layer in Fig. 1. The SyF free layer consists of two ferromagnetic layers separated by a paramagnetic metal, where the magnetizations of two ferromagnetic layers are set to anti-parallel alignment. In this MTJ structure, the magnetization in the pinned layer is along z-axis. The magnetization of ferromagnetic layer near the pinned layer, which is defined as F1 in the SyF layer, is in-plane with the angle θ relative to the magnetization in the pinned layer. And F2 is defined as the other ferromagnetic layer in the SyF layer. The typical MTJ structure is described in nm unit as pin(5)/insulator(1)/F1(2)/normal(0.8)/F2(2). Our calculation is restricted to the bias that causes the electron flowing from the pinned to the SyF layers. This bias direction is expected to exert the torque on the magnetization of F1

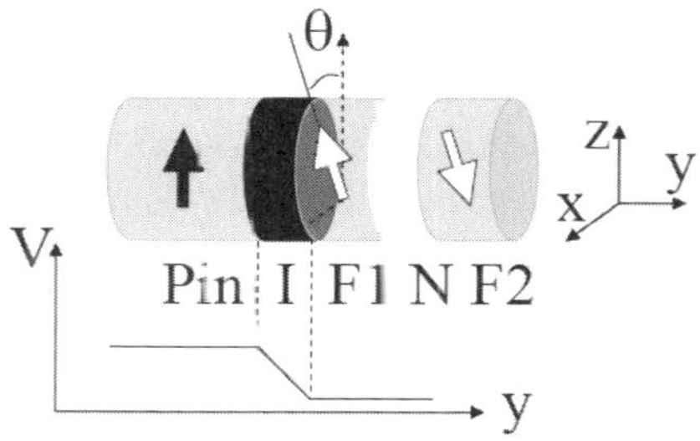

Fig. 1. Schematic picture of the MTJ with the SyF free layer and the bias voltage dependence in the MTJ. Pin and I denoted the pinned and insulating layers, respectively. F1/N/F2 is the SyF free layer, where N is non-magnetic layer.

parallel to the magnetization of the pinned layer, from the result of the MTJ with the single free layer.

Figure 2 shows the STT of F1, F2, and SyF free layer in the above typical MTJ structure, with barrier height U=0.7eV, bias voltage V=0.5V, and polarization of the all ferromagnetic layers $p = (k_\uparrow - k_\downarrow)/(k_\uparrow + k_\downarrow) = 0.6$, where k_σ is the Fermi wave number of spin σ. The sign changes in x-component of STT for F1 and F2 mean the unidirectional rotation, which shows the tendency to align the magnetization of F1 parallel to that of pinned layer, and the magnetization of F2 anti-parallel to that of pinned layer, respectively. The all components of STT of F1 and F2 show the opposite direction of torque each other. Especially, the opposite torques in x- and z-components mean the same direction of rotation. It is considered that this same direction of rotation induces the cooperative motion of magnetization in F1 and F2 via the interlayer exchange coupling in the SyF layer. This cooperative motion in the SyF layer suggests the reduction of the critical current for the magnetization reversal in MTJs.

Here, we try to interpret intuitively the cooperative rotation of magnetization in F1 and F2. At first, we focus on the torque on the magnetization of F1. As we noted in the bias direction, the rotational direction of magnetization in F1 is understood as the Slonczewski's torque in the single free layer, which is caused by the conservation of angular momentum between the magnetization in F1 and the polarization of spin current parallel to the magnetization of the pinned layer.

Next, we focus on the torque on the magnetization of F2. It can be considered that the effective polarization of spin current exerting torque on the magnetization of F2 is parallel to the magnetization of the pinned layer which is not filtered by the magnetization of F1, because there are no torque from the spin current aligned *just* anti-parallel to the magnetization of F1. However, this leads to the same direction of torque exerted on F2 as F1, that means the opposite direction of rotation to F1. This intuitive speculation fails to interpret the torque on the MTJ with the SyF. On the other hand, it can be considered that the multiple reflection of spin current in the MTJ is important since the filtering of spin current is enhanced by the SyF layer. Taking account of the effect of the multiple reflection, the pinned layer reflects the spin current polarizing anti-parallel to the magnetization of the pinned layer to the SyF layer. This polarization of the spin current contributes to the opposite direction of torque exerted on F2 to F1, that leads the unidirectional rotation of the magnetizations of F1 and F2. It is considered that the reflection of the spin current from the pinned layer plays a dominant role for the STT on F2.

Regarding the SyF free layer as an *effective* single free layer, the STT shown in Fig.2 (c) is reduced relative to the STT on F1 shown in Fig.2 (a). This is consistent with the unidirectional rotation of the magnetizations of F1 and F2. From the view point of an *effective* single free layer, the ideal *symmetric* SyF layer with the same thickness, the same amplitude and the opposite direction of magnetizations of F1 and F2, which has no net magnetization, can be expected no STT. However, due to the internal degree of freedom in the SyF free layer, the STT on F1 and F2 contribute to the unidirectional rotation.

It should be noted that we find rarely the in-phase STT between F1 and F2 when the bias voltage is relatively high as 1V. We also note that the normal torque (the y component in Fig.2) is the comparable order of magnitude with the in-plane torque. This result is consistent with the case of single free layer MTJ [7].

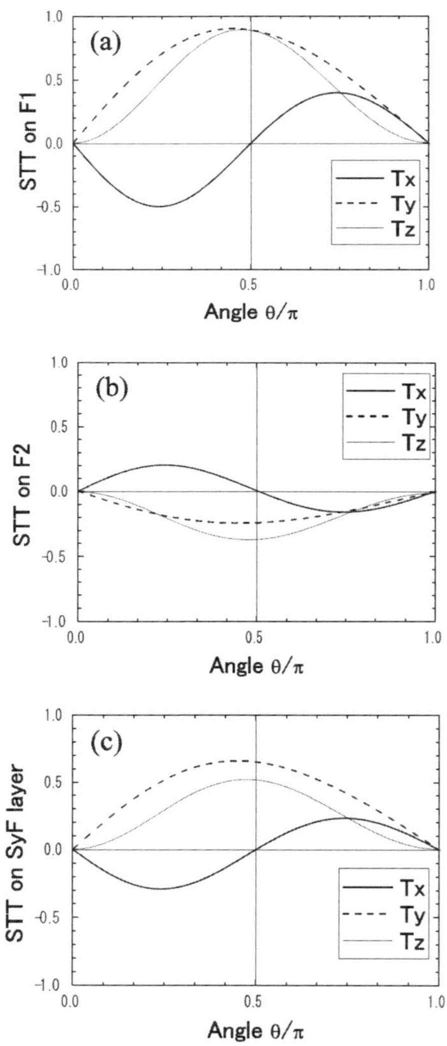

Fig. 2. Angular dependence of the STT on F1(a), F2(b), and SyF free layer(c).

4. Summary

We study spin and charge transports in the MTJs with the SyF free layers in the ballistic regime, based on the spin-polarized free-electron model. It is shown that the STT exerted on the magnetizations of two ferromagnetic layers in the SyF free layers shows the tendency of the same direction of rotation in the realistic junctions. It is suggested that, via the interlayer exchange coupling in the SyF free layer, this rotation induces the cooperative motion of magnetizations in two ferromagnetic layers, and this motion reduces the critical current for the magnetization reversal in the SyF free layer.

Acknowledgments

This work is supported by Next Generation Supercomputing Project, Nanoscience Program, the Ministry of Education, Culture, Sports, Science and Technology, Japan.

References

1. C. Slonczewski, J. Magn. Magn. Mater. **159**, L1 (1996).
2. Berger, Phys. Rev. B **54**, 9353 (1996).
3. Theodonis *et al.*, Phys. Rev. Lett. **97**, 237205 (2006).
4. Kubota *et al.*, Nat. Phys. **4**, 37 (2008).
5. Hayakawa *et al.*, J. Jpn. Appl. Phys. **45**, L1057 (2006).
6. Imamura, N. Kobayashi, S. Takahashi, and S. Maekawa, Phys. Rev. Lett. **84**, 1003 (2000).
7. Wilczyński, J. Barnaś, and R. Świrkowicz, Phys. Rev. B **77**, 054434 (2008).

GAPLESS CHIRALITY EXCITATIONS IN ONE-DIMENSIONAL SPIN-1/2 FRUSTRATED MAGNETS

SHUNSUKE FURUKAWA[1], MASAHIRO SATO[1], YASUHIRO SAIGA[2], AND SHIGEKI ONODA[1]

[1] *Condensed Matter Theory Laboratory, RIKEN, Wako, Saitama 351-0198, Japan*

[2] *Center for Educational Assistance, Shibaura Institute of Technology, Minuma-ku, Saitama 337-8570, Japan*

We theoretically study one-dimensional spin-1/2 multiferroic systems. We focus on the chiral ordered phase with gapless excitations that appears under easy-plane anisotropy. The in-plane chirality dynamics contains gapless solitons associated with local flips of the chirality. These excitations contribute to the low-frequency dielectric response through the magnetoelectric coupling.

Keywords: multiferroics, vector spin chirality, quantum spin systems, dielectric response

1. Introduction

The vector spin chirality in Mott insulators can induce a ferroelectric polarization through the spin-orbit coupling.[1,2,3] This new type of multiferroic behavior, originally discovered in a spiral magnet $TbMnO_3$,[4] has received considerable attentions for both its fundamental importance and its potential applications. In particular, recently discovered multiferroic behavior in quasi-one-dimensional (1D) frustrated magnets $LiCuVO_4$[5,6] and $LiCu_2O_2$[7,8] has raised an intriguing issue of quantum fluctuations in multiferroics, because of the low dimensionality and the spin-1/2 nature.

A remarkable aspect of multiferroics is that the dynamics of the spin chirality can be probed from the dielectric functions $\varepsilon^{\alpha\alpha}(\omega)$ through the magnetoelectric coupling,[9,10] as performed for RMnO$_3$ (R=Tb,Dy).[11,12] Here, a local flip of the chirality induces the transverse charge dynamics. In particular, in 1D systems, a pair of such defects may propagate as elementary excitations.

Here, we uncover the gapless dielectric response originating from chirality excitations in 1D spin-1/2 multiferroics.[13]

2. Model and ground-state properties

A minimal model for the quasi-1D spin-1/2 multiferroics, especially for $LiCuVO_4$,[14] is given by a simple spin Hamiltonian

$$H = \sum_{\rho=1}^{2}\sum_{j} J_\rho \left[\mathbf{S}_j \cdot \mathbf{S}_{j+\rho} + (\Delta - 1) S_j^z S_{j+\rho}^z\right], \quad (1)$$

where $\boldsymbol{S}_j$ represents the spin at the site $j(=1,\dots,L)$ and $\Delta(\le 1)$ is an easy-plane exchange anisotropy. We focus on the case $J_2 > 0$. Then the model contains a geometrical frustration irrespective of the sign of J_1. We introduce the Fourier component of the vector spin chirality by

$$\mathcal{O}_\kappa^\alpha(q) \equiv \frac{1}{L}\sum_j e^{-iqj}(\boldsymbol{S}_j \times \boldsymbol{S}_{j+1})^\alpha, \ \alpha = y, z. \quad (2)$$

The uniform component couples with the ferroelectric polarization via the magnetoelectric coupling,[1,2,3] $P^\alpha \propto \epsilon_{\alpha\beta x}\mathcal{O}_\kappa^\beta(q=0)$, with $\epsilon_{\alpha\beta\gamma}$ being the fully antisymmetric tensor.

In the classical limit, a spiral ground state is realized for $|J_1|/J_2 < 4$. This state has a uniform chirality $\kappa^z = \langle \mathcal{O}_\kappa^z(q=0)\rangle \neq 0$, leading to the ferroelectricity $P^y \neq 0$. In the quantum case with spin-1/2, the chiral ordering competes with the valence bond ordering driven by quantum fluctuations. A bosonization analysis for $|J_1|/J_2 \ll 1$ has predicted the appearance of a chiral ordered phase with a quasi-long-range spiral spin correlation.[15] This "gapless chiral phase" has been identified numerically[16] around the XY case $\Delta \approx 0$ for $J_1 > 0$. For $J_1 < 0$, i.e., a relevant case to multiferroic cuprates, the stability of the chiral phase has not been examined in detail.

In the rest of this section, we show that the competition with the valence bond orders becomes weaker for $J_1 < 0$ under weak easy-plane anisotropy $\Delta \sim 1$, by calculating the spin gap and the incommensurability by Lanczos diagonalization. Minimization of the spin gap under the twisted boundary condition results in a smooth dependence on the system size L even in the presence of an incommensurate correlation.[17] The obtained gaps for $L = 12, 16, 20, 24$, and 28 are extrapolated to $L \to \infty$ via a linear fitting.[18] Figure 1 shows the intensity plot of the extrapolated spin gap. For $J_1 > 0$, a gapped region characterized by a singlet dimer order extends largely from $\Delta = 0$ to 1, which indicates the suppression of the chiral ordering. For $J_1 < 0$, another gapped phase with a different dimer $|\uparrow\downarrow\rangle + |\downarrow\uparrow\rangle$ appears under

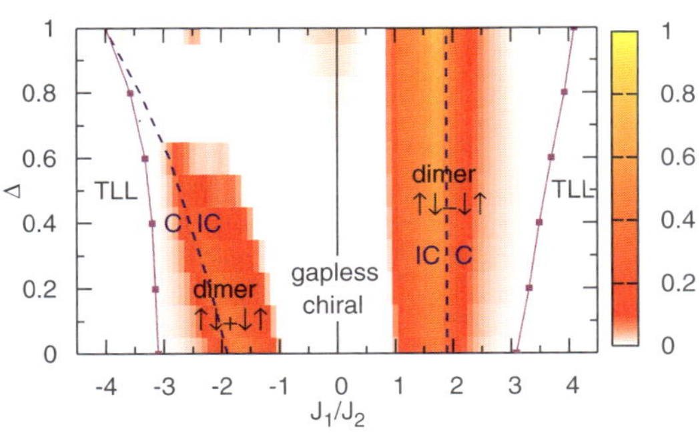

Fig. 1. The intensity plot of the spin gap of the model (1). The purple solid lines are the phase boundaries between Tomonaga-Luttinger liquids (TLL) and dimer phases determined in Refs. 19 and 20. The blue broken lines indicate Lifshitz transitions.

strong easy-plane anisotropy but is destabilized around $\Delta \sim 1$.

From the value of flux that minimizes the spin gap, one can determine the pitch angle Q of the spin correlation.[17] The Lifshitz transitions, where the spin correlation change its character from incommensurate (IC) to commensurate (C), are drawn by blue broken lines in Fig. 1. While the incommensurate region for $J_1 > 0$ terminates at $J_1/J_2 \simeq 2$, that for $J_1 < 0$ persists more largely under weak anisotropy $\Delta \sim 1$. This reflects the suppression of the dimer order for $J_1 < 0$.

3. Chirality dynamics

We discuss the dynamical chirality correlations in the gapless chiral phase. They are defined by

$$K^\alpha(q,\omega) = L\sum_\lambda |\langle\lambda|\mathcal{O}^\alpha_\kappa(q)^\dagger|\mathrm{GS}\rangle|^2\delta(\omega-\omega_\lambda), \tag{3}$$

where $|\mathrm{GS}\rangle$ is the ground state, and $|\lambda\rangle$ is an excited state with an excitation energy ω_λ. Through the magnetoelectric coupling, the in-plane and out-of-plane components, K^z and K^y, contribute to the dielectric functions $\mathrm{Im}\,\varepsilon^{yy}$ and $\mathrm{Im}\,\varepsilon^{zz}$, respectively.

Now we introduce the biquadratic Dzyaloshinskii-Moriya (DM) interaction:

$$H_{\mathrm{BDM}} = -V_z\sum_j (\boldsymbol{S}_j\times\boldsymbol{S}_{j+1})^z(\boldsymbol{S}_{j+2}\times\boldsymbol{S}_{j+3})^z. \tag{4}$$

This arises from the coupling of spins to transverse optical phonons through the DM interaction.[21] For small $|J_1|/J_2$, this term can amplify the chiral order while keeping the pitch angle almost commensurate ($Q \approx \pi/2$).

Figure 2 shows the lowest-energy levels contributing to $K^z(0,\omega)$, $K^z(\pi,\omega)$, and $K^y(0,\omega)$. We have taken $J_1/J_2 = -0.1$, $\Delta = 0.9$, and $V_z/J_2 = 3$ so that the system is in the gapless chiral phase. One can observe that the out-of-plane component $K^y(0,\omega)$ corresponding to electromagnons[9] is gapful. On the other hand, the in-plane

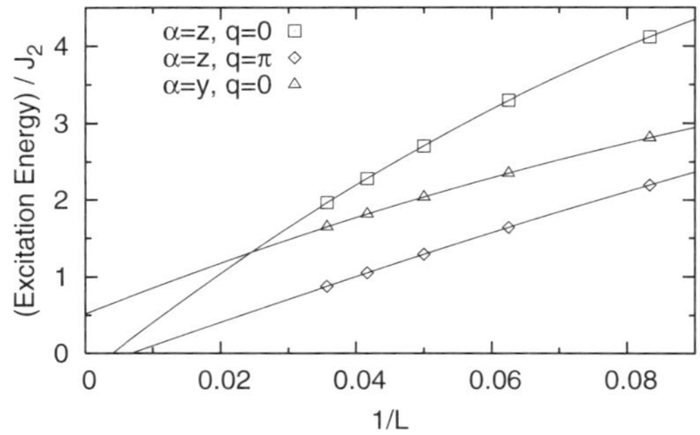

Fig. 2. The lowest-energy levels ω_λ contributing to $K^z(0,\omega)$, $K^z(\pi,\omega)$, and $K^y(0,\omega)$, plotted against $1/L$. The lines show the extrapolations of the levels to $L \to \infty$ via quadratic fittings.

components $K^z(0,\omega)$ and $K^z(\pi,\omega)$ eventually become gapless in the thermodynamic limit, leading to a gapless dielectric response in the Mott insulator. Also, the gapless chiral phase possesses the staggered-chiral quasi-long-range order, in addition to the chiral long-range order.

Now we clarify the nature of the gapless chirality excitations in terms of the bosonization formulation in Ref. 15. Therein, they have introduced two pairs of dual boson fields $(\Phi_\pm, \Theta_\pm)$, which represent the average $(+)$ and relative $(-)$ polar angles for the neighbouring spins. If the system enters the gapless chiral phase, the field Θ_- is locked to produce the uniform vector chirality $\kappa^z \propto \langle \sin(\sqrt{4\pi}\Theta_-)\rangle \neq 0$, and the excitations associated with (Φ_-, Θ_-) acquire a finite energy gap. In contrast, those associated with (Φ_+, Θ_+) are described by a gapless Tomonaga-Luttinger liquid (TLL).

In this framework, the chirality operators are represented as

$$\mathcal{O}^z_\kappa(0) = \int \frac{dx}{La_0}[c_1 \sin(\sqrt{4\pi}\Theta_-) + c_2 a_0^2 \kappa^z (\partial_x \Theta_+)^2 + \ldots], \quad (5a)$$

$$\mathcal{O}^z_\kappa(\pi) = \int \frac{dx}{La_0}\, c_3 \kappa^z \cos(\sqrt{4\pi}\Phi_+) + \cdots, \quad (5b)$$

where c_{1-3} are constants of the order of unity and a_0 is a lattice spacing. The second term $(\partial_x\Theta_+)^2$ in Eq. (5a) gives rise to a nonsingular gapless contribution $K^z(0,\omega) \sim \omega^2$. This gapless mode turns to two-phason excitations if a 3D incommensurate spiral magnetic LRO appears due to a weak inter-chain coupling. In realistic systems containing defects that act as pinning centers, single phason excitation can also appear in $K^z(0,\omega)$.[22]

Similarly, the term $\cos(\sqrt{4\pi}\Phi_+)$ in the staggered part $\mathcal{O}^z_\kappa(\pi)$ also yields a gapless contribution. This gapless mode has a solitonic character as we explain in the following. The term $\int dx \cos(\sqrt{4\pi}\Phi_+)$ corresponds to the leading term in the bosonized form of $\sum_j S^z_j S^z_{j+1}$, which rotates the two neighbouring spins by π about the z axis. We operate this term on the background of a quasi-long-range spiral order with the positive (L) chirality, as shown in Fig. 3. The flips of two neighbouring spins creates a pair of kinks with a negative (R) chirality (Fig. 3 (b)). Then, via the exchange of two neighboring chiral spin pairs encircled by broken lines, the kinks propagate without changing the sign of the chirality of each domain (Fig. 3 (c)). Since this soliton carries the opposite chirality to the background, it induces a transverse charge displacement. This contrasts with the longitudinal nature of the well-known charge

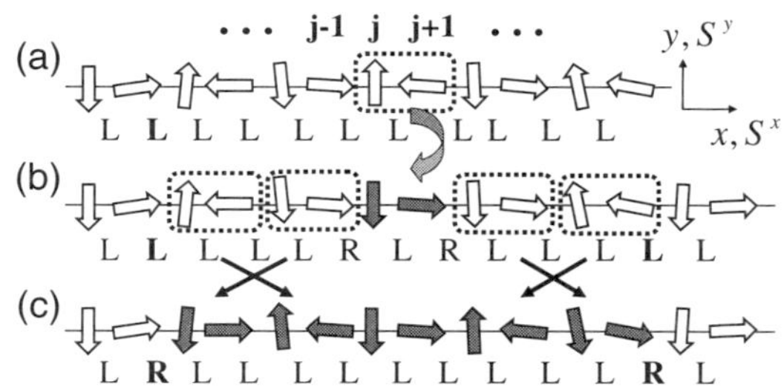

Fig. 3. Schematic pictures on the creation and propagation of the chiral soliton pair in the background of quasi-long-range spiral order with a pitch angle $\pi/2$. The symbol L (R) denotes a positive or *left-handed* (negative or *right-handed*) chirality at each bond.

soliton in the polyacetylene.

In the 1D multiferroics like $LiCuVO_4$ where the unit cell is doubled, the chiral solitons observed in the staggered channel can also appear in the uniform channel. If the $U(1)$ symmetry is weakly broken by spin anisotropy or an in-plane magnetic field, the soliton can acquire a small gap. It will be interesting to look for the soliton-pair excitation by low-frequency electrical or optical probes.[23]

Acknowledgement

The authors thank N. Nagaosa, Y. Tokura, S.-W. Cheong, T. Rõõm, M. Mostovoy for stimulating discussions. This work was partly supported by Grants-in-Aid for Scientific Research (No. 17071011, No. 20029006, and No. 20046016) from MEXT of Japan. Numerical calculations were partly performed on the RIKEN Super Combined Cluster System.

References

1. H. Katsura, N. Nagaosa, and A.V. Balatsky, *Phys. Rev. Lett.* **95**, 057205 (2005).
2. I.A. Sergienko and E. Dagotto, *Phys. Rev.* **B73**, 094434 (2006).
3. C. Jia, S. Onoda, N. Nagaosa, and J.H. Han, *Phys. Rev.* **B76**, 144424 (2007).
4. T. Kimura, T. Goto, H. Shintani, K. Ishizaka, T. Arima, Y. Tokura, *Nature* **426**, 55 (2003).
5. M. Enderle, C. Mukherjee, B. Fåk, R.K. Kremer, J.-M. Broto, H. Rosner, S.-L. Drechsler, J. Richter, J. Malek, A. Prokofiev, W. Assmus, P. Pujol, J.-L. Raggazzoni, H. Rakoto, M. Rheinstädter, and H.M. Rønnow, *Europhys. Lett.* **70**, 237 (2005).
6. Y. Naito, K. Sato, Y. Yasui, Y. Kobayashi, Y. Kobayashi, and M. Sato, *J. Phys. Soc. Jpn.* **76**, 023708 (2007).
7. S. Park, Y. J. Choi, C. L. Zhang, and S-W. Cheong, *Phys. Rev. Lett.* **98**, 057601 (2007).
8. S. Seki, Y. Yamasaki, M. Soda, M. Matsuura, K. Hirota, and Y. Tokura, *Phys. Rev. Lett.* **100**, 127201 (2008).
9. H. Katsura, A.V. Balatsky, and N. Nagaosa, *Phys. Rev. Lett.* **98**, 027203 (2007).
10. H. Katsura, S. Onoda, J.H. Han, and N. Nagaosa, *Phys. Rev. Lett.* **101**, 187207 (2008).
11. A. Pimenov, A.A. Mukhin, V.Yu. Ivanov, V.D. Travkin, A.M. Balbashov, and A. Loidl, *Nature Physics* **2**, 97 (2006).
12. N. Kida, Y. Ikebe, Y. Takahashi, J.P. He, Y. Kaneko, Y. Yamasaki, R. Shimano, T. Arima, N. Nagaosa, and Y. Tokura, *Phys. Rev.* **B78**, 104414 (2008).
13. S. Furukawa, M. Sato, Y. Saiga, and S. Onoda, *J. Phys. Soc. Jpn.* **77**, 123712 (2008).
14. The neutron scattering results in ref. [5] give $J_1 \sim -19K$ and $J_2 \sim 45K$ for $LiCuVO_4$.
15. A.A. Nersesyan, A.O. Gogolin, and F.H.L. Eßler, *Phys. Rev. Lett.* **81**, 910 (1998).
16. T. Hikihara, M. Kaburagi, and H. Kawamura, *Phys. Rev.* **B63**, 174430 (2001).
17. A.A. Aligia, C.D. Batista, and F.H.L. Eßler, *Phys. Rev.* **B62**, 3259 (2000).
18. The linear extrapolation sometimes produces small negative values. In Fig. 1, we have replaced such unphysical negative values by zero.
19. K. Nomura and K. Okamoto, *J. Phys.* **A27**, 5773 (1994).
20. R. D. Somma and A. A. Aligia, *Phys. Rev.* **B64**, 024410 (2001).
21. S. Onoda and N. Nagaosa, *Phys. Rev. Lett.* **99**, 027206 (2007).
22. H. Fukuyama and P.A. Lee, *Phys. Rev.* **B17**, 535 (1978).
23. T. Rõõm *et al.*, private communication.

ELECTRONIC STATES OF GRAPHENE AND ITS MULTI-LAYERS

TSUNEYA ANDO* AND MIKITO KOSHINO
Department of Physics, Tokyo Institute of Technology,
2-12-1 Ookayama, Meguro-ku, Tokyo 152-8551, Japan
**E-mail: ando@phys.titech.ac.jp*

A brief review is given on exotic electronic properties of monolayer, bilayer, and multilayer graphenes from a theoretical point of view. In multi-layer graphenes, the Hamiltonian is decomposed into those of monolayer and bilayer graphenes.

Keywords: graphite; carbon nanotube; Berry phase; interlayer interaction; diamagnetic susceptibility; neutrino, Dirac Fermion.

1. Introduction

In an effective-mass approximation, an electron in a graphite monolayer or graphene is described by Weyl's equation for a massless neutrino.[1,2,3,4] Electronic properties in such an exotic electronic structure are quite intriguing, and the conductivity with/without a magnetic field including the Hall effect,[5,6] quantum corrections to the conductivity,[7] and the dynamical transport[8] were investigated theoretically. The results show that the system exhibits various characteristic behaviors different from conventional two-dimensional systems.[9] Quite recently, this single layer graphene was fabricated using the so-called scotch-tape technique[10] and the magnetotransport was measured including the integer quantum Hall effect, demonstrating the validity of the neutrino description of the electronic states.[11,12] Since then the graphene became the subject of extensive theoretical and experimental study. The purpose of this paper is to give a brief review on such electronic properties of monolayer, bilayer, and multilayer graphenes from a theoretical point of view.

Later experiments in higher magnetic fields showed spin splitting for several low-lying Landau levels and additional splitting for level $n = 0$.[13] Inter-band magneto-absorptions were reported for the first time[14] in graphenes epitaxially grown on SiC substrates.[15,16] The cyclotron resonance was reported also in scotch-tape graphene.[17,18] The local density-of-states and carrier concentrations were measured using scanning single-electron-transistor.[19] The negligibly small negative magnetoresistance was observed,[20] a weak positive magnetoresistance was reported,[21] and sensitivity to scatterers was demonstrated.[22] There were reports on optical phonons using the Raman scattering,[23,24] spin transport,[25,26] angle-resolved photoemission spectroscopy,[27,28] and so on. There has also been a rapid development in the theory of graphene.[29]

2. Monolayer Graphene

The structure of 2D graphite sheet is shown in Fig. 1. A unit cell contains two carbon atoms which are denoted by A and B. We have the vectors connecting nearest neighbor carbon atoms $\boldsymbol{\tau}_l$ ($l=1,2,3$) with length $|\boldsymbol{\tau}_l| = a/\sqrt{3}$, where the lattice constant is given by $a=2.46$ Å. The first Brillouin zone is given by a hexagon with two corner points K and K' with corresponding wave vectors $\mathbf{K}$ and $\mathbf{K}'$, respectively.

Let $\phi(\mathbf{r})$ be a p_z orbital localized around the origin. In a nearest-neighbor tight-binding model, the wave function is written

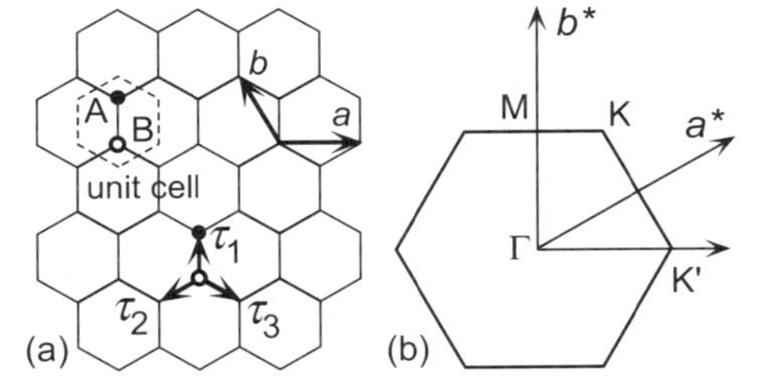

Fig. 1. The lattice structure (a) and the first Brillouin zone (b) of graphene. The unit cell contains two carbon atoms denoted by A and B. Two primitive translation vectors are denoted by **a** and **b** and three vectors connecting neighboring A and B atoms are $\boldsymbol{\tau}_l$ ($l = 1,2,3$). Reciprocal lattice vectors are given by $\mathbf{a}^*$ and $\mathbf{b}^*$.

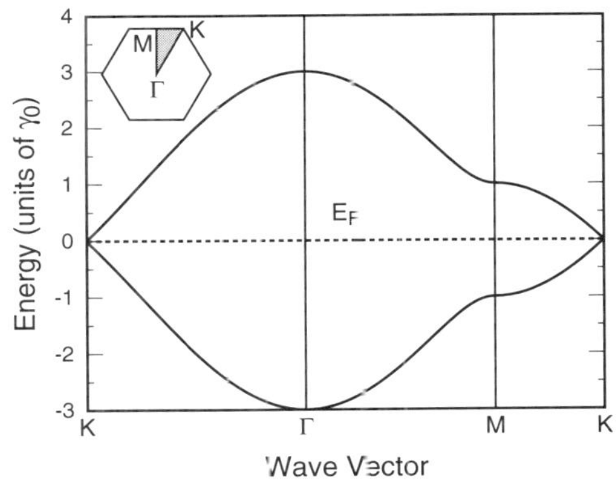

Fig. 2. The band structure obtained in a tight-binding model.

as

$$\psi(\mathbf{r}) = \sum_{\mathbf{R}_A} \psi_A(\mathbf{R}_A)\phi(\mathbf{r}-\mathbf{R}_A) + \sum_{\mathbf{R}_B} \psi_B(\mathbf{R}_B)\phi(\mathbf{r}-\mathbf{R}_B), \quad (1)$$

where, $\psi_A(\mathbf{R}_A)$ and $\psi_B(\mathbf{R}_B)$ are the amplitudes at sites $\mathbf{R}_A$ and $\mathbf{R}_B$. Taking into account only the hopping integral $-\gamma_0$ ($\gamma_0 \sim 3$ eV)[30] between nearest-neighbor atoms and neglecting overlapping integrals, we have

$$\begin{aligned} \varepsilon\psi_A(\mathbf{R}_A) &= -\gamma_0 \sum_l \psi_B(\mathbf{R}_A - \boldsymbol{\tau}_l), \\ \varepsilon\psi_B(\mathbf{R}_B) &= -\gamma_0 \sum_l \psi_A(\mathbf{R}_B + \boldsymbol{\tau}_l), \end{aligned} \quad (2)$$

where we have chosen the energy origin at the level of an isolated p_z orbital.

Setting $\psi_A(\mathbf{R}_A) = f_A(\mathbf{k})\exp(i\mathbf{k}\cdot\mathbf{R}_A)$ and $\psi_B(\mathbf{R}_B) = f_B(\mathbf{k})\exp(i\mathbf{k}\cdot\mathbf{R}_B)$, we have

$$\begin{pmatrix} 0 & h_{AB}(\mathbf{k}) \\ h_{AB}(\mathbf{k})^* & 0 \end{pmatrix} \begin{pmatrix} f_A(\mathbf{k}) \\ f_B(\mathbf{k}) \end{pmatrix} = \varepsilon \begin{pmatrix} f_A(\mathbf{k}) \\ f_B(\mathbf{k}) \end{pmatrix}, \quad (3)$$

$$h_{AB}(\mathbf{k}) = -\gamma_0 \sum_l \exp(-i\mathbf{k}\cdot\boldsymbol{\tau}_l). \quad (4)$$

Then, the energy bands become $\varepsilon_\pm(\mathbf{k}) = \pm|h_{AB}(\mathbf{k})|$. We have $\varepsilon_\pm(\mathbf{K}) = \varepsilon_\pm(\mathbf{K}') = 0$, showing that there is no gap at K and K' points. Figure 2 shows the obtained π bands.

To see the behavior in the vicinity of the K point, for example, we rewrite **k** as $\mathbf{K}+\mathbf{k}$ and expand $h_{AB}(\mathbf{k})$ in terms of $|\mathbf{k}|a$. To the lowest order, we have

$$\gamma(\boldsymbol{\sigma}\cdot\mathbf{k})\tilde{\mathbf{f}}(\mathbf{k}) = \varepsilon\tilde{\mathbf{f}}(\mathbf{k}), \quad \hat{\mathbf{f}}(\mathbf{k}) = \begin{pmatrix} \tilde{f}_A(\mathbf{k}) \\ \tilde{f}_B(\mathbf{k}) \end{pmatrix}, \quad (5)$$

with $\gamma = (\sqrt{3}/2)a\gamma_{\rm C}$, $f_A(\mathbf{K}+\mathbf{k}) = \tilde{f}_A(\mathbf{k})$ and $f_B(\mathbf{K}+\mathbf{k}) = -\omega\tilde{f}_B(\mathbf{k})$, $\omega = \exp(2\pi i/3)$, and $\boldsymbol{\sigma} = (\sigma_x, \sigma_y)$, where σ_x and σ_y are Pauli spin matrices. In a similar manner, we obtain the equation for the K' point, for which $\boldsymbol{\sigma}$ is replaced with $\boldsymbol{\sigma}^*$.

Thus, in the vicinity of the K and K' points, the energy dispersion is given by $\varepsilon_s(\mathbf{k}) = s\gamma|\mathbf{k}|$ ($s = \pm 1$) and the density of states becomes

$$D(\varepsilon) = \frac{g_{\rm v} g_{\rm s}|\varepsilon|}{2\pi\gamma^2}, \quad (6)$$

where $g_{\rm s} = 2$ is the spin degeneracy and $g_{\rm v} = 2$ is the valley degeneracy corresponding to the K and K' points. These results are illustrated in Fig. 3. The density of states vanishes at $\varepsilon = 0$ and therefore graphene is often called a zero-gap semiconductor. However, the energy-dependence of the conductivity indicates that graphene is a metal rather than a semiconductor.[5,29,31]

In an effective-mass approximation or a $\mathbf{k}\cdot\mathbf{p}$ scheme near the K or K' point, the electron motion is described by a Schrödinger equation in which **k** is replaced by $\hat{\mathbf{k}} = -i\vec{\nabla}$.

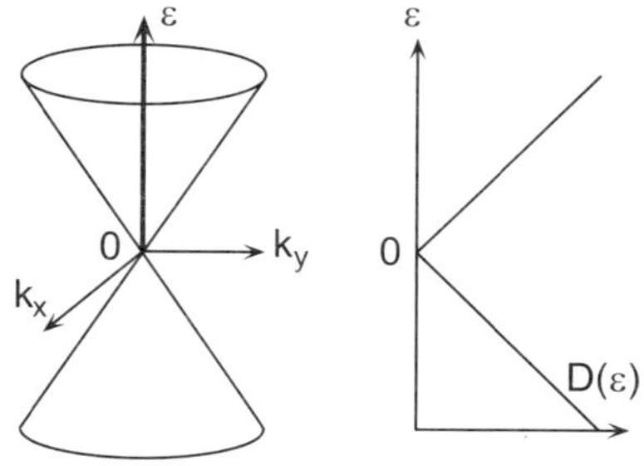

Fig. 3. The energy dispersion and density of states in the vicinity of K and K' points obtained in a $\mathbf{k}\cdot\mathbf{p}$ scheme.

We have

$$\mathcal{H}_0\mathbf{F}(\mathbf{r}) = \varepsilon\mathbf{F}(\mathbf{r}), \tag{7}$$

with

$$\mathcal{H}_0 = \gamma(\boldsymbol{\sigma}\cdot\hat{\mathbf{k}}) = \begin{pmatrix} 0 & \gamma\hat{k}_- \\ \gamma\hat{k}_+ & 0 \end{pmatrix}, \tag{8}$$

where $\hat{k}_\pm = \hat{k}_x \pm i\hat{k}_y$ and $\mathbf{F}(\mathbf{r})$ is a two-component wave function. This is exactly the same as Weyl's equation for a neutrino, except that the velocity, given by $v = \gamma/\hbar$, is much smaller than the light velocity (about 1/300). The equation for the K' point is given by replacing $\boldsymbol{\sigma}$ with $\boldsymbol{\sigma}^*$.

The wave function near the K point is written as

$$\mathbf{F}_{s\mathbf{k}}(\mathbf{r}) = \frac{1}{L}\exp(i\mathbf{k}\cdot\mathbf{r})\,\mathbf{F}_{s\mathbf{k}}, \tag{9}$$

where $\mathbf{F}_{s\mathbf{k}}$ is the spin function in which the spin is in the direction of $\mathbf{k}$ or $-\mathbf{k}$,

$$\mathbf{F}_{s\mathbf{k}} = \exp[i\phi_s(\mathbf{k})]\,R^{-1}[\theta(\mathbf{k})]\,|s). \tag{10}$$

Here, $\phi_s(\mathbf{k})$ is an arbitrary phase, $\theta(\mathbf{k})$ is the angle between $\mathbf{k}$ and the k_y axis, i.e., $k_x + ik_y = +i|\mathbf{k}|e^{i\theta(\mathbf{k})}$ and $k_x - ik_y = -i|\mathbf{k}|e^{-i\theta(\mathbf{k})}$, $R(\theta)$ is the spin rotation operator around the z direction perpendicular to graphene sheet, and $|s)$ is the spin function with $\mathbf{k}$ in the k_y axis. The spin rotation operator is given by $R(\theta) = \exp(i\theta\sigma_z/2)$ with σ_z being a Pauli spin matrix. The rotation operator satisfies $R(\theta_1)R(\theta_2) = R(\theta_1+\theta_2)$ and $R(-\theta) = R^{-1}(\theta)$. Further, we have $R(\theta \pm 2\pi) = -R(\theta)$ and $R(-\pi) = -R(+\pi)$. Therefore, for a phase $\phi_s(\mathbf{k})$ independent of the direction of $\mathbf{k}$, wave function $\mathbf{F}_{s\mathbf{k}}$ changes its sign under the 2π rotation of $\mathbf{k}$ or it has opposite sign under $+\pi$ and $-\pi$ rotation of $\mathbf{k}$.

The sign of the wave function itself depends on the choice of the phase. We can choose $\phi_s(\mathbf{k}) = -\theta(\mathbf{k})/2$, for example. Then, we have

$$\mathbf{F}_{s\mathbf{k}} = \frac{1}{\sqrt{2}}\begin{pmatrix} -i\,s\,e^{-i\theta(\mathbf{k})} \\ 1 \end{pmatrix}, \tag{11}$$

which does not change the sign under $\theta(\mathbf{k}) \to \theta(\mathbf{k}) \pm 2\pi$. However, this wave function changes its sign if we consider Berry's phase under the wave-vector rotation.[32,33] In fact, Berry's phase $e^{-i\eta}$ under the wave-vector rotation is calculated as

$$\eta = -i\int_0^T dt\Big\langle s\mathbf{k}(t)\Big|\frac{d}{dt}\Big|s\mathbf{k}(t)\Big\rangle = -\pi, \tag{12}$$

where $\mathbf{k}(t)$ moves along a closed contour around $\mathbf{k} = 0$ between time $t = 0$ and T. Therefore, the wave function acquires phase $-\pi$ when $\mathbf{k}$ is rotated around the origin.[4,34,35] The sign change occurs only when the closed contour encircles the origin $\mathbf{k} = 0$ but not when the contour does not contain $\mathbf{k} = 0$, showing the presence of a topological singularity at $\mathbf{k} = 0$. This topological singularity is the origin of the absence of backward scattering in metallic carbon nanotubes.[4,34,35]

A singularity at $\varepsilon = 0$ manifests itself in magnetic fields even in classical mechanics. The equation of motion in magnetic field $\mathbf{B}$ perpendicular to the system is given by

$$\hbar\dot{\mathbf{k}} = -\frac{e}{c}\mathbf{v}\times\mathbf{B}. \tag{13}$$

This gives the cyclotron frequency $\omega_c = eBv^2/c\varepsilon$, which diverges and changes its sign at $\varepsilon = 0$, showing that $\varepsilon = 0$ is singular in a magnetic field.[4,6,29,31]

In quantum mechanics $\hat{k}_x$ and $\hat{k}_y$ satisfy the commutation relation $[\hat{k}_x, \hat{k}_y] = -i/l^2$, where l is the magnetic length given by $l =$

$\sqrt{c\hbar/eB}$. Semiclassically, the Landau levels can be obtained by the condition

$$\oint k_x dk_y = \pm\frac{2\pi}{l^2}\Big(|n|+\frac{1}{2}\Big), \qquad (14)$$

as $\varepsilon_n=\mathrm{sgn}(n)\sqrt{|n|+1/2}\,\hbar\omega_B$ with integer n, where $\hbar\omega_B=\sqrt{2}\gamma/l$ and $\mathrm{sgn}(n)=n/|n|$ for $n\neq 0$ and 0 for $n=0$. Because of the uncertainty principle, $k^2=0$ is not allowed and there is no Landau level at $\varepsilon=0$. However, a full quantum mechanical treatment gives

$$\varepsilon_n=\mathrm{sgn}(n)\sqrt{|n|}\,\hbar\omega_B, \qquad (15)$$

leading to the formation of Landau levels at $\varepsilon=0$.[1] This can be understood by the cancellation of factor 1/2 in Eq. (14) by Berry's phase corresponding to the rotation in the $\mathbf{k}$ space.

The most direct manifestation of this zero-mode anomaly is likely to be the singular susceptibility given by

$$\chi(\varepsilon_{\mathrm{F}})=-\frac{g_{\mathrm{v}}g_{\mathrm{s}}\gamma^2}{6\pi}\Big(\frac{e}{c\hbar}\Big)^2\delta(\varepsilon_{\mathrm{F}}). \qquad (16)$$

with ε_{F} being the Fermi energy. This was obtained first by McClure[1] and later by many others.[29,36,37,38] Singularities at $\varepsilon=0$ manifest themselves in various transport quantities, such as the zero-field conductivity,[5] the dynamical conductivity,[8] the half-integer quantum Hall effect,[6] etc. Further, the system has a special time reversal symmetry $\mathbf{F}^S=K\mathbf{F}^*$ with $K=-i\sigma_y$ and therefore belongs to the so-called symplectic universality class without spin-orbit interaction.[4,39] This symmetry and associated crossover play important roles in transport in carbon nanotubes[34,35,39,40,41,42,43] and in graphene.[7,44]

3. Bilayer Graphene

We consider a bilayer graphene which is arranged in the AB (Bernal) stacking as shown in Fig. 4. The upper layer is denoted as 1 and the lower layer denoted as 2. In each layer, the unit cell contains two carbon atoms denoted by A_1 and B_1 in layer 1 and A_2 and B_2 in layer 2. For the inter-layer coupling, we include only the coupling γ_1 between vertically neighboring atoms. Then, electronic states are described by the $\mathbf{k}\cdot\mathbf{p}$ equation:[45,46]

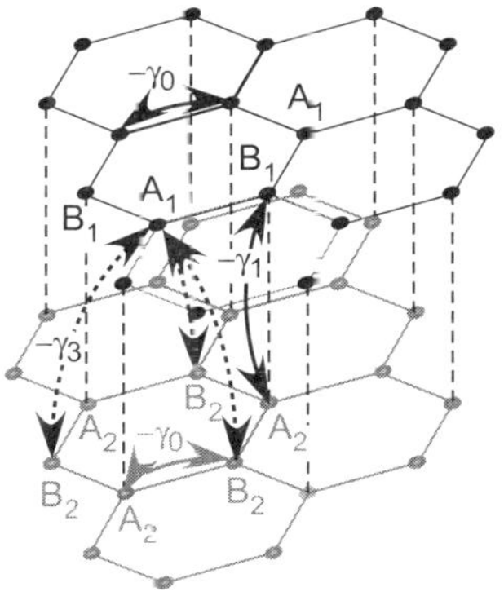

Fig. 4. The lattice structure of a bilayer graphene. Some representative hopping integrals are shown.

$$\mathcal{H}_0\mathbf{F}(\mathbf{r})=\varepsilon\mathbf{F}(\mathbf{r}), \qquad (17)$$

with

$$\mathcal{H}_0=\begin{array}{c} \\ \end{array}\!\!\begin{array}{c}\begin{array}{cccc}\mathrm{A}_1 & \mathrm{B}_1 & \mathrm{A}_2 & \mathrm{B}_2\end{array}\\ \begin{pmatrix} 0 & \gamma\hat{k}_- & 0 & 0\\ \gamma\hat{k}_+ & 0 & \gamma_1 & 0\\ 0 & \gamma_1 & 0 & \gamma\hat{k}_-\\ 0 & 0 & \gamma\hat{k}_- & 0\end{pmatrix}\end{array}. \qquad (18)$$

Figure 5 shows the corresponding energy bands.

We have $\gamma_1\approx 0.39$ eV.[47] The coupling γ_3 between vertically neighboring atoms $A_1\leftrightarrow B_2$ as shown in Fig. 4, estimated as $\sim$0.315 eV,[48] is known to cause trigonal warping and four band-touching points at zero energy in the energy scale of a few meV.[45,46] Such a structure can easily be washed out in actual bilayer graphenes with disorder[46] and will be neglected in the following. The parameter γ_4 giving coupling $A_1\leftrightarrow A_2$ and $B_1\leftrightarrow B_2$ is of the same order of magnitude as γ_3. However, this does not change the qualitative feature of the low-energy spectrum and is not important.[49,50]

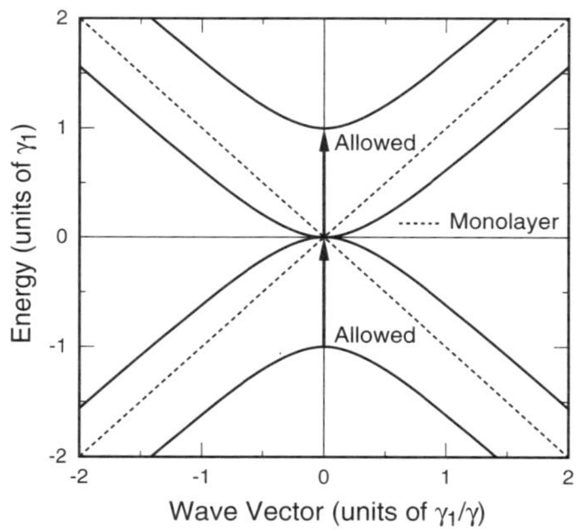

Fig. 5. The energy dispersion of a bilayer graphene. The dotted lines show that of a monolayer graphene. The allowed optical transitions at $k=0$ are denoted by the arrows.

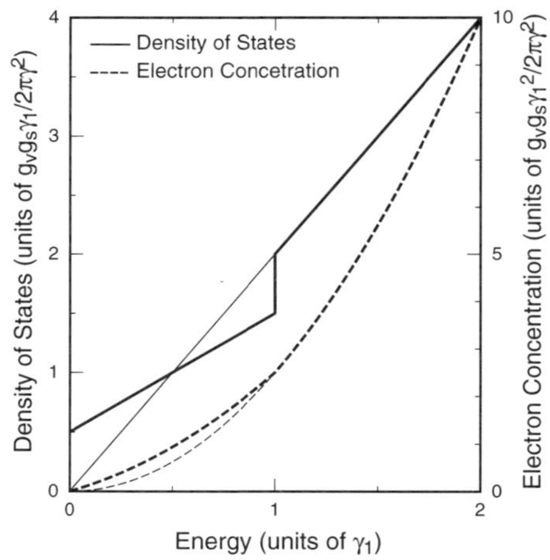

Fig. 6. The density of states and the electron concentration versus energy of a bilayer graphene. The thin lines show those in a monolayer graphene.

The density of states is given by

$$D(\varepsilon) = \frac{g_v g_s}{2\pi\gamma^2}\Big[\frac{\Delta}{2}+|\varepsilon| + \theta(|\varepsilon|-\Delta)\Big(|\varepsilon|-\frac{\Delta}{2}\Big)\Big], \tag{19}$$

where $\theta(t)$ is the step function defined by $\theta(t)=1$ for $t>0$ and 0 for $t<0$. The density of states is shown in Fig. 6.

In the vicinity of $\varepsilon=0$, the Hamiltonian is reduced to a (2,2) form for the basis set (A_1, B_2),[45,46]

$$\mathcal{H} \approx \frac{\hbar^2}{2m^*}\begin{pmatrix} 0 & \hat{k}_-^2 \\ \hat{k}_+^2 & 0 \end{pmatrix}, \tag{20}$$

with $m^*=\hbar^2\gamma_1/2\gamma^2$. We have the energy dispersion $\varepsilon(k)=\pm\hbar^2k^2/2m^*$. The corresponding Landau levels become

$$\varepsilon_{\pm,n} = \pm\hbar\omega_c\sqrt{n(n+1)}, \quad (n=0,1,\ldots), \tag{21}$$

with $\omega_c=eB/m^*c$. The Landau level $n=0$ is doubly degenerate. Because of this degeneracy, the diamagnetic susceptibility has a logarithmic singularity at $\varepsilon=0$,[38,51]

$$\chi(\varepsilon_F) = \frac{g_v g_s}{4\pi}\frac{e^2\gamma^2}{\hbar^2}\frac{\theta(\gamma_1-|\varepsilon_F|)}{\gamma_1}\Big(\ln\frac{|\varepsilon_F|}{\gamma_1}+\frac{1}{3}\Big). \tag{22}$$

There has been theoretical development on how this singularity is removed in the presence of disorder.[52,53]

4. Multi-Layer Graphene

We consider a multilayer graphene composed of N graphene layers arranged in the AB (Bernal) stacking. For the simplicity we include the nearest-neighbor intra-layer coupling parameter γ_0, and the inter-layer coupling γ_1 between A and B atoms located vertically with respect to the layer plane. We can show that the Hamiltonian of a multilayer graphene can be decomposed into those equivalent to monolayer or bilayer, which allows us to systematically study the dependence of various quantities on layer numbers.

Let $|A_j\rangle$ and $|B_j\rangle$ be the Bloch functions at the K point, corresponding to the A and B sublattices, respectively, of layer j. For convenience we divide carbon atoms into two groups as

$$\begin{aligned}\text{Group I}: &\ B_1, A_2, B_3, \cdots\\ \text{Group II}: &\ A_1, B_2, A_3, \cdots\end{aligned}$$

The atoms of group I are arranged along vertical columns normal to the layer plane, while those in group II are above or below the center of hexagons in the neighboring layers.

If the basis is taken as $|A_1\rangle, |B_1\rangle$; $|A_2\rangle, |B_2\rangle$; $\cdots$; $|A_N\rangle, |B_N\rangle$, the Hamiltonian for the multilayer graphene around the K point becomes

$$\mathcal{H} = \begin{pmatrix} H_0 & V & & & \\ V^\dagger & H_0 & V^\dagger & & \\ & V & H_0 & V & \\ & & \ddots & \ddots & \ddots \end{pmatrix}, \quad (23)$$

with

$$H_0 = \begin{pmatrix} 0 & \gamma\hat{k}_- \\ \gamma\hat{k}_+ & 0 \end{pmatrix}, \quad V = \begin{pmatrix} 0 & 0 \\ \gamma_1 & 0 \end{pmatrix}. \quad (24)$$

The effective Hamiltonian for K' is obtained by exchanging $\hat{k}_+$ and $\hat{k}-$.

The Hamiltonian (23) can be decomposed into smaller subsystems for the basis appropriately chosen.[54,55] First, we define the orthonormal sets

$$\phi_l^{(\mathrm{I})} = \psi_l^{(1)}|B_1\rangle + \psi_l^{(2)}|A_2\rangle + \psi_l^{(3)}|B_3\rangle + \cdots,$$
$$\phi_l^{(\mathrm{II})} = \psi_l^{(1)}|A_1\rangle + \psi_l^{(2)}|B_2\rangle + \psi_l^{(3)}|A_3\rangle + \cdots, \quad (25)$$

where

$$\psi_l^{(j)} = \sqrt{\frac{2}{N+1}}\sin j\kappa_l, \quad \kappa_l = \frac{\pi}{2} - \frac{l\pi}{2(N+1)}, \quad (26)$$

with

$$l = -(N-1),\ -(N-3),\ \ldots,\ N-1. \quad (27)$$

Here, l is an odd integer when N is even, while l is even when N is odd, and therefore $l=0$ is allowed only for odd N.

Next, for $m>0$, we take the basis

$$\{(\phi_m^{(\mathrm{II})}+\phi_{-m}^{(\mathrm{II})})/\sqrt{2}, \quad (\phi_m^{(\mathrm{I})}+\phi_{-m}^{(\mathrm{I})})/\sqrt{2},$$
$$(\phi_m^{(\mathrm{I})}-\phi_{-m}^{(\mathrm{I})})/\sqrt{2}, \quad (\phi_m^{(\mathrm{II})}-\phi_{-m}^{(\mathrm{II})})/\sqrt{2}\}. \quad (28)$$

For $m=0$, we take the basis $\{\phi_0^{(\mathrm{II})}, \phi_0^{(\mathrm{I})}\}$. Then, the Hamiltonian has no off-diagonal elements between different m's. For $m>0$, the sub-Hamiltonian within the basis Eq. (28) becomes

$$\mathcal{H}_m = \begin{pmatrix} 0 & \gamma\hat{k}_- & 0 & 0 \\ \gamma\hat{k}_+ & 0 & \lambda_m\gamma_1 & 0 \\ 0 & \lambda_m\gamma_1 & 0 & \gamma\hat{k}_- \\ 0 & 0 & \gamma\hat{k}_+ & 0 \end{pmatrix}, \quad (29)$$

with

$$\lambda_m = 2\cos\kappa_m, \quad (30)$$

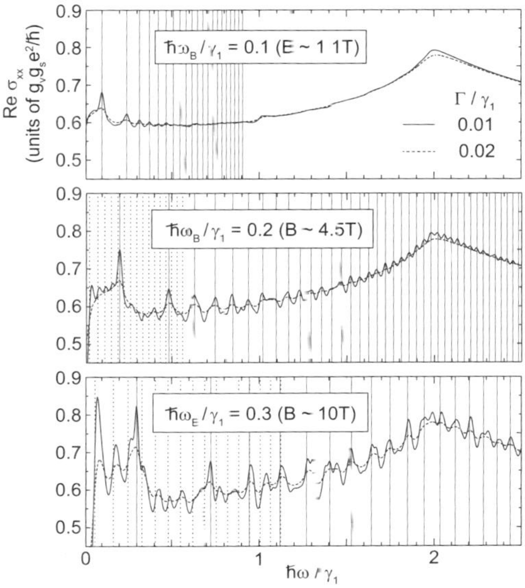

Fig. 7. Examples of the real part of the dynamical conductivity averaged over the layer numbers from $N=1$ to 20, for magnetic fields with $\hbar\omega_B/\gamma_1=0.1$, 0.2 and 0.3. Vertical solid and dashed lines represent transition energies for monolayer-type ($\kappa=\pi/2$) and bilayer-type ($\kappa=0$) subbands, respectively.

which is equivalent to the Hamiltonian of bilayer graphene, while the inter-layer coupling γ_1 is multiplied by λ_m. For $m=0$, we have the Hamiltonian identical to that of the monolayer graphene. These subsystems are labeled as $m=0, 2, 4, \cdots, N-1$ for odd N and $m=1, 3, 5, \cdots, N-1$ for even N.

The eigenstate of a finite-layer graphene can be regarded as a part of a standing wave in bulk graphite, which is a superposition of opposite traveling waves with $\pm k_z$. The quantity κ $(=\kappa_m)$ in our representation corresponds to a wave number via $\kappa=|k_z|c_0/2$, where $c_0/2$ is the interlayer distance. Thus the monolayer-type subband $\kappa=\pi/2$ is related to an H point in the Brillouin zone of graphite, while no states correspond to $k_z=0$ since κ never becomes zero.

Thus, the Hamiltonian has been decomposed into those equivalent to the monolayer or bilayer bands. The monolayer-like band exists only in odd-layered graphenes,

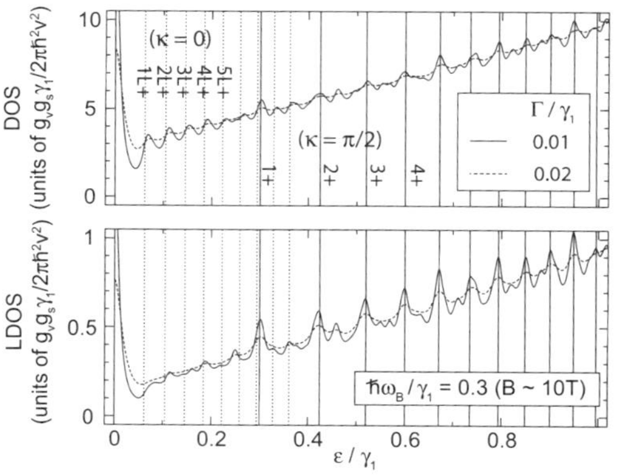

Fig. 8. (Top) Density of states averaged over the layer numbers $N = 1$ to 20, at $\hbar\omega_B/\gamma_1 = 0.3$. (Bottom) Corresponding plot for the local density of states on the top layer. Vertical solid and dashed lines indicate the ideal Landau level energies for the monolayer-type ($\kappa = \pi/2$) and the bilayer-type ($\kappa = 0$) subbands, respectively.

while the bilayer-like bands always exist. This decomposition can be used for the analysis of diamagnetic response of multilayer graphenes.[54] In fact, in odd-layered graphenes, the monolayer-like band gives a strong diamagnetic peak at $\varepsilon_F = 0$, while the bilayer-like bands present a broad peak. With the increase of the layer number, the susceptibility is dominated by that of bilayer-like bands.

This decomposition rule can also be used for the dynamical conductivity describing optical absorption and the local density of states observed by scanning tunneling spectroscopy.[55] Figure 7 shows the dynamical conductivity averaged over the samples $N = 1, 2, \cdots, 20$. Surprisingly, we still see the series of peaks in the monolayer graphene $\hbar\omega \propto \sqrt{B}$. This comes from the monolayer-type subbands which appears in every odd layered graphene.

In Fig. 8, the top panel shows the total density of states and the bottom panel shows the corresponding plot for the local density of states on the top layer. In the total density of states, we observe the several peaks coming from the monolayer-type subband similarly to the optical absorption spectra. In the local density of states, interestingly, the peaks of the monolayer-type subband are much more pronounced.

5. Summary and Conclusion

In summary, the electron motion in monolayer graphene is equivalent to that of a massless Dirac particle, characterized by k-linear dispersion and Berry's phase π. In bilayer graphene the dispersion is proportional to k^2 with Berry's phase 2π. In multilayer graphenes, the Hamiltonian is decomposed into those of a monolayer and bilayers with renormlized interlayer coupling.

Acknowledgments

This work was supported in part by Grants-in-Aid for Scientific Research and for Scientific Research on Priority Area "Carbon Nanotube Nanoelectronics" from the Ministry of Education, Culture, Sports, Science and Technology, Japan.

References

1. J. W. McClure, *Phys. Rev.* **104**, 666 (1956).
2. J. C. Slonczewski and P. R. Weiss, *Phys. Rev.* **109**, 272 (1958).
3. T. Ando, *Semicond. Sci. Technol.* **15**, R13 (2000); *Solid State Commun.* **12**, 69 (2003); *Physica E* **20**, 24 (2003); *Physics E* **22**, 656 (2004).
4. T. Ando, *J. Phys. Soc. Jpn.* **74**, 777 (2005).
5. N. H. Shon and T. Ando, *J. Phys. Soc. Jpn.* **67**, 2421 (1998).
6. Y. Zheng and T. Ando, *Phys. Rev. B* **65**, 245420 (2002).
7. H. Suzuura and T. Ando, *Phys. Rev. Lett.* **89**, 266603 (2002).
8. T. Ando, Y. Zheng, and H. Suzuura, *J. Phys. Soc. Jpn.* **71**, 1318 (2002).
9. T. Ando, A. B. Fowler, and F. Stern, *Rev. Mod. Phys.* **54**, 437 (1982) and references cited therein.
10. K. S. Novoselov, A. K. Geim, S. V. Morozov, D. Jiang, Y. Zhang, S. V. Dubonos, I. V. Grigorieva, and A. A. Firsov, *Science* **306**, 666 (2004).

11. K. S. Novoselov, A. K. Geim, S. V. Morozov, D. Jiang, M. I. Katsnelson, I. V. Grigorieva, S. V. Dubonos, and A. A. Firsov, *Nature* **438**, 197 (2005).
12. Y. Zhang, Y.-W. Tan, H. L. Stormer, and P. Kim, *Nature* **438**, 201 (2005).
13. Y. Zhang, Z. Jiang, J. P. Small, M. S. Purewal, Y.-W. Tan, M. Fazlollahi, J. D. Chudow, J. A. Jaszczak, H. L. Stormer, and P. Kim, *Phys. Rev. Lett.* **96**, 136806 (2006).
14. M. L. Sadowski, G. Martinez, M. Potemski, C. Berger, and W. A. de Heer, *Phys. Rev. Lett.* **97**, 266405 (2006).
15. C. Berger, Z. Song, T. Li, X. Li, A. Y. Ogbazghi, R. Feng, Z. Dai, A. N. Marchenkov, E. H. Conrad, P. N. First, and W. A. de Heer, *J. Phys. Chem. B* **108**, 19912 (2004).
16. J. Hass, R. Feng, T. Li, X. Li, Z. Zong, W. A. de Heer, P. N. First, E. H. Conrad, C. A. Jeffrey, and C. Berger, *Appl. Phys. Lett.* **89**, 143106 (2006).
17. J. Jiang, R. Saito, G. G. Samsonidze, A. Jorio, S. G. Chou, G. Dresselhaus, and M. S. Dresselhaus, *Phys. Rev. B* **75**, 035407 (2007).
18. R. S. Deacon, K.-C. Chuang, R. J. Nicholas, K. S. Novoselov, and A. K. Geim, *Phys. Rev. B* **76**, 081406 (2007).
19. J. Martin, N. Akerman, G. Ulbricht, T. Lohmann, J. H. Smet, K. von Klitzing, and A. Yacoby, *Nat. Phys.* **4**, 144 (2008).
20. S. V. Morozov, K. S. Novoselov, M. I. Katsnelson, F. Schedin, L. A. Ponomarenko, D. Jiang, and A. K. Geim, *Phys. Rev. Lett.* **97**, 016801 (2006).
21. X.-S. Wu, X.-B. Li, Z.-M. Song, C. Berger, and W. A. de Heer, *Phys. Rev. Lett.* **98**, 136801 (2007).
22. F. V. Tikhonenko, D. W. Horsell, R. V. Gorbachev, and A. K. Savchenko, *Phys. Rev. Lett.* **100**, 056802 (2008).
23. J. Yan, Y. Zhang, P. Kim, and A. Pinczuk, *Phys. Rev. Lett.* **98**, 166802 (2007).
24. S. Pisana, M. Lazzeri, C. Casiraghi, K. S. Novoselov, A. K. Geim, A. C. Ferrari, and F. Mauri, *Nat. Mat.* **6**, 198 (2007).
25. N. Tombros, C. Jozsa, M. Popinciuc, H. T. Jonkman, and B. J. van Wees, *Nature* **448**, 571 (2007).
26. M. Ohishi, M. Shiraishi, R. Nouchi, T. Nozaki, T. Shinjo, and Y. Suzuki, *Jpn. J. Appl. Phys.* **46**, L605 (2007).
27. T. Ohta, A. Bostwick, T. Seyller, K. Horn, and E. Rotenberg, *Science* **313**, 951 (2006)
28. T. Ohta, A. Bostwick, J. L. McChesney, T. Seyller, K. Horn, and E. Rotenberg, *Phys. Rev. Lett.* **98**, 206802 (2007).
29. T. Ando, *Physica E* **40**, 213 (2007) and references cited therein.
30. W. W. Toy, M. S. Dresselhaus, and G. Dresselhaus, *Phys. Rev. B* **15**, 4077 (1977).
31. T. Ando, *J. Phys. Soc. Jpn.* **75**, 074716 (2006).
32. M. V. Berry, *Proc. Roy. Soc. London* **A392**, 45 (1984).
33. B. Simon, *Phys. Rev. Lett.* **51**, 2167 (1983).
34. T. Ando and T. Nakanishi, *J. Phys. Soc. Jpn.* **67**, 1704 (1998).
35. T. Ando, T. Nakanishi, and R. Saito, *J. Phys. Soc. Jpn.* **67**, 2857 (1998).
36. S. A. Safran and F. J. DiSalvo, *Phys. Rev. B* **20**, 4889 (1979).
37. J. Blinowski and C. Rigaux, *J. Phys. (Paris)* **45**, 545 (1984).
38. R. Saito and H. Kamimura, *Phys. Rev. B* **33**, 7218 (1986).
39. T. Ando and H. Suzuura, *J. Phys. Soc. Jpn.* **71**, 2753 (2002).
40. T. Ando, *J. Phys. Soc. Jpn.* **73**, 1273 (2004).
41. T. Ando and K. Akimoto, *J. Phys. Soc. Jpn.* **73**, 1895 (2004).
42. K. Akimoto and T. Ando, *J. Phys. Soc. Jpn.* **73**, 2194 (2004)
43. T. Ando, *J. Phys. Soc. Jpn.* **75**, 054701 (2006).
44. E. McCann, K. Kechedzhi, V. I. Falko, H. Suzuura, T. Ando, and B. L. Altshuler, *Phys. Rev. Lett.* **97**, 146805 (2006).
45. E. McCann and V. I. Falko, *Phys. Rev. Lett.* **96**, 086805 (2006).
46. M. Koshino and T. Ando, *Phys. Rev. B* **73**, 245403 (2006).
47. A. Misu, E. Mendez, and M. S. Dresselhaus, *J. Phys. Soc. Jpn.* **47**, 199 (1979).
48. R. E. Doezema, W. R. Datars, H. Schaber, and A. Van Schyndel, *Phys. Rev. B* **19**, 4224 (1979).
49. J. W. McClure, *Phys. Rev.* **108**, 612 (1957).
50. J. W. McClure, *Phys. Rev.* **119**, 606 (1960).
51. S. A. Safran, *Phys. Rev. B* **30**, 421 (1984).
52. H. Fukuyama, *J. Phys. Soc. Jpn.* **76**, 043711 (2007).
53. M. Koshino and T. Ando, *Phys. Rev. B* **75**, 235333 (2007)
54. M. Koshino and T. Ando, *Phys. Rev. B* **76**, 085425 (2007).
55. M. Koshino and T. Ando, *Phys. Rev. B* **77**, 115313 (2008).

INTER-LAYER MAGNETORESISTANCE IN MULTILAYER MASSLESS DIRAC FERMIONS SYSTEM α-(BEDT-TTF)$_2$I$_3$

N. TAJIMA*

RIKEN, Hirosawa 2-1, Wako-shi, Saitama, 351-0198, Japan.
*E-mail: taji@riken.jp

S. SUGAWARA, Y. NISHIO and K. KAJITA

Department of Physics, Toho University - Miyama 2-2-1, Funabashi-shi, Chiba 274-8510, Japan

Zero-gap state with the Dirac cone type energy dispersion has been found in an organic conductor α-(BEDT-TTF)$_2$I$_3$ under high hydrostatic pressures. This is the first two-dimensional zero-gap state discovered in bulk crystals with layered structure and exhibits transport phenomena characteristic to electrons on the Dirac cone type energy structure. We find a characteristic transport feature of bulk system in inter-layer resistance. In the zero-gap system, n=0 Landau level called zero-mode always appear at the contact point of Dirac cone under the magnetic field and the degeneracy increases proportional to the strength of the magnetic field. The effects of zero-mode conduces the remarkable negative inter-layer magnetoresistance at low temperatures.

Keywords: α-(BEDT-TTF)$_2$I$_3$; massless Dirac Fermions system; zero-mode Landau level, inter-layer resistivity

1. Introduction

Recently, Novoselov *et al.*[1] and Zhang *et al.*[2] have experimentally demonstrated that a graphene (a single layer of graphite) is a zero-gap-semiconductor with massless Dirac particles. In this paper, we report the discovery of a bulk zero-gap-semiconductor. It was realized in a quasi-two dimensional organic conductor α-(BEDT-TTF)$_2$I$_3$ under high pressures.[3-8] This is the first bulk material with zero gap energy band and exhibits transport phenomena characteristic to electrons on the Dirac cone type energy structure. We have no other example. Thus, α-(BEDT-TTF)$_2$I$_3$ provides us a testing ground for the transport of multilayer massless Dirac fermions system. In this paper, we demonstrate that this system shows strong negative inter-layer magnetoresistance at low temperatures. This is a characteristic feature on transport in multilayer Dirac Fermions system.

A crystal of α-(BEDT-TTF)$_2$I$_3$ consists of conductive layers of BEDT-TTF molecules and insulating layers of I_3^- anions (Fig. 1).[9] Since the conductive layers are separated by insulating layers, carriers in this system have strongly two-dimensional (2D) nature. The ratio of the in-plane conductivity to the out-of-plane conductivity is about 10^3. Therefore, each conductive layer is expected to be almost

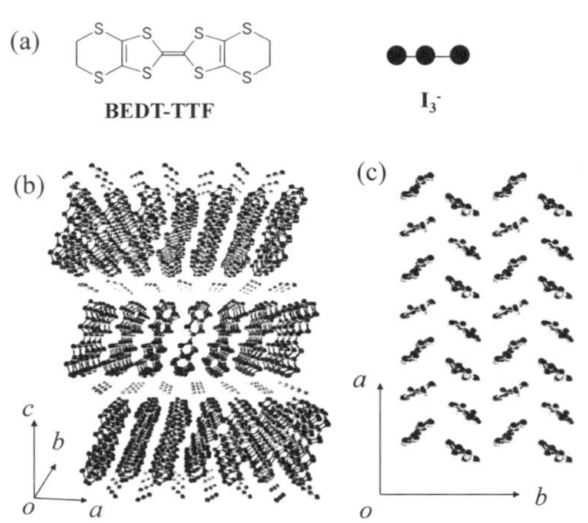

Fig. 1. (a) BEDT-TTF molecule and I_3^- molecule. (b) Crystal structure of α-(BEDT-TTF)$_2$I$_3$ viewed from *b*-axis. (c) Crystal structure viewed from *c*-axis. (After ref. 4)

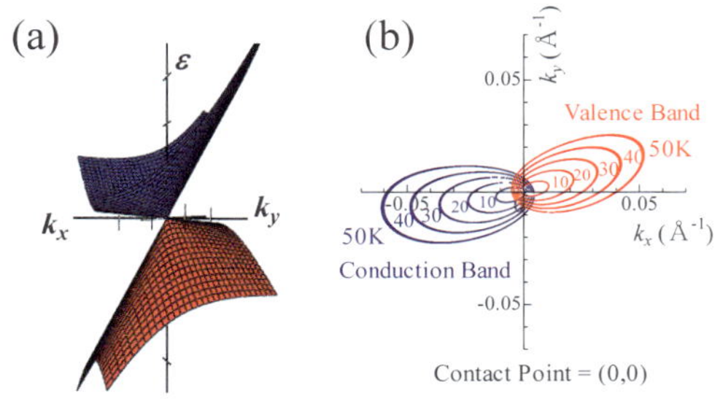

Fig. 2. (a) Band structure and (b) energy contours near a contact point of Dirac cone. They are calculated using the parameters for p=0.6 GPa in ref. 8. Note that the origins of the axes are taken at the position of the contact point. (After ref. 6)

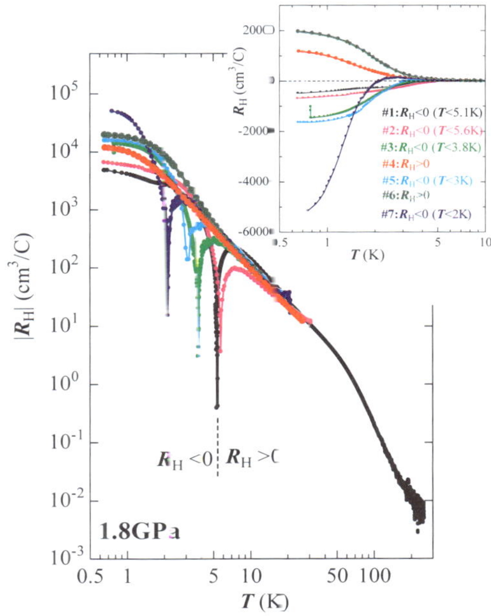

Fig. 3. Sample dependence of the Hall coefficient. When cooled, some samples cause a sign inversion of Hall coefficient from positive to negative at dip.

independent. Under the ambient pressure, this material behaves as a metal down to 135 K where it undergoes a phase transition to a charge-ordered insulator. In the low temperature phase, horizontal charge stripes pattern for +1 e and 0 is formed.[10-13] When the crystal is placed under a high hydrostatic pressure above 1.5 GPa, the metal-insulator transition is suppressed and the metallic region expands to low temperatures.[14-16] This change in the electronic state is accompanied by the disappearance of the charge ordering state as shown by the Raman experiment.[13] Note that in the high pressure phase, the resistance is almost constant over the temperature region from 300 K to 2 K. It looks like a dirty metal with high density of impurities or lattice defects. In dirty metals, mobilities of carriers are low and depend on temperature only weakly, and therefore, resistance is also independent of temperature. In the present case, however, the situation is different. Carriers in this material were found to have a very high mobility about 10^5 cm^2/V.s at low temperatures.[3, 4] It indicates that the crystal is clean. In our investigation of the magnetoresistance and Hall effect from 1.5 K to 300 K, both the carrier density (hole) and the mobility change by about six orders of magnitude so that the effect just cancel out giving rise to the temperature independent conductivity. We claimed that this is a new type of conductor, which behaves as a semiconductor with an extremely narrow gap less than 1 meV.[3, 4]

Recently, Kobayashi *et al.* published a paper that supports our conclusion. They performed an energy band calculation of α-(BEDT-TTF)$_2$I$_3$ and found that this material under high pressures is in the zero-gap state.[6,7] According to refs. 7 and 8, the bottom of the conduction band and the top of the valence band contact to each other at two points (We call them as "contact points") in the first Brillouin zone as shown in Fig. 2. These results were supported by first principle band calculations.[17]

In the picture of zero-gap system, remarkable transport phenomena are naturally understood as follows. The carrier density written as $n \propto T^2$ is a characteristic feature of

2D zero-gap structure.[4, 6] The resistivity per layer (sheet resistance R_S) is given as $R_S = h/e^2$ which is temperature independent[6] is characteristic transport of zero-gap conductors.[18] Recently, Kobayashi *et al.* clearly explained the origin of anomalous Hall effect as shown in Fig. 3, which some samples cause a sign inversion of Hall coefficient from positive to negative at low temperatures.[19] According to their theory, lack of I_3-molecules gives rise to the effect of electron doping in a crystal so that the strong sample dependence of the Hall coefficient dose exist as shown in Fig. 3. The strange temperature dependences of Hall coefficient are also clear signatures of the existence of zero gap state.

One problem that remains to be solved is anomalous transport phenomena under the magnetic field at low temperatures.[4, 14, 15] Transport phenomena of this system are very sensitive to the magnetic field at low temperatures. Three stepwise changes in the in-plane resistivity were observed, when the magnetic field was increased or the temperature were decreased.[4] Both the quantum oscillations and the quantum Hall effect, on the other hand, were not observed in the magnetic field up to 15T.

In this paper, we explain the electronic structure of this zero-gap system in the magnetic field.

2. Experiments

In order to understand the electronic structure in the magnetic field, Inter-layer magnetotransport was investigated at low temperatures.

A sample on which four electrical leads were attached was encased in a Teflon capsule filled with pressure medium (Idemitsu DN-oil 7373). The capsule was set in a clamp-type pressure cell made of MP35N hard alloy and then hydrostatic pressure was applied up to 1.7 GPa. The resistance measurements were carried out using a conventional DC method with the electrical current in the *c*-crystal axis, which is normal to the 2D plane under the magnetic field below 3 T.

3. Electronic state in the magnetic field

Figure 4 shows the magnetic field dependences of inter-layer magentoresistance at 4 K. The behavior of resistivity changes depending on the direction of the magnetic field. In the magnetic field parallel to 2D plane, a gradual growth of resistivity was observed. In the field normal to the 2D plane, on the other hand, the resistivity also increases in low fields below 0.2T. Then, it turns to decreases. The drop of the resistivity is large. In the field of 3 T, the resistivity is about 1/3 of that at 0.2 T. In-plane magnetoresistance of this system, on the other hand, shows positive behavior in the magnetic field normal to 2D plane.[4,14,15] At

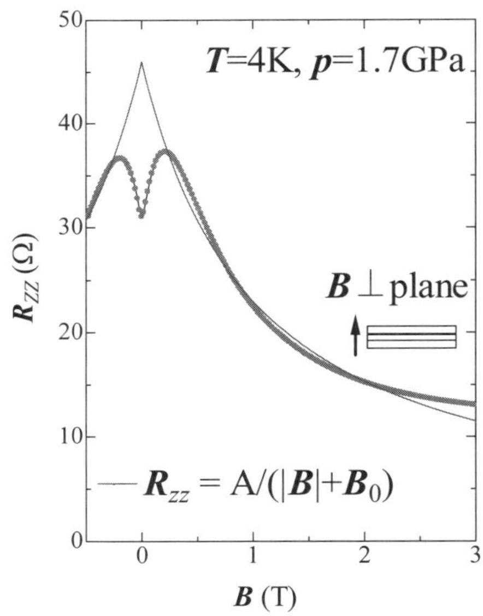

Fig. 4. Magnetic field dependence of the inter-layer resistance at 4 K. When the magnetic field is applied along normal to 2D plane, remarkable nagative magnetoresistance are observed. As for negative magnetoresistance, the fitting is done by the formura R_{zz}=A/(|B|+B_0).

4 K, it rapidly increases by about one order of magnitude from 0 to 1 T. To our knowledge, this is the first material which shows those remarkable magnetoresistance. It is very hard to explain those magnetoresistance behaviors by the picture of conventional semiclassical Boltzmann magnetotransport. We demonstrate that the negative inter-layer magnetoresistance as shown in Fig. 4 is one of the characteristic transport features in multilayer Dirac Fermions system.

Important difference of zero gap conductors from usual conductors is the appearance of a Landau level at the contact points when magnetic fields are applied normal to the conductive layer. Since the energy of this level is E_F irrespective of the field strength, the Fermi distribution function is always 1/2. It means that a half of Landau states degenerated in zero-mode are occupied. Landau degeneracy is proportional to the strength of magnetic field B and therefore it can be a high value in high magnetic fields. This effect is expected to be detected in the inter-layer resistance in the transverse magnetic field as pointed out by Osada.[20] The field parallel to the current do not exert the Lorentz force on carriers, so that the motion of carriers in the current direction is not affected by the magnetic fields. Hence, the conductivity in this direction is proportional to the carrier density. The effect of magnetic field appears only if the change in the carrier density occurs.

To understand the effect of zero mode Landau carriers, we should examine them quantitatively. The energy of Landau levels in zero-gap systems are expressed as $E_{nLL} = \pm(2e\hbar v_F^2|n|B)^{1/2}$, where v_F is the Fermi velocity and n is the Landau index.[20] In drawing Fig. 5, we used $v_F=10^7$ cm/s estimated from the band calculation by Katayama *et al.*[8] According to the band calculation, the Dirac cone in this system has highly anisotropic Fermi velocity v_F (ϕ).[7,8] Therefore, the present v_F is some average of v_F (ϕ).

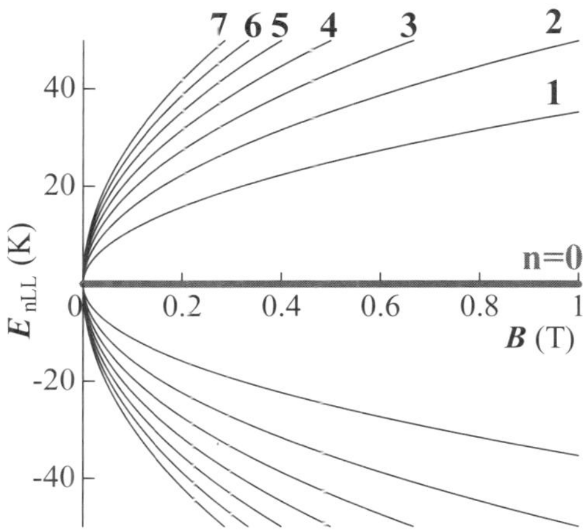

Fig. 5. Landau levels of zero-gap system α-$(BEDT\text{-}TTF)_2I_3$.

At low temperatures, the carriers on the Landau level with index n=0 (zero-mode) dominants the transport phenomena in the magnetic field because other Landau levels are far from the Fermi energy as shown in Fig. 5. At 1 T, for example, the energy gap between the zero-mode level and the n=1 level is about 30 K. If the experiment were performed at temperatures below 30 K and at magnetic fields above 1 T, most of the mobile carriers would be in the zero-mode Landau level. Such a situation expressed as $k_BT < E_{1LL}$ is called the "quantum limit". Important fact is that the degeneracy of zero-mode is increased in proportional to the magnetic field. Thus, change in the density of zero-mode carriers is the origin of the negative magnetoresistance, as shown in Fig. 4.

Quite recently, Osada gave an analytical formula of the inter-layer magnetoresistance in the multilayer Dirac Fermions system.[20] It is expressed as $R_{zz} = A/(|B|+B_0)$ for B normal to 2D plane, where A was assumed as a field-independent constant in the case of pure system and B_0 was fitting parameter. Important is that this formula shows strongly negative magnetoresistance proportional to $1/|B|$. When we assumed B_0 = 0.7 T, this simple formula reproduces well the field dependences of the resistance except for the low field region below 0.2 T as shown by solid lines in Fig. 4.

To conclude this discussion, this experiment evidences the existence of zero-mode Landau carriers in α-(BEDT-TTF)$_2$I$_3$ under high pressures.

An apparent discrepancy of the data from the formula is seen at low magnetic fields, because the model is oversimplified. This formula was derived based on the quantum limit picture that only the zero-mode Landau level exists at the Fermi level. Since each Landau level is broadened by scattering, zero-mode Landau level is sure to overlap with other Landau levels at low magnetic field. In such a region the formula loses its validity. We can recognize this region in Fig. 4 below 0.2 T where positive magnetoresistance is observed. This critical magnetic field B_p shifts to a lower field with decreasing temperature, as shown in Fig. 6. The overlap between zero-mode and other Landau levels will be sufficiently small above the critical magnetic field, as a result there the negative magnetoresistance is observed. Figure 5 support this conclusion. At 0.2 T and 4 K, the energy of n=1 Landau level of about 12 K is sufficiently large compared with the thermal energy $2k_BT$.

4. Conclusions

In conclusion, we succeeded in detecting the zero-mode Landau level in massless Dirac Fermions system α-(BEDT-TTF)$_2$I$_3$ under high hydrostatic pressure. The characteristic feature of zero-mode Landau carriers was clearly seen in the inter-layer magnetoresistance. Increase of zero-mode degeneracy by the magnetic field gives rise to strong negative inter-layer magnetoresistance proportional to $1/|B|$ at low temperatures. This is a characteristic feature on transport in multilayer massless Dirac Fermions systems.

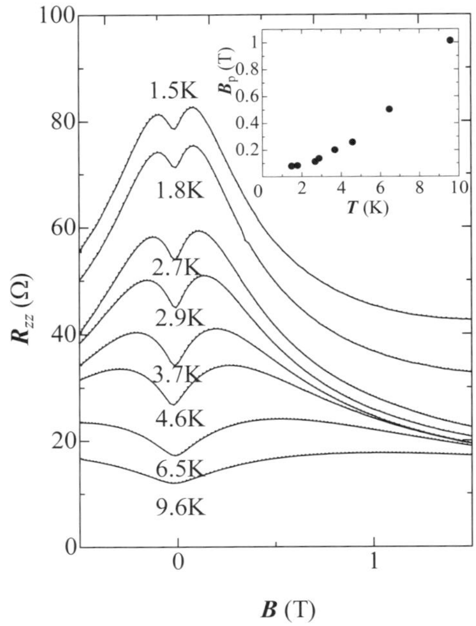

Fig. 6. Magnetic field dependence of inter-layer resistance for the magnetic field normal to the 2D plane at several temperatures from 1.5 K to 9.6 K. Inset is the temperature dependence of the critical field B_p.

Acknowledgments

We thank Prof. T. Osada, Dr. A. Kobayashi, Dr. S. Katayama, Prof. Y. Suzumura, Dr. R. Kondo, Dr. T. Morinari, Prof. T. Tohyama and Prof. H. Fukuyama for valuable discussions. This work was supported by Grants-in-Aid for Scientific Research (No. 19540387) from the Ministry of Education, Culture, Sports, Science and Technology, Japan.

References

1. K. S. Novoselov, A. K. Geim, *et al.*, *Nature* **438,** 197 (2005).
2. Y. Zhang, Y. W. Tan, *et al.*, Nature **438**, 201 (2005).
3. N. Tajima, *et al.*, *J. Phys. Soc. Jpn.* **69**, 543 (2000).
4. N. Tajima, *et al.*, *J. Phys. Soc. Jpn.* **75**, 051010 (2006).

5. N. Tajima, *et al.*, *J. Phys. Soc. Jpn.* **71**, 1832 (2002).
6. N. Tajima, *et al.*, *Europhys. Lett.* **80**, 47002 (2007).
7. A. Kobayashi, *et al.*, *J. Phys. Soc. Jpn.* **73**, 3135 (2004).
8. S. Katayama, *et al.*, *J. Phys. Soc. Jpn.* **75**, 054705 (2006).
9. K. Bender, *et al.*, *Mol. Cryst. Liq. Cryst.* **108**, 359 (1984).
10. H. Kino and H. Fukuyama, *J. Phys. Soc. Jpn.* **64**, 1877 (1995).
11. H. Seo, *J. Phys. Soc. Jpn.* **69**, 805 (2000).
12. Y. Takano, *et al.*, *J. Phys. Chem. Solid* **62**, 393 (2001).
13. R. Wojciechowski, *et al.*, *Phys. Rev. B* **67**, 224105 (2003).
14. T. Ojiro, *et al.*, *Synth. Met.* **56**, 2268 (1993).
15. K. Kajita, *et al.*, *J. Phys. Soc. Jpn.* **61**, 23 (1993).
16. T. Mishima, *et al.*, *Synth. Met.* **70**, 771 (1995).
17. H. Kino and T. Miyazaki, *J. Phys. Soc. Jpn.* **75**, 034704 (2006).
18. N. H. Shon and T.Ando, *J. Phys. Soc. Jpn.* **67**, 2421 (1998).
19. A. Kbayashi, *et al.*, *J. Phys. Soc. Jpn* **77**, 064718 (2008).
20. T. Osada, *J. Phys. Soc. Jpn.* **77**, 084711 (2008).

THEORY ON ELECTRONIC PROPERTIES OF GAPLESS STATES IN MOLECULAR SOLIDS, α-(BEDT-TTF)$_2$I$_3$

A. KOBAYASHI*

Institute for Advanced Research, Nagoya University, Furo-cho, Chikusa-ku, Nagoya, 464-8602, Japan

**E-mail: akito@slab.phys.nagoya-u.ac.jp*

Y. SUZUMURA

Department of Physics, Nagoya University, Furo-cho, Chikusa-ku, Nagoya, 464-8602, Japan

H. FUKUYAMA

Department of Applied Physics, Faculty of Science, Tokyo University of Science, 1-3 Kagurazaka, Shinjuku-ku, Tokyo 162-8601, Japan

In molecular solids α-(BEDT-TTF)$_2$I$_3$, it has been clarified that the gapless state originates from extremely anisotropic massless Dirac fermions described by the tilted Weyl equation. The inter-band effects of magnetic filed and extremely small amount of electron doping give rise to anomalous transport phenomena, *i.e.* the sharp and continuous reversal of the sign of the Hall coefficient at low temperatures. Among materials having Dirac fermions, α-(BEDT-TTF)$_2$I$_3$ has a unique feature of layered bulk structure, where motions of the fermions are highly two-dimensional but have weak inter-layer hopping. In order to study how this unique feature affects the electronic properties, we examine roles of the weak inter-layer hopping in the conductivity, the Hall conductivity, and the orbital susceptibility. It is shown how the sharpness of the sign reversal of the Hall coefficient is reduced by the inter-layer hopping. The sharp sign-reversal observed in α-(BEDT-TTF)$_2$I$_3$ can be explained when the inter-layer hopping is comparable or smaller than the effective spectrum broadening.

Keywords: massless Dirac fermion; tilted Weyl equation; zerogap state; α-(BEDT-TTF)$_2$I$_3$; inter-band effects of magnetic field; Hall effect, inter-layer hopping

1. Introduction

It has been clarified theoretically that the gapless state in molecular solid α-(BEDT-TTF)$_2$I$_3$ comes from the unusual zero-gap fermions.[1] Those fermions are elucidated to be extremely anisotropic massless Dirac fermions described by the tilted Weyl equation.[2] This theoretical finding is expected to resolve long standing mysteries of experimental findings of the weak temperature (T) dependence of resistivity with the sheet resistivity close to h/e^2 and the strong increase of Hall coefficient with decreasing T.[3,4]

In the previous study,[5] we calculated the conductivity, the Hall conductivity, and the orbital susceptibility in the tilted Weyl equation. By taking into account the T-dependence of the chemical potential given by the 1st principle calculation,[6] it is found that the Hall coefficient is enhanced strongly with decreasing T. It is also proposed that the sharp but continuous reversal of the sign of the Hall coefficient is possible if the extremely small amount of electron doping exists.[5] These results can explain the anomalous features of the Hall coefficient observed in α-(BEDT-TTF)$_2$I$_3$.[7] In addition, it is shown that the orbital diamagnetism should have a peak at the temperatures where the Hall coefficient changes sign. These temperature dependences originate from the inter-band effects of magnetic field.[8]

The layered structure is one of unique features of α-(BEDT-TTF)$_2$I$_3$ among materials having Dirac fermions such as graphene.[9] The inter-layer hopping has been estimated to be smaller than 10^{-3} eV,[6] although the accurate value has not been determined. In the present study, the roles of the inter-layer hopping for the conductivity, the Hall conductivity and the orbital susceptibility are investigated in order to study how the layered structure affects the transport phenomena and the orbital diamagnetism.

2. Tilted Weyl Equation

Figure 1 shows the energy dispersions of α-(BEDT-TTF)$_2$I$_3$ in the vicinity of the crossing point, where the unit of energy is electron volt (eV). The upper and lower cones correspond to the conduction and valence bands, respectively. The chemical potential coincides with the crossing point at $T = 0$. The linear dispersions exist within 100 meV around the crossing point. These characteristic band structures have been theoretically concluded[1] by using the transfer integrals obtained from structural analysis,[10] and then these have been verified by the first principle calculation.[6,11]

It has been proposed that quasi-particles around the crossing point in α-(BEDT-TTF)$_2$I$_3$ are described by the tilted Weyl equation[2]

$$\mathcal{H} = \sum_{\rho=0}^{3} \mathbf{k} \cdot \mathbf{v}_\rho \sigma_\rho, \tag{1}$$

which is rigorous in the vicinity of the crossing point, where $\sigma_{1,2,3}$ are the Pauli matrices and σ_0 is the identity matrix. We take the Planck constant $\hbar$ and the lattice constant a as $\hbar = a = 1$. The band energy dispersions $\xi_{\mathbf{k}}^{\pm}$ are obtained by diagonalization,

$$\xi_{\mathbf{k}}^{\pm} = \mathbf{k} \cdot \mathbf{v}_0 \pm \sqrt{\sum_\nu (\mathbf{k} \cdot \mathbf{v}_\nu)^2}, \tag{2}$$

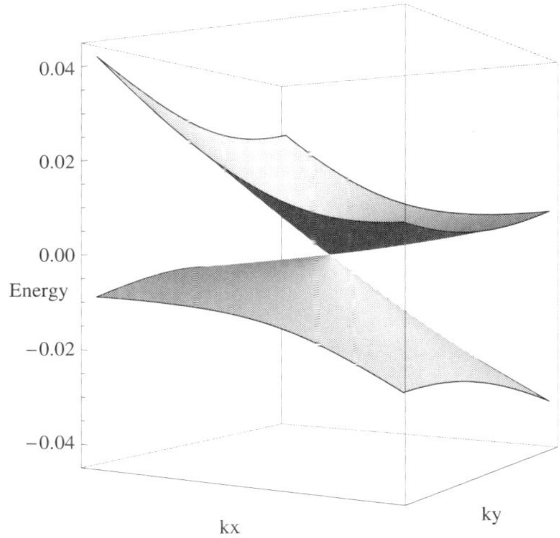

Fig. 1. The zerogap fermion in α-(BEDT-TTF)$_2$I$_3$. The vertical axis represents the band energy in unit of eV, on the k_x-k_y plane in the vicinity of the crossing point ($k_x^0 - 0.1\pi/a \le k_x \le k_x^0 + 0.1\pi/a$ and $k_y^0 - 0.1\pi/a \le k_y \le k_y^0 + 0.1\pi/a$).

where $\nu = 1, 2, 3$. Based on the transfer integrals given by the first principle calculation[6], we take $\mathbf{v}_0 = (v_0, 0)$, $\mathbf{v}_1 = (v_c, 0)$, $\mathbf{v}_2 = \mathbf{0}$ and $\mathbf{v}_3 = (0, v_c)$ approximately for analytical calculations in the following, where $v_c = 1.0 \times 10^5$m/s and $v_0 = 0.8 \times 10^5$m/s.

3. Conductivity, Hall Conductivity and Orbital Susceptibility

Electronic states in periodic potential of crystal are described by the Bloch bands, $\epsilon_{\mathbf{k}}^n$, where n and $\mathbf{k}$ are the band index and the wave vector, respectively. One may naively think that the description of the electronic states in the presence of magnetic field is given by the substitution $\mathbf{k} \to \mathbf{k} + e/c\mathbf{A}$ in $\epsilon_{\mathbf{k}}^n$, where $\mathbf{A}$ is the vector potential. However, this substitution is not justified, because $\mathbf{A}$ has matrix elements between different Bloch bands. This is because the Bloch's functions $\phi_{\mathbf{k}}^n(\mathbf{r}) = \exp(i\mathbf{kr})u_{\mathbf{k}}^n(\mathbf{r})$ are not plane waves but contain periodic functions, $u_{\mathbf{k}}^n(\mathbf{r})$ depending $\mathbf{k}$. On the other hand, in the Luttinger-Kohn representation,[12] the

k-dependences of the basis of the wave functions are the same as plane waves, since the periodic functions $u^n_{\mathbf{k}}(\mathbf{r})$ are fixed on a constant momentum $\mathbf{k} = \mathbf{k}_c$. Then the substitution of $\mathbf{k} \to \mathbf{k} + e/c\mathbf{A}$ is justified in the Luttinger-Kohn representation, as in the present case of the tilted Weyl equation which is represented at the crossing point, $\mathbf{k}_c = \mathbf{k}_0$. Thus the results in the Luttinger-Kohn representation include the contributions of the inter-band effects of magnetic field exactly.

The conductivity and the Hall conductivity and are given by Kubo formula,

$$\sigma_{\mu\nu}(\mathbf{q},\omega) = [\Phi_{\mu\nu}(\mathbf{q},\omega+i\delta) - \Phi_{\mu\nu}(\mathbf{q},i\delta)]/i\omega,$$
$$\Phi_{\mu\nu}(\mathbf{q},i\omega_\lambda) = \frac{1}{\beta}\int_0^\beta\int_0^\beta d\tau d\tau' \exp^{i\omega_\lambda(\tau-\tau')}$$
$$\times\langle TJ_\mu(\mathbf{q},\tau)J_\nu(0,\tau')\rangle, \quad (3)$$

where μ and ν denote x or y, and $\omega_\lambda = 2\lambda\pi T$ is the Matsubara frequency. The current **J** is given by $\mathbf{J_q} \equiv -c\partial\mathcal{H}_H/\partial\mathbf{A_q} = -e\sum_\rho \mathbf{v}_\rho\sigma_\rho$, where $\mathcal{H}_H$ is the Hamiltonian with external magnetic field, $\mathcal{H}_H = \sum_\rho(\mathbf{k}+e\mathbf{A_q}/c)\cdot\mathbf{v}_\rho\sigma_\rho$. The vector potential is included as a plane wave $\mathbf{A}(\mathbf{r}) = \exp(i\mathbf{qr})\mathbf{A_q}$.

The conductivity is given by

$$\sigma_{xx}(\omega) = (L_{xx}(\omega+i\delta) - L_{xx}(i\delta))/i\omega$$
$$L_{xx}(i\omega_\lambda) = T\sum_n\sum_{\mathbf{k}}\mathrm{Tr}[J_xG(+)J_xG], \quad (4)$$

where G is the Green's function defined by $G(i\epsilon_n - \mathcal{H}) = 1$ with the Matsubara frequency $\epsilon_n = (2n+1)\pi T$ and $G(+)$ is the Green's function with $\epsilon_n \to \epsilon_n + \omega_\lambda$. The effective spectrum broadening due to elastic impurity scatterings at $T = 0$, Γ_0, is taken into account phenomenologically here by $i\epsilon_n \to i\tilde{\epsilon}_n = i\epsilon_n + i\Gamma_0\mathrm{sgn}(\epsilon_n)$.

The Hall conductivity σ_{xy} with weak magnetic field is given by the first order terms of the vector potential and the momentum **q**. The gauge-invariant formula of the Hall conductivity in the Luttinger-Kohn representation is given by

$$\sigma_{xy}(\omega) = (L_{xy}(\omega+i\delta) - L_{xy}(i\delta))/i\omega$$
$$L_{xy}(i\omega_\lambda) = iHT\sum_n\sum_{\mathbf{k}}$$
$$\mathrm{Tr}[-J_xG(+)J_xG(+)J_yGJ_yG$$
$$-J_xG(+)J_yGJ_yGJ_xG$$
$$-J_xG(+)J_yG(+)J_xG(+)J_yG$$
$$+J_xG(+)J_xG(+)J_yG(+)J_yG$$
$$+J_xG(+)J_yG(+)J_yGJ_xG$$
$$+J_xG(+)J_yGJ_xGJ_yG]/ce. \quad (5)$$

The orbital susceptibility is defined by

$$\chi = -[\frac{1}{H}\frac{\partial\Omega}{\partial H}]_{H=0}$$
$$\Omega = -k_{\mathrm{B}}T\,ln\mathrm{Tr}\exp[-\beta(\mathcal{H}-\mu N)]. \quad (6)$$

Fukuyama formula for the orbital susceptibility in weak magnetic field[13] is given by

$$\chi = \frac{e^2}{c^2}T\sum_n\sum_{\mathbf{k}}\mathrm{Tr}G\gamma_xG\gamma_yG\gamma_xG\gamma_y. \quad (7)$$

In the Luttinger-Kohn representation, we obtain $\gamma_x = v_0\sigma_0 + v_c\sigma_1$ and $\gamma_y = v_c\sigma_3$ for the tilted Weyl equation.

The chemical potential dependences of σ_{xx}, σ_{xy}, and χ have been shown in Refs. 5 and 14.

4. Dimension Crossover on Hall Effect and Orbital Diamagnetism

In order to investigate the T-dependence of the Hall coefficient in α-(BEDT-TTF)$_2$I$_3$, we have taken the T-dependence of the chemical potential μ into account. The T-dependence of μ based on tight binding model with transfer integrals fitted to the 1st principle calculation[6] is shown by the dashed line in Fig. 2.[5] The possible T-dependence of the effective spectrum broadening due to the inelastic scattering has also been taken into account phenomenologically, *i. e.* , Γ_0 is replaced by $\Gamma = \Gamma_0 + \theta T$, where θ is the coefficient of the linear term of T.[5] The val-

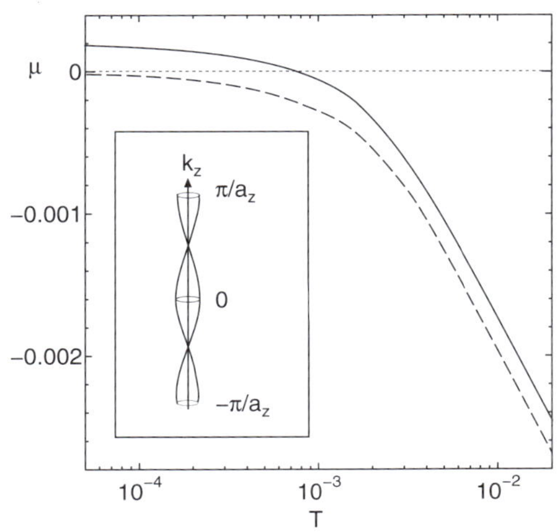

Fig. 2. T-dependences of μ in the presence of electron doping 1.6ppm with $\Delta\mu = 2.0 \times 10^{-4}$ (the solid line), and in the absence of the carrier doping (the dashed line). The inset shows the spindle-shaped Fermi pockets with electron and hole along the crossing line due to weak inter-layer hopping.

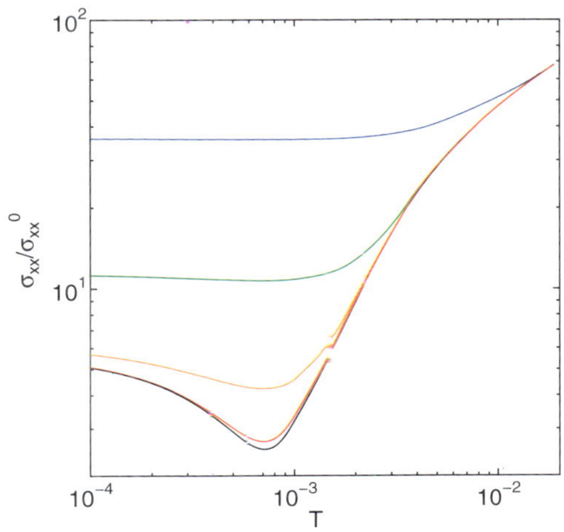

Fig. 3. T-dependences of the conductivity with $t_z = 10^{-5}$ (the black line), 3×10^{-5} (the red line), 10^{-4} (the orange line), 3×10^{-4} (the green line), and 10^{-3} (the blue line).

ues of $\Gamma_0 = 10^{-5}$eV and $\theta = 2 \times 10^{-3}$ (dimensionless) are chosen hereafter. Three-dimensionality has been taken into account by adding a term $-2t_z cos k_z a_z$ in eq. (1), where t_z is the inter-layer hopping and a_z is the layer spacing 1.74×10^{-9}m.[5] The inter-layer hopping t_z makes the spindle-shaped Fermi pockets with electrons and holes along the crossing line when $\mu = 0$ as schematically shown in the inset of Fig. 2.

In order to explain the sign reversal of the Hall coefficient, we have assumed the existence of electron doping which is possibly induced by the lack of I_3^- ions.[5] The solid line in Fig. 2 shows the T-dependence of μ in the presence of electron doping 1.6ppm with $\Delta\mu = 2.0 \times 10^{-4}$, where $\mu = \Delta\mu$ at $T = 0$. When electrons are doped, μ crosses zero at the finite temperature T_r. The value of $\Delta\mu = 2.0 \times 10^{-4}$ is chosen hereafter.

Figure 3 shows the T-dependences of the conductivity with $t_z = 10^{-5}$, 3×10^{-5}, 10^{-4}, 3×10^{-4}, and 10^{-3}, where $\sigma_{xx}^0 \equiv e^2/2\pi^2$. When $t_z = 10^{-3}$, the conductivity exhibits weak T-dependence but the value at $T \to 0$ is much higher than σ_{xx}^0. When t_z is decreased, the T-dependence becomes strong. The conductivity decreases with decreasing T in high temperature region, and it has a minimum at T_r and then weakly increases at lower temperature than T_r. The minimum value of the conductivity is close to $e^2/2\pi^2$ at $t_z = 10^{-5}$, which has been studied as the quantum transport in graphene.[14,15,16]

Figure 4 shows the T-dependences of the Hall conductivity with $t_z = 10^{-5}$, 3×10^{-5}, 10^{-4}, 3×10^{-4}, and 10^{-3}, where $\sigma_{xy}^0 \equiv e^3 v_c^2 H/4\pi^2 c\Gamma_0^2$. When $t_z = 10^{-3}$, the Hall conductivity exhibits weak T-dependence and the values at low temperatures are close to zero. When t_z is decreased, the T-dependence becomes strong and exhibits sharp sign-reversal at T_r.

Figure 5 shows the the T-dependences of the Hall coefficient R with $t_z = 10^{-5}$, 3×10^{-5}, 10^{-4}, 3×10^{-4}, and 10^{-3}, where $R = \sigma_{xy}/H\sigma_{xx}^2$ with $R_0 = \pi^2 v_c^2/ec\Gamma_0^2$. The Hall coefficient is strongly enhanced with decreasing T from room temperature to T_r. The sign of the Hall coefficient reverses con-

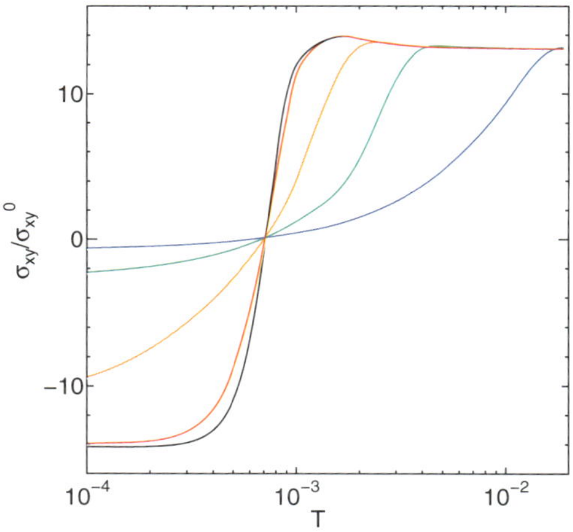

Fig. 4. T-dependences of the Hall conductivity with $t_z = 10^{-5}$ (the black line), 3×10^{-5} (the red line), 10^{-4} (the orange line), 3×10^{-4} (the green line), and 10^{-3} (the blue line).

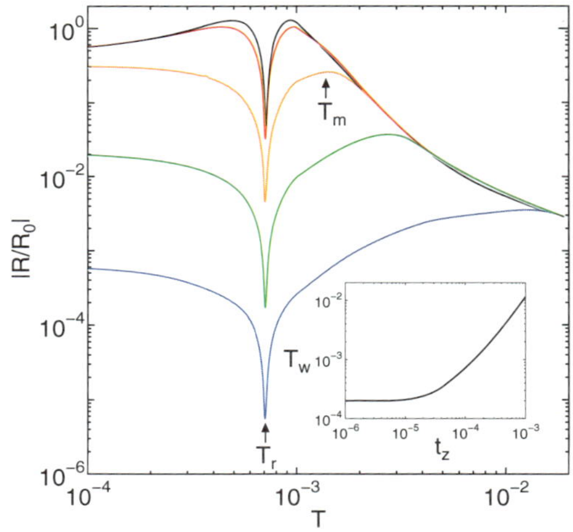

Fig. 5. T-dependences of the Hall coefficient with $t_z = 10^{-5}$ (the black line), 3×10^{-5} (the red line), 10^{-4} (the orange line), 3×10^{-4} (the green line), and 10^{-3} (the blue line). The inset is t_z-dependence of $T_W = T_m - T_r$.

tinuously at T_r, since μ crosses zero. The sharpness of the reversal increases with decreasing t_z. The maximum value of the Hall coefficient at T_m also increases with decreasing t_z, and then the maximum value is 10^3 times larger that that of room temperature when t_z is smaller than 10^{-4}, which is in qualitative agreement with experimental observation in α-(BEDT-TTF)$_2$I$_3$.[7] The inset shows t_z-dependence of the half-width of the reversal, $T_W \equiv T_m - T_r$. When t_z is larger than $\Gamma \sim 10^{-5}$, T_W decreases with decreasing t_z. When t_z is smaller than Γ, on the other hand, the T-dependences of the Hall coefficient are ruled by Γ, and then T_W becomes independent of t_z.

Figure 6 shows the T-dependences of the orbital susceptibility with $t_z = 10^{-5}$, 3×10^{-5}, 10^{-4}, 3×10^{-4}, and 10^{-3}, where $\chi_0 \equiv -e^2 v_c^2/3\pi^2 c^2 \Gamma_0$. The T-dependence becomes strong with decreasing t_z. When t_z is smaller than 10^{-4}, a characteristic peak emerges at T_r.

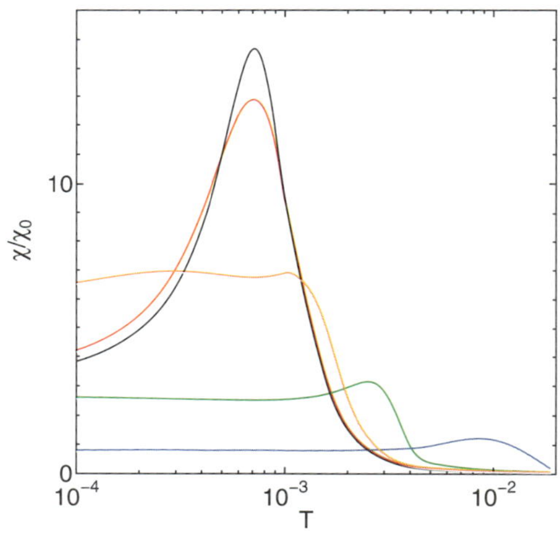

Fig. 6. T-dependences of the orbital susceptibility with $t_z = 10^{-5}$ (the black line), 3×10^{-5} (the red line), 10^{-4} (the orange line), 3×10^{-4} (the green line), and 10^{-3} (the blue line).

5. Summary and Discussion

The roles of the inter-layer hopping were investigated in the conductivity, the Hall conductivity, and the orbital susceptibility in order to study how the layered structure of α-(BEDT-TTF)$_2$I$_3$ affects the anomalous transport phenomena and the orbital diamagnetism. It was found that the continuous reversal of the sign of the Hall coefficient increases the sharpness with decreasing the inter-layer hopping. The inter-layer hopping is required to be comparable or smaller than the effective spectrum broadening, expected as 10^{-5} eV, so as to explain very sharp sign-reversal observed in α-(BEDT-TTF)$_2$I$_3$. Such a small inter-layer hopping is also required so that the characteristic peak of the orbital diamagnetism emerges.

We emphasize that the sensitivities of the Hall coefficient and the orbital diamagnetism as functions of temperatures are essentially due to the inter-band effects of the magnetic field. However the present results indicate that high two-dimensionality is required to the appearance of characteristic phenomena due to the two-dimensional zerogap fermions. The inter-layer hopping could be controlled by the uniaxial pressure perpendicular to the conducting plane, and then experiments for transport phenomena and orbital diamagnetism are expected with varying the uniaxial pressure along the c-axis in α-(BEDT-TTF)$_2$I$_3$ in order to observe the dimension-crossover.

Acknowledgments

The authors are thankful to N. Tajima and Y. Fuseya for fruitful discussions. This work has been financially supported by a Grant-in-Aid for Special Coordination Funds for Promoting Science and Technology (SCF) from the Ministry of Education, Culture, Sports, Science and Technology in Japan, and for Scientific Research on Priority Areas of Molecular Conductors (No. 15073213) and Scientific Research (No. 19740205) from the Ministry of Education, Culture, Sports, Science and Technology in Japan.

References

1. S. Katayama, A. Kobayashi and Y. Suzumura: J. Phys. Soc. Jpn. **75** (2006) 054705.
2. A. Kobayashi, S. Katayama, Y. Suzumura and H. Fukuyama: J. Phys. Soc. Jpn. **76** (2007) 034711.
3. K. Kajita, T. Ojiro, H Fujii, Y. Nishio, H. Kobayashi, A. Kobayashi and R. Kato: J. Phys. Soc. Jpn. **61** (1992) 23.
4. N. Tajima, S. Sugawara, M. Tamura, Y. Nishio and K. Kajita: J. Phys. Soc. Jpn. **75** (2006) 051010.
5. A. Kobayashi, Y. Suzumura, and H. Fukuyama: J. Phys. Soc. Jpn. **77** (2008) 064718.
6. H. Kino and T. Miyazaki: J. Phys. Soc. Jpn. **75** (2006) 034704.
7. N. Tajima, S. Sugawara, M. Tamura, R. Kato, Y. Nishio and K. Kajita: private communication.
8. R. Kubo and H. Fukuyama: Proc. 10th Int. Conf. Physics of Semiconductors, Cambridge, Masachussets, 1969 (United States Atomic Energy Commission, (1970) 551).
9. K. S. Novoselov, A. K. Geim, S. V. Morozov, D. Jiang, M. I. Katsnelson, I. V. Grigorieva, S. V. Dubonos, and A. A. Firsov: Nature **438** (2005) 197.
10. R. Kondo, S. Kagoshima and J. Harada: Rev. Sci. Instrum. **76** (2005) 093902.
11. S. Ishibashi, T. Tamura, M. Kohyama and K. Terakura: J. Phys. Soc. Jpn. **75** (2006) 015005.
12. J. M. Luttinger and W. Kohn: Phys. Rev. **97** (1955) 869.
13. H. Fukuyama: Prog. Theor. Phys. **45** (1971) 704.
14. H. Fukuyama: J. Phys. Soc. Jpn. **76** (2007) 043711.
15. T. Ando: J. Phys. Soc. Jpn. **74** (2005) 777.
16. K. Nomura and A. H. MacDonald: Phys. Rev. Lett. **98** (2007) 076602.

HALL EFFECT AND DIAMAGNETISM OF BISMUTH

Y. FUSEYA* and M. OGATA
Department of Physics, University of Tokyo, Hongo, Bunkyo-ku, Tokyo 113-0033, Japan
**E-mail: fuseya@hosi.phys.s.u-tokyo.ac.jp*

H. FUKUYAMA
Department of Applied Physics, Faculty of Science, Tokyo University of Science, 1-3 Kagurazaka, Shinjuku-ku, Tokyo 162-8601, Japan

Hall effects of bismuth have been studied theoretically with a special reference to the diamagnetic susceptibility. The conductivity, the Hall conductivity, and the orbital susceptibility are calculated on the basis of the Kubo formula for the 4 × 4 Dirac fermions in three dimension. The Hall coefficient exhibits unexpected peaks at the band-edge and sign change at the center of the band gap. Implications of the present results to bismuth alloys are discussed.

Keywords: Bismuth; Hall effect; Diamagnetism; Dirac fermion; Kubo formula.

1. Introduction

Bismuth has played a distinguished role in the solid-state physics. It will be enough to note that such phenomena as the Nernst effect, the high magnetoresistance, the Schubnikov-de Haas effect, the de Haas-van Alphen effect, were all first discovered in Bi. Not only old days, but also recently, quantum Nernst effects[1] and three-dimensional fractional quantum Hall effects[2] were firstly discovered in Bi. Also, the quantum spin-Hall effects in Bi are lively discussed[3,4,5].

In order to understand these fascinating phenomena, it is indispensable to understand the motion of electrons and holes of Bi in a magnetic field. As for Bi, however, even the very fundamental galvanomagnetic effect, the Hall effect in a weak magnetic field, has not been understood so far in detail. There are some old measurements of the Hall effect on Bi[6,7], but they are not consistent with each other. So the Hall effect of Bi has been controversial for a long time. Also, there has been no theoretical analysis. In the following, we study the Hall effect in a three-dimensional Dirac electron system as a effective model of Bi. We discuss implications of the results of Hall effect and orbital susceptibility to Bi alloys.

2. Theory

The fact that low-energy properties of Bi can be described by Dirac electrons was first introduced by Cohen and Blount[8], and then by Wolff in an elegant way[9]. For pure Bi, electrons locate near the L-point in the Brillouin zone and holes near the T-point. Previous studies concentrated on the L-point electrons, since their contribution is dominant due to much larger velocity and much smaller band gap of electrons than those of holes at the T-point. Based on this model, anomalous behaviors of Bi, such as a large g-factor[8], magneto-optics[9], large diamagnetism[10], have been clarified. The Hamiltonian for electrons and holes is given by

$$\mathscr{H} = \Delta\beta + \mathrm{i}v \sum_{\mu} k_{\mu}\beta\alpha_{\mu}, \qquad (1)$$

where 2Δ is the band gap, β and α_i are the 4×4 matrices that appear in the Dirac theory. Here, we omit the k^2-term in the original two-band Dirac model since the effect of this term is negligibly small[10]. We also assume that the velocity of electrons, v, are isotropic for simplicity. This "relativistic" Hamiltonian already includes spin-orbit interactions[9,10], which are very strong in Bi contrary to graphene or α-ET_2I_3.

We calculate various response functions, such as the conductivity, σ_{xx}, the Hall conductivity, σ_{xy}, and the orbital susceptibility, χ, on the basis of the Kubo formula in a gauge-invariant expression[11,12]. Here we have introduced a finite damping, Γ, which is assumed to be constant in order to make our argument as simple and transparent as possible[13,14], although Shon and Ando have indicated that Γ somewhat depends on energy and momentum[15]. Due to this constant Γ, we can obtain the analytic result as Eqs. (2)-(8), where $f(\varepsilon)$ is the Fermi distribution function.

3. Results

The obtained results, Eqs. (2)-(8), seem to be rather complicated, but their chemical potential, μ, dependences are simple. They are basically given as $\sigma_{xx} \propto \mu^2$, $\sigma_{xy} \propto -\mu$, for the metallic region, $|\mu| > \Delta$. For the insulating region, $|\mu| < \Delta$, of course, they almost vanish. However, their Γ-dependences are not so simple. Figure 1 (a) and (b) show the Γ-dependences of σ_{xx} and σ_{xy}, respectively. In the metallic region, e.g., $\mu/\Delta = 3.0$, $\sigma_{xx} \propto \Gamma^{-1}$ and $\sigma_{xy} \propto \Gamma^{-2}$ for $\Gamma < \Delta$. These results are consistent with the well known results for free electrons, $\sigma_{xx} = ne^2\tau/m_e$ and $\sigma_{xy} = ne^2\omega_c\tau^2/m_e$, where ω_c being the cyclotron frequency, and $\tau = \Gamma^{-1}$. Therefore, the Hall coefficient, which is given by $R_H = \sigma_{xy}/\sigma_{xx}^2 H$ does not depends on Γ in the metallic region (Fig. 1 (c)). On the other hand, in the insulating region ($\mu/\Delta = 0.1$) or at the band-edge ($\mu/\Delta = 1.0$), this relation is broken. In the insulating region (at the band-edge), $\sigma_{xx} \propto \Gamma^2$ ($\sigma_{xx} \propto \Gamma^{0.75}$) and $\sigma_{xy} \propto \Gamma^{2.75}$ ($\sigma_{xy} \propto \Gamma^{-0.5}$), so that $R_H \propto \Gamma^{-1.25}$ ($R_H \propto \Gamma^{-2}$). These results

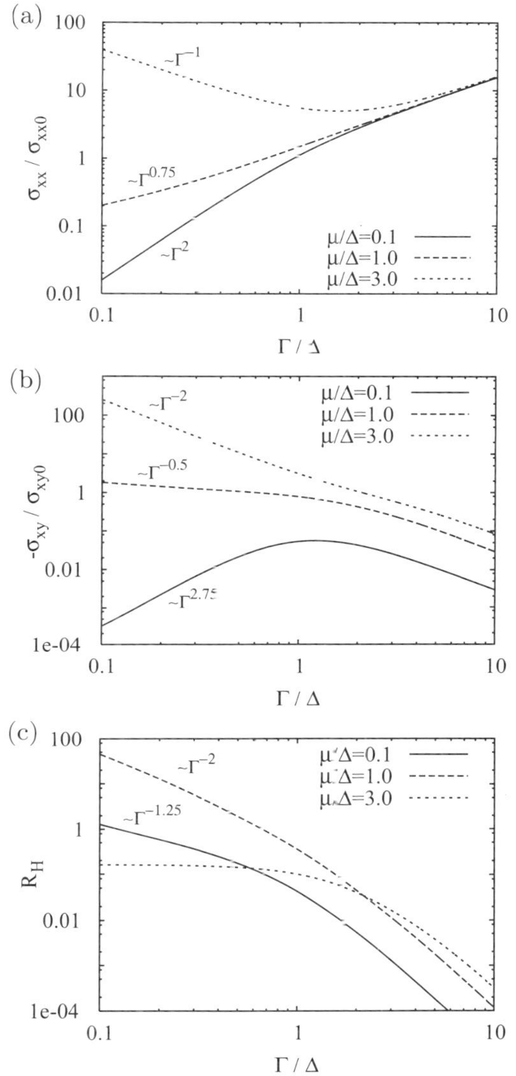

Fig. 1. The Γ-dependences of (a) the conductivity σ_{xx}, (b) the Hall conductivity σ_{xy}, (c) the Hall coefficient R_H at $T = 0$ for the metallic ($\mu/\Delta = 3.0$), the band-edge ($\mu/\Delta = 1.0$), and the insulating ($\mu/\Delta = 0.1$) regions. The normalization factors are taken to be $\sigma_{xx0} = e^2/\pi^2 v$, $\sigma_{xy0} = e^3 v/12\pi^2 c$, and $R_{H0} = \sigma_{xy0}/\sigma_{xx0}^2$.

indicate that the well known formula for free electrons, $R_H^{free} = 1/(nec)$, is not valid any more for $|\mu| \leq \Delta$; R_H strongly depends on Γ. R_H at the band-edge depends on Γ more strongly than that in the insulating region, so that peak structures appear at the band-

$$\sigma_{xx} = -\frac{e^2}{\pi^3 v}\int_{-\infty}^{\infty} d\varepsilon f'(\varepsilon-\mu)\int_0^{\infty} dX\left[F_1(\varepsilon,X)-F_2(\varepsilon,X)\right], \quad (2)$$

$$F_1(\varepsilon,X) = \frac{X^2(\varepsilon^2+\Gamma^2-\frac{1}{3}X^2-\Delta^2)}{\{(\varepsilon+\mathrm{i}\Gamma)^2-X^2-\Delta^2\}\{(\varepsilon-\mathrm{i}\Gamma)^2-X^2-\Delta^2\}}, \quad (3)$$

$$F_2(\varepsilon,X) = \frac{X^2\left\{(\varepsilon+\mathrm{i}\Gamma)^2-\frac{1}{3}X^2-\Delta^2\right\}}{2\left\{(\varepsilon+\mathrm{i}\Gamma)^2-X^2-\Delta^2\right\}^2} + \text{c.c.}, \quad (4)$$

$$\sigma_{xy} = \frac{e^3 vH}{12\pi^2 c}\int_{-\infty}^{\infty} d\varepsilon\left[F_3(\varepsilon)f(\varepsilon-\mu)+F_4(\varepsilon)f'(\varepsilon-\mu)\right]\mathrm{sgn}(\varepsilon), \quad (5)$$

$$F_3(\varepsilon) = \frac{\varepsilon+\mathrm{i}\Gamma}{\{(\varepsilon+\mathrm{i}\Gamma)^2-\Delta^2\}^{3/2}} + \text{c.c.}, \quad (6)$$

$$F_4(\varepsilon) = \frac{-2\Gamma^4-\Gamma^2\Delta^2+(\Delta^2-\varepsilon^2)^2+2\mathrm{i}\Gamma^3\varepsilon-\mathrm{i}\Gamma\mu(\Delta^2-\varepsilon^2)}{2\Gamma^2\varepsilon^2\sqrt{\varepsilon^2-\Gamma^2-\Delta^2+2\mathrm{i}\Gamma\varepsilon}} + \text{c.c.}, \quad (7)$$

$$\chi = \frac{e^2 v}{c^2\pi^2}\frac{4}{15}\int_{-\infty}^{\infty} d\varepsilon f(\varepsilon)\mathrm{sgn}(\varepsilon)\left[\frac{1}{\sqrt{(\varepsilon+\mathrm{i}\Gamma)^2-\Delta^2}} + \text{c.c.}\right]. \quad (8)$$

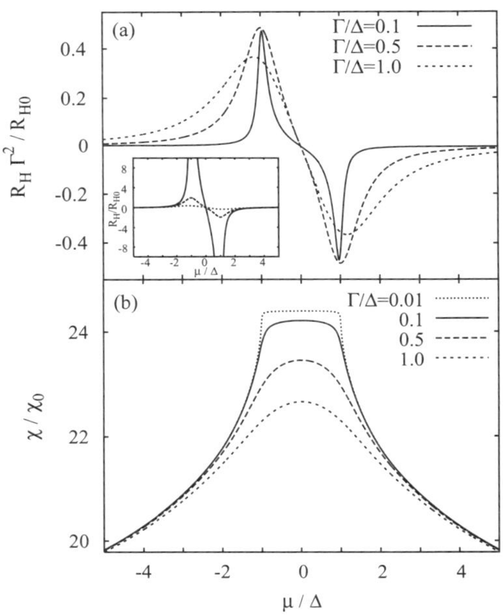

Fig. 2. The chemical-potential dependences of (a) the Hall coefficient, and (b) the orbital susceptibility.

edge for clean system as is shown in Fig. 2 (a). Note that there is no carriers at $T = 0$ of $|\mu| < \Delta$ in clean systems. It is rather surprising that R_{H} does not diverge but has small values in this region (esp. $R_{\mathrm{H}} = 0$ at $\mu = 0$).

4. Discussions & Conclusions

Here, we discuss the implications of the present results to Bi. Some old measurements of R_{H} in Bi have already exhibited the signs of peak structures and sign changes with respect to μ[6,7]. They are, however, not consistent with each other. This would be due to the purity of samples or the inhomogeneity of either chemical or external pressure. So in this paper, we propose guidelines for the measurements of R_{H} in Bi.

First, it will be rather difficult to find the location of band-edge within the measurements of R_{H} when the peak is very sharp. This will be resolved by a simultaneous measurement of χ together with that of R_{H}. A clear kink of χ at $\mu \simeq \pm\Delta$ (Fig. 2 (b)) has been actually observed experimentally[16]. Thus, the kink in χ can be used to identify the band-edge, and then we can expect the peak of R_{H} in the same region.

Next, when Bi alloys are used, it will be important to consider the contributions of not only electrons at the L-points but also holes at T-points. The characteristic features — a peak of R_{H} and a kink of χ at $\mu = 0$, a sign change at $\mu = 0$ — will ap-

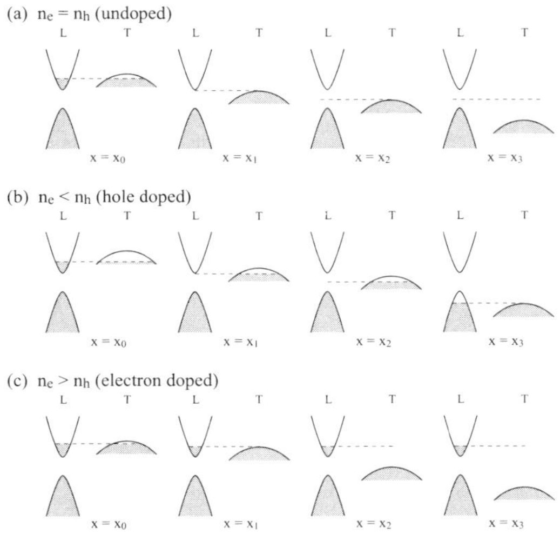

Fig. 3. Schematic band structure around L- and T-points of Bi and its alloys for (a) undoped ($n_e = n_h$), (b) hole doped ($n_e < n_h$), and (c) electron doped ($n_e > n_h$) case. The dashed lines indicate the level of the chemical potential.

pear differently following three experimental situations: (a) undoped ($n_e = n_h$), (b) hole doped ($n_e < n_h$), and (c) electron doped Bi ($n_e > n_h$). The band structure around the L- and T-points for each case is illustrated in Fig. 3. Here, x denotes some controlling parameter which changes the relative position of electron- and hole-band, for example, external pressure or alloy concentration. For undoped case, a clear band-edge property, a peak in R_H and a kink in χ, can be seen, since n_e and n_h vanish simultaneously at $x = x_1$. In this case, the sign change in R_H will not be observed, since μ is always positive for electrons. For the hole doped case, on the other hand, μ of electrons can be negative for $x > x_2$, so that the sign change in R_H is possible. However, the band-edge property in R_H will not be so clear due to the finite contribution from the holes at T-point, even though it will be clear in χ. For the electron doped case (Fig. 3 (c)), neither the band-edge effect nor the sign change will be hardly seen.

In conclusion, we have investigated the Hall effect of Bi and its alloys. The R_H exhibits unexpected peak structures at the band-edges and sign change at the center of the band gap. We have proposed the simultaneous measurement of χ and R_H for clarifying the experimental situations. These predictions will help to resolve the longstanding and puzzling reports on R_H of Bi.

Acknowledgments

We thank A. Kobayashi for helpful discussions. Y.F. is supported by JSPS Research Fellowships for Young Scientists.

References

1. K. Behnia, M.-A. Méasson, and Y. Kopelevich, *Phys. Rev. Lett.* **98**, 166602 (2007).
2. K. Behnia, L. Balicas, and Y. Kopelevich, *Science* **317**, 1729 (2007).
3. S. Murakami, *Phys. Rev. Lett.* **97**, 236805 (2006).
4. Liang Fu, C. L. Kane, and E. J. Mele, *Phys. Rev. Lett.* **98**, 106803 (2007); Liang Fu and C. L. Kane, *Phys. Rev. B* **76**, 045302 (2007).
5. D. Hsieh *et al.*, *Nature* **452**, 970 (2008).
6. A. L. Jain, *Phys. Rev.* **114**, 1518 (1959)
7. N. B. Brandt and Y. G. Ponomarev, *Sov. Phys. JETP* **28**, 635 (1969).
8. M. H. Cohen and E. I. Blount, *Phil. Mag.* **5**, 115 (1960).
9. P. A. Wolff, *J. Phys. Chem. Solids* **25**, 1057 (1964).
10. H. Fukuyama and R. Kubo, *J. Phys. Soc. Jpn.* **28**, 570 (1970).
11. H. Fukuyama, H. Ebisawa, and Y. Wada, *Prog. Theor. Phys.* **42**, 494 (1969); H. Fukuyama, *ibid.* **42**, 1284 (1969).
12. H. Fukuyama, *Prog. Theor. Phys.* **45**, 704 (1971).
13. H. Fukuyama, J. Phys. Soc. Jpn. **76**, 043711 (2007).
14. A. Kobayashi, Y. Suzumura, and H. Fukuyama, *J. Phys. Soc. Jpn.* **77**, 064718 (2008).
15. N. H. Shon and T. Ando J. Phys. Soc. Jpn. **67**, 2421 (1998).
16. L. Wehrli, *Phys. Kondens. Materie* **8**, 87 (1968).

QUANTUM NERNST EFFECT IN A BISMUTH SINGLE CRYSTAL

MARI MATSUO*

Graduate School of Integrated Sciences, Ochanomizu University, Otsuka, Bunkyo, Tokyo 112-8610, Japan
**E-mail: matsuo@iis.u-tokyo.ac.jp*

AKIRA ENDO

Institute for Solid State Physics, University of Tokyo, Kashiwanoha, Kashiwa, Chiba 277-8581, Japan

NAOMICHI HATANO

Institute of Industrial Science, University of Tokyo, Komaba, Meguro, Tokyo 153-8505, Japan

HIROAKI NAKAMURA

Department of Simulation Science, National Institute for Fusion Science, Oroshi-cho, Toki, Gifu 509-5292, Japan

RYŌEN SHIRASAKI

Department of Physics, Faculty of Engineering, Yokohama National University, Tokiwadai, Hodogaya-ku, Yokohama, Kanagawa 240-8501, Japan

KO SUGIHARA

1-40-6-506 Shibayama, Funabashi, Chiba, 274-0816, Japan

We report a theoretical calculation explaining the quantum Nernst effect observed experimentally in a bismuth single crystal. Generalizing the edge-current picture in two dimensions, we show that the peaks of the Nernst coefficient survive in three dimensions due to a van Hove singularity. We also evaluate the phonon-drag effect on the Nernst coefficient numerically. Our result agrees with the experimental result for a bismuth single crystal.

Keywords: Bismuth; quantum Hall effect; quantum Nernst effect; phonon-drag effect.

1. Introduction

The Nernst effect is a thermoelectric phenomenon which yields a transverse voltage when there is a magnetic field perpendicular to a temperature gradient. In 2005, Nakamura *et al.*[1] predicted the *quantum* Nernst effect of the two-dimensional electron gas in a semiconductor heterojunction under a strong magnetic field, using the edge current picture of the quantum Hall effect.[2] In the quantum Nernst effect, the Nernst coefficient shows sharp peaks when a Landau level crosses the Fermi level and the thermal conductance in the direction of the temperature gradient is quantized. Shirasaki *et al.*[3] showed that impurities do not change the conclusions much.

In 2007, Behnia *et al.*[4,5] have reported quantum oscillation of the Nernst coefficient in a bismuth single crystal, which was similar to the prediction by Nakamura *et al.*[1] It is remarkable that the strong quantum effect is observable in three dimensions; the quantized plateaus of the Hall conductance can be hardly identified in three dimensions.

In the present study we extend the theoretical prediction of the quantum Nernst effect in two dimensions[1] to a three-dimensional system, motivated by the experiment in a bismuth single crystal. We find that the peaks of the Nernst coefficient survive in three dimensions because of a van Hove singularity. A simple calculation explains Behnia *et al.*'s experimental result qualitatively. Consideration of the electron-phonon interaction leads even to quantitative agreement with the experimental result.

2. Edge-current picture of the quantum Nernst effect

We consider the adiabatic Nernst effect. The magnetic field B is applied in the direction of the trigonal axis, which we set to the z axis. The system is electrically and thermally insulated on all surfaces except for the two surfaces attached to heat baths, which induce a temparature bias ΔT in the x direction. Then the Nernst voltage V_N is generated in the y direction. The adiabatic Nernst coefficient is then defined by

$$N = -\frac{V_N/W}{B\Delta T/L}, \tag{1}$$

where L and W are the lengths of the crystal in the x and y directions, respectively.

Nakamura *et al.*[1] assumed that, in a two-dimensional clean sample at low temperatures, an edge current circulates around the system ballistically when the chemical potential is in between two Landau levels. The edge current that departs the colder side maintains the Fermi distribution $f(T_-,\mu_-)$ of the colder bath until it reaches the hotter side and then maintains the Fermi distribution $f(T_+,\mu_+)$ of the hotter bath until it reaches the colder side. The difference of the chemical potential generates the Nernst voltage as $V_N = (\mu_+ - \mu_-)/|e|$.

The Hamiltonian of the two-dimensional electron gas yields energy levels $E(n,k_x)$, where $n = 0,1,2,\cdots$ represents a Landau level and the label $-k_m \le k_x \le k_m$ represents a state in the Landau level with k_m determined by the confining potential in the y direction. After some algebra they obtained the electric current in the x direction as

$$I = \frac{e}{\pi\hbar}\left(A_0\Delta\mu - A_1 k_B \Delta T\right), \tag{2}$$

where $\Delta T \equiv T_+ - T_-$, $\Delta\mu \equiv \mu_+ - \mu_-$ and

$$A_\nu(\mu) = \sum_{n=0}^{\infty}\int_{X_0(n)}^{X_1(n)} \frac{x^\nu dx}{4\cosh^2(x/2)} \tag{3}$$

with $X_i(n) \equiv (E_i(n)-\mu)/(k_BT)$, $E_0(n) \equiv E(n,k_x=0)$ and $E_1(n) \equiv E(n,k_x=k_m)$. Setting $I=0$ and $e=|e|>0$, they obtained the Nernst coefficient as

$$N(B,T) = \frac{k_B}{|e|B_z}\frac{L}{W}\frac{A_1}{A_0}, \tag{4}$$

which showed sharp peaks when the chemical potential is equal to a Landau level. This quantum behavior basically comes from the oscillation of the coefficient A_1.

We now extend the above to three dimensions. This extension introduces an additional argument k_z in the energy level as $E_i(n,k_z) = E_i(n) + (\hbar k_z)^2/(2m_z)$ and hence in $X_i(n,k_z)$. Equation (3) now has integration over k_z as

$$\tilde{A}_\nu(\mu) = \sum_{n=0}^{\infty}\int\frac{dk_z}{2\pi}\int_{X_0(n,k_z)}^{X_1(n,k_z)}\frac{x^\nu dx}{4\cosh^2(x/2)} = \int_0^\infty D(E_z)A_\nu(\mu - E_z)dE_z, \tag{5}$$

where we have changed the integration variable from k_z to $E_z = (\hbar k_z)^2/(2m_z)$. This variable transformation introduces the density of states $D(E_z) \equiv [2\pi(dE_z/dk_z)]^{-1}$, which has a van Hove singularity at $E_z = 0$.

This van Hove singularity is the essence of the survival of the quantum oscillation in three dimensions. The overlap of the peaks of $A_1(\mu)$ with the van Hove singularity leads to oscillatory behavior of the coefficient $\tilde{A}_1(\mu)$ and hence of the Nernst coefficient.

Figure 1 presents the quantum Nernst effect thus obtained in three dimensions. This

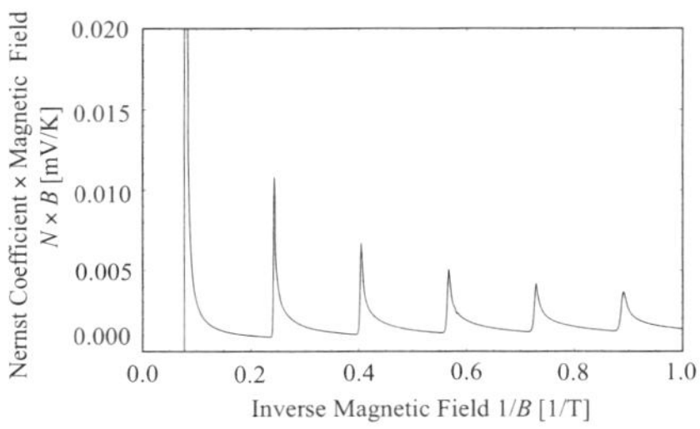

Fig. 1. The Nernst coefficient N times the magnetic field, which is equal to the thermomagnetic power S_{xy}, in three dimensions at $T = 0.28$ K against the inverse magnetic field $1/B$ [1/T], after Eq. (4). We used the following parameter values: the effective mass is $m = 0.06289m_0$ and $m_z = 0.6667m_0$[6] where m_0 is the bare mass of the electron; the size of the Hall bar is $L = 4.0$mm[4,5] and $W = 2.2$mm[4,5]; the confining potential at the edges is $V_0 = 4.22$ eV[7]; the chemical potential at equilibrium is $\mu = 11.4$meV[8].

agrees with the experimental result qualitatively, though the magnitudes of the peaks is much smaller in our calculation than in the experimental result.

3. Phonon-drag effect on the Nernst coefficients

Since the above argument does not quantitatively agree with the experimental result, we now consider the phonon-drag effect. We basically follow Sugihara's theory[9,10] of the thermoelectric power in bismuth. The difference here is that we treat the Fermi and Bose distribution functions without any approximations and evaluate the magnetic-field dependence of the thermomagnetic power numerically.

The thermomagnetic power $S_{xy} = NB$ is given by

$$S_{xy} = -\frac{Q_y}{TF}\rho_{xx}, \qquad (6)$$

where Q_y denotes the heat current in the y direction, F denotes the electric field in the x direction, ρ_{xx} denotes the diagonal resistivity, and we assumed $\rho_{xy} \ll \rho_{xx}$ in bismuth. At low temperatures we may neglect all excitations except acoustic phonons having the energy $\hbar\omega_q$ and the wave vector $\boldsymbol{q}$, which are generated through deformation coupling. The heat current in the y direction is then given by

$$Q_y = \int \frac{d\boldsymbol{q}}{(2\pi)^3}\hbar\omega_q v_s \frac{q_y}{q}\left(N_q - N_q^{(0)}\right), \qquad (7)$$

where v_s is the group velocity of the phonons. The term $N_q - N_q^{(0)}$ represents the displacement of the phonon distribution N_q from its equiribrium Bose distribution $N_q^{(0)}$.

In order to estimate the displacement, we use the Boltzmann equation in the steady state

$$\left(\frac{\partial N_q}{\partial t}\right)_{\text{carrier}} + \left(\frac{\partial N_q}{\partial t}\right)_{\text{relaxation}} = 0. \qquad (8)$$

The first term represents the change of the phonon distribution due to the interaction with carriers and the second term represents that due to other interactions such as boundary scattering, phonon-phonon interaction and impurity scattering. In the steady state these two terms are balanced. In the second term, we assume $(\partial N_q/\partial t)_{\text{relaxation}} = -(N_q - N_q^{(0)})/\tau_{\text{r}}(q)$ in the relaxation-time approximation; the electron-phonon interaction replaces the phonon distribution, but the relaxation effects make the nonequilibrium phonon distribution to the equilibrium one in time τ_{r}.

The theoretical result is shown in Fig. 2 (a) and the experimental result is shown in Fig. 2 (b) at three temperatures, respectively. In Fig. 2 (c) we compared the theoretical and experimental result at T=0.28[K]. The magnitude of the peaks of the theoretical result, after considering the phonon-drag effect, is consistent with the experimental result. Note that there are no adjustable fitting parameters.

4. Summary

We have shown that the quantum Nernst effect survive in three dimensions due to a van

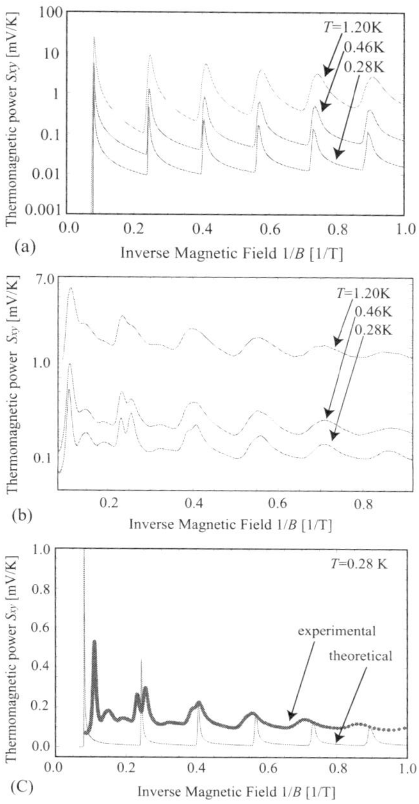

Fig. 2. The thermomagnetic power in three dimensions at $T=$ 0.28 K, 0.46 K, 1.20 K against the inverse magnetic field $1/B$[1/T]: (a) Our theoretical result with the phonon-drag effect; (b) The experimental result by Behnia *et al.*[4,5]; (c) Comparison of the results at $T=$ 0.28 K. In the calculation we used the following parameter values: the deformation potential $D = 1.2\text{eV}$[11]; the electric resistivity $\rho_{xx} = 0.35\Omega\text{cm}$[4,5]; the group velocity of the phonons 2×10^3m/s; the bismuth density 9.75×10^3kg/m^3. The parameter values for the effective mass m and m_z and the chemical potential at equilibrium μ are the same as in Fig. 1.

Hove singularity. We have presented an extension of the edge-current picture in two dimensions to three dimensions, in order to explain Behnia *et al.*'s[4] experiment for a bismuth single crystal. The naive extension reproduced the quantum oscillation of the experimental result qualitatively. Furthermore, consideration of the phonon-drag effect led us to a theoretical result quantitatively consistent with the experimental result. The peaks are sharper than those in the experiment probably because we neglected impurity scattering of carriers. For simplicity, we also assumed the Fermi surface of bismuth is ellipsoid and considered only the hole contribution. Going beyond these approximations would be an interesting future problem.

Acknowledgments

We thank Dr Y. Hasegawa for constructive comments. We also thank Dr. Behnia for valuable comments and providing experimental data. The work is supported partly by the Murata Science Foundation as well as by the National Institutes of Natural Sciences undertaking Forming Bases for Interdisciplinary and International Research through Cooperation Across Fields of Study and Collaborative Research Program (No. NIFS08KEIN0091) and Grants-in-Aid for Scientific Research (No. 17340115 and No. 20340101) from the Ministry of Education, Culture, Sports, Science and Technology.

References

1. H. Nakamura, N. Hatano, R. Shirasaki, *Solid State Comm.* **135** 510 (2005)
2. B. I. Halperin, *Phys. Rev. B* **25**, 2185 (1982)
3. R. Shirasaki, H. Nakamura, N. Hatano, *e-J. Surf. Sci. Nanotech.* **3**, 518 (2005)
4. K. Behnia, M.A. Measson, Y. Kopelevich, *Phys. Rev. Lett.* **98**, 166602 (2007)
5. K. Behnia, L. Balicas, Y. Kopelevich, *Science* **317**, 1729 (2007)
6. The Japan Institute of Metals (ed), *Handoutai to Hankinzoku (in Japanese)* (Agne Gijutsu Center, Tokyo, 1990)
7. H. B. Michaelson, *J. Appl. Phys.* **48**, 4729 (1977)
8. H. Kawamura, *Kotai Purazuma (in Japanese)* (Asakura, Tokyo, 1972)
9. K. Sugihara, *J Phys. Soc. Jpn.* **27**, 356 (1969)
10. K. Sugihara, *J. Phys. Soc. Jpn.* **27**, 362 (1969)
11. K. Walther, *Phys. Rev.* **174**, 782 (1968)

KONDO EFFECT AND SUPERCONDUCTIVITY IN SINGLE InAs QUANTUM DOTS CONTACTED WITH SUPERCONDUCTING LEADS

S. TARUCHA[a,b,c], C. BUIZERT[a], K. SHIBATA[d,e], Y. KANAI[a], R.S. DEACON[a], K. HIRAKAWA[d,e], and A. OIWA[a,b,e]

[a]*Department of Applied Physics and QPEC, Univ. of Tokyo, Bunkyo-ku, Tokyo 113-8656, Japan*
[b]*Quantum Spin Information Project, ICORP, JST, Atsugi-shi, Kanagawa 243-0198, Japan*
[c]*Institute for Nano Quantum Information Electronics, Univ. of Tokyo, Meguro-ku, Japan*
[d]*Institute of Industrial Science, Univ. of Tokyo, Meguro-ku ,Tokyo 153-8505, Japan*
[e]*CREST JST, Kawaguchi, Saitama 332-0012, Japan*
[*]*E-mail: tarucha@ap.t.u-tokyo.ac.jp*

We study the transport properties of a single InAs self-assembled quantum dot contacted with superconducting leads. The charging energy E_c of the quantum dot is much larger than the superconducting gap energy Δ. For the dot whose tunnel coupling Γ to the lead is much larger than Δ but smaller than E_c, we observe enhancement of first-order Andreev reflections by the Kondo effect. We find that the zero-bias conductance measured for various Δ's and Kondo temperature T_K's collapses onto a single curve with Δ/T_K as the only relevant energy scale, providing experimental evidence for universal scaling in this system. On the other hand, for the dot with Γ comparable to E_c we observe a supercurrent flowing through the dot, reflecting the charge fluctuation sufficiently greater that one.

Keywords: InAs, quantum do, superconductivity, Kondo effect, Andreev reflection, supercurrent, proximity effect, Coulomb blockade, spin correlation

1. Introduction

Electron transport through a pair of electrodes with a quantum dot (QD) in between is predominantly influenced by the feature of the QD [1]. This has been used to study the electronic properties of QD itself [2] and also manipulate the charge and spin degrees of freedom in the conduction [3]. Particularly, associated with the latter, novel spin-correlation phenomena, such as Pauli effect [4] and the Kondo effect [5-7], have been revealed. Use of superconductor instead of normal metal for making contacts to QDs can further expand the freedom of spin correlation in the electron transport. Transport properties of such QDs have been studied for carbon nanotubes [8-10] and InAs nanowires [11-13].

In this work we perform electronic spectroscopy measurements on single InAs self-assembled QDs contacted with superconducting leads to explore features of the Kondo effect and superconductivity. The transport characteristics are pretty much different from sample to sample, depending on the parameters, particularly dot-lead tunnel coupling, Γ, and Coulomb gap, E_c. For samples having relatively large Γ but smaller than E_c, we observe the interplay between the Kondo effect and Andreev reflection but no supercurrent, while we observe supercurrent for samples having large Γ comparable to E_c. Note neither of the above features is observed for samples having small Γ.

2. Experimental

2.1 Samples

Samples are prepared using uncapped self-assembled InAs QDs whose diameter is 80 to 90 nm. A pair of contact electrodes with a 20 to 30 nm gap is directly placed on the QD. An n-GaAs layer is embedded 0.3 μm below the surface with an i-AlGaAs layer in between, and used as a back gate to tune the QD electrostatic potential [14,15]. The electrode metal is 5 nm Ti/200 nm Al [16,17]. We previously observed the spin-half Kondo effect for a similar InAs QD but coupled to normal metal (Ti-Au) electrodes [4].

In this study we use two samples A and B prepared in the same way but actually having different tunnel couplings Γ: relatively large for Sample A and large for Sample B. The charging energy E_c, is a few meV for both samples, which is evaluated from the Coulomb diamonds for the normal state (N-state) contacts, i.e. measured at a magnetic field B larger than the critical field B_c (~100 mT). The superconducting gap Δ measured from the stability diagram is 140 μeV, and 180 μeV for Sample A, and B, respectively. The critical temperature T_c = 0.9 K, and 1.2 K for Sample A, and B, respectively.

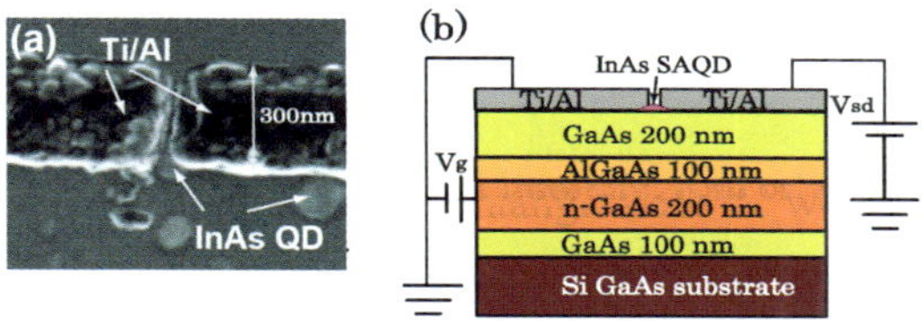

Fig. 1 (a) Typical scanning electron microscope image of InAs dot with superconducting nanogap electrodes. (b) Cross-sectional illustration of the device.

2.2 Interplay between Kondo effect and Andreev reflection

Figure 2 shows the differential conductance dI/dV_{sd} vs. gate voltage V_g and source-drain voltage V_{sd} (a) and dI/dV_{sd} vs. V_g at V_{sd} = 0 mV (b), measured for Sample A in the N-state. We observe the zero-bias anomaly inside the

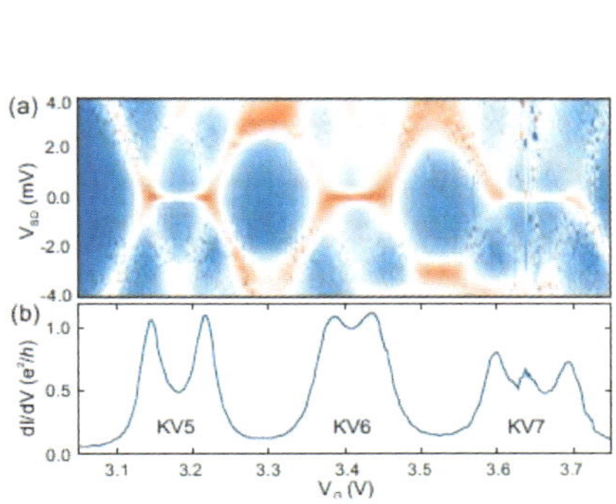

Fig. 2(a) Differential conductance dI/dV_{sd} as a function of V_g and V_{sd} measured at 35 mK and 100 mT. (b) dI/dV_{sd} vs. V_g at V_{sd} = 0 mV in units of e^2/h.

Coulomb blockade (CB) regime (KV5, 6, and 7) holding odd number of electrons, due to the Kondo effect. From Fig. 2(a) E_c of ~2.5 meV and Γ of 0.5 to 0.6 meV are evaluated. The Kondo temperature evaluated from the width of the zero-bias anomaly is 1.04, 1.65, and 1.18 K for KV5, 6, and 7, respectively.

We then turned off the magnetic field and measured the same as in Fig. 2(a). The result is shown in Fig. 3(a). The two parallel lines around zero bias are clearly resolved in the non-Kondo Coulomb valleys. They mark the onset of direct quasi-particle tunneling at $|V_{sd}| = 2\Delta/e$ (Fig. 3(b)). Only the first order Andreev reflection at $|V_{sd}| = \Delta/e$ is faintly visible as a shoulder to the 2Δ feature.

On the other hand, for the Kondo valley, we observe coexistence of first order Andreev reflection (peaks at $eV_{sd}=\pm\Delta$) and the Kondo effect (zero-bias peak) (Fig. 3(c)). The Andreev process is enhanced by the Kondo effect in such a way that the zero-bias peak develop side-peaks at $eV_{sd}=\pm\Delta$. We then measured the zero-bias conductance G^{SC} for various Δ's and Kondo temperature T_K's, and plot the data of G^{SC}/G^N_0 in Fig. 4, where G^N_0 is

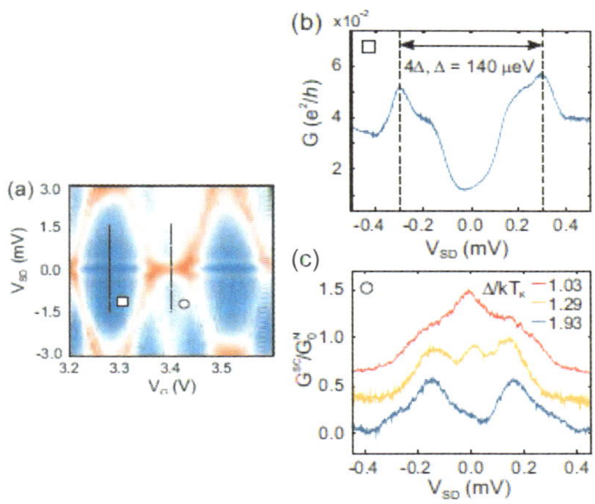

Fig. 3(a) dI/dV_{sd} vs. V_g and V_{sd}, with the leads in the SC-state (B = 0 T). One Kondo valley (KV6 in Fig. 2(a)) and two non-Kondo Coulomb valleys are indicated. (b) Scan of the gap structure in the CB regime, as indicated by the square in (a). (c) Scan of the gap structure, as indicated by the circle in (a), for three different Kondo valleys with Δ/kT_K = 1.03, 1.29, and 1.93, respectively. The curves are normalized by their N-state zero-bias conductance G^N_0. Offset for clarity.

the value measured for the N-state leads. We find that all data collapse on a single curve with Δ/T_K as the only relevant energy scale. This can represent a new type of Kondo universality.

In this sample we observed no signature of supercurrent as shown in Fig. 5(b), probably due to the small dot-lead coupling as compared to the large Coulomb gap. We, therefore, selected another sample (Sample B) with a large dot-lead coupling to observe supercurrent. The large tunnel coupling is helpful in raising the proximity effect, because the charge fluctuation in the dot can be more than one.

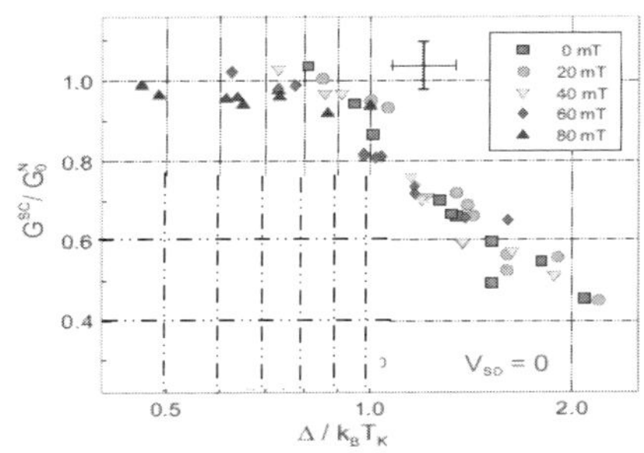

Fig. 4 Semi-log plot of $G^{SC}/G^N{}_0$ obtained for various values of Δ/kT_K. For these data, Δ is varied as a function of magnetic field. The measurement uncertainty is given by the bars.

2.3 Proximity effect

Figure 5(a) shows Coulomb oscillations in the normal state measured at 150 mT. The peaks are rather broad and overlapped with each other in the V_g range between 0.4 and 0.7 V, indicating the strong dot-lead coupling Γ comparable to E_c (= 1 to 2 meV). The supercurrent characteristics are only observed in this gate voltage range with current-bias measurement (Fig. 5(b)). When the current bias is swept from zero to critical current I_c, dissipationless supercurrent flows though the dot. The observed hysteretic behavior suggests an underdamped Josephson junction.

The critical current is a maximum near Coulomb peaks and is significantly smaller in the CB regime. In this condition, the Kondo anomaly is not well identified. The observed I_c is very small compared with the theoretical value for single-mode junction probably because the tunnel-junction is not transparent enough to compensate for the effect of Coulomb gap.

Note in most samples the zero-bias peak due to supercurrent is observed in the CB region holding even number of electrons in the dot. This reflects influence of singlet correlation to suppress the supercurrent flow through the dot.

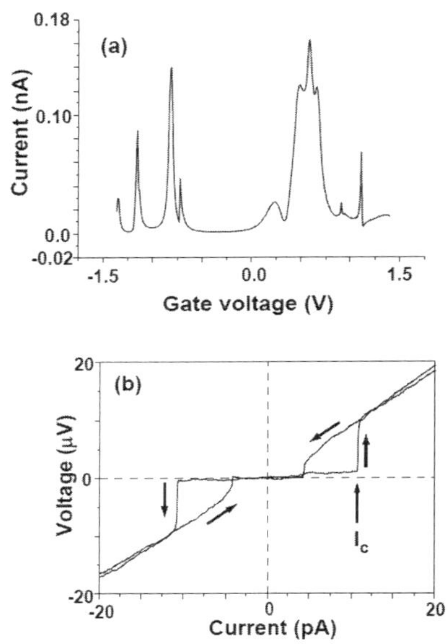

Fig. 5(a) Coulomb oscillations in the N-state of Sample B measured at 130 mK and 150 mT. V_{sd} is less than 100 μV. (b) Typical supercurrent characteristic measured at 130 mK and 0 T.

3. Conclusions

Transport properties of single InAs self-assembled quantum dots contacted with superconducting leads are studied. The observed transport features strongly depend on the dot-lead tunnel coupling Γ and charging energy E_c. For the dot with $\Delta < \Gamma < E_c$, enhancement of Andreev reflections due to the Kondo effect is observed. We find that the zero-bias Kondo conductance measured for various Δ's and Kondo temperature T_K's collapses onto a single curve with Δ/T_K as the only relevant energy scale, providing experimental evidence for universal scaling in this system. On the other hand, for the dot with $\Gamma \sim E_c$ a supercurrent flowing through the dot is observed.

Acknowledgments

We acknowledge K. Yoshida for assistance in device fabrication. Part of this work is financially supported by the Grant-in-Aid for Scientific Research S (No. 19104007) and B (No. 18340081).

References

1. Mesoscopic Electron Transport, edited by L.L. Sohn, *et al.*: Kluwer, Dordrecht, Netherlands, Series E, vol. **345** (1997) 105.
2. S. Tarucha et al. Phys. Rev. Lett., **77,** 3631 (1996).
3. Concept of Spin Electronics, edited by S. Maekawa: Oxford University Press, Oxford, New York (2006).
4. K. Ono et al. Science **297**, 1313 (2002).
5. D. Goldharber Gordon et al. Nature, **391**, 156 (1998).
6. S.M. Cronenwett et al. *Science* **281**, 540 (1998).
7. S. Sasaki et al. Nature, **405**,764 (2000).
8. M. R. Buitelaar et al. Phys. Rev. Lett., **89** 256801 (2002).
9. J.–P. Cleuziou et al. Nature Nanotechnology, **1**, 53 (2006).
10. P. Jarillo-Herrero et al. Nature, **439**, 953 (2006).
11. T. Sand-Jseperson et al. Phys. Rev. Lett., **99**, 126603 (2007).
12. Y.-J. Doh et al. Sceince **309**, 272 (2005).
13. J. A. Van Dam et al. Nature **442**, 667 (2006).
14. M. Jung et al. App. Phys. Lett., **86**, 33106 (2005).
15. Y. Igarashi et al. Phys. Rev. B (RC), **76**, 081303 (2007) .
16. C. Buizert et al. Phys. Rev. Lett., **99**,136806 (2007).
17. K. Shibata et al. Appl. Phys. Lett., **91**, 112102 (2007) .

ELECTRON TRANSPORT THROUGH A LATERALLY COUPLED TRIPLE QUANTUM DOT FORMING AHARONOV-BOHM INTERFEROMETER

T. KUBO*[,1], Y. TOKURA[1,2], S. AMAHA[1], T. HATANO[1], and S. TARUCHA[1,3]
[1] *Quantum Spin Information Project, ICORP, JST, Kanagawa, Atsugi-shi 243-0198, Japan*
[2] *NTT Basic Research Laboratories, NTT Corporation, Kanagawa, Atsugi-shi 243-0198, Japan*
[3] *Department of Applied Physics, University of Tokyo, Tokyo, Bunkyo-ku 113-8656, Japan*
**E-mail: kubo@tarucha.jst.go.jp*

We theoretically discuss coherent charge transport through an Aharonov-Bohm interferometer formed by a triple quantum dot using the nonequilibrium Green's function method. In particular, we introduce the coherent indirect coupling parameter α, which characterizes the indirect coupling strength via the reservoir between two quantum dots. When $\alpha = 0$, the period of the Aharonov-Bohm oscillations for the linear conductance is h/e. For $\alpha \neq 0$, the period of the AB oscillations becomes $3h/e$ associated with the three small triangular paths of electrons, which are formed by the triple quantum dot and the reservoir.

Keywords: laterally coupled triple quantum dot; Aharonov-Bohm oscillation; coherent indirect coupling.

1. Introduction

Quantum phase coherence in mesoscopic systems had been attracting considerable interest of many physicists. Quantum phase coherence is often probed by measuring interference oscillations of the conductance as a function of magnetic flux, chemical potential or electric field. In interference experiments with an Aharonov-Bohm (AB) ring containing a quantum dot (QD), quasiperiodic modulation of the tunneling current as a function of magnetic flux threading through the ring was demonstrated [1]. This result demonstrates that phase coherence is maintained during the tunneling process through a QD. Subsequently, for a parallel coupled double QD system, AB oscillations were observed in a tunneling current through the double QD between the common source and drain contacts [2]. Recently we have succeeded in fabricating laterally coupled vertical triple QD (TQD) devices with a triangular arrangement, and studied the transport properties [3]. It is very important to understand the basic properties of TQD systems since TQD systems can provide novel electronic effects such as ratchet effect [4], correlation effects [5,6], and charge detection [7].

In this article, we examine coherent charge transport through an AB interferometer formed by a TQD, as schematically shown in Fig. 1. In particular, we take into account the coherent indirect coupling via the reservoirs between QDs [8]. We calculate the linear conductance using the nonequilibrium Green's function method. The coherent indirect coupling affects the AB oscillation period of the linear conductance.

2. Model

We consider a laterally coupled TQD shown in Fig. 1. We model this system with the Hamiltonian

$$H = H_R + H_{TQD} + H_T, \qquad (1)$$

where H_R describes Fermi seas of non-interacting electrons in two reservoirs:

$$H_R = \sum_{\nu \in \{U,L\}} \sum_k \epsilon_{\nu k} {c_{\nu k}}^\dagger c_{\nu k}. \qquad (2)$$

Here $\epsilon_{\nu k}$ is the energy of conduction electrons with wave number k in the reservoir ν, and

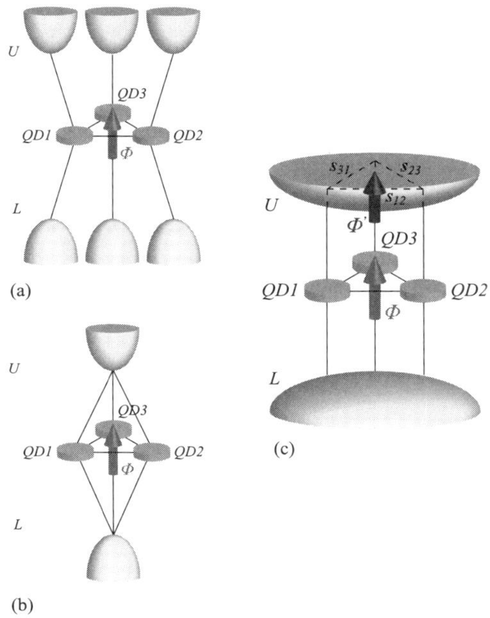

Fig. 1. Schematic diagram of a laterally coupled TQD forming AB interferometer. Φ denoted by an arrow is the magnetic flux threading through the triangle formed by three QDs, and Φ' is the magnetic flux threading through the triangle formed by electron's propagation in the reservoirs. s_{ij} is the propagation length of electrons in the reservoirs between i-th QD and j-th QD. (a) $\alpha = 0$. (b) $\alpha = 1$. (c) $0 < |\alpha| < 1$.

the operator $c_{\nu k}$ ($c_{\nu k}{}^{\dagger}$) annihilates (creates) an electron in the reservoir ν. H_{TQD} is the Hamiltonian of the isolated TQD:

$$\begin{aligned} H_{TQD} &= \sum_{i=1}^{3} \epsilon_0 d_i{}^{\dagger} d_i \\ &\quad + \Big(t_{12} e^{i\phi/3} d_1{}^{\dagger} d_2 + t_{23} e^{i\phi/3} d_2{}^{\dagger} d_3 \\ &\quad + t_{31} e^{i\phi/3} d_3{}^{\dagger} d_1 + \text{h.c.} \Big) \\ &\equiv \sum_{i=1}^{3} \epsilon_0 d_i{}^{\dagger} d_i \\ &\quad + \Big[t_{12}(\phi) d_1{}^{\dagger} d_2 + t_{23}(\phi) d_2{}^{\dagger} d_3 \\ &\quad t_{31}(\phi) d_3{}^{\dagger} d_1 + \text{h.c.} \Big] . \end{aligned} \tag{3}$$

Here ϵ_0 is the single-particle energy level of three QDs, and d_i ($d_i{}^{\dagger}$) annihilates (creates) an electron in the ith QD. In the second term, t_{ij} represents the direct inter-dot tunnel coupling between the ith QD and jth QD. The factors $e^{\pm i\phi/3}$ indicate the effect of the magnetic flux ($\phi = 2\pi\Phi/\Phi_0$ is an AB phase in the triangle formed by three QDs, where Φ is the magnetic flux threading through the triangular loop as shown in Fig. 1, and $\Phi_0 = h/e$ is the magnetic flux quantum). The electron spin and inter-dot Coulomb interaction are disregarded. H_T is the tunneling Hamiltonian between reservoirs and QDs:

$$H_T = \sum_{\nu \in \{U,L\}} \sum_k \sum_{i=1}^{3} \left[t_{\nu k}^{(i)} c_{\nu k}{}^{\dagger} d_i + \text{h.c.} \right], \tag{4}$$

where $t_{\nu k}^{(i)}$ is real. In this model, we assume that the linewidth functions are

$$\mathbf{\Gamma}^U = \mathbf{\Gamma}^L = \Gamma \begin{pmatrix} 1 & \alpha_{12} & \alpha_{31} \\ \alpha_{12} & 1 & \alpha_{23} \\ \alpha_{31} & \alpha_{23} & 1 \end{pmatrix} . \tag{5}$$

Here the boldface notation indicates the 3×3 matrices, and we introduced the coherent indirect coupling between QDs via the reservoir [8]. In general, α_{ij} between i-th QD and j-th QD is a function of the propagation length s_{ij} of electrons in the reservoirs. For simplicity, we assume that $\alpha_{ij} = \alpha$. As shown in Fig. 1(a), when $\alpha = 0$ (s is infinite), there is no coherence via the reservoirs between QDs. In contrast, when the propagation length s of electrons in the reservoirs is zero, the coherence via the reservoirs between QDs is maximal (see Fig. 1(b)). Most of experimental conditions correspond to the situation ($0 < |\alpha| < 1$) shown in Fig. 1(c). Then, if all direct inter-dot tunnel couplings are the same, $t_{ij} = t_c$, we can simultaneously diagonalize the TQD Hamiltonian H_{TQD} and the linewidth function matrices $\mathbf{\Gamma}^U$ and $\mathbf{\Gamma}^L$. The eigenvalues of the linewidth function matrices have to be non-negative since those are related to the lifetime. In this system, the eigenvalues of the linewidth functions are $1 + 2\alpha$ and $1 - \alpha$ (degenerate). Therefore, $-\frac{1}{2} \leq \alpha \leq 1$. This is in contrast to the situation of double QDs where the condition is $|\alpha| \leq 1$ [8]. Depending on the situations, therefore, we have to be careful about the physically adequate range of the coherent indirect coupling parameter α. In general, we can determine the range of the coherent indirect coupling parameters from the condition, that the poles of the retarded Green's function have to be in the lower half-plane of the complex ϵ plane.

Moreover, as shown in Fig. 1(c), in fact, we have to take into account the magnetic flux dependence of the coherent indirect coupling parameter α since there exists finite magnetic flux Φ' threading through the trajectry formed by s_{12}, s_{23}, and s_{31}.

3. Formulation

The tunneling current is given by [9]

$$I = -\frac{e}{2\pi}\int \frac{d\epsilon}{\hbar}[f_U(\epsilon) - f_L(\epsilon)] \times \mathrm{Im}\left[\mathrm{Tr}\left\{\frac{\mathbf{\Gamma}}{2}\boldsymbol{G}^r(\epsilon)\right\}\right], \quad (6)$$

where $\mathbf{\Gamma} = \mathbf{\Gamma}^U + \mathbf{\Gamma}^L$, and the Fermi-Dirac distribution function of the reservoir ν is $f_\nu(\epsilon) = 1/[1 + e^{(\epsilon-\mu_\nu)/k_BT}]$. $\boldsymbol{G}^r(\epsilon)$ is the Fourier transform of the retarded Green's function matrix $\boldsymbol{G}^r(t,t') = -i\theta(t-t')\left\langle\{d_i(t), {d_j}^\dagger(t')\}\right\rangle$.

From the equation-of-motion (EOM) for the retarded Green's function $G^r_{ij}(t,t') = -i\theta(t-t')\langle\{d_i(t), {d_j}^\dagger(t')\}\rangle$, we obtain the following Fourier transform of $\boldsymbol{G}^r(t,t')$

$$\boldsymbol{G}^r(\epsilon) = \left[\{\boldsymbol{g}^r(\epsilon)\}^{-1} + \frac{i}{2}\mathbf{\Gamma}\right]^{-1}, \quad (7)$$

where $\boldsymbol{g}^r(\epsilon)$ is the retarded Green's function of an isolated TQD given as

$$\boldsymbol{g}^r(\epsilon) = \begin{pmatrix} \frac{\epsilon-\epsilon_0}{\hbar} & -\frac{t_{12}(\phi)}{\hbar} & -\frac{t_{31}(\phi)}{\hbar} \\ -\frac{t_{12}(\phi)}{\hbar} & \frac{\epsilon-\epsilon_0}{\hbar} & -\frac{t_{23}(\phi)}{\hbar} \\ -\frac{t_{31}(\phi)}{\hbar} & -\frac{t_{23}(\phi)}{\hbar} & \frac{\epsilon-\epsilon_0}{\hbar} \end{pmatrix}^{-1}. \quad (8)$$

4. Results

In the following, we focus on the situation when $t_{ij} = t_c$ and zero temperature. The following discussion about the period of AB oscillations is still valid even under finite temperature. When $\alpha = 0$, we observe simple AB oscillations with the period of h/e. When $\epsilon_0/\hbar\Gamma = -1$ and $t_c/\hbar\Gamma = -2$, the linear conductance is shown by solid line in Fig. 2. However, for a finite coherent indirect coupling, as an example we consider the situation for $\alpha = 1$, the period of the AB oscillation for the linear conductance is $3h/e$ (see dashed line of Fig. 2). When $\alpha = 0$, tunneling processes enclosing the magnetic flux are only the triangular AB ring formed by the TQD. Although $\Phi' = 0$ under this condition (see Fig. 1(b)), for $0 < |\alpha| < 1$, we have to consider the effect of Φ' [10]. For $\alpha = 1$, however, as shown in Fig. 3, magnetic flux threads through a side surface formed by two QDs and one reservoir. This provides the period $3h/e$ of the AB oscillation for the linear conductance.

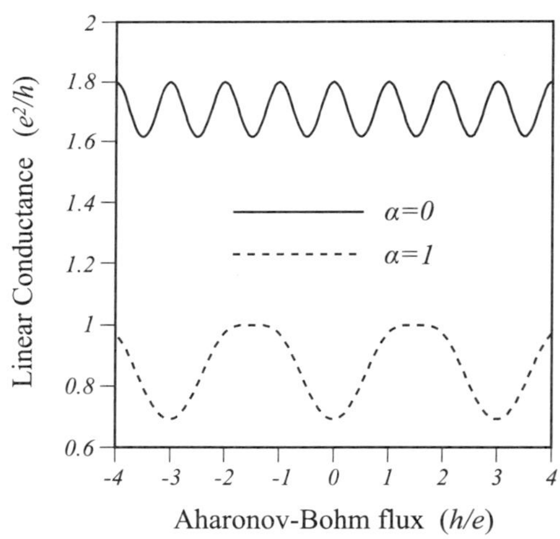

Fig. 2. AB oscillations for the linear conductance at zero temperature for $\epsilon_0/\hbar\Gamma = -1$ and $t_c/\hbar\Gamma = -2$. Solid and dashed lines indicate the situations for $\alpha = 0$ and $\alpha = 1$, respectively.

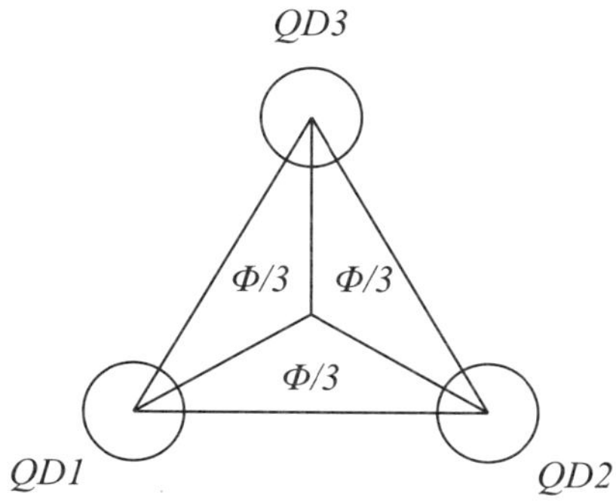

Fig. 3. Schematic view from the top for $\alpha = 1$. The magnetic flux $\Phi/3$ threads through each side surface formed by two QDs and one reservoir.

5. Conclusions

We examined the electron transport through a laterally coupled TQD forming AB interferometer with a coherent indirect coupling parameter α using the nonequilibrium Green's function method. In the TQD systems, it is noted that we have to be careful about the range of the coherent indirect coupling parameter. Moreover, we discussed the period of the AB oscillations for the tunneling current through a TQD. While, for $\alpha = 0$, it is h/e, the period of the AB oscillations becomes $3h/e$ when $\alpha \neq 0$.

Acknowledgments

We would like to thank S. Sasaki and S. Teraoka for useful discussions. Part of this work is financially supported by a Grant-in-Aid for Scientific Research S (No. 19104007), B (No. 18340081), SORST-JST, and Special Coordination Funds for Promoting Science and Technology, MEXT.

References

1. A. Yacoby, M. Heiblum, D. Mahalu, and H. Shtrikman, Phys. Rev. Lett. **74**, 4047 (1995).
2. T. Hatano, M. Stopa, W. Izumida, T. Yamaguchi, T. Ota, and S. Tarucha, Physica E **22**, 534 (2004).
3. S. Amaha, T. Hatano, T. Kubo, Y. Tokura, D. G. Austing, and S. Tarucha, Physica E **40**, 1322 (2008).
4. M. Stopa, Phys. Rev. Lett. **88**, 146802 (2002).
5. T. Kuzmenko, K. Kikoin, and Y. Avishai, Phys. Rev. Lett. **96**, 046601 (2006).
6. N. J. Craig, J. M. Taylor, E. A. Lester, C. M. Marcus, M. P. Hanson, and A. Gossard, Science **304**, 565 (2004).
7. D. S. Saraga and D. Loss, Phys. Rev. Lett. **90**, 166803 (2003)
8. T. Kubo, Y. Tokura, T. Hatano, and S. Tarucha, Phys. Rev. B **74**, 205310 (2006).
9. Y. Meir and N. S. Wingreen, Phys. Rev. Lett. **68**, 2512 (1992).
10. in preparation.

AHARONOV-BOHM OSCILLATIONS IN PARALLEL COUPLED VERTICAL DOUBLE QUANTUM DOT

T. HATANO[a, *], T. KUBO[a], S. AMAHA[a], S. TERAOKA[a], Y. TOKURA[a,b] and S. TARUCHA[a,c]
[a]*Quantum Spin Information Project, ICORP, JST, Atsugi-shi, Kanagawa 243-0198, Japan*
[b]*NTT Basic Research Laboratories, NTT Corporation, Atsugi-shi, Kanagawa 243-0198, Japan*
[c]*Department of Applied Physics, Univ. of Tokyo, Hongo, Bunkyo-ku, Tokyo 113-8656, Japan*
**E-mail: hatano@tarucha.jst.go.jp*

We used a parallel coupled vertical double quantum dot device consisting of two laterally tunnel-coupled dots and a common source and drain electrodes to study the Aharanov-Bohm (AB) effects in the vertically flowing single electron tunneling current. In the charging diagram measured at a finite source-drain voltage a bonding state as a ground state, and an antibonding state as an excited state were distinguished for the one-electron state. We observed periodic magnetic oscillations for the current through the bonding state, due to the AB effect for the two tunnel-coupled dots.

Keywords: parallel coupled vertical double quantum dot; bonding and antibonding states; Aharonov-Bohm oscillations.

1. Introduction

Quantum dots are expected to be applied for novel devices, for example, in the field of spintronics and quantum information processing because both charge states and spin states can be fully controlled [1]. For these applications, a number of quantum dots should work collectively. Particularly for applications to quantum information devices, it is important to conserve coherence in the total system, including quantum dots and outer environments. Aharonov-Bohm (AB) effect [2] is a useful probe to investigate the quantum coherence in such collective systems [3]. For systems of two quantum dots, AB effects were experimentally studied using an AB ring with one quantum dot in each path [4, 5]. Theory predicted that there should be a specific phase shift in the AB oscillations for the current through the two tunnel-coupled states: a spin singlet state and a spin triplet state [6], and also that the AB oscillations and Fano effects of the bonding and antibonding states [7]. Particularly, we introduced the coherent indirect coupling parameter, which characterizes the strength of the indirect coupling via the reservoirs between two quantum dots, and found that the AB oscillations and Fano effects for the current flowing through the bonding and antibonding states are affected by this indirect coupling [7]. However, it has still been difficult to experimentally observe the AB oscillations in the current through the tunnel-coupled states, probably because of difficulty in the device fabrication. Previously, we observed AB oscillations for a vertical AB ring including one vertical dot in each arm but for the dots very weakly coupled [5].

In this paper, we fabricated a parallel coupled vertical double quantum dot device (See Fig. 1) and studied the transport properties. We clearly identified a set of a bonding state and an anti-bonding state in the excitation spectrum for the one-electron state. We observed AB oscillations in the current flowing through the bonding state.

2. Experimental results

Figure 1(a) schematically shows the device configuration. Two quantum dots are vertically confined by a double-barrier heterostructure (DBH) and laterally defined by four split Schottky gates [8]. The DBH is made of an undoped 10 nm GaAs well and undoped AlGaAs barriers of thicknesses 8.0nm and 7.0nm (the thicker is closest to the substrate). Two of the gates (side gates) are used to tune the number of electrons in each dot independently, and the remaining two gates (center gates) are used to tune the inter-dot tunnel coupling. Current vertically flows between the common source and drain through the parallel coupled two dots (Fig. 1(b)). The line mesas split the surrounding gate metal.

Measurements were carried out for the device placed in a dilution refrigerator at 20mK with an in-plane magnetic field as a parameter. Because the two dots are tunnel-coupled to each other, the magnetic field penetrates the two closed current paths (upper and lower paths) with more or less the same area. The respective closed area is defined by the inter-dot distance and the distance between the coupled dots and the source or drain, and it is approximately given by S∼350 nm × 15 nm $=5.3\times10^{-15}$ m^{-2}. Therefore, the expected period of AB oscillations is about 0.77T.

Figure 2(a) shows the conductance of the device measured as a function of the left and the right gate voltages, V_{sL} and V_{sR}. This figure is called charging diagram [8], in which the charge state in each honeycomb-like structure is labeled (N_L, N_R), where N_L (N_R) is the number of electrons in the left (right) dot. The applied source-drain voltage V_{sd} is small enough that the ground state of the double dot is only probed as the dark line or Coulomb peak in this figure. No Coulomb peaks were observed for $V_{sL} \lesssim -1.7$ V and $V_{sR} \lesssim -0.1$ V. Therefore, in Fig. 2(a) the double dot state changes from the empty state (0,0) to the charged state (1,1) via the state (1,0) or (0,1) in the solid circle.

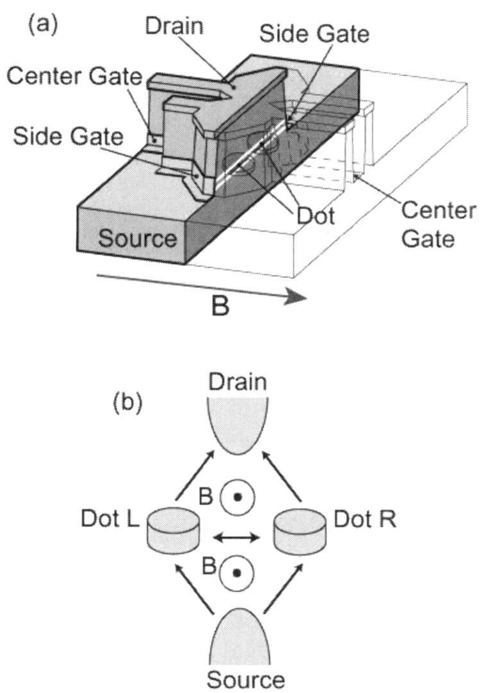

Fig. 1 (a) Structure of the double quantum dot device. (b) Schematics of parallel coupled double quantum dot. Arrows indicate the current flow.

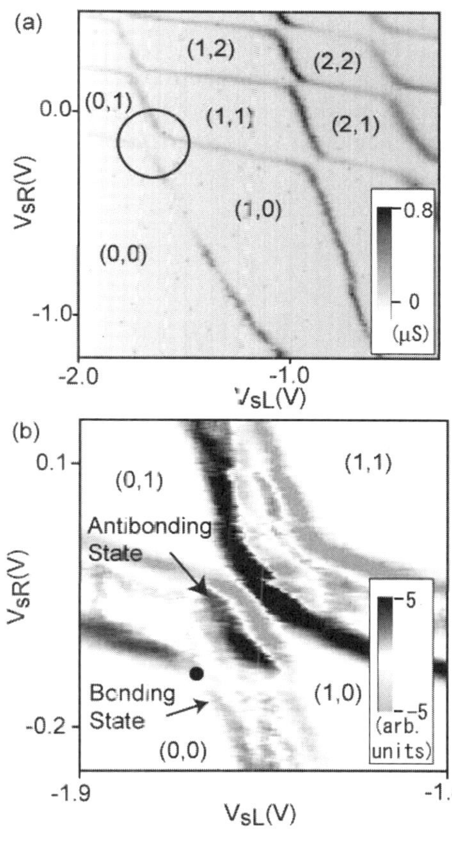

Fig. 2 (a) Conductance as a function of the left and right side gate voltages, V_{sL} and V_{sR} at V_{sd}=50 μV and V_c=-1.20 V. (b) Average values of the transconductance $(dI_{sd}/dV_{sL}+dI_{sd}/dV_{sR})/\sqrt{2}$ as a function of V_{sL} and V_{sR} at V_{sd}=800μV in the region with just one electron in the double dot.

Figure 2(b) shows the average value of the transconductance $(dI_{sd}/dV_{sL}+dI_{sd}/dV_{sR})/\sqrt{2}$ as a function of V_{sL} and V_{sR} at V_{sd}=800μV in the solid circle of Fig. 2(a). The two anti-crossing stripes are the Coulomb peaks widened by the finite V_{sd}, corresponding to addition of one extra electron to the empty state, and one more electron to the one-electron state, respectively. The region in black or dark gray indicates the positive derivative and highlights either the ground state or excited states. In the lower left Coulomb stripe, we observe the ground state (left lower dark line) and the excited state (right upper dark line), which are repulsive from each other. This feature arises from the inter-dot tunnel coupling, reflecting a bonding ground state and an antibonding excited state [9].

The magnetic field dependence of the current through the bonding state, which is indicated by the solid circle in Fig.2 (b), is shown in Fig. 3(a). This current shows clear periodic oscillations. The fast Fourier transform (FFT) of the oscillations is shown in Fig. 3(b). The dashed vertical line indicates the inverse magnetic field of the maximum FFT component. The inverse of the value is 0.81T, corresponding to the period of the AB oscillations in Fig. 3(a). This period is very close to the calculated value, 0.77T, which is described before.

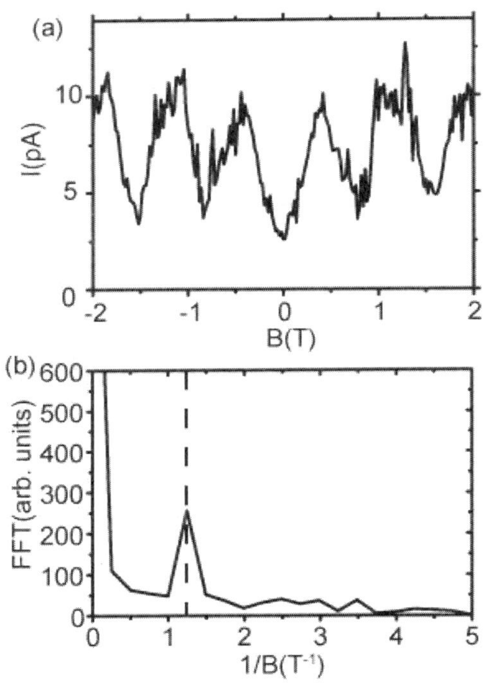

Fig. 3 (a) Magnetic field dependence of current through the bonding state at V_c=-1.20V and V_{sd}=800μV. (b) Fast Fourier transform of the current oscillation of (a).

As compared with our previous experiment with a similar geometry but having a very weak inter-dot tunnel coupling [5], we found two different points. First, the AB oscillations have a minimum at B=0 T in Fig. 3(a). On the other hand, in our previous experiment [5], the AB oscillations had a maximum at B=0 T. Since both devices used in the present and previous experiments have mirror symmetry in the geometry, we expect the current path between the source and drain is almost identical between the two dots. Therefore, if we neglect the inter-dot tunnel coupling, the current path forms a single loop made out of two symmetric branches each containing one dot, which brings about a peak in the AB oscillations at B=0 T. We only speculate here that the observed dip in the AB oscillation at B=0 T can be due to additional interference paths associated with the inter-dot tunnel coupling.

The visibility, ~60% of the AB oscillations in Fig. 3(a) is higher than that (~15%) of our previous experiment, probably due to the difference in the measurement temperature between 20 mK and 1.5K [10]. Note the visibility of our experiment is also higher than that (~20%) previously observed for an AB ring with a single dot embedded in one of the two branches [11]. This is probably because the AB ring used in our experiment is about ten times smaller than that used in [11].

Summary

Single electron tunneling characteristics were studied using a parallel coupled vertical double quantum dot. For the double dot having just one electron, the ground state as the bonding state, and the excited state as the antibonding state were distinguished in the charge diagram with a finite source-drain voltage. The current through the bonding state showed periodic oscillations with magnetic field, indicating AB oscillations associated with the bonding state.

Acknowledgments

We thank Y. -S. Shin, A. Shibatomi and S. Sasaki for experimental help and useful discussions. Part of this work is financially supported by the Grant-in-Aid for Scientific Research S (No. 19104007), B (No. 18340081), SORST-JST, Special Coordination Funds for Promoting Science and Technology, MEXT.

References

[1]S. Tarucha, D. G. Austing, T. Honda, R. J. van der Hage, L. P. Kouwenhoven, Phys. Rev. Lett. **77**, 3613 (1996).
[2]Y. Aharonov and D. Bohm, Phys. Rev. **115** (1959) 485.
[3]T. Bergsten, T. Kobayashi, Y. Sekine, and J. Nitta, Phys. Rev. Lett. **97**, 196803 (2006).
[4]A. W. Holleitner, C. R. Decker, H. Qin, K. Eberl and R. H. Blick, Phys. Rev. Lett. **87**, 256802(2001).
[5]T. Hatano, M. Stopa, W. Izumida, T. Yamaguchi, T. Ota and S. Tarucha, Physica E **22**, 534 (2004).
[6]D. Loss and E. V. Sukhorukov, Phys. Rev. Lett. **84**, 1035 (2000).
[7]T. Kubo, Y. Tokura, T. Hatano, and S. Tarucha, Phys Rev. B **74**, 205310 (2006).
[8] W. G. van der Wiel, S. De Franceschi, J. M. Elzerman, T. Fujisawa, S. Tarucha and L. P. Kouwenhoven, Rev. Mod. Phys. **75**, 1 (2003).
[9]T. Hatano, Y. Tokura S. Amaha, T. Kubo, Y. Nishi, Y. Hirayama and S. Tarucha, Phys. Rev. B 77, 241301(R)(2008).
[10]A.E. Hansen, A. Kristensen, S. Pedersen, C.B. Sørensen, and P.E Lindelof, Phys. Rev. B **64**, 045327 (2001).
[11]A. Yacoby, M. Heiblum, D. Mahalu, and H. Shtrikman, Phys. Rev. Lett. **74**, 4047 (1995).

LATERALLY COUPLED TRIPLE SELF-ASSEMBLED QUANTUM DOTS

S. AMAHA*, T. HATANO, T. KUBO, S. TERAOKA, A. SHIBATOMI,

Quantum Spin Information Project, ICORP, JST, Atsugi-shi, Kanagawa, 243-1098, Japan
**E-mail: amaha@tarucha.jst.go.jp*

Y. TOKURA

Quantum Spin Information Pjoject, ICORP, JST and NTT Basic Research Laboratories, NTT Corporation, Atsugi-shi, Kanagawa, 243-0198, Japan

S. TARUCHA

Quantum Spin Information Pjoject, ICORP, JST and *Institute for Nano Quantum Information Electronics, University of Tokyo* and *Applied Physics, School of Engineering, University of Tokyo, Hongo, Bunkyo-ku, 113-8656, Japan*

We investigate the electronic properties of a sub-micron vertical resonant tunneling structure containing three self-assembled InAs quantum dots (SADs) surrounded by four gate electrodes. The stability diagram is obtained by measuring Coulomb oscillation peaks as a function of the four gate voltages, which are used to modulate the electro-chemical potential of each SAD differently. We assign the charge states in the diagram by identifying three sets of Coulomb oscillation lines. We observe specific anti-crossing behaviors for the two Coulomb oscillation lines in the different sets, reflecting parallel couplings between three SADs in the conduction channel.

Keywords: Quantum dots; triple quantum dots; self-assembled quantum dots.

1. Introduction

Multiple quantum dots (QDs) are attracting interests due to the enhanced degrees of freedom associated with the inter-dot couplings. Especially for triple QDs several unique effects have been predicted, such as charge ratchet effect [1], RKKY interaction [2] and Nagaoka ferro magnetism, depending on the configuration of three dots [3]. However, just a few experiments have been presented for triple QDs so far due to the difficulty in fabrication using a top-down technique [4, 5, 6]. On the other hand, self-assembled InAs QD (SAD) structures, which are typical examples of a system made by a bottom-up technique, have been actively studied not only for optical [7] and electrical properties [8] but also for applications to optical devices [9]. Novel electronic properties, including the Kondo effect [10], spin injection [11], and interplay between the Kondo effect and Andreev reflection [12] have been well studied for single SADs directly contacted to metal electrodes. Another attention has been directed to the properties of multiply coupled SADs. The effects of inter-dot coupling on the optical [13] and electrical properties [14] have been exclusively studied for *vertically* coupled double SAD structures. Technically *laterally* coupled SADs can be realized by adjusting the density of SADs in the epitaxial growth, but investigation on their electronic properties is difficult because individual coupled SADs cannot be distinguished in a controlled manner. Therefore, no work has ever been presented to manipulate the charge states in *laterally* coupled SADs.

Here we investigate the electronic properties of a vertical resonant tunnel diode containing three SADs with four gate electrodes. These gate electrodes act not only to squeeze the conduction channel of the tunnel diode but also manipulate electro-chemical potential of each SAD differently. Therefore, the stability diagram of the triple

SADs can be obtained by sweeping particular gate voltage pairs of the four gates [15]. We then assign three sets of Coulomb oscillation lines in the stability diagram, related to the charging of the three QDs, and observe a couple of anti-crossings between two Coulomb oscillation lines in the different sets. These anti-crossings indicate the existence of capacitive and/or quantum mechanical tunnel coupling between relevant two SADs.

2. Experimental

2.1 *Device structure*

The material used for making the devices was grown by molecular beam epitaxy on a semi-insulating GaAs substrate. The average dot diameter and height are estimated to be ~20 and 3-5 nm, respectively [15]. The SAD sheet density derived from the AFM image (not shown) is about $\sim 2\times 10^{10}/\mathrm{cm}^2$. The density is high enough to expect the existence of two or three SADs inside the squeezed conduction channel of the cross-shaped mesa ($\sim (0.1\mu\mathrm{m})^2$).

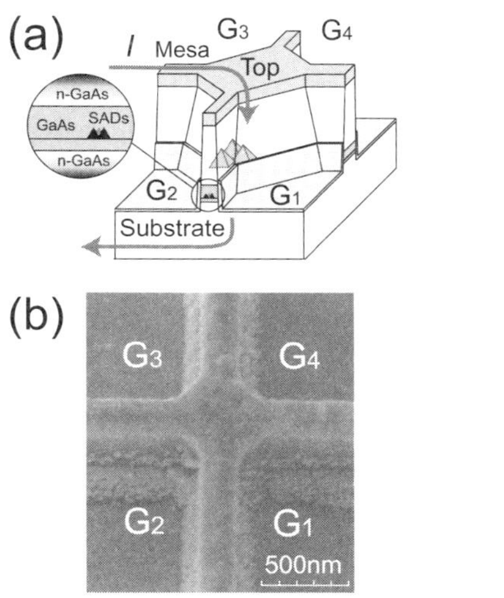

Fig.1. (a) Schematic diagram of four-gated vertical resonant tunneling device. Thin line mesas are connected to the square mesa and separate the gate electrodes (G1, G2, G3, G4) as four. We measure the current I through the square mesa by sweeping the four gate voltages (V_{g1}, V_{g2}, V_{g3}, V_{g4}) on (G1, G2, G3, G4). (b) SEM image of a typical four-gated vertical resonant tunneling device.

The SADs with a wetting layer are embedded in a thin un-doped GaAs layer between two GaAs layers with graded n-type doping. Figure 1(a) schematically shows the sub-micron vertical resonant tunneling structure with four separate gate electrodes fabricated by a typical technique of making vertical QD structures [16].

Figure 1 (b) is a scanning electron microscope (SEM) image of the device fabricated. The gate electrodes are separated from each other by the four thin line mesas. These line mesas are sufficiently thin to be fully depleted. Therefore, the current flows along the metal on top of the line mesas and then downwards through several SADs in the central square mesa. We apply different gate voltages (V_{g1}, V_{g2}, V_{g3}, V_{g4}) to the four gate electrodes (G1, G2, G3, G4) (see Fig.1 (a)), and measure the tunneling current under the source-drain voltage V_{sd} at ~10mK using a general DC measurement technique.

2.2 *Measurement results*

Figure 2 (a) shows the intensity plot of measured current as a function of $V_{g1}(=V_{g2})$ and V_{g3} with constant voltages of V_{g4}= +0.30V and V_{sd}=0.3mV. The obtained stability diagram is not similar to so called "honeycomb" structure of double QDs. To indentify the charge states in the obtained stability diagram, we show schematic diagrams of Figs. 2 (b) and (c). In Fig. 2 (b), solid thick lines labeled $▲_i$ (i=1,2) and solid thin lines labeled $■_i$ (i=1,2) form the honeycomb-like shape. The thick lines are induced by the charging of one dot (Dot1) and the thin lines by the charging of the other dot (Dot2). To consider the charging effect of the third QD, we draw a dotted line with different slope from the other solid lines in Fig. 2 (b). In the realistic case, each crossing of the solid and dotted line should be repulsive due to the inter-dot Coulomb interaction and quantum mechanical tunneling between the relevant two QDs. Therefore, we can build up the stability diagram of the triple QD as shown in Fig. 2 (c).

The experimentally obtained stability diagram shown in Fig. 2 (a) is consistent with the schematic diagram of Fig. 2(c). We can classify three sets of Coulomb oscillation lines, as labeled ▲$_i$ (i=1,2), ■$_i$ (i=1,2) and ●$_i$ (i=1), respectively in Fig. 2 (a), depending on the slopes of the Coulomb oscillation lines in the stability diagram. This indicates that three QDs are located in the conduction channel. Each Coulomb oscillation line labeled ▲$_i$ (i=1,2), ■$_i$ (i=1,2) or ●$_i$ (i=1) is associated with the charging of the respective QD (Dot1, Dot2 or Dot3). Note that the slope of each Coulomb oscillation line in the stability diagram is associated with the gate capacitance to the corresponding dot, which depends on the location of the dot. No Coulomb oscillation lines are observed in the region of $V_{g1}(=V_{g2})<$ -0.55V and $V_{g3}<$0.05V, indicating that all three dots are empty, so that we can assign the electron numbers of Dot1 (N_1), Dot2 (N_2) and Dot3 (N_3) in each Coulomb blockade region as shown by (N_1, N_2, N_3) in Fig. 2 (a). We observe a weak but definite rounding at the anti-crossing between two dark lines, which are highlighted by the dotted circles (α), and (β) formed by the ▲$_1$ and ●$_1$ lines, and by the ■$_1$ and ▲$_1$ lines, respectively in Fig. 2(a). Such anti-crossing indicates existence of inter-dot electrostatic and/or tunnel coupling between Dot1 and Dot3 for (α), and between Dot2 and Dot3 for (β), respectively. In case of (γ) and (δ), the Coulomb oscillation line ●$_1$ closely approaches the round anti-crossing of the two Coulomb oscillation lines ■$_2$ and ▲$_2$, where the electronic states in Dot3 should be coupled to the molecular state spread in Dot1 and Dot2. Therefore, we refer to the two anti-crossings (γ) and (δ) as "quadruple point". These anti-crossings suggest that all three dots might be coupled to each other in the triangular arrangement. The molecular like states formed near the gate voltage condition of (γ) and (δ) should be interesting to investigate the spin states and Aharonov-Bohm effects in triple QDs under a finite magnetic field [17, 18].

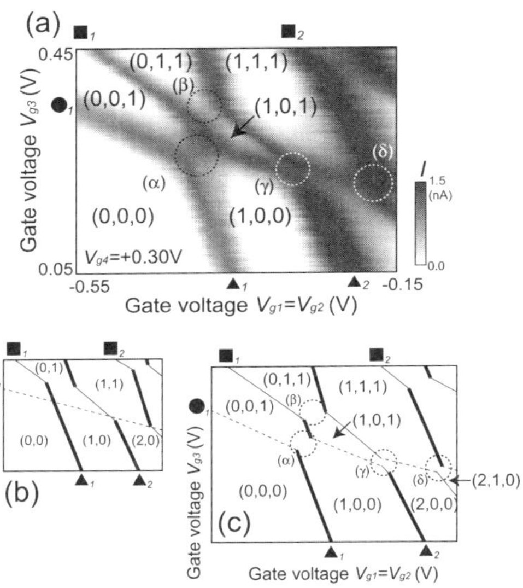

Fig. 2. (a) Measured current levels as a function of V_{g1}(= V_{g2}) and V_{g3} under constant V_{g4}=+0.30V and V_{sd}=0.3mV. (b) Schematic diagram of double QD. Solid thick and solid lines are induced by the charging of Dot1 and Dot2, labeled by ▲$_i$ (i=1,2) and ■$_i$ (i=1,2), respectively. A dotted line is guideline of the additional Coulomb oscillation line (●$_i$ (i=1)) to understand the stability diagram of (c). (c) Schematic diagram of triple QD. Solid heavy lines, solid lines and dotted lines are induced by the charging of Dot1, Dot2 and Dot3, respectively. The fact that Fig. 2(c) has agreement for (a) indicates that triple QD is formed in our device. (α)-(δ) are anti-crossings of Coulomb oscillation lines caused by inter-dot Coulomb interaction and/or quantum mechanical tunneling between relevant two QDs.

3. Conclusion

We investigated the electronic properties of four gated resonant tunnel diode containing parallel coupled triple SADs. In the stability diagram obtained by tuning the particular gate voltage pairs, we identified three sets of Coulomb oscillation lines in the obtained current intensity plot as a function of the two gate voltages. Any two of the Coulomb oscillation lines in the different sets anti-cross with each other. Therefore we concluded that the related two of the three SADs are capacitively and/or quantum mechanically coupled to each other. This system will be useful to explore the physics in triple QDs coupled in the triangular arrangement.

Acknowledgments

The authors thank D. G. Austing, T. Sato, T. Ota, T. Kodera, K. Ono and T. Maruyama for fruitful comments. Part of this work was financially supported by a Grant-in-Aid for Scientific Research S (No. 19104007), B (No. 18340081), and Special Coordination Funds for Promoting Science and Technology, MEXT. ST acknowledges the support from DARPA-QuEST (BAA-0824).

References

1. M. Stopa, *Phys. Rev. Lett.* **88**, 146802 (2002).
2. N. J. Craig *et al.*, Science **304**, 565 (2004).
3. M. Korkusinski *et al.*, *Phys. Rev.* B **75**, 115301 (2007).
4. A. Vidan *et al.*, *Appl. Phys. Lett.* **85**, 3602 (2004).
5. L. Gaudreau *et al.*, *Phys. Rev. Lett.*, **97**, 036807 (2006).
6. S. Amaha *et al.*, Physica E **40**, 1322 (2008).
7. R.J. Warburton *et al.*. Nature **405**, 926 (2000).
8. M. Jung *et al.*, *Appl. Phys. Lett.*, **87**, 203109 (2005).
9. Y. Arakawa and H. Sakaki, *Appl. Phys. Lett.*, **40**, 939 (1982).
10. Y. Igarashi *et al.*, *Phys. Rev. B.* **76**, 081303 R (2007).
11. K. Hamaya *et al.*, *Appl. Phys. Lett.* **90**, 053108 (2007).
12. C. Buizert *et al.*, *Phys. Rev. Lett.* **99**, 136806 (2007).
13. M. Bayer *et al.*, Science **291**, 451 (2001).
14. T. Ota *et al.*, *Phys. Rev. Lett.* **95**, 236801 (2005).
15. S. Amaha *et al.*, *Appl. Phys. Lett.* **92**, 202109 (2008).
16. D. G. Austing *et al.*, *Jpn. J. Appl. Phys.* **36**, 4151 (1997).
17. F. Delgado *et al.*, *J. Phys.: Condens. Matter* **20**, 315207 (2008).
18. T. Kubo *et al.*, to be published (2009).

SPECTROSCOPY OF CHARGE STATES OF A SUPERCONDUCTING SINGLE-ELECTRON TRANSISTOR IN AN ENGINEERED ELECTROMAGNTIC ENVIRONMENT

E. ABE*, Y. KIMURA, Y. HASHIMOTO, Y. IYE and S. KATSUMOTO

The Institute for Solid State Physics, The University of Tokyo, 5-1-5 Kashiwanoha, Kashiwa, Chiba 277-8581, Japan

**E-mail: e-abe@issp.u-tokyo.ac.jp*

We fabricate superconducting single-electron transistors on a semiconductor heterostructure substrate, and examine the energy levels of their charge states by means of microwave spectroscopy. Light illumination induces the carriers at the heterointerface, which work as a dissipative electromagnetic environment. We observe prohibition of tunneling due to dissipation in a device with smaller tunnel coupling, while the other device with larger tunnel coupling preserves its quantum nature in spite of dissipation.

Keywords: Cooper pair box; Single-electron transistor; Quantum coherence; Dissipation.

1. Introduction

Experiments on macroscopic quantum coherence seek for the boundary between classical and quantum worlds. A model system is a double-well potential with one level in each well, where tunnel coupling Δ between the left and right levels leads to the formation of the bonding and anti-bonding states. When the dissipation (parameterized as α) is considered, the Caldeira-Leggett theory predicts that it can prohibit tunneling and 'classicalize' the system [1,2]. Such dissipation can be viewed as a leakage of system's information into its environment. The environment then 'sees' particle's location, which prevents a further movement of the particle.

Figure 1-(a) illustrates a qualitative phase diagram in the (α, Δ) parameter space obtained from a scaling method [2,3]. An ultimate goal of this line of research is to observe the quantum-classical crossover by controlling the parameters and to map out the phase diagram. In this contribution, we focus on the charge states of superconducting single-electron transistors (SETs), and present experimental results toward this goal. Dissipation is introduced by preparing the devices on a heterostructure wafer that is designed to be insulating at low-temperatures. The carriers are induced by light illumination and the two-dimensional electron gas (2DEG) layer becomes conductive, which is sustained long after the illumination is terminated. This realizes switching from the complete absence of dissipation to the presence. We observe prohibition of tunneling due to dissipation in a device with smaller

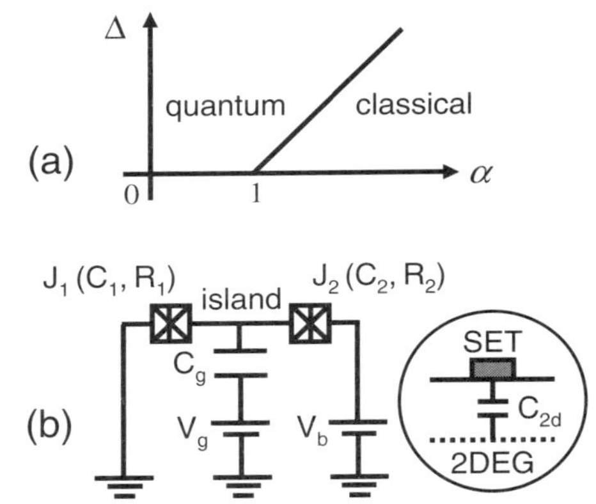

Fig. 1. (a) Cartoon illustrating the quantum-classical transition in the (α, Δ) parameter space. (b) Circuit diagram of our device. When the 2DEG is present, it capacitively couples with the device through capacitance C_{2d}.

tunnel coupling, while the other device with larger tunnel coupling preserves its quantum nature in spite of dissipation.

2. Device and method

The circuit diagram of the device we fabricated is shown in Fig. 1-(b). The Josephson junction J_1 and the gate electrode together form a Cooper pair box, the Hamiltonian of which is written as

$$H = \sum_{n \in Z} \Big[E_C (n - n_g)^2 |n\rangle\langle n| + \frac{E_J}{2} (|n\rangle\langle n+1| + |n+1\rangle\langle n|) \Big]. \quad (1)$$

Here $E_C = (2e)^2/(2C_\Sigma)$ is the Cooper pair Coulomb energy with C_Σ the total capacitance of the island, E_J is the Josephson energy of J_1, $n_g = (C_g Vg + Q_b)/(2e)$ is the gate charge including the background charge Q_b, and n is the number of 'excess' Cooper pairs in the island. $|n\rangle$ is the associated eigenstate.

For experimental convenience, J_1 is split into two to form a SQUID that allows for tuning of E_J. By imposing the flux bias Φ on the SQUID, E_J is controlled as $E_{J0} \cos(\pi\Phi/\Phi_0)$ (*i.e.*, E_J is maximum at $\Phi/\Phi_0 = 0$ and minimum at 1/2.), where Φ_0 is the flux quantum $h/2e$. The degeneracy of the two lowest levels at $n_g = n+1/2$ when $E_J = 0$ is lifted by a finite E_J. When $E_J \ll E_C$, the eigenstates are written as linear superposition of $|n\rangle$ and $|n+1\rangle$. This is equivalent to the case of the double-well system. Not to mention, E_J corresponds to Δ.

The exited state is experimentally addressable by means of microwave spectroscopy. By attaching the probe junction J_2 to the island and thus making the device an SET, microwave irradiation and subsequent observation of photon-assisted (PA) Josephson quasiparticle (JQP) current allow us to map out the energy dispersion relation of the system [4]. Under an appropriate bias condition, the JQP current peak is observed around $n_g = n + 1/2$, since the JQP cycle requires injection of one Cooper pair across J_1 followed by two sequential quasiparticle tunnelings across J_2. When the photon energy matches the energy difference of the two lowest levels, the Cooper pair is injected into the island, leading to the PAJQP process. In order not to disturb coherence at J_1, the condition $R_1 \ll R_2$ is requested, which is achieved by a triple-angle shadow evaporation technique.

We prepared two devices (Device 1 and 2), which share the same lithographic design but have different junction resistances. The devices were cooled with a dilution refrigerator down to 60 mK. Microwaves were sent through a coaxial cable that finally reaches the gate electrode.

3. Results and discussion

3.1. *DC measurement*

The device parameters are deduced from the DC measurement as a function of V_b and V_g. E_C before light illumination are estimated as 26.5 GHz for Device 1 and 16.2 GHz for Device 2. E_J at the respective junctions are hard to estimate because only the series resistance $R_1 + R_2$ is measurable.

When the carriers were introduced, they screened the electric field and made the gate electrode ineffective. This is due to the 2DEG beneath the gate electrode. The necessary voltage had to be applied from the 2DEG side via Ohmic contacts on the surface. The carriers capacitively couple with the SET through capacitance C_{2d}, which reduces E_C to 22.0 GHz for Device 1 and 14.3 GHz for Device 2.

The I-V_b curves near zero-bias for Device 1 at $\Phi/\Phi_0 = 0$ are shown in Fig. 2, and a qualitatively similar result is also obtained from Device 2. It is observed that the current due to Cooper pair tunneling increases after illumination. The reduction of E_C due to C_{2d} means that the Josephson tunneling effect becomes relatively more important. More-

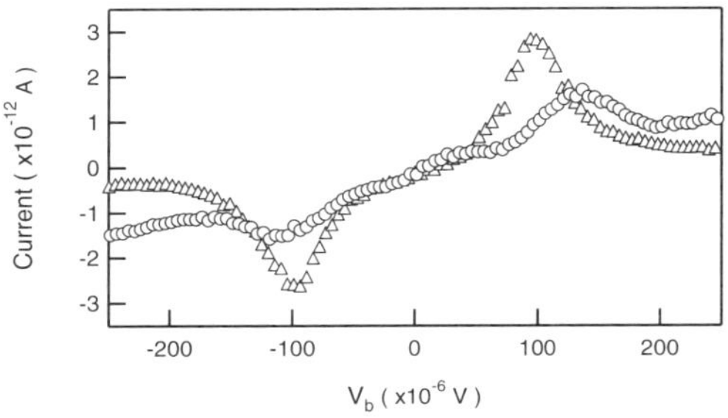

Fig. 2. I-V_b curves for Device 1 at $\Phi/\Phi_0 = 0$ before (circles) and after (triangles) illumination.

over, dissipation arising from the charge fluctuation of 2DEG makes the superconducting phase more localized. These lead to the enhancement of Cooper pair tunneling. On the other hand, this tunneling current is always suppressed at $\Phi/\Phi_0 = 1/2$ (not shown).

3.2. *Microwave spectroscopy*

Figure 3 is the result of spectroscopic measurement on Device 1. Here, unlike DC voltages, microwaves were applied through the gate electrode even after illumination. Although higher power was required to observe the PAJQP peaks, the spectroscopy was still possible. In Fig. 3-(a), spectroscopy at $\Phi/\Phi_0 = 1/2$ is shown. The dispersion relations are linear within the accuracy of the measurements because E_J is close to zero and the coherence is destroyed by the quasiparticle tunneling at the probe junction. The slopes then yield $2E_C$, and the obtained E_C are 26.7 and 22.7 GHz before and after illumination, respectively, which are close to those estimated from the DC measurements.

Figure 3-(b) shows spectroscopy at $\Phi/\Phi_0 = 0$. To facilitate comparison, the vertical axis is normalized by E_C of the respective conditions. The dispersion relations show qualitatively different behaviors with or without 2DEG. Without 2DEG, a clear deviation from linear behavior, indicative of the formation of the bonding and anti-bonding states, is observed. The data points fall on

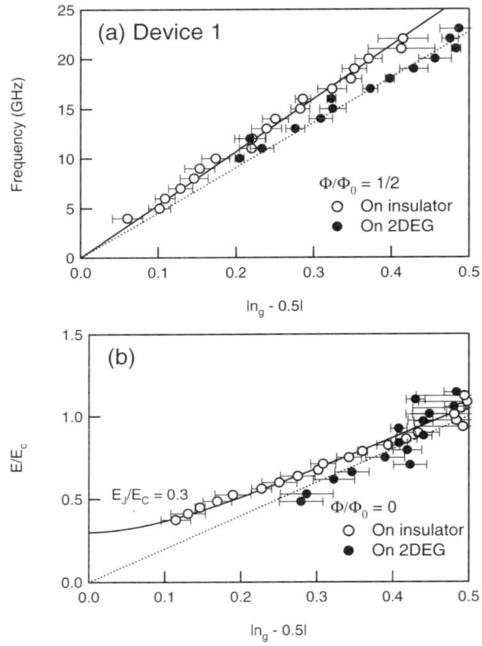

Fig. 3. Energy dispersion relations of Device 1. (a) At $\Phi/\Phi_0 = 1/2$ before (○) and after (●) light illumination. The solid (dashed) line is a linear fit to the data ○ (●). (b) At $\Phi/\Phi_0 = 0$. The vertical axis is given in units of E_C. The solid curve is the numerical solution of Eq. (1) for $E_J/E_C = 0.3$. The dashed line has a slope of $2E_C$. The positions of the PAJQP peaks are determined from the average of the absolute values of the photon-absorption and -emission peaks when the both peaks were observed, and the absolute value is plotted when only either of them was observed.

the theoretical curve calculated for $E_J/E_C = 0.3$, and E_J is estimated as about 8 GHz. In the presence of 2DEG, it is seen that the data points fall mostly on the line with the slope $2E_C$. Since the value of E_J itself should not be modified by the presence of 2DEG, this result suggests tunneling between the two charge states is forbidden. In other words, this is a strong indication of the dissipation-driven quantum-to-classical transition.

We note that the positions of the low-frequency PAJQP peaks after illumination were difficult to determine, for they are broader than those before illumination and overlap with the central JQP peaks. The

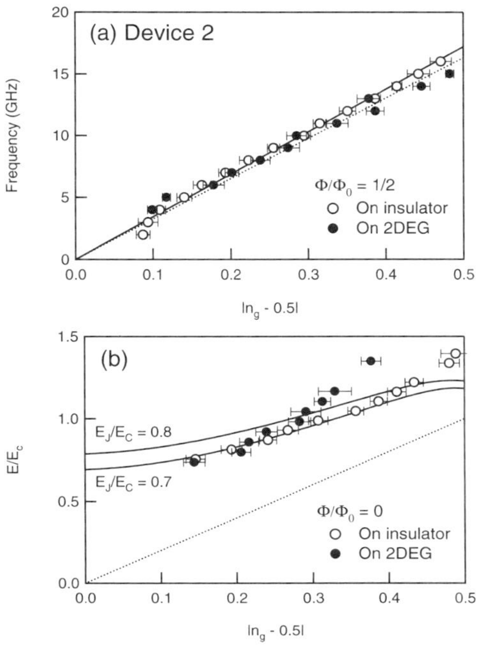

Fig. 4. Energy dispersion relations of Device 2. The solid curves in (b) are the numerical solutions of Eq. (1) for $E_J/E_C = 0.7$ and 0.8.

broader PAJQP peaks may reflect the level broadenings due to dissipation. but we have not identified a definitive reason.

Figure 4 shows the result on Device 2. As before, spectroscopy at $\Phi/\Phi_0 = 1/2$ was carried out to check E_C. The obtained E_C are 17.2 and 16.3 GHz before and after illumination, respectively. Although the reduction of E_C is observed, the amount is not as large as expected from the DC measurements. On the other hand, it is clear from Fig. 4-(b) that the energy gap is observed even in the presence of 2DEG and, thus, tunneling between the two charge states is still allowed. E_J/E_C is around 0.75 ($E_J \approx 13$ GHz).

The striking difference between the behaviors of the two devices would lie in the difference in E_J. Before illumination, the parameter α is small (expected to be close to zero) and the two devices are both in the quantum regime irrespective of E_J. Light illumination increases α, the values of which will not be much different between the two devices because we used the same heterostructure substrate and indeed C_{2d} was nearly the same. Therefore, E_J determines whether the system remains quantum or not. If E_J is relatively large, the device remains in the quantum regime. Otherwise, the device will fall on the classical regime. It is conceivable that Device 2 is the former case and Device 1 the latter.

4. Conclusion

In conclusion, we investigated the charge states of the SETs in the absence or presence of 2DEG by means of microwave spectroscopy. The 2DEG worked as a dissipative electromagnetic environment, and we found, in the device with smaller E_J, an indication of prohibition of tunneling due to dissipation. The device with larger E_J exhibited an energy gap even after dissipation is introduced.

Currently, we have not been able to quantitatively estimate α due to the lack of detail knowledge on how the 2DEG is modeled as dissipation. Experimentally, it is certainly desired that the carrier density be controlled systematically, using top metallic gates or a back gate. Then, the quantum-classical crossover may be observed as continuously increasing the carrier density.

Acknowledgments

This work is supported by Grant-in-Aid for Scientific Research from MEXT of Japan and also by Special Coordination Funds for Promoting Science and Technology.

References

1. A. O. Caldeira and A. J. Leggett, *Phys. Rev. Lett.* **46**, 26 (1981)
2. S. Chakravarty, *Phys. Rev. Lett.* **49**, 681 (1982)
3. P. W. Anderson, G. Yuval and D. R. Hamann, *Phys. Rev B* **1**, 4464 (1970)
4. Y. Nakamura, C. D. Chen and J. S. Tsai, *Phys. Rev. Lett.* **79**, 2328 (1997)

NUMERICAL STUDY OF THE COULOMB BLOCKADE IN AN OPEN QUANTUM DOT

YUJI HAMAMOTO* and TAKEO KATO
Institute for Solid State Physics, University of Tokyo, Kashiwa, Chiba 277-8581, Japan
**E-mail: hamamoto@issp.u-tokyo.ac.jp*

Resonant tunneling through an open quantum dot in the Coulomb blockade regime $T \ll e^2/C_0$ is studied numerically using the path-integral Monte Carlo method. The quantum dot is connected to two bulk leads by single-mode point contacts. We consider on-site repulsion between electrons and analyze its influence on the resonant conductance. Our numerical results exhibit power-law dependence of the peak width and the phase transition between perfect conductor and insulator, which is consistent with the analytical results.

Keywords: Coulomb blockade; quantum dot; Tomonaga-Luttinger liquid.

1. Introduction

In quantum dots, interaction between electrons has a great influence on the transport property. When tunneling of an electron into or out of the dot increases the charging energy, such a tunneling process is generally suppressed in the Coulomb blockade regime $T \ll e^2/C_0$, where C_0 is the dot-gate capacitance. An exception occurs when the average number of the dot charge is equal to a half-integer $n+1/2$, which is realized by tuning the gate voltage V_g. Degenerate charge states ne and $(n+1)e$ mediate electron tunneling with the charging energy unchanged. As a result of the enhanced charge fluctuation, observables such as conductance have periodic resonant peaks as a function of V_g.

It has become possible to experimentally realize one-dimensional quantum dot systems in quantum wires[1] and carbon nanotubes,[2] and the characteristic behaviors of the resonant peaks observed in such systems are considered to derive from Tomonaga-Luttinger liquids. The system we study in this paper is an *open* quantum dot, where the dot and the leads are strongly connected and one can treat electrons one-dimensionally. Such system have been treated analytically,[3,4] but this approach can not give a direct prediction about the non-perturbative region of interest experimentally. To overcome this difficulty, Hügle and Egger have applied to a similar system the path-integral Monte Carlo (PIMC) method based on the *real*-time formalism.[5] However, the power-law behaviors of the peak width they have observed contradicts analytical predictions,[3,4] which presumably derives from the essential difficulty due to the sign problem. In this paper, we apply the *imaginary*-time PIMC method, which is numerically easer to handle, to our system employing new update schemes recently developed in the field of Josephson junctions.[6] The framework of our PIMC method is the same as in Ref. 7.

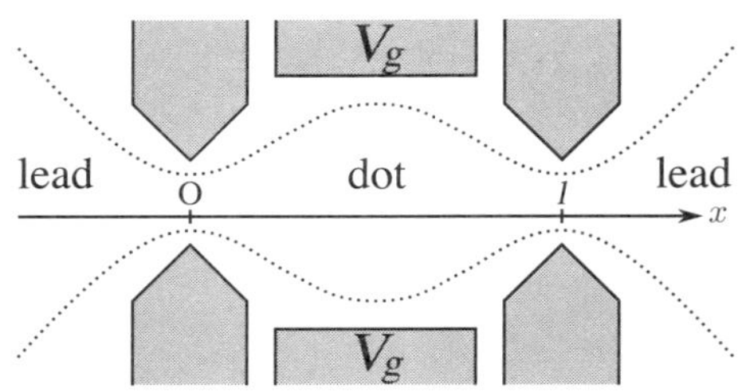

Fig. 1. Schematic figure of an open quantum dot. Two-dimensional electron gas is defined within the dotted lines by applying negative voltage to the gates (gray regions).

2. The open quantum dot system

Bosonization is a useful technique to analyze the open quantum dot system shown in Fig. 1. Two-dimensionality of the dot and the leads does not matter, since the transport behavior is dominated by the one-dimensional electrons near the two point contacts. If we assume that the dot is connected to each lead by a single mode point contact, we can model the system with an infinite Tomonaga-Luttinger liquid (TLL). In the following, we assume spinless electrons and symmetric barriers on the point contacts for simplicity, although it is easy to extend our approach to the spinful case or asymmetric barriers. If we define the bosonization of the annihilation operator of an electron as $\psi_r(x) \propto e^{irk_F x - i[r\phi(x)+\theta(x)]}$ ($r = +$ for a right-mover and $r = -$ for a left-mover), the system is effectively described by only two bosons $\phi_L \equiv \phi(0)$ and $\phi_R \equiv \phi(l)$ on the left and right constrictions, respectively. This is because both the charging effect of the dot and the reflection on the point contacts can be expressed by these bosons. Integrating out the other bosons in the dot and the leads, we obtain the effective action[4]

$$\begin{aligned} S = &\frac{1}{2\pi K\beta}\sum_{\omega_n}\frac{|\omega_n|}{1+e^{-|\omega_n|/\varepsilon}}|\tilde{\phi}_I(\omega_n)|^2 \\ &+\frac{1}{2\pi K\beta}\sum_{\omega_n}\frac{|\omega_n|}{1-e^{-|\omega_n|/\varepsilon}}|\tilde{\phi}_Q(\omega_n)|^2 \\ &+U\int d\tau[\phi_Q(\tau)-\pi N]^2 \\ &+V\int d\tau\cos\phi_I(\tau)\cos[\phi_Q(\tau)-k_F l]. \quad (1) \end{aligned}$$

Here new bosonic fields $\phi_I = \phi_R + \phi_L$ and $\phi_Q = \phi_R - \phi_L$ are defined, [a] and $\tilde{\phi}_I$ and $\tilde{\phi}_Q$ denote their Fourier components.

The transport property of this quantum dot system is determined by five constants K, $k_F l$, U, V, and N. The Luttinger parameter K controls on-site interaction in the TLL, and we here focus on repulsive interaction ($K < 1$). $\varepsilon \equiv v_F/Kl$ is the level spacing of the order of the crossover temperature between coherent and incoherent regimes. Note that in the incoherent limit $\varepsilon \ll T$, the system at the degeneracy points is mapped to the multi-channel Kondo problem.[8] The charging energy is bosonized as the third term in Eq. (1) with the coefficient $U \equiv e^2/2\pi^2 C_0$ and the dimensionless gate voltage $N \equiv -C_0 V_g/e$. The last term in Eq. (1) describes reflection on the double barrier with strength V, which is proportional to the barrier height.

3. Results and discussion

Linear conductance characterizes transport through the quantum dot. This quantity can be calculated using the Kubo formula;

$$G = \frac{e^2}{h}\lim_{i\omega_n\to 0} g(i\omega_n), \quad (2)$$

$$g(i\omega_n) = \frac{2|\omega_n|}{\pi}\int_0^\beta d\tau\langle\phi_L(\tau)\phi_R(0)\rangle e^{i\omega_n\tau}. \quad (3)$$

For a typical set of parameters, $g(i\omega_n)$ behaves quadratically in the small ω_n region. Therefore, we carry out the extrapolation $\omega_n \to 0$ by fitting the first five points with a parabolic curve.

In the rest of this paper we present our numerical results obtained for $V = 4$, $U = 1$ and $\Delta\tau = 0.25$. [b] First of all, the line shape of the resonant peak for $k_F l = \pi$ and $K = 0.5$ is shown in Fig. 2. Since the level spacing in the quantum dot is lifted by the charging effect, the peak appears when the resonant condition $eV_g + (e^2/C_0 + \pi\varepsilon/K)(n+1/2) = 0$ with an integer n is satisfied. One can see the peak sharpen monotonically with decreasing temperature (increasing β). This is in marked contrast to the non-interacting case, where the peak width is almost inde-

[a] $\dot{\phi}_I$ and ϕ_Q are propotional to the average current and the dot charge, respectively.

[b] $\Delta\tau$ is the width of the discretized time steps.

pendent of temperature if thermal fluctuation is smaller than the level broadening. [c] The analytical calculation predicts[3] that the peak width decreases in a power-law fashion $\propto T^{1-K}$. To examine temperature dependence of the peak width, we estimate the full width at half maximum (FWHM) of the peak for each temperature using interpolation, and plot it in Fig. 3 as a function of the inverse temperature. $k_F l = \pi$ is fixed, and the results for $K = 0.3$ and 0.5 are shown. The data points for each K scale as T^{1-K} in consistent with the analytical prediction,[3] which suggests that the power-law dependence of the FWHM holds in the intermediate region of tunneling amplitude.

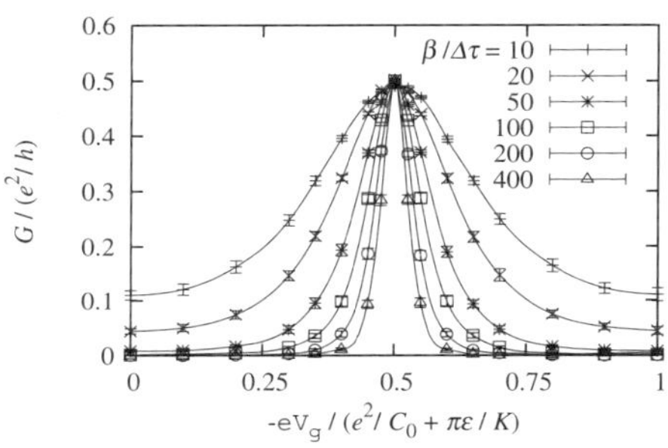

Fig. 2. Line shape of the resonant peak for $k_F l = \pi$ and $K = 0.5$. As temperature is lowered, the peak width decreases with the peak height unchanged.

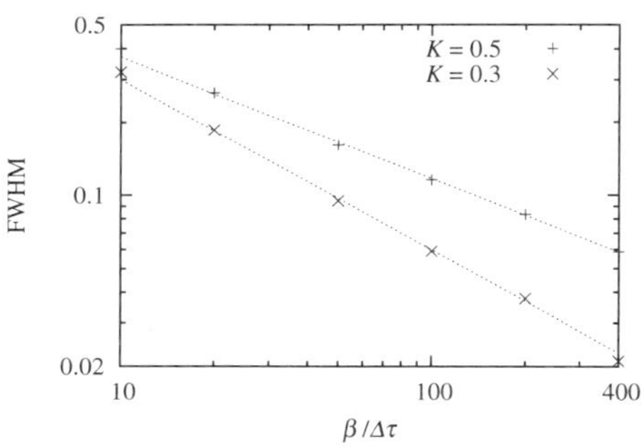

Fig. 3. FWHMs of resonant peaks for $k_F l = \pi$. For each K, a dotted line $\propto T^{1-K}$ is also shown as an eye guide.

Next, we turn to a more repulsive region, where if K crosses a critical value K^*, the resonant peak vanishes as $T \to 0$, even though the resonant condition is satisfied. This process can be observed by measuring the temperature dependence of the resonant conductance. In Fig. 4, the resonant conductance for $k_F l = 20\pi$ is plotted as a function of the inverse temperature for various values of K. For a quantum dot of finite size, the level spacing ε acts as an important energy scale, across which a coherent-incoherent crossover occurs. Therefore the conductance temporarily increases as $T \to 0$ when T is of the order of ε, reflecting the recovery of coherence. Note that the conductance through an incoherent double barrier is roughly half of the coherent value. As the temperature is lowered, the behavior of conductance undergoes a drastic change. In this regime, the temperature dependence of the conductance depends on value of K. With decreasing temperature, the conductance (I) increases monotonically for $K > 0.25$ (perfect conductor phase); (II) decreases monotonically for $K < 0.25$ (insulator phase); (III) apparently remains unchanged for $K = 0.25$ (marginal).

By repeating the similar analyses for larger values of V, we would be able to draw a phase diagram in the K-V plane. In

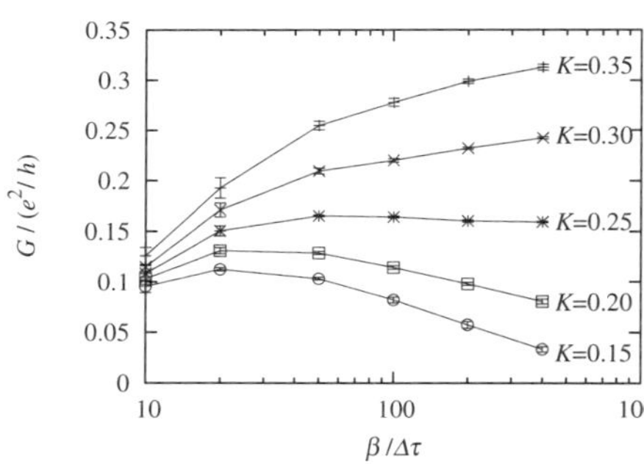

Fig. 4. Temperature dependence of the resonant conductance for $k_F l = 20\pi$. The system undergoes the phase transition between perfect conductor and insulator at $K^* \simeq 0.25$.

---
[c] We can check this temperature-independence using the PIMC simulation for $U = 0$ and $K = 1$.

that region, conductance at zero temperature drastically changes showing the Kosterlitz–Thouless transition, as have been predicted analytically.[3] We will discuss this problem elsewhere.

4. Summary

We have demonstrated the PIMC simulation of an open quantum dot connected to two bulk leads and calculated the resonant conductance. The line shape of the resonant peak is strongly influenced by the on-site repulsion, and our results are consistent with the analytical predictions in the strong tunneling limit. The phase transition between perfect conductor and insulator is also observed.

Acknowledgements

The computation in this work has been done using the facilities of the Supercomputer Center, Institute for Solid State Physics, University of Tokyo.

References

1. O. M. Auslaender, A. Yacoby, R. de Picciotto, K. W. Baldwin, L. N. Pfeiffer, and K. W. West, Phys. Rev. Lett. **84**, 1764 (2000).
2. H. W. Ch. Postma, T. Teepen, Z. Yao, M. Grifoni, and C. Dekker, Science, **293**, 76 (2001).
3. C. L. Kane and M. P. A. Fisher, Phys. Rev. B, **46**, 7268 (1992); *ibid.* **46**, 15233 (1992).
4. A. Furusaki and N. Nagaosa, Phys. Rev. B **47**, 3827 (1993).
5. S. Hügle and R. Egger, Europhys. Lett. **66**, 565 (2004).
6. N. Kimura and T. Kato, Phys. Rev. B **69**, 0102504 (2004); P. Werner and M. Troyer, Phys. Rev. Lett. **95**, 060201 (2005).
7. Y. Hamamoto and T. Kato, Phys. Rev. B **77**, 245325 (2008).
8. A. Furusaki and K. A. Matveev, Phys. Rev. Lett. **75**, 709 (1995); Phys. Rev. B **52**, 16676 (1995).
9. J. M. Kosterlitz and D. J. Thouless, J. Phys. C **6**, 1181 (1981).

SYMMETRY IN THE FULL COUNTING STATISTICS, THE FLUCTUATION THEOREM AND AN EXTENSION OF THE ONSAGER THEOREM IN NONLINEAR TRANSPORT REGIME

Y. UTSUMI[1] AND K. SAITO[2]

[1] *Institute for Solid State Physics, University of Tokyo, Kashiwa, Chiba 277-8581, Japan*
[2] *Graduate School of Science, University of Tokyo, 113-0033, Japan*

We study the full counting statistics for multi-terminal quantum dots. We show that the microscopic reversibility naturally results in a symmetry of the cumulant generating function, which generalizes the fluctuation theorem in the context of the coherent quantum transport. Using this symmetry, we derive the fluctuation-dissipation theorem and the Onsager-Casimir relation in the linear transport regime and the universal relations among nonlinear transport coefficients.

Keywords: Full counting statistics; Fluctuation theorem; Onsager reciprocal theorem

1. Introduction

Full counting statistics (FCS) has become an active topic in the mesoscopic physics [1,2,3,4]. FCS addresses the probability distribution of the transmitted electrons. It can characterize statistical properties of quantum transport in the far-from-equilibrium regime. Since the first paper by Levitov and Lesovik [1], many theories have clarified various aspects of FCS. Recently, experiments to measure third current cumulants [5] and the probability distributions of current [6,7] have been conducted.

In the last decade, the relations referred to as the *fluctuation theorem* (FT) have been a major issue of the nonequilibrium statistical physics [8]. The FT predicts that the probability distribution of entropy production $P(\Delta S)$ satisfies,

$$P(\Delta S)/P(-\Delta S) = \mathrm{e}^{\Delta S}\,. \tag{1}$$

The equality is important since it is one of a few exact relations in the far-from-equilibrium regime. Moreover, it leads the second law of thermodynamics, that the average of entropy production is always positive, $\langle \Delta S \rangle = \int d\Delta S\, P(\Delta S)\, \Delta S \geq 0$. (It is derived from a trivial transformation of Eq. (1), $\langle \mathrm{e}^{-\Delta S} \rangle = 1$, where we used the normalization condition $\int d\Delta S\, P(\Delta S) = 1$). Moreover, as we will show later, the FT can leads the linear response theory, i.e., the fluctuation-dissipation theorem and the Onsager reciprocal theorem [9]. The experimental confirmation of the FT has been performed by measuring the work done by dragging an colloidal particle in water using an optical tweezer [10].

Transport phenomena inevitably produces entropy. Therefore, FT and FCS are considered to be closely related. Recently a link between FT and FCS was studied with regard to classical mesoscopic electron transport, i.e., the sequential tunneling in the quantum dots (QDs) [11]. In this paper, we will derive a symmetry in the FCS, which clarifies the general link between FT and FCS with respect to coherent electron transports [12]. Importantly, the symmetry enables us to derive the universal relations among nonlinear transport coefficients beyond the Onsager reciprocal theorem [13,14].

2. Characteristic function

We consider interacting QDs connected to m electron reservoirs [Fig. 1] [12]. The total Hamiltonian consists of the reservoirs H_r ($r=1,\cdots,m$), the QDs H_d, interaction $H_{\rm int}$, and tunneling H_T,

$$H = \sum_{r=1}^{m} H_r + H_d + H_{\rm int} + H_T. \qquad (2)$$

Hamiltonians for the reservoir r and the QDs are described as $H_r = \sum_{k\sigma} \varepsilon_{rk} a^{\dagger}_{rk\sigma} a_{rk\sigma}$, and $H_d = \sum_{ij\sigma} t_{ij}\, d^{\dagger}_{i\sigma} d_{j\sigma}$, where $d_{i\sigma}$ annihilates an electron with spin σ at site i. In the reservoir r, $a_{rk\sigma}$ annihilates an electron with spin σ and the wave vector k. The Coulomb interaction in the QD region is described as, $H_{\rm int} = \sum_{ij\sigma\sigma'} (U_{i\sigma j\sigma'}/2)\, d^{\dagger}_{i\sigma} d^{\dagger}_{j\sigma'} d_{j\sigma'} d_{i\sigma}$. The tunnel couplings between the reservoirs and the QDs are described by, $H_T = \sum_{rki\sigma} t_{rki}\, d^{\dagger}_{i\sigma} a_{rk\sigma}$ + H.c.. When a magnetic field B is applied, the elements of the hopping and tunneling matrix acquire the phases, which are odd functions of the magnetic field $\phi(-B) = -\phi(B)$, as $t_{ij} = |t_{ij}| \exp(i\,\phi_{ij})$ and $t_{rki} = |t_{rki}| \exp(i\,\phi_{ri})$.

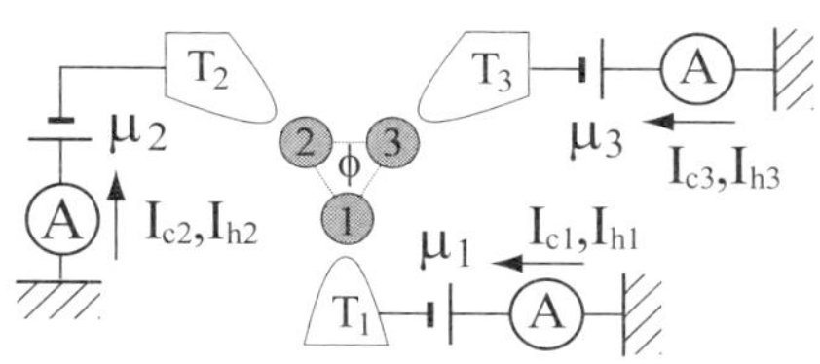

Fig. 1. Schematic picture of interacting quantum dots connected to m reservoirs ($m = 3$). The magnetic filed applied to the dot region is accounted for by the Aharonov-Bohm phase ϕ.

The proper definition of the characteristic function is nontrivial and requires a fictitious spin-1/2 galvanometer [1]. Here we adopt another scheme [15] almost equivalent to the 'two-time measurement protocol' [16]. At the initial state, the leads and the QDs are decoupled. Then at time $t = 0$, the tunneling is turned on and we allow the system to evolve during the 'measurement time' $t = \tau$. In the initial and final states, the electron number and the energy are good quantum numbers in each region, $H_s|n\rangle_s = E_s^{(n)}|n\rangle_s$ and $N_s|n\rangle_s = N_s^{(n)}|n\rangle_s$, where $s = d$ for the QDs and $s = 1, \cdots, m$ for the leads. An eigen state of the total system is the tensor product $|n\rangle = |n\rangle_d \otimes |n\rangle_1 \otimes \cdots \otimes |n\rangle_m$. At the initial time, the density matrix reads, $\rho_0 = \prod_s \rho_s$, $(s = 1, \cdots, m, d)$,

$$\rho_s = \frac{\exp[-\beta_s(H_s - \mu_s N_s)]}{Z_{s0}}, \qquad ({\rm Tr}\rho_s = 1).$$

Here $T_s = 1/\beta_s$ is the temperature ($\hbar = k_{\rm B} = {\rm e} = 1$) and μ_s is the chemical potential. The joint probability to find a state $|i\rangle$ at the initial time and a state $|f\rangle$ at the final time reads,

$$P_{\tau;|i\rangle\to|f\rangle} = \left|\langle f|{\rm e}^{-iH\tau}|i\rangle\right|^2 \langle i|\rho_0|i\rangle.$$

The joint probability distribution for q_{cs} electrons and q_{hs} energy flowing out of a lead $s = 1, \cdots, m$ and the QD region $s = d$ is,

$$P(\{q_{\alpha s}\}) = \sum_{if} P_{\tau;|i\rangle\to|f\rangle} \prod_{s'} \delta_{q_{cs'}, N_{s'}^{(f)} - N_{s'}^{(i)}} \times \delta\left(q_{hs'} - E_{s'}^{(f)} + E_{s'}^{(i)}\right),$$

where $\alpha = c, h$. After the Fourier transform, we obtain the characteristic function

$$\mathcal{Z} = \sum_{q_{cd}} \sum_{q_{c1}} \cdots \sum_{q_{cm}} \int dq_{hd}\, dq_{h1} \cdots dq_{hm} \times P(\{q_{\alpha s}\}; B) \exp\left(i \sum_{\alpha s} q_{\alpha s}\, \chi_{\alpha s}\right).$$

Since $|f\rangle$ and $|i\rangle$ are eigen vectors of H_s and N_s, we immediately obtain,

$$\mathcal{Z}(\{\chi_{\alpha s}\}; B) = \left\langle V^{-\dagger} e^{iH\tau} V^2 e^{-iH\tau} V^{\dagger} \right\rangle, \quad (3)$$

where $\langle \cdots \rangle$ is performed in terms of the initial density matrix ρ_0. The unitary operator is given as,

$$V = \prod_s V_s, \quad V_s = \exp\left(\frac{\chi_{hs} H_s + \chi_{cs} N_s}{2i}\right).$$

Note that Eq. (3) satisfies the normalization condition, $\mathcal{Z}(\{0\}; B) = 1$.

3. Microscopic reversibility

So far several symmetries are known. For instance, $\mathcal{Z}$ is a 2π-periodic function of χ_{cs} as a consequence of the discreteness of charge.

In the following, we will show a symmetry related to the microscopic reversibility. Let Θ be the time reversal operator, which reverses the linear and angular momentum operators and spin operator [17]. The time reversal operator is antilinear and antiunitary;

$$\langle n|n'\rangle = \langle \tilde{n}'|\tilde{n}\rangle,$$
$$\Theta\left(c_1\,|n\rangle + c_2\,|n'\rangle\right) = c_1^*\,|\tilde{n}\rangle + c_2^*\,|\tilde{n}'\rangle,$$

where c_1 and c_2 are arbitrary complex numbers and $|\tilde{n}\rangle = \Theta|n\rangle$. Then we can show the key equality,

$$\begin{aligned}&\mathrm{Tr}\left[V^\dagger e^{iH\tau} V^2 e^{-iH\tau} V^\dagger \rho_0\right]\\ &= \mathrm{Tr}[(\Theta\rho_0\Theta^{-1})(\Theta V\Theta^{-1})\,(\Theta e^{iH\tau}\Theta^{-1})\\ &\times (\Theta {V^\dagger}^2\Theta^{-1})(\Theta e^{-iH\tau}\Theta^{-1})(\Theta V\Theta^{-1})]\\ &= \mathrm{Tr}\left[\rho_0 V^\dagger\, e^{-iH\tau} V^2\, e^{iH\tau} V^\dagger\right]\big|_{B\to -B},\end{aligned}$$

where it should be noticed that, $\Theta H_d\Theta^{-1} = H_d|_{B\to-B}$ and $\Theta H_T\Theta^{-1} = H_T|_{B\to-B}$. Further we notice that the density matrix in the initial state ρ_s is the squire of the unitary operator V_s^2 with replacing $\chi_{hs}\to -i\beta_s$ and $\chi_{cs}\to i\beta_s\mu_s$. Then from the definition (3),

$$\mathcal{Z}(\{\chi_{\alpha s}\}; B) = \mathcal{Z}(\{-\chi_{\alpha s} + iA_{\alpha s}\}; -B), \quad (4)$$

where $A_{cs} = \beta_s\mu_s$ and $A_{hs} = -\beta_s$.

In the quantum transport problem, we are interested in the stationary state reached after long measurement time. The cummulant generating function (CGF) at the stationary state is,

$$\mathcal{F}(\{\chi_{\alpha r}\}; B) = \lim_{\tau\to\infty} \frac{\ln \mathcal{Z}(\{\chi_{\alpha r}\}; B)}{\tau}.$$

At the stationary state, no extra charge and heat accumulate inside the QDs. Then the current and charge conservations imply that CGF depends only on the differences between the counting fields of the reservoirs. Namely, the symmetry (4) is reduced to,

$$\mathcal{F}(\{\boldsymbol{\chi}_{\alpha r}\}; B) = \mathcal{F}(\{-\boldsymbol{\chi}_{\alpha r} + i\mathcal{A}_{\alpha r}\}; -B), (5)$$

where $\boldsymbol{\chi}_{\alpha r} = \chi_{\alpha r} - \chi_{\alpha m}$ and $\mathcal{A}_{\alpha r} = A_{\alpha r} - A_{\alpha m}$, which is the affinity. It is possible to prove this conjecture based on the perturbative expansion in powers of U [12].

Equation (5) is regarded as a steady-state quantum FT. Let us consider a two-terminal setup ($m=2$) and define the entropy production, $\Delta S = \mathcal{A}_{c1}q_{c1} + \mathcal{A}_{h1}q_{h1}$. Then by fixing the counting fields as $\boldsymbol{\chi}_{c1}\,\mathcal{A}_{c1}^{-1} = \boldsymbol{\chi}_{h1}\,\mathcal{A}_{h1}^{-1} = \chi$ and by performing the inverse Fourier transform, we obtain Eq. (1).

4. Beyond the Onsager reciprocal theorem

We will show that the symmetry (5) predicts general relations among nonlinear transport coefficients. In the following we will concentrate on the isothermal case $\beta_r = \beta$ and we will consider the situation that only the chemical potentials μ_1 and μ_2 are varied and the current is measured at terminals 1 and/or 2 [Fig. 1]. The general form of cumulant of the currents is

$$\langle\langle I_{c1}^{k_1} I_{c2}^{k_2}\rangle\rangle = \frac{\partial^{k_1+k_2}\mathcal{F}(\{0\}; B)}{\partial\,(i\boldsymbol{\chi}_{c1})^{k_1}\,\partial\,(i\boldsymbol{\chi}_{c2})^{k_2}}.$$

The transport coefficients are introduced as,

$$\langle\langle I_{c1}^{k_1} I_{c2}^{k_2}\rangle\rangle = \sum_{\ell_1,\ell_2=0}^{\infty} \frac{L_{\ell_1\ell_2}^{k_1k_2}(B)}{\ell_1!\ell_2!}\mathcal{A}_{c1}^{\ell_1}\mathcal{A}_{c2}^{\ell_2}.$$

We symmetrize the coefficients as

$$L_{\ell_1\ell_2,\pm}^{k_1k_2} = L_{\ell_1\ell_2}^{k_1k_2}(B) \pm L_{\ell_1\ell_2}^{k_1k_2}(-B).$$

Then by taking the derivatives with respect to the affinities and counting fields for both sides of Eq. (5), we obtain the general relations among the coefficients as

$$\begin{aligned}L_{\ell_1\ell_2,\pm}^{k_1k_2} &= \pm\sum_{n_1=0}^{\ell_1}\sum_{n_2=0}^{\ell_2}\binom{\ell_1}{n_1}\binom{\ell_2}{n_2}\\ &\times (-1)^{n_1+n_2+k_1+k_2}\,L_{\ell_1-n_1\,\ell_2-n_2,\pm}^{k_1+n_1\,k_2+n_2},\end{aligned}$$

where $L_{\ell_1\ell_2,\pm}^{0\,0} = 0$ is also satisfied from the normalization condition. They connect the relations among the nonlinear transport coefficients satisfying $k_1+k_2+\ell_1+\ell_2 = \mathcal{N}$. Therefore, it is possible to classify the equalities using an inter $\mathcal{N}$. For $\mathcal{N}=1$, we obtain trivial relations $L_{0\,0}^{1\,0}(B) = L_{0\,0}^{0\,1}(B) = L_{1\,0}^{0\,0}(B) = L_{0\,1}^{0\,0}(B) = 0$. For $\mathcal{N} = 2$ we reproduce the linear response theory, the fluctuation-dissipation theorem,

$$L^{1\,0}_{1\,0}(B) = L^{2\,0}_{0\,0}(B)/2,$$
$$L^{1\,1}_{0\,0}(B) = L^{0\,1}_{1\,0,+} = L^{1\,0}_{0\,1,+},$$

and the Onsager-Casimir relations [14],

$$L^{1\,0}_{0\,1}(B) = L^{0\,1}_{1\,0}(-B), \quad L^{1\,0}_{1\,0}(B) = L^{1\,0}_{1\,0}(-B),$$
$$L^{0\,1}_{0\,1}(B) = L^{0\,1}_{0\,1}(-B).$$

Relations beyond linear response regime can be also obtained. For $\mathcal{N} = 3$, we obtain

$$L^{3\,0}_{0\,0,+} = 0, \quad L^{1\,0}_{2\,0,+} = L^{2\,0}_{1\,0,+},$$
$$L^{1\,0}_{2\,0,-} = L^{2\,0}_{1\,0,-}/3 = L^{3\,0}_{0\,0,-}/6, \quad L^{2\,1}_{0\,0,+} = 0,$$
$$L^{0\,1}_{2\,0,+} = L^{1\,1}_{1\,0,+} = 2\,L^{1\,0}_{1\,1,+} - L^{2\,0}_{0\,1,+},$$
$$L^{2\,0}_{0\,1,-} = L^{1\,1}_{1\,0,-} = L^{2\,1}_{0\,0,-}/2 = L^{0\,1}_{2\,0,-} + 2\,L^{1\,0}_{1\,1,-},$$

This extension of the Onsager reciprocal theorem relays solely on the microscopic reversibility and are insensitive to the setup details. We expect that the relations can be measured in the presently available experimental setups [18].

5. Conclusion

We derived the symmetry (5) in full counting statistics from the microscopic reversibility. It can be regarded as the quantum version of the fluctuation theorem under a magnetic filed. We derive the linear response theory and relations among the nonlinear transport coefficients beyond the Onsager reciprocal theorem.

References

1. L. S. Levitov and G. B. Lesovik, JETP Lett. **58**, 230 (1993); L. S. Levitov, H.-W. Lee, and G. B. Lesovik, J. Math. Phys., **37**, 4845 (1996).
2. *Quantum Noise in Mesoscopic Physics*, Vol. 97 of *NATO Science Series II: Mathematics, Physics and Chemistry* edited by Yu. V. Nazarov (Kluwer Academic Publishers, Dordrecht/Boston/London, 2003).
3. D. A. Bagrets, Y. Utsumi, D. S. Golubev, G. Schön, Fortschritte der Physik **54**, 917-938 (2006).
4. K. Saito and A. Dhar, Phys. Rev. Lett **90**, 180601 (2007).
5. B. Reulet *et al.*, Phys. Rev. Lett. **91**, 196601 (2003); Yu. Bomze *et al.*, Phys. Rev. Lett. **95**, 176601 (2005).
6. S. Gustavsson *et al.*, Phys. Rev. Lett. **96**, 076605 (2006).
7. T. Fujisawa *et al.*, Science **312**, 1634 (2006).
8. D. J. Evans, E G. D. Cohen, and G. P. Morriss, Phys. Rev. Lett. **71**, 2401 (1993), G. Gallavotti and E. G. D. Cohen, Phys. Rev. Lett. **74**, 2694 (1995), J. L. Lebowitz and H. Spohn, J. Stat. Phys. **95**, 333 (1999).
9. G. Gallavotti, Phys. Rev. Lett. **77**, 4334 (1996).
10. G. M. Wang, E. M. Sevick, E. Mittag, D. J. Searles, D. Evans, Phys. Rev. Lett. **89**, 050601 (2002).
11. D. Andrieux and P. Gaspard, J. Stat. Mech. P01011 (2006); J. Stat. Phys. 107 (2007); M. Esposito, U. Harbola, and S. Mukamel, Phys. Rev. B **75**, 155316 (2007).
12. K. Saito and Y. Utsumi, arXiv:0709.4128.
13. L. Onsager, Phys Rev. 37, 405, (1931); H. B. G. Casimir, Rev. Mod. Phys. 17, 343, (1945).
14. M. Büttiker, Phys. Rev. Lett. **57**, 1761 (1986).
15. Yu. Makhlin, G. Schön, and A. Shnirman, Phys. Rev. Lett. **85**, 4578 (2000); M.-S. Choi, F. Plastina, and R. Fazio, Phys. Rev. Lett. **87**, 116601 (2001); M.-S. Choi, F. Plastina, and R. Fazio, Phys. Rev. B **67** 045105 (2003).
16. J. Kurchan, cond-mat/0007360; H. Tasaki, cond-mat/0009244; C. Jarzynski and D. K. Wójcik, Phys. Rev. Lett. **92**, 230602 (2004).
17. J. J. Sakurai, *Modern Quantum Mechanics* (Benjamin/Cummings. Menlo Park, California, 1985).
18. A. Löfgen *et al.* Phys. Rev. Lett. **92** 046803 (2004); C. A. Marlow *et al.*, Phys. Rev. Lett. **96** 116801 (2006); R. Leturcq *et al.*, Phys. Rev. Lett. **96**, 126801 (2006).

SINGLE-ARTIFICIAL-ATOM LASING AND ITS SUPPRESSION BY STRONG PUMPING

J.R. JOHANSSON[1], S. ASHHAB[1,2], A.M. ZAGOSKIN[1,3] AND FRANCO NORI[1,2]

[1] *Advanced Science Institute, The Institute of Physical and Chemical Research (RIKEN), Wako-shi, Saitama, Japan*

[2] *Center for Theoretical Physics, Physics Department, The University of Michigan, Ann Arbor, Michigan, USA*

[3] *Department of Physics, Loughborough University, Loughborough, United Kingdom*

We analyze single-atom lasing in a system composed of an artificial atom coupled to a cavity mode. The population inversion of the artificial atom is modeled as a reversed relaxation process that takes the atom from its ground state to its excited state. In this setup there is no lasing threshold below which lasing does not occur, but lasing can be suppressed if the 'relaxation' rate, i.e. the atom pumping rate, is larger than a certain threshold value. Using transition-rate equations, we derive analytic expressions for the lasing suppression condition and the photon-distribution of the cavity, both in the lasing and suppressed-lasing regimes.

Keywords: artificial atom, cavity, lasing

1. Introduction

Recently there has been an increased interest in superconducting electronic circuits and the quantum phenomena that these circuits can exhibit. With such circuits it is possible to design and engineer quantum systems, which makes them particularly interesting for applications in quantum information processing and as artificial atoms in experiments analogous to those in quantum optics.[1] For instance, systems with an artificial atom coupled to a cavity in the form of a harmonic-oscillator circuit element have received considerable attention lately.[2,3,4,5] These circuit-QED systems makes it possible to study various quantum-optics phenomena in a highly controllable and easily tunable setting, as well as explore parameter regimes that are inaccessible using natural atoms. An example of such a phenomenon is lasing,[6] which, in the context of superconducting circuit-QED systems, recently has been the topic of both theoretical work[7,8] and experimental demonstrations.[9,10] Here[11] we analyze a model of such a circuit-QED lasing setup, inspired by recent experimental results.[9] A similar model has also been studied in previous work.[12,13]

Using simple transition-rate relations, we derive the counterintuitive result that if the pumping of the atom from the lower to the upper level becomes stronger than a certain critical value, lasing action is lost. These results agree with the results of previous work based on a quantum-optical master-equation approach.[12] We also analyze the state of the cavity both in the lasing state and in the suppressed-lasing regime, where the cavity is in a thermal state.

2. Model

We consider a system composed of an atom coupled to a single-mode cavity. The atom is modeled as a quantum two-level system, and the cavity-mode is modeled as a harmonic oscillator. The Hamiltonian for this system can be expressed as

$$\hat{H} = \frac{\hbar\omega_a}{2}\left(\sin\theta\hat{\sigma}_x + \cos\theta\hat{\sigma}_z\right) + \hbar\omega_0\hat{a}^\dagger\hat{a} + \hbar g_0\hat{\sigma}_x\left(\hat{a} + \hat{a}^\dagger\right), \quad (1)$$

where ω_a and ω_0 are the frequencies of the atom and cavity, respectively, θ is the atom's deviation from the its degeneracy point, and g_0 is the atom-cavity interaction strength. The $\hat{\sigma}_x$ and $\hat{\sigma}_y$ operators are the Pauli operators acting on the atomic state, and $\hat{a}$ and $\hat{a}^\dagger$ are the creation and annihilation operators acting on the cavity state. In the following we shall assume that the atom and the cavity are in resonance, i.e. $\omega_a = \omega_0$, and that this is the largest frequency in the problem.

The population inversion of the atom is accounted for by a reversed relaxation with rate Γ, i.e. a pumping process that excites the atom from its ground state to its excited state. The cavity is assumed to have a decay rate κ, which has to be smaller than Γ for lasing to occur.

This situation can be described by the Lindblad master equation[14]

$$\frac{d\rho}{dt} = -\frac{i}{\hbar}\left[\hat{H},\rho\right] + \frac{\kappa}{2}\left(2\hat{a}\rho\hat{a}^\dagger - \hat{a}^\dagger\hat{a}\rho - \rho\hat{a}^\dagger\hat{a}\right) + \frac{\Gamma}{2}\left(2\hat{\sigma}_+\rho\hat{\sigma}_- - \hat{\sigma}_-\hat{\sigma}_+\rho - \rho\hat{\sigma}_-\hat{\sigma}_+\right), \quad (2)$$

where $\hat{\sigma}_\pm$ are the raising and lowering operators that act on the atomic state. Below we will analyze this model by numerically solving Eq. (2), and by studying the rate-equations for the number of excitations in the atom-cavity system. For the latter we shall use the notation $|n_a, n\rangle$ to describe the states of the system, where $n_a = 0$ for the atomic ground state and $n_a = 1$ for the atomic excited state, and n represents the number of photons in the cavity.

3. Photon emission and loss rates

The model described above contains a mechanism for photon emission into the cavity and a mechanism for photon loss from the cavity. The loss rate of photons from the cavity ($|n_a, n\rangle \to |n_a, n-1\rangle$) is given by

$$\Gamma_{\text{loss}} = n\kappa, \quad (3)$$

where n is the number of photons in the cavity. Assuming that the atom initially is in its excited state, and that the cavity contains a small number of photons, $n-1$, the atom-cavity coupling (with matrix element $g\sqrt{n}$, where $g = g_0\cos\theta$) induces dynamics between the states $|1, n-1\rangle$ and $|0, n\rangle$. Since $\Gamma \gg g$, this process can be described by an incoherent transfer of excitations from the atom to the cavity, resulting in the photon emission rate ($|1, n-1\rangle \to |0, n\rangle \to |1, n\rangle$)

$$\Gamma_{\text{emission}} = \frac{4ng^2}{\Gamma}. \quad (4)$$

The photon emission rate therefore increases linearly with n for small values of n. As n increases, eventually $g\sqrt{n}$ becomes comparable to or larger than Γ, and the $|1, n-1\rangle \leftrightarrow |0, n\rangle$ transitions must be treated as coherent oscillations. In the limit of large n the system effectively spends half the time in the two states ($|1, n-1\rangle$ and $|0, n\rangle$), resulting in an asymptotic photon emission rate of $\Gamma/2$.

4. Lasing condition and possible steady states

Using the photon emission and loss rates as functions of photon number n, one can obtain the probability distribution of photon number states in the cavity. In the following analysis we treat the atom's relaxation rate Γ as a tunable parameter, while g and κ are kept fixed. By combining Eqs. (3) and (4), we find that if

$$\frac{4g^2}{\Gamma} > \kappa, \quad (5)$$

the photon emission rate is larger than the photon loss rate, assuming a small photon number in the cavity. Starting with a small photon number, the number increases exponentially in time until the photon emission rate saturates and reaches a balance with the photon loss rate. As Γ is increased, the initial photon emission rate decreases, but the final number of photons in the cavity is increased [cf. Fig. 1(a) and 1(b)], as expected for an increased atom pumping rate.

A change in behavior is observed when Γ is increased enough that Eq. (5) is not satisfied. The loss rate will then always be higher than the emission rate, and lasing will not occur [cf. Fig 1(d)]. Equation (5) can therefore be considered a threshold condition for lasing in this model.

We now consider the situation where the lasing condition (Eq. 5) is satisfied, and we analyze the probability distribution of the photon number in the cavity. Deep in the lasing regime, we can assume that the emission rate is well approximated by $\Gamma/2$. The loss rate is still given by Eq. (3). The peak in the photon-number probability distribution therefore occurs at

$$n_{\text{peak}} = \frac{\Gamma}{2\kappa}. \tag{6}$$

The width of the probability distribution can be calculated by using the photon emission and loss rates in detailed-balance relations.[11] The resulting photon number distribution is

$$P_n = P_0 \exp\left\{ -\frac{(n - n_{\text{peak}})^2}{2n_{\text{peak}}} \right\}, \tag{7}$$

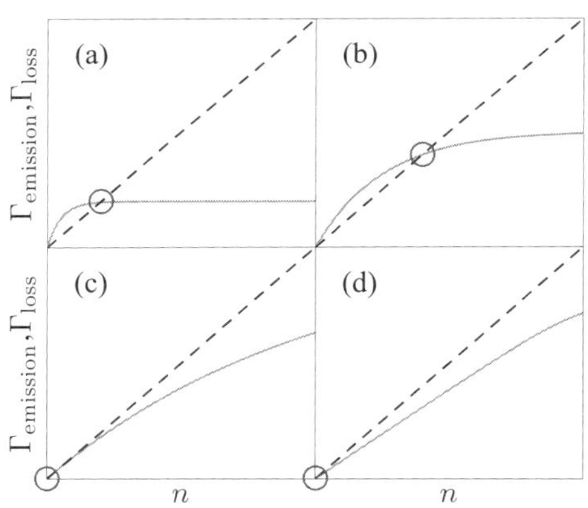

Fig. 1. Schematic plot of the photon emission rate Γ_{emission} (solid line) and the photon loss rate Γ_{loss} (dashed line) as functions of the photon number in the cavity n. The intersection point (circle) determines the value of n in the photon-number probability distribution with the highest probability. Going from (a) to (d), Γ is increased, while κ is kept fixed.

and the width is thus of the order of $\sqrt{n_{\text{peak}}}$, as would be expected for the lasing state.

We now turn to the situation where lasing is suppressed, i.e. when $4g^2 < \Gamma\kappa$. In the linear regime (i.e. when n is small), we can write simple detailed balance equations for the probabilities P_n;

$$\frac{P_{n+1}}{P_n} = \frac{\Gamma_{\text{emission}}(n)}{\Gamma_{\text{loss}}(n+1)} = \frac{4g^2}{\Gamma\kappa}. \tag{8}$$

This equation can be identified as the detailed-balance equation for a cavity in thermal equilibrium at effective temperature

$$T_{\text{eff}} = \frac{\hbar\omega_0}{k_B}\left[\log\left\{\frac{\Gamma\kappa}{4g^2}\right\}\right]^{-1}, \tag{9}$$

and the Bose-distribution formula gives the corresponding average photon cavity number

$$\bar{n} = \left(\frac{\Gamma\kappa}{4g^2} - 1\right)^{-1}. \tag{10}$$

5. Numerical results

By solving Eq. (2) numerically for different values of the atom-pumping rate Γ, and for fixed g_0 and κ, we get a complete picture of the atom-cavity system's response to different pumping rates. The results agree with the analytical calculations in both the lasing and suppressed-lasing regimes, and show the behavior of the system in the intermediate regime, see Fig. 2.

6. Conclusions

We have analyzed the lasing behavior of a single artificial atom in a cavity. Although increased pumping strength initially results in a larger photon population in the cavity, increasing the pumping rate beyond a certain point starts to suppress the number of photons in the lasing state. When the pumping rate reaches a critical threshold value, lasing action is completely lost and a thermal state of the cavity is formed. We have analyzed the properties of both the lasing and suppressed-lasing (thermal) states. Our

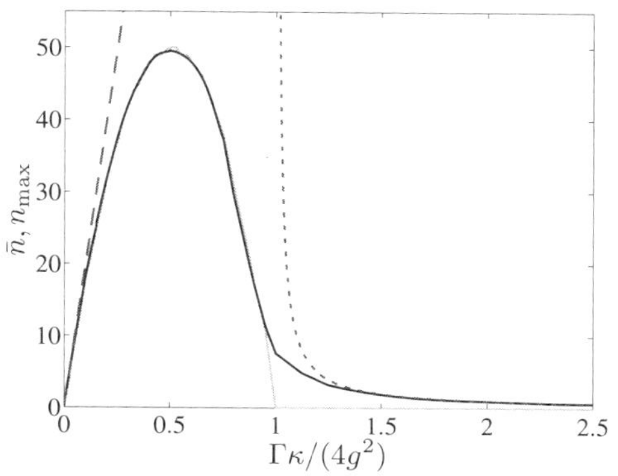

Fig. 2. The average photon number $\bar{n}$ (black solid line) and maximum-probability photon number $n_{\max}$ (gray solid line) in the cavity as functions of the parameter $\Gamma\kappa/(4g^2)$. Note that $n_{\max}$ corresponds to circles in Fig. 1. The values $g/\omega_0 = 8 \times 10^{-3}$ and $\kappa/\omega_0 = 5 \times 10^{-3}/(2\pi)$ were used in the calculations. The dashed line shows the predictions of Eq. (6) in the lasing regime, and the dotted line shows the predictions of Eq. (10) in the thermal regime.

analysis and results are very relevant to recent experiments,[9] suggesting that experimental tests of this phenomenon should be possible in the near future.

Acknowledgments

This work was supported in part by the National Security Agency (NSA), the Laboratory for Physical Sciences (LPS), the Army Research Office (ARO), the National Science Foundation (NSF), the JSPS-RFBR and Core-to-Core (CTC) program supported by the Japan Society for Promotion of Science (JSPS).

References

1. J.Q. You and F. Nori, Phys. Today **58** (11), 42 (2005); G. Wendin and V. Shumeiko, in *Handbook of Theoretical and Computational Nanotechnology*, ed. M. Rieth and W. Schommers (ASP, Los Angeles, 2006).
2. I. Chiorescu, P. Bertet, K. Semba, Y. Nakamura, C. J. P. M. Harmans, and J. E. Mooij, Nature **431**, 159 (2004).
3. A. Wallraff, D. I. Schuster, A. Blais, L. Frunzio, R. S. Huang, J. Majer, S. Kumar, S. M. Girvin, and R. J. Schoelkopf, Nature **431**, 162 (2004);
4. J. Johansson, S. Saito, T. Meno, H. Nakano, M. Ueda, S. Semba, and H. Takayanagi, Phys. Rev. Lett. **96**, 127006 (2006).
5. E. Il'ichev, N. Oukhanski, A. Izmalkov, Th. Wagner, M. Grajcar, H.-G. Meyer, A. Yu. Smirnov, A. Maassen van den Brink, M. H. Amin, and A. M. Zagoskin, Phys. Rev. Lett. **91**, 097906 (2003).
6. P. W. Milonni and J. H. Eberly, *Lasers* (Wiley, New York, 1988); M. O. Scully and M. S. Zubairy, *Quantum Optics* (Cambridge University Press, 1997); L. Davidovich, Rev. Mod. Phys. **68**, 127 (1996); W. E. Lamb, W. P. Schleich, M. O. Scully, and C. H. Townes, Rev. Mod. Phys. **71**, S242 (1999).
7. J. Q. You, Y.-X. Liu, C. P. Sun, and F. Nori, Phys. Rev. B **75**, 104516 (2007).
8. J. Hauss, A. Fedorov, C. Hutter, A. Shnirman, and G. Schön, Phys. Rev. Lett. **100**, 037003 (2008).
9. O. Astafiev, K. Inomata, A. O. Niskanen, T. Yamamoto, Yu. A. Pashkin, Y. Nakamura, and J. S. Tsai, Nature **449**, 588 (2007).
10. M. Grajcar, S. H. W. van der Ploeg, A. Izmalkov, E. Il'ichev, H.-G. Meyer, A. Fedorov, A. Shnirman, G. Schoen, Nature Physics **4**, 612 (2008); F. Nori, Nature Physics **4**, 589 (2008).
11. S. Ashhab, J.R. Johansson, A.M. Zagoskin and F. Nori, arXiv:0803.1209v2.
12. Y. Mu and C. M. Savage, Phys. Rev. A **46**, 5952 (1992).
13. H.-J. Briegel and B.-G. Englert, Phys. Rev. A **47**, 3311 (1993); C. Ginzel, H.-J. Briegel, U. Martini, B.-G. Englert, and Z. Schenzle, Phys. Rev. A **48**, 732 (1993).
14. C. W. Gardiner and P. Zoller, *Quantum Noise* (Springer, Berlin, 2000).

PHOTONIC ENTANGLEMENT IN QUANTUM COMMUNICATION AND QUANTUM COMPUTATION

A. ZEILINGER*

Institute for Quantum Optics and Quantum Information IQOQI, , Austrian Academy of Sciences, and Faculty of Physics, University of Vienna, A-1090 Vienna, Austria

E-mail: anton.zeilinger@univie.ac.at

Since ISQM 05, entangled photons have been used in our laboratory for a number of novel experiments, both in the foundations of quantum mechanics and in quantum information. Because of the high quality of entanglement of photons, precision tests of a non-local realistic theory proposed by Leggett were possible. These experiments indicate that the concept to be abandoned in quantum mechanics is most likely that of objectivism or realism. In parallel experiments, many schemes of one-way quantum computation were performed. There, because of active feed-forward, one can realize an efficient quantum computer and eliminate the effect of the randomness of individual events. The most significant properties of all-optical quantum computation are the very short cycle time, which can easily be below 100 ns and the high fidelity of the operation. Finally, experiments on quantum entanglement over distances of up to 144 km not only confirmed that quantum communication with satellites is possible, they also allowed novel tests of Bell's inequality.

Keywords: Entanglement; quantum computation; quantum communication.

1. Introduction

The current ISQM Tokyo 08 may be an opportunity to look back 25 years to the *First International Symposium on the Foundations of Quantum Mechanics in the Light of New Technology* which took place in Tokyo in 1983. Back at that time, I was still interested in neutron interferometry, and ISQM 1983 gave us the opportunity to present some results in neutron interferometry where we searched for new physics beyond standard quantum mechanics [1].

While many of these ideas of neutron interferometry are now used routinely in experiments with cold atoms and Bose-Einstein-condensates, it is still most challenging to come up with good experiments testing quantum mechanics with matter waves. Indeed, it is still very difficult to design tests which do not run immediately into inconsistencies.

Back in 1983, this was witnessed by a comment by Professor Y. Nambu after my presentation, where he questions whether it is possible at all to form a consistent theory which would allow for one of the tests we did, namely, the search for a new gauge theory coupling to iso-spin. Luckily, Prof. C. N. Yang came to the rescue, pointing out that our experiment was setting a limit on any gauge theory which contains the iso-spin SU(2) independent of the structure of the entire group.

The reason I point this out is to reflect on the one hand on the enormous progress which has been made in the field, with numerous new experiments in matter wave interferometry of many different systems far beyond the neutron and on the other hand on the fact that, to date, no test theory exists which would allow us to perform broad classes of quantum mechanics without immediately running into internal contradictions or into contradictions with existing experiments.

Obviously, it is a great advantage that experimental progress since then has put many more tools into our hands for performing fundamental experiments on quantum mechanics. One of them is entanglement, which has achieved such a maturity that it can now be applied to various quantum communication and quantum computation protocols.

2. Entanglement - Fundamentals

When in 1935 Einstein, Podolsky and Rosen discussed entanglement for the first time [2], they already presented the apparent conflict with relativity theory because of the instantaneous collapse of the wave function for entangled systems irrespective of distance.

They could actually have referred to Isaac Newton [3] "that Entanglement should be innate, inherent an essential to Matter, so that one Body may act upon another at a Distance thro' a Vacuum, without the Mediation of any thing else … is to me so great an Absurdity that I believe no Man who has in philosophical Matters, a competent Faculty of Thinking can ever fall into it". All I did here was to replace Newton's originally used word "Gravity" by "Entanglement".

While Schrödinger already in 1935 pointed out [4] about entanglement "I would not call that one, but rather the characteristic trait of quantum mechanics, the one that enforces its entire departure from classical lines of thought." Despite this clear indication that entanglement is something very significant, the field lay essentially dormant until John Bell [5] discovered that there is an inherent contradiction between the predictions of quantum mechanics and those of any local realistic theory. This prediction is most striking for the case of entanglement between three or more qubits [6]. There, the situation may arise for the three-photon entangled state

$$|GHZ\rangle = \frac{1}{\sqrt{2}}\left(|H\rangle_1|H\rangle_2|H\rangle_3 + |V\rangle_1|V\rangle_2|V\rangle_3\right). \quad (1)$$

This state describes three entangled photons, each of them either horizontally (H) or vertically (V) polarized.

There, based on two measurements in a complementary basis, a local realist predicts for the third photon to be, say, vertically polarized while quantum mechanics predicts it to be horizontally polarized. This prediction was confirmed within experimental uncertainty

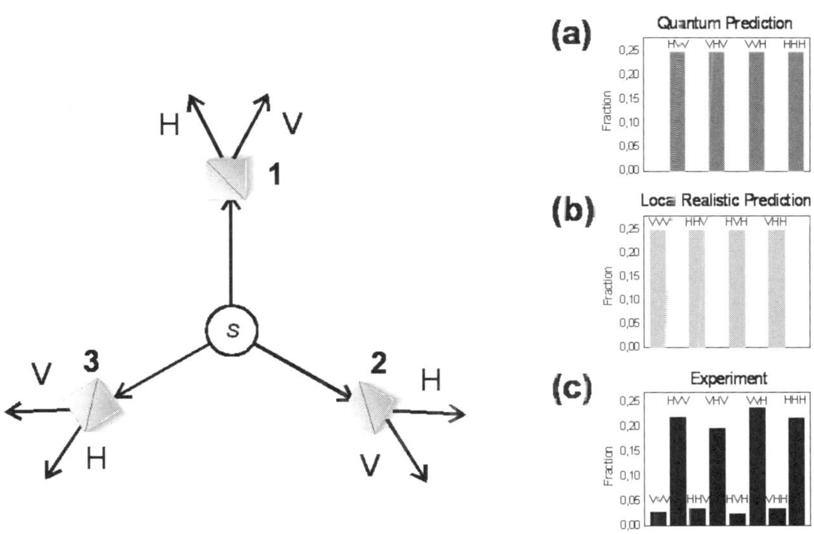

Fig. 1. Test of local realism [7] using three-particle correlations, so-called GHZ states. Principle of the experimental arrangement (left): a source emits three photons entangled in the state given by equation 1. For observation in a conjugate basis rotated by 45°, quantum mechanics predicts that only three kinds of correlations can occur (a). Local realism predicts exactly those four correlations which quantum mechanics does not predict (b). The experiment confirms the quantum mechanical prediction within experimental accuracy (c).

[7] (Fig. 1). It might be interesting to remark that, while such states and also other multi-particle entangled states, for example the so-called W-states [8] were invented just for pointing out fundamental features of quantum mechanics, they have now become central in many quantum computation protocols.

3. Entanglement and Reality

In their 1935 paper, EPR take issue with entanglement and, as they say at the end of the paper, "we believe, however, that such a theory is possible", by which they mean a theory which offers a more complete description of physical reality than quantum mechanics, in particular, it describes a realistic description of entanglement.

The starting point of their argument is the introduction of "elements of physical reality" which are defined in their famous statement "if, without in any way disturbing a system, we can predict with certainty (i.e. with probability equal to unity) the value of a physical quantity, then there exists an element of physical reality corresponding to this physical quantity." It is therefore amusing to note that states exist where entanglement dismissed by Einstein as "spooky" is even an element of reality itself in the EPR sense. An example for such a state is

$$\left|\Phi^-\right\rangle_{a,b} = \frac{1}{\sqrt{2}}\left(\left|H\right\rangle_a\left|\phi^-\right\rangle_{b_1,b_2} - \left|V\right\rangle_a\left|\psi^+\right\rangle_{b_1,b_2}\right) \quad (2)$$

In this three-photon state, photon A is either horizontally or vertically polarized while photons B1 and B2 are in one of the Bell states

$$\begin{aligned}\left|\Phi^\pm\right\rangle_{12} &= \frac{1}{\sqrt{2}}\left(\left|0\right\rangle_1\left|0\right\rangle_2 \pm \left|1\right\rangle_1\left|1\right\rangle_2\right)\\ \left|\Psi^\pm\right\rangle &= \frac{1}{\sqrt{2}}\left(\left|0\right\rangle_1\left|1\right\rangle_2 \pm \left|1\right\rangle_1\left|0\right\rangle_2\right)\end{aligned} \quad (3)$$

where the $\left|0\right\rangle$ and $\left|1\right\rangle$ stand for two orthogonal states. In our experiment [9] they represent again horizontal and vertical polarization respectively of a photon. The state of equation 2 introduces entanglement as an element of reality because, depending on whether the polarization of photon a is horizontal or vertical, one may predict with certainty the entangled state for the other two photons. The experiment [9] not only confirms state 2 but also allows to observe the violation of a CHSH inequality for the case where for photon a, the measurement observable was linear polarization while for photons b1 and b2, it was a superposition of two different Bell states.

The recent experiments would not have been possible without significant progress in the development of sources for entanglement. An important step was the development of a source for collinear entangled photons using Sagnac geometry [10]. Recent development [11] allows to produce more than 1 million pairs per second for laser pump powers of a few tens of milliwatts.

4. Testing Nonlocal Realism

The fact that, as just mentiones, recent sources produce entanglement is of both very quality and intensity was of crucial importance for a test of a nonlocal hidden variable theory proposed by Leggett [12]. These crypto non-local hidden variable theories, as Leggett called them, go significantly beyond the local hidden variable theories ruled out by Bell's inequality. The individual measurement result on one particle in an entangled pair are allowed to not only depend on the local measurement parameter, but also on which measurement is performed at its distant entangled twin brother.

Leggett was able to show that it is possible to construct such theories in a way that they obey the impossibility condition to send a signal faster than the speed of light. Therefore, such theories are not in conflict with special relativity. Furthermore, they can be constructed such that they are in agreement with the results of all known tests of Bell's inequality and certainly with all experiments on individual particles. As in local realistic theories, the total

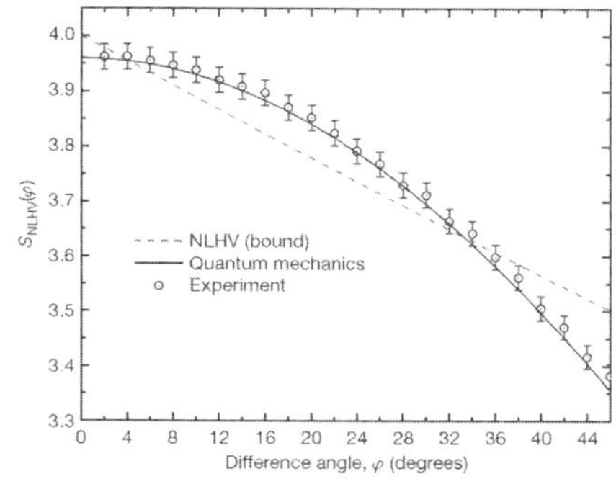

Fig. 2. Test of nonlocal quantum theory of the class proposed by Leggett [12]. The graph (from [13]) shows a specific parameter, a sum of correlations generalized from the original CHSH inequalities. The difference angle is that angle by which one has to go with the measurement of elliptical polarization of one photon out of the plane defined by the linear polarization of the other photon in the Poincaré sphere. The limit for a nonlocal hidden variable scheme is given by the dashed line. The experiment clearly violates this limit and thus rules out the theories of the class tested. It should be mentioned that in the same experiment, a violation of a CHSH inequality was also observed, thus ruling out not only the new nonlocal theories, but also local realism in the usual way.

ensemble of particles is divided into sub-ensembles characterized by distributions of hidden variable.

In a specific, physically rather reasonable type of theory, physical states are statistical mixtures of sub-ensembles with definite polarizations such that the expectation values for each sub-ensemble obey Malus' Law. In the experiment [13], the polarization-correlations between two photons were observed.

The new feature of these experiments compared to earlier Bell tests was that in order to test the non-local realistic theories as proposed by Leggett, one has to observe correlations between photons which are not within one plane of the Bloch (or Poincaré) sphere, that is, one observes for one photon, say, linear polarizations and for the other photon both linear and elliptical polarization.

The experimental results are in clear agreement with quantum mechanics (Fig. 2). This confirms that certain specific nonlocal variants of a realistic theory proposed by Leggett are not viable. While the theory ruled out was an intuitively rather reasonable alternative to quantum mechanics, the question is left open which more, possibly less intuitive nonlocal variants might also be ruled out in future experiments.

5. Quantum Computation with Entangled Photons

It is well known that due to the lack of effective photon-photon interactions, the implementation of quantum computation with photons only provides a significant challenge. As has been shown in the seminal paper by Knill, Laflamme and Milburn [14], one nevertheless can implement quantum computation using linear optical elements only. The essential idea here is to exploit the fact that the measurement process in quantum mechanics is outside the linear and unitary quantum evolution. Yet, the intrinsic randomness of quantum measurement implies a significant limit to scalability.

Raussendorf and Briegel [15] suggested an intrinsically non-unitary scheme for quantum computation where this can be overcome. In that approach, one starts with so-called cluster states, and the program is implemented by a sequence of measurements on that state. Most significantly, randomness is compensated by active feed-forward in the sense that, depending on earlier measurement results on some qubits of the state, unitary operations are set changing effectively the measurement bases of future measurements on the remaining, still higly enttangled qubits of that state. In the experiment [16], we used coherent superposition of two photon pairs created in double passage through a down-conversion crystal (Fig. 3). We were able to demonstrate [16] some of the most fundamental concepts of one-way quantum computing including single and two-qubit operations. The intrinsic

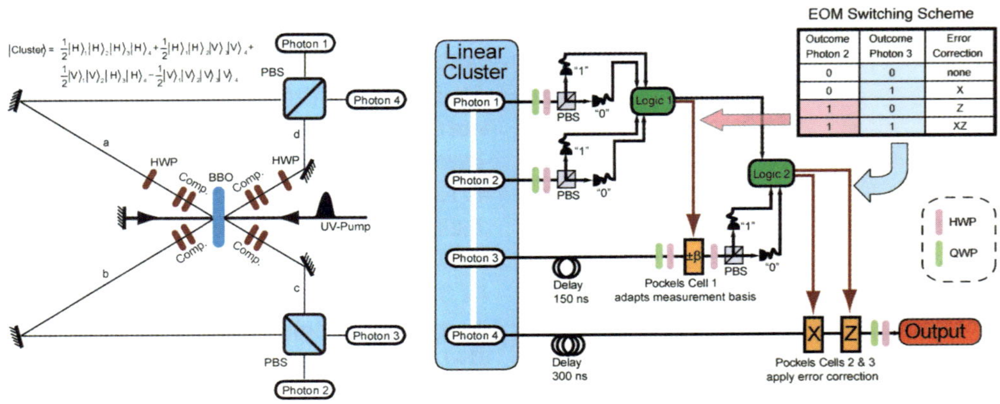

Fig. 3. Experimental one-way quantum computation. Left: A UV pulse passing twice through a nonlinear BBO crystal creates probabilistically two pairs of entangled photons. Coherent superposition of two photons each at the two polarizing beam splitters PBS create the cluster state shown under the condition that one photon each is detected in one of the four detectors. This set-up requires interferometric stability over a long time. The proper state is set by half-wave plates (HWP) at appropriate angles. Right: Linear optics quantum computation with active feed-forward. Depending on the measurement results of photon 1 and photon 2, the measurement basis for photon 3 is adapted using a Pockels cell which can implement a polarization rotation by the angle $\pm\beta$. Finally, depending on the then obtained results on all three photons, Pauli errors can be corrected by applying the proper spin rotations, again with Pockels cells on photon 4. These two operations are the most basic building blocks for linear optics quantum computation with active feed-forward. The insert shows the logic of the switching scheme (from [16] and [18]).

quantum nature of the procedure was demonstrated by violating a cluster-state Bell inequality [17].

In a more recent experiment [18], active feed-forward was implemented. In the experiment, both rotations of the measurement basis and Pauli error corrections could be achieved using fast detectors, rapid electronics and Pockels cells. It should be noted that one such cycle from measurement all the way to the setting of the new parameter took less than 150 ns.

Our one-way quantum computation scheme was applied to a number of basic quantum computation protocols. These include the realization of Deutsch's algorithm [19], decoherence-free quantum information transfer [20], the realization of a quantum version of prisoner's dilemma [21] and of Grover's search algorithm [16]. Future challenges of all-optical quantum computation certainly include the development of efficient sources for multi-photon entangled states, of detectors with both higher efficiency and the ability to discriminate photon numbers, and of feed-forward schemes with even shorter time constants. All these goals seem to be within reach of technology in reasonable time.

6. Long-Distance Quantum Communication

In the time since the last ISQM, a significant part of the work of our group was to extend quantum communication both to larger distances and to higher data rates, higher fidelity and towards more technical applications. In terms of distance, we are routinely operating a quantum communication link between the Canary Islands of Tenerife and La Palma [22].

Typically, on La Palma, we implement a source generating entangled photons, and on Tenerife we utilize the telescope of the Optical Ground Station (OGS) operated by the European Space Agency ESA. As an example, we might discuss the transmission of one of two entangled photons. A source on La Palma generates an entangled pair, of which one photon is measured by Alice on La Palma and the other photon is transmitted through an optical telescope over to Tenerife. In parallel to the quantum link, we operate tracking lasers to correct for the atmospheric instability. This is done by readjusting both the sending and the receiving telescope continuously in real time. In the experiment mentioned, both photons are measured using polarizing beam splitters, one rotated by 45° relative to each other. This makes it possible to implement the BBM protocol of quantum cryptography.

Using this scheme, it was both possible to violate the Bell inequality for the photon locally measured and the transmitted photon as well as to establish a quantum cryptography key. It should be noted that the attenuation across the link between La Palma and Tenerife is comparable to the typical attenuation between a ground station and a satellite in low-earth orbit. In the long run, we intend to establish an entangled source on a satellite and transmit either one or both photons to ground stations on Earth [23].

7. Concluding Comments

After having been a philosophical curiosity in the 1930s, entanglement has significantly matured over the last 20 years as a resource for quantum communication and quantum computation. Photonic entanglement will certainly remain one of the most essential resources for long-distance quantum communication. Besides that, all-optical one-way quantum computation offers a very exciting perspective, paralleling quantum computation with other means.

Acknowledgments

This work was supported by the Austrian Science Funds FWF, by the European Commission (projects QAP and EMALI), by the IARPA-funded U.S. Army Research Office (QCCM) and by the European Space Agency.

References

1. A. Zeilinger, M.A. Horne and C.G. Shull, in *Proceedings of the International Symposium on Foundations of Quantum Mechanics in the Light of New Technology*, (S. Kamefuchi et al., Eds.), Phys. Soc. Japan, Tokyo, 289-293 (1983).
2. A. Einstein, B. Podolsky and N. Rosen, Phys. Rev. **47**, 777 (1935).
3. Newton to Bentley, February 25, 1692 / 3.
4. E. Schrödinger, Proceedings of the Cambridge Philosophical Society **31**, 555 (1935).
5. J.S. Bell, Physics **1**, 195 (1964).
6. D.M. Greenberger, M.A. Horne, A. Zeilinger, in *Bell's Theorem, Quantum Theory, and Conceptions of the Universe*, (M. Kafatos, Ed.), Kluwer, Dordrecht, 69 (1989); arXiv: quant-ph 0712.0921.
7. J.W. Pan, D. Bouwmeester, M. Daniell, H. Weinfurter, A. Zeilinger, Nature **403**, 515 (2000).
8. A. Zeilinger, M.A. Horne, D. Greenberger, in *Proceedings of the NASA Conference „Squeezed States and Quantum Uncertainty"*, NASA Conference Publication 3135, 73-81 (1992).
9. P. Walther, K. J. Resch, C. Brukner and A. Zeilinger, Phys. Rev. Lett. **97,** 020501 (2006).
10. T. Kim, M. Fiorentino and F. N. Wong, Phys. Rev. A **73**, 012316 (2006).
11. A. Fedrizzi, T. Herbst, A. Poppe, T. Jennewein and A. Zeilinger, Optics Express **15**, 15377-15386 (2007).
12. A.J. Leggett, Found. Phys. **33**, 1469 (2003).
13. S. Gröblacher, T. Paterek, R. Kaltenbaek, C. Brukner, M. Zukowski, M. Aspelmeyer and A. Zeilinger, Nature **446**, 871-875 (2007).
14. E. Knill, R. Laflamme and G. J. Milburn, Nature **409**, 46 - 52 (2001).

15. R. Raussendorf and H. J. Briegel, Phys. Rev. Lett. **86**, 5188 - 5191 (2001).
16. P. Walther, K. J. Resch, T. Rudolph, E. Schenck, H. Weinfurter, V. Vedral, M. Aspelmeyer and A. Zeilinger, Nature **434**, 169 - 176 (2005).
17. P. Walther, M. Aspelmeyer, K.J. Resch, and A. Zeilinger, Phys. Rev. Lett. **95**, 020403 (2005).
18. R. Prevedel, P. Walther, F. Tiefenbacher, P. Böhi, R. Kaltenbaek, T. Jennewein and A. Zeilinger, Nature **445**, 65 – 69 (2007).
19. M. S. Tame, R. Prevedel, M. Paternostro, P. Böhi, M. S. Kim and A. Zeilinger, Phys. Rev. Lett. **98**, 140501 (2007).
20. R. Prevedel, M. S. Tame, A. Stefanov, M. Paternostro, M. S. Kim and A. Zeilinger, Phys. Rev. Lett. **99**, 250503 (2007).
21. R. Prevedel, A. Stefanov, P. Walther and A. Zeilinger, New Journal of Physics **9**, 205 (2007).
22. R. Ursin, F. Tiefenbacher, T. Schmitt-Manderbach, H. Weier, T. Scheidl, M. Lindenthal, B. Blauensteiner, T. Jennewein, J. Perdigues, P. Trojek, B. Ömer, M. Fürst, M. Meyenburg, J. Rarity, Z. Sodnik, C. Barbieri, H. Weinfurter and A. Zeilinger, Nature Physics **3**, 481 - 486 (2007).
23. J. Perdigues Armengol, B. Furch, C. J. d. Matos, O. Minster, L. Cacciapuoti, M. Pfennigbauer, M. Aspelmeyer, T. Jennewein, R. Ursin, T. Schmitt-Manderbach, G. Baister, J. G. Rarity, W. Leeb, C. Barbieri, H. Weinfurter and A. Zeilinger, Acta Astronautica **63,** 165 – 178 (2008).

QUANTUM NON-DEMOLITION MEASUREMENT OF A SUPERCONDUCTING FLUX QUBIT

J.E. MOOIJ

Kavli Institute of Nanoscience, Delft University of Technology, Delft, The Netherlands

E-mail: j.e.mooij@tudelft.nl

Measurements have been performed on superconducting flux qubits using a Josephson bifurcation amplifier (JBA). The qubit states are identified with the two stable dynamic states of a strongly driven non-linear oscillator that contains a SQUID sensor. Readout with high fidelity is achieved. Correlations in repeated measurements demonstrate that this readout is fully projective.

Keywords: Superconducting qubit, quantum measurement.

1. Introduction

In recent years, superconducting nanocircuits have been developed that behave as single quantum particles of which the properties can be engineered in fabrication [1]. It was not a priori obvious that the laws of quantum mechanics would apply for such macroscopic objects. Superconducting quantum bits (qubits) may in the long term provide a scalable technology for quantum information processing. Superconducting qubits can be driven into a controlled quantum superposition state, and interacting qubits can be entangled. Here we discuss a well-controlled non-destructive readout method of qubits. By repeated measurements on qubits in superposition states we show that this readout is fully projective. Also in this respect, superconducting quantum bits behave as true quantum systems.

The method of readout employs a non-linear oscillator that is strongly driven below its resonance frequency [2]. This so-called Duffing oscillator has two stable dynamic states with low (l) and high (h) amplitude of oscillation. In the detection method, these two oscillator states are correlated with the two basic states of the qubit $|0\rangle$ and $|1\rangle$. The bifurcation of the oscillator response has been employed for qubit readout for the first time by Siddiqi *et al.* [3].

The experiments that are described in this paper were performed by A. Lupaşcu, C.J.P.M. Harmans, R.N. Schouten, C.J.M. Verwijs, E.F.C. Driessen, L. Roschier, S. Saito, T. Picot and P.C. de Groot. They have been published in references [4-7].

2. System

2.1. *Qubit*

The qubit consists of a superconducting loop with three Josephson junctions, biased at a magnetic flux near $0.5\Phi_o$, where Φ_o is the flux quantum h/2e. The basic states have circulating currents of about 0.5 μA that run clockwise or anti-clockwise. At the symmetry point of exactly $\Phi_o/2$ a symmetric superposition is created with energy splitting Δ. Away from the symmetry point the Hamiltonian can be written, using Pauli spin matrices, as

$$H = -\frac{1}{2}(\varepsilon\sigma_z + \Delta\sigma_x)$$
$$\varepsilon = 2I_p(\Phi - \Phi_o/2) \qquad (1)$$

Here Φ is the applied flux; I_p is the maximum circulating current. The energy splitting is $E_1 - E_0 = (\Delta^2 + \varepsilon^2)^{1/2}$. Additional states exist at large distances. The circulating current is

$$I_{circ} = I_p \frac{\varepsilon}{\sqrt{\varepsilon^2 + \Delta^2}} \qquad (2)$$

This current induces a magnetic flux that is measured by the SQUID. At the symmetry point $\varepsilon=0$ the qubit is insensitive to flux noise. However, as I_{circ} is zero there, no linear measurement can be performed. It is possible to prepare the qubit at $\varepsilon=0$, followed by a fast but adiabatic flux shift to a point $\varepsilon\neq0$ for readout.

2.2. *SQUID circuit*

The SQUID consists of a parallel circuit of two Josephson junctions. The maximum supercurrent through the SQUID depends on the flux through the junction loop. Readout for flux qubits is usually performed by ramping up the transport current through the SQUID until it switches to a finite voltage. This switching induces complete decoherence, even in neighboring other qubits. An alternative method is to use the fact that for transport currents below the critical value the SQUID behaves as a nonlinear inductance. That inductance is inversely proportional to the critical current and is also flux-dependent. It is measured by shunting the SQUID with a capacitor and measuring the resonance frequency of the resultant oscillator. The qubit signal results in a positive or negative shift of the resonance. It can be read out with a transmission line, in transmission or reflection.

A higher sensitivity can be reached with the bifurcation oscillator method, usually referred to as the Josephson bifurcation amplifier (JBA). The SQUID/capacitor resonator is driven with an RF signal of high amplitude at a frequency slightly below the resonance. The SQUID acts as the nonlinear element in the Duffing oscillator (the inductance is lower for high driving amplitude). As shown in Fig. 1a, for intermediate driving amplitudes the oscillator

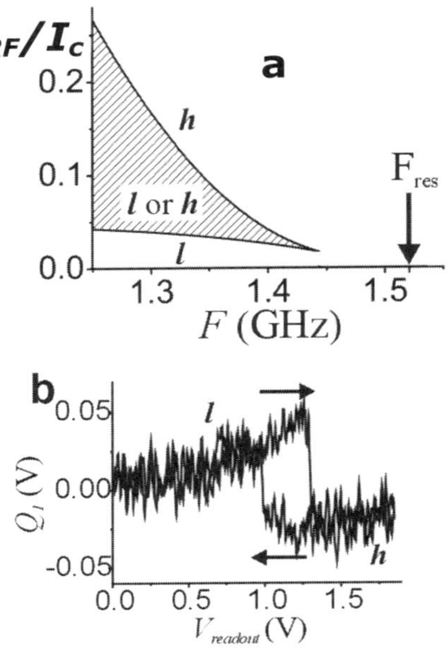

Fig.1. Regimes of the driven nonlinear SQUID oscillator.
a. SQUID response for driving at frequency F and amplitude IRF/Ic. The regimes with high (h) and low (l) oscillation amplitude overlap in a hysteretic region.
b. Measured hysteretic response.

has two possible stable dynamic responses with low or high amplitude. Transitions are hysteretic, as illustrated in Fig. 1b which shows the measured response of a SQUID. Whether the SQUID oscillates in the high or low mode can be determined with very high fidelity. The measurement procedure is to apply a short RF pulse (6 ns) that either induces the oscillator to switch to the *h* mode or not (outcome *l*). Directly afterwards, the RF amplitude is reduced to a value in the middle of the hysteretic regime during 100 ns. No switching is induced in this holding plateau, during which the electronics read out the state. The procedure is repeated 5000 times, each of which gives an outcome *l* or *h*. The switching probability *P*(*h*) of switching to the *h* mode is used as the measurement signal.

The qubit effectively shifts the resonance frequency, producing two hysteretic regions that are shifted slightly in the horizontal direction of Fig. 1a. The probe pulse is given amplitude such that switching to the h mode will only occur for one state of the qubit. For the other qubit state the SQUID oscillator remains in the l mode. In the experiments described here, the h mode of the oscillator corresponded to the ground state of the qubit, while the l mode indicated the excited state. The fidelity of qubit readout was 0.87.

3. Experimental Results

Measurements have been made on a qubit that was prepared in an arbitrary superposition state. The switching probability was measured as a function of the amplitude of the RF measurement pulse. The derivative of this curve gives the probability that the oscillator has switched before that value of the probe amplitude was reached. In Fig. 2 this derivative is shown. The black line gives the distribution for the qubit in the ground state. A peak around amplitude 1.3 is found. The dark gray line gives the result when beforehand a π-pulse has been applied to the qubit, bringing it into its excited state. Now a slightly lower peak is observed around amplitude 1.7 as well as a low bump at the previous value of 1.3. When the qubit is prepared in a symmetric superposition,

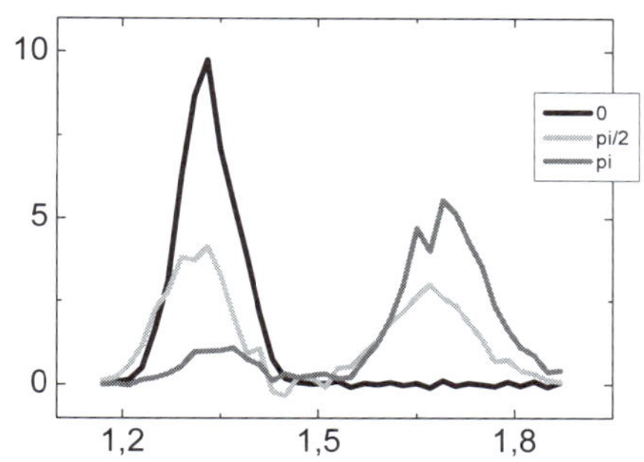

Fig. 2. Measurement results on a qubit that is prepared in the ground state (black), in a symmetric superposition (light gray) and in the excited state (dark gray).
Plotted is the differential probability that the SQUID has switched to its h mode before the driving amplitude reaches the indicated value of the RF driving amplitude (horizontal scale). Units are arbitrary.

two well-separated peaks of almost equal height are found. Clearly the ground state of the qubit is found near amplitude 1.3, the excited state near 1.7 and nothing in the region in the middle. This is consistent with projective measurement. In the excited state some relaxation takes place as will be discussed later, shifting some of the weight to the ground state.

The readout process with the bifurcation method takes about 100 ns. The relaxation time is significantly longer and it is possible to perform two consecutive measurements on the same qubit. Beforehand the qubit can be prepared in the ground or excited state, or in a superposition. If measurement is projective, the result of the second measurement should be identical to that of the first measurement, irrespective of the initial state of the qubit. Fig. 3 gives the result of such measurements.

The measuring scheme is indicated in Fig. 3.a. The qubit is prepared in an initial state by a resonant microwave pulse with length t_0. One readout is performed, followed by a second readout. In between a second microwave pulse with length t_1 can change the qubit state in a coherent fashion. The holding plateau of the first measurement is reduced to bring the two readouts closer together. This gives only a negligible loss in fidelity.

It is important to note that determination of the probability of finding the h mode of the oscillator requires thousands of individual readouts. For each consecutive single shot readout, however, both the outcomes of the first and of the second measurement are stored. That allows the determination of P(h) using only the first measurement, but also of correlations between the various outcomes of the two measurements. P($h|h$) is the conditional probability that the second readout yields h when the first readout had given h. Similarly P($l|l$) correlates outcome l after an initial result l. A perfect, fully projective measurement should always give P($h|h$)= P($l|l$)=1.

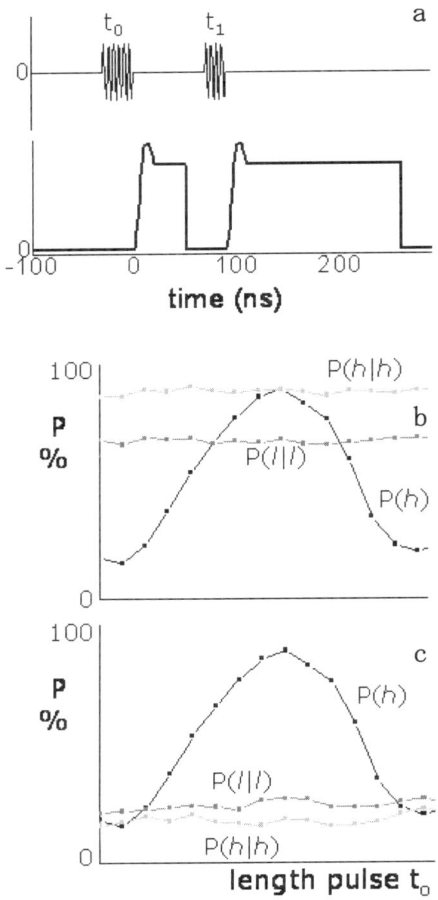

Fig. 3. Repeated measurements on the same qubit. The initial state of the qubit varies with the length of pulse to. a. readout scheme. b. Pulse *t*1 has length 2p. c. Pulse *t*1 has length p. Outcomes *(l* or *h*) in the two readouts are correlated. See text.

Fig. 3.b gives the measurement results when the initial state is the ground state. The single first readout P(*h*) shows the expected Rabi oscillation of the qubit. One should remember that oscillator mode *h* indicates the ground state of the qubit. Zero length for t_o is outside the range of the figure, but clearly a peak of P*(h*) is reached in the middle of the figure. There the pulse t_o induces a rotation of 2π, so the qubit is back in the ground state before the first readout. Where P*(h*) is minimal, the qubit is initially in the excited state. One finds that P(*h*|*h*) is very close to 1, even when the initial qubit state is near the excited state, as indicated by P(*h*). The values of the conditional probabilities for the excited state P(*l*|*l*) are slightly lower, but still there is a high correlation. Some relaxation occurs between the readouts, partially induced by the first measurement.

Fig. 3.c shows the corresponding results when in between the two readouts a π-pulse is applied. This should turn full correlation into full anti-correlation. Indeed this is found within the limitations imposed by relaxation.

The overall result of these repeated measurements indicates that the readout method employing the dispersive bifurcation method is non-destructive. After the first readout, the qubit is in a well-defined state (either $|0\rangle$ or $|1\rangle$). Readout is consistent with projection, as quantum mechanics textbooks like to teach, but few real experiments manage to achieve. Is this measurement of the quantum non-demolition type (QND)? The results indicate yes, with high fidelity. When the fidelity is defined as $P_{\rm QND}$=(P(*h*|*h*)+ P(*l*|*l*))/2, one finds $P_{\rm QND}$=88% for our experiments.

True QND requires that the quantum system under study commutes with the measurement system. When the qubit Hamiltonian of Eq.(1) is considered in connection with the SQUID readout which is in the σ_z-direction, one sees that only when the qubit is far away from the symmetry point do the SQUID and the qubit commute. In our experiments Δ/h is equal to 4.6 GHz, while the qubit splitting is 14.2 GHz. The cosine of the angle between the qubit and the SQUID is 0.95 under these conditions.

The influence of relaxation on the readout process as well as the influence of the readout on qubit relaxation was studied in detail. It was found that relaxation in the qubit is significantly faster when the detector oscillator is excited, in particular when the oscillator is in the *h* mode. The circulating current of the

SQUID varies in time and couples to the qubit. As a result, in the h mode the qubit energy splitting varies over several GHz, at the oscillator frequency of 1.4 GHz. During this sweep the qubit may either directly relax due to coupled dynamics, or interact with neighboring two-level defect states. In Fig.4. the decay of the qubit excited state is shown while the detector oscillates in the l or the h mode. A significant difference is observed. See reference [7] for details.

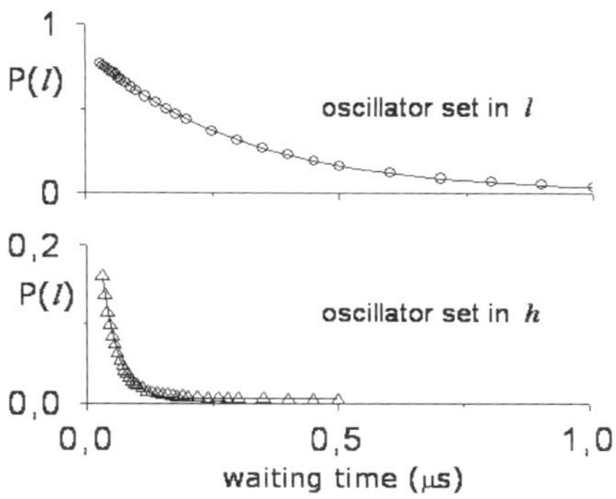

Fig.4. Relaxation of the qubit while the SQUID oscillator is in the low amplitude l mode (top) or the high amplitude h mode (bottom).

4. Conclusions

A readout scheme has been developed for superconducting flux qubits, using a SQUID flux detector as a nonlinear element in a Duffing oscillator. The fidelity of readout is around 87%. With repeated measurement it is found that this readout is fully projective, with a quantum non-demolition fidelity of 88%.

Acknowledgments

Support for this work was received from FOM, the Netherlands Program for Nanotechnology NanoNed and the European Union FET project EuroSQIP.

References

1. G. Wendin and V.S. Shumeiko, *Low Temp. Phys.* **33**, 724 (2007).
2. M.I. Dykman and M.A. Krivoglaz, *Physica* **104A**, 480 (1980).
3. I. Siddiqi *et al.*, *Phys. Rev. Lett.* **93**, 207002 (2004).
4. A. Lupaşcu *et al.* *Phys. Rev. Lett.* **93**, 177006 (2004).
5. A. Lupaşcu *et al.* *Phys. Rev. Lett.* **96**, 127003 (2006).
6. A. Lupaşcu *et al.* *Nature Phys.* **3**, 119 (2007).
7. T. Picot *et al.*, *Phys. Rev. B* **78**, 132508 (2008)

ATOMIC PHYSICS AND QUANTUM INFORMATION PROCESSING WITH SUPERCONDUCTING CIRCUITS

FRANCO NORI*

Advanced Science Institute, The Institute of Physical and Chemical Research (RIKEN), Wako-shi, Saitama 351-0198, Japan
CREST, Japan Science and Technology Agency, Kawaguchi, Saitama 332–0012, Japan
Department of Physics, Center for Theoretical Physics, The University of Michigan, Ann Arbor, Michigan 48109-1040, USA
**E-mail: nori@umich.edu*

Artificial atoms, making transitions between discrete energy levels, can be made using superconducting circuits. Such circuits can used to conduct atomic-physics experiments on a silicon chip and test quantum mechanics at macroscopic scales.

Keywords: Superconducting qubits; quantum information; quantum optics and atomic physics on a chip.

1. Bridge between Atomic and Condensed Matter Physics

Quantum information processing provides an exciting natural bridge between atomic and condensed matter physics. Quantum information processing involves preparing, manipulating, and reading out the quantum states of a many-qubit system. Therefore, it is desirable to have qubits that can be individually controlled. Moreover, they should be scalable; that is, simply adding more qubits should create a larger circuit capable of more complex calculations. Several types of qubits satisfy these requirements. This very brief overview of the field will focus on one type of qubit: superconducting quantum circuits [1-3]. These are currently being studied by many research groups. Additional information and many more references to other work can be found in Refs. [1-3].

2. Quantum Circuits as Artificial Atoms

The electrons confined on small superconducting island collectively behave much like a single natural atom; so they can be interpreted as a single "artificial atom" [1-3]. Natural atoms and artificial atoms both have discrete energy levels and exhibit coherent quantum oscillations between those levels—so-called Rabi oscillations. But whereas natural atoms are usually driven using visible photons that excite electrons from one state to another, the artificial atoms in the circuits are driven by currents, voltages, and microwave photons. The resulting electric and magnetic fields control the tunneling of electrons between the superconducting island and nearby electrodes. The effects of those external fields on the circuits are analogues of the more familiar ways that natural atoms respond to electric and magnetic fields (the so-called Stark and Zeeman effects).

One of the differences between quantum circuits and natural atoms is that atoms couple weakly to their environment whereas circuits couple strongly. Another distinction is that the energy scales of the two systems differ.

One of the advantages of artificial atoms over naturally occurring atoms is that they can be lithographically designed to have specific characteristics, such as a large dipole moment or particular transition frequencies.

3. Superconducting Electronics

Quantum circuits made of Josephson junctions provide good qubits [1-3]. Josephson junctions—superconducting grains or electrodes separated by an insulating oxide—act like nonlinear inductors in a circuit. The nonlinearity ensures an unequal spacing between energy levels, so that the lowest levels can be accessed using external fields. Two important energy scales determine the quantum mechanical behavior of a Josephson-junction circuit: the Josephson coupling energy and the electrostatic Coulomb energy for a single Cooper pair. In analogy to the usual position–momentum duality in quantum mechanics, the phase ϕ of the Cooper-pair wave function and the number n of Cooper pairs are conjugate variables and obey the Heisenberg uncertainty relation.

There are several kinds of superconducting qubits realized in different regimes. The charge qubit is in the "charge regime" where the number n of Cooper pairs is well defined and the phase ϕ fluctuates strongly. The so-called flux and phase qubits are both in the phase regime, in which the phase is well defined and n fluctuates strongly. And the charge–flux qubit lies in the intermediate regime in which charge and phase degrees of freedom play equally important roles.

4. Circuits Mimicking Atoms

These type of superconducting circuits can behave like atoms, including exhibiting the so-called Rabi oscillations. Let us briefly review Rabi oscillations of an atom in a strong laser field. Consider an electromagnetic wave resonant with an atom transition frequency. First, it excites the atom from the ground state to the excited state. Therefore the probability amplitude of finding the atom in the excited state increases over time. Afterwards, when the atom reaches the excited state, the wave de-excites the atom, also known as stimulated emission. This periodic cycle of absorption-emission is called Rabi oscillations, and it occurs at a frequency that is proportional to the strength of the applied electric field driving these transitions. During this cycle the atom is usually in a superposition of ground and excited state. Over the past decade, these oscillations have been observed in macroscopic artificial atoms embedded in superconducting circuits.

5. Different Circuits Mimicking a Variety of Artificial Atoms

5.1. *Charge qubit or Cooper-pair Box*

Figure 1(a) shows a charge qubit, known also as a Cooper-pair box (CPB; shown as a dashed blue square). This is driven by an applied voltage V_g (green) through the gate capacitance C_g to induce an offset charge $2e\, n_g = C_g\, V_g$. A Josephson junction, the barrier denoted by the ×, connects the box to a wire lead. Each junction has a capacitance C and a Josephson coupling energy E_J. The electrostatic energy of the CPB is equal to $E_C\ (n - n_g)^2$, where the charging energy $E_C = (2e)^2/2C$, is plotted as a function of the number n of excess electron pairs. The lowest energy states, $|0\rangle$ and $|1\rangle$ (shown in red on the right side of Fig. 1), are degenerate when $n_g = 0.5$, and are used as the qubit state basis. Those states are coupled via the junction energy E_J, which controls the tunneling between them.

5.2. *Magnetic Flux Loop (RF-SQUID)*

Figure 1(b) shows a magnetic-flux "box", or more precisely, a loop enclosing magnetic flux, which is the "magnetic analogue" of the electrostatic CPB shown in Fig. 1(a). A magnetic bias now replaces the electric bias: A current-driven magnetic field pierces the loop with a strength given by a mutual inductance

M. While an electric field prompts stored electron pairs to tunnel into or out of the CPB, a magnetic field pushes magnetic flux quanta Φ_0 into or out of the superconducting quantum interference device (SQUID) loop. The adjacent potential energy diagram (energy versus "phase drop ϕ along the barrier") plots a Josephson energy term (proportional to $\cos\phi$) and an additional superimposed inductive energy term—proportional to $(\phi - \phi_{ext})^2 / 2L$, where L is the SQUID's inductance—as a function of the phase ϕ of the junction. The lowest energy states (shown in red) are superpositions of the clockwise and counterclockwise supercurrent states, $|\downarrow\rangle$ and $|\uparrow\rangle$, that flow in the SQUID loop; Δ here is the tunneling energy between the supercurrent states. Those energy states are degenerate when the externally applied magnetic flux $\phi_{ext} = \pi$.

5.3. *Flux Qubit: Magnetic Flux Loop with three junctions*

Figure 1(c) shows a three-junction flux qubit. This loop with three junctions works like a magnetic flux "box" in Fig. 1(b), except that one of the three junctions has a slightly smaller (~25% smaller) capacitance and coupling energy. The contour plot on the right of Fig. 1(c) shows the potential energy as a function of two junctions' phases. The two red dots inside the potential wells on the right correspond to the qubit basis states $|\downarrow\rangle$ and $|\uparrow\rangle$. Having several junctions, here three instead of one, provide a higher level of control over the potential energy of the circuit. Versions of this circuit using four junctions are now used by several research groups, since these are more symmetric and offer more control over the potential energy of the circuit.

5.4. *Magnetic Flux Loop (RF-SQUID)*

Figure 1(d) shows a so-called phase qubits where a current source biases a junction. Logic operations can be achieved by driving the qubit with a microwave field at frequency $(E_1 - E_0)/h$. Pulsing the qubit with a microwave field at a frequency $(E_2 - E_1)/h$ produces a transition from $|1\rangle$ to $|2\rangle$. One can then read the qubit's state by measuring the occupation probability of state $|2\rangle$.

6. Future Challenges

One of the major obstacles to quantum computing is decoherence due to noise. This effect degrades the information carried by the qubit. Another central issue in future developments will be the efficient and nondissipative readout of qubit states. It is still too early to say which type of qubit will do best in addressing these challenges, and so win the race to realize a quantum computer in practice.

For superconducting qubits the status is that all quantum states on the so-called Bloch sphere—a geometrical representation of the states of a two-level system—can be reached; spin-echo techniques, borrowed from nuclear magnetic resonance, can reduce the effect of $1/f$ noise; and readout efficiency greater than 99% and a coherence quality factor of approximately 10^6 can be achieved. When techniques for manipulating two or three qubits become well established, the next step will be to build circuits with a larger number of qubits, increased readout efficiency, and lower decoherence. Such conditions would allow quantum computing with superconducting qubits.

7. Quantum Mechanical Circuitry

Even if no quantum computing is ever achieved using superconducting circuits, they still provide researchers with tools to test fundamental quantum mechanics in novel ways (see, e.g., [1-3], and the many references therein). For example, these artificial atoms can be used to simulate atomic physics using quantum circuits; researchers have already observed Rabi oscillations and interference patterns that are manifest during the phase evolution of a superconducting qubit. Moreover, the devices can also do basic tests of quantum mechanics, and study the striking quantum non-localities of quantum mechanical wave-functions (discussed in the celebrated Einstein-Podolsky-Rosen paper).

This very short overview of the field has touched on only a few of the many fundamental physics questions that can be addressed by the quantum control of macroscopic entangled states, not to mention the potential for technological applications. This research is extremely active worldwide.

Acknowledgments

We thank our collaborators, especially J.Q. You, S. Ashhab, Y.X. Liu, and R. Johansson, for their valuable contributions to earlier versions of this manuscript. This work was supported in part by the US National Security Agency, Laboratory of Physical Sciences, Army Research Office, and the National Science Foundation.

References

[1] J.Q. You, F. Nori, "Superconducting circuits and quantum information", *Physics Today* **58**, No. 11, 42-47 (2005).

[2] F. Nori, "Atomic physics with a circuit", *Nature Physics* **4**, 589 (2008).

[3] our papers on this subject are available online at http://dml.riken.jp/publications.php and also here http://www.umich.edu/~nori/

[4] Y.X. Liu, J.Q. You, L.F. Wei, C.P. Sun, F. Nori, "Optical selection rules and phase-dependent adiabatic state control in a superconducting quantum circuit", *Phys. Rev. Lett.* **95**, 087001 (2005).

Table 1. Transitions between energy levels in atoms obey selection rules, like the electric-dipole selection rule. Artificial atoms have different types of selections rules [4].

Transitions between energy levels: flux qubit circuits and natural atoms

Properties / Atoms	Dipole moment or its analog	Parity	Symmetry of potential	Selection rules
Natural atoms	$e\vec{r}$	Odd	Well-defined symmetry	Yes
Flux qubit circuits $f=1/2$	$\sin(2\varphi_m)$	Odd	Well-defined symmetry	Yes
Flux qubit circuits $f\neq 1/2$	$\sin(2\varphi_m+2\pi f)$	No parity	Broken symmetry	No

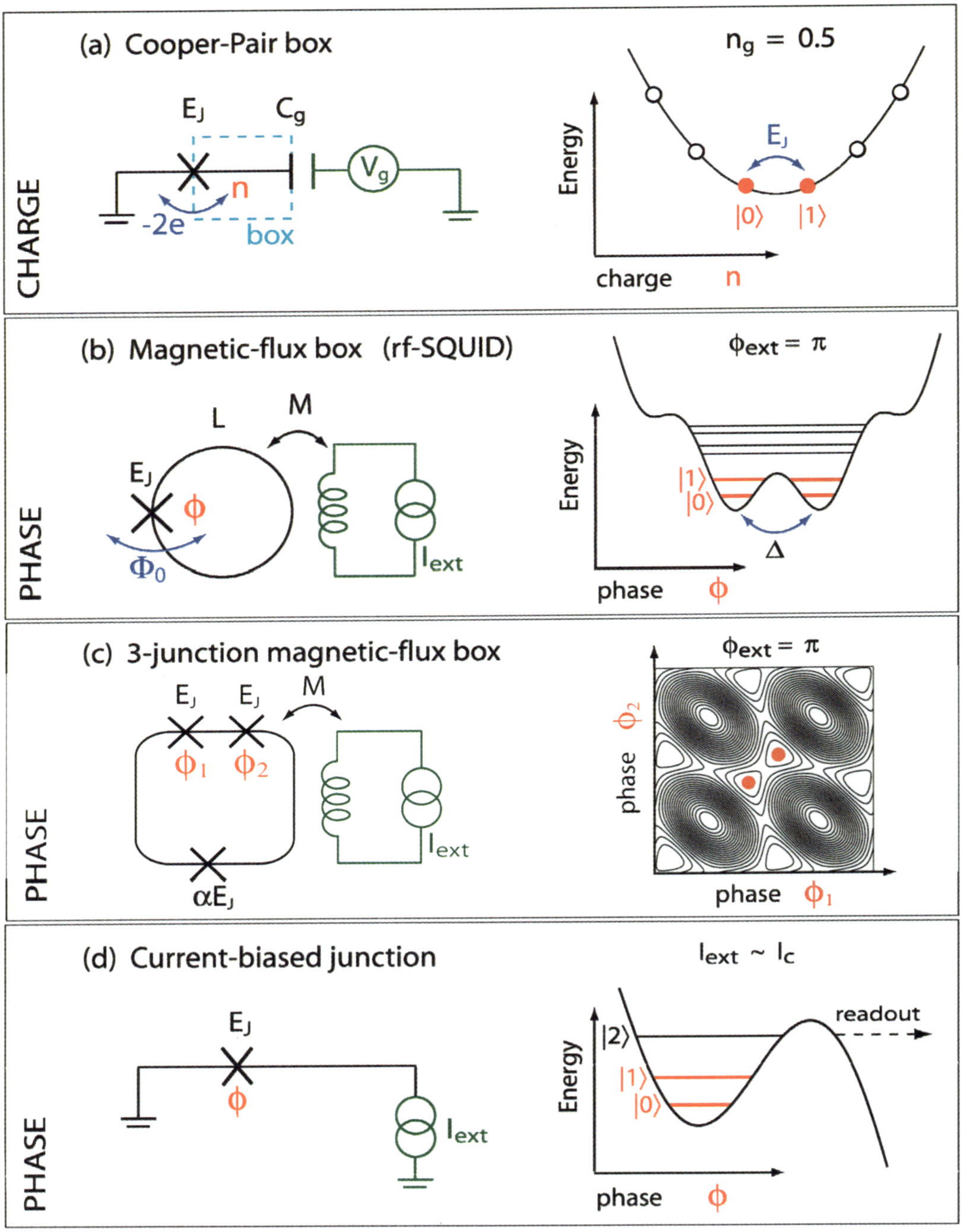

Fig. 1. Schematic diagrams of circuits (left side) and their corresponding energy levels (right side). These superconducting circuits can behave like atoms making transitions between discrete energy levels. Such circuits can used to conduct atomic-physics experiments on a silicon chip and test quantum mechanics at macroscopic scales.

THEORY OF MACROSCOPIC QUANTUM DYNAMICS IN HIGH-T_C JOSEPHSON JUNCTIONS

SHIRO KAWABATA

Nanotechnology Research Institute (NRI), National Institute of Advanced Industrial Science and Technology (AIST), Tsukuba, Ibaraki, 305-8568, Japan and

CREST, Japan Science and Technology Corporation (JST), Kawaguchi, Saitama 332-0012, Japan

E-mail: s-kawabata@aist.go.jp

We have theoretically investigated macroscopic quantum tunneling (MQT) and the influence of nodal quasiparticles and zero energy bound states (ZES) on MQT in s-wave/ d-wave hybrid Josephson junctions. In contrast to d-wave/d-wave junctions, the low-energy quasiparticle dissipation resulting from nodal quasiparticles and ZES is suppressed due to a quasiparticle-tunneling blockade effect in an isotropic s-wave superconductor. Therefore, the inherent dissipation in these junctions is found to be weak. This result suggests high potential of s-wave/d-wave hybrid junctions for applications in quantum information devices.

Keywords: Josephson Effect; Macroscopic Quantum Tunneling; High-T_c Superconductor; Path-integral Method; Quantum Dissipation.

1. Introduction

Since the experimental observations of macroscopic quantum tunneling (MQT) in YBCO grain boundary[1,2] and Bi2212 intrinsic[3,4,5,6] Josephson junctions, the macroscopic quantum dynamics of high-T_c d-wave junctions has become a hot topic in the field of superconductor quantum electronics and quantum computation. Recently, the effect of low-energy quasiparticles, e.g., the nodal-quasiparticles and the zero energy Andreev bound states (ZES)[7,8] on MQT in d-wave junctions have been theoretically investigated.[9,10,11] It was found that, in c-axis type junctions, the suppression of MQT resulting from the nodal-quasiparticles is very small.[9,10] This result is consistent with recent experimental observations.[3,4,5,6] In the case of in-plane type d-wave junctions, however, the ZES give a strong dissipative effect.[11] Therefore, it is important to avoid the formation of ZES in order to observe MQT with a high quantum-to-classical crossover temperature T^*.

On the other hand, recently, quiet qubits consisting of a superconducting loop with an s-wave/d-wave (s/d) hybrid junction have been proposed.[12] In the quiet qubits, a quantum two level system is spontaneously generated and therefore it is expected to be robust to decoherence caused by fluctuations of the external magnetic field.[12,13] However, influence of the low-energy quasiparticle dissipation due to nodal-quasiparticles and ZES on s/d quiet qubits is not yet understood. Therefore, it is important to investigate such intrinsic dissipation effects on the macroscopic quantum dynamics in order to realize quiet qubits.

In this paper, motivated by the above studies, we have investigated investigate MQT in s/d hybrid Josephson junctions [see Fig. 1(a)].[14] In contrast with d/d junctions,[9,11] we have shown that the influence of the low-energy quasiparticle dissipation is suppressed by a small but finite isotropic gap in the s-wave superconductor. Therefore, weak quasiparticle dissipation is expected in such junctions.

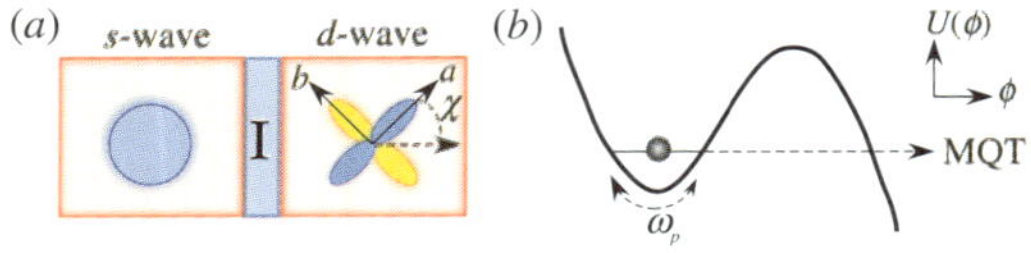

Fig. 1. (a) Schematic of an in-plane s-wave/d-wave hybrid Josephson junction. a and b denote the crystalline axes of the d-wave superconductor, and χ is the mismatch angle between the normal of the insulating barrier (I) and the a-axis. (b) Potential $U(\phi)$ v.s. the phase difference ϕ between two superconductors and ω_p is the Josephson plasma frequency of the junction.

2. Effective action

In the following we derive the effective action for s/d junctions with a clean insulating barrier and without extrinsic dissipation, e.g., an ohmic dissipation in a shunt resistance. The partition function of a junction can be described by a functional integral over the macroscopic variable (the phase difference ϕ),[15] i.e., $\mathcal{Z} = \int \mathcal{D}\phi(\tau) \exp\left(-\mathcal{S}_{\rm eff}[\phi]/\hbar\right)$. We consider the high barrier limit, where $z_0 \equiv m w_0/\hbar^2 k_F \gg 1$. Here m is the electron mass, w_0 is the height of the delta function-type insulating-barrier I (see Fig. 1(a)), and k_F is the Fermi wave length. In this limit, the effective action $\mathcal{S}_{\rm eff}$ is given by $\mathcal{S}_{\rm eff} = \mathcal{S}_0 + \mathcal{S}_\alpha$, where

$$\mathcal{S}_0[\phi] = \int_0^{\hbar\beta} d\tau \left[\frac{M}{2}\left(\frac{\partial\phi(\tau)}{\partial\tau}\right)^2 + U(\phi)\right],$$

$$\mathcal{S}_\alpha[\phi] = -\int_0^{\hbar\beta} d\tau d\tau' \alpha(\tau-\tau') \times \cos\frac{\phi(\tau)-\phi(\tau')}{2}. \qquad (1)$$

In this equation, $\beta = 1/k_B T$, $M = C\,(\hbar/2e)^2$ is the mass (C is the capacitance of the junction) and the potential $U(\phi)$ can be described by

$$U(\phi) = \frac{\hbar}{2e}\left[\int_0^1 d\lambda\ \phi I_J(\lambda\phi) - \phi\ I_{ext}\right], \quad (2)$$

where I_J is the Josephson current and I_{ext} is the external bias current. The dissipation kernel $\alpha(\tau)$ is related to the quasiparticle current $I_{\rm qp}(V)$ under the constant bias voltage V by $\alpha(\tau) = (\hbar/e)\int_0^\infty d\omega/(2\pi)\exp(-\omega\tau) \times I_{\rm qp}(V=\hbar\omega/e)$ at zero temperature.

In order to investigate effects of the nodal-quasiparticles and the ZES on MQT, we derive effective actions for two types of s/d_χ junctions, i.e., $\chi = 0$ and $\pi/4$, with χ being the mismatch angle between the normal of the insulating barrier (I) and the crystalline axis of the d-wave superconductor (see Fig. 1(a)). In the case of s/d_0 junctions, ZES are completely absent.[7] On the other hand, in $s/d_{\pi/4}$ junctions, ZES are formed near the interface between I and the d-wave superconductor $d_{\pi/4}$.

By solving the Bogoliubov-de Gennes equation,[7] we obtain analytical expressions for the potential as

$$U(\phi) \approx \begin{cases} -\dfrac{\hbar I_{C_1}}{2e}\left(\cos\phi + \eta\phi\right) & \text{for } s/d_0 \\[2ex] -\dfrac{\hbar I_{C_2}}{2e}\left(-\dfrac{\cos 2\phi}{2} + \eta\phi\right) & \text{for } s/d_{\pi/4} \end{cases} \quad (3)$$

where I_{C_1} (I_{C_2}) is the Josephson critical current for the s/d_0 ($s/d_{\pi/4}$) junction, and $\eta \equiv I_{\rm ext}/I_{C_1(C_2)}$. The dissipation kernel $\alpha(\tau)$ is given by

$$\alpha(\tau) \approx \begin{cases} \dfrac{3\hbar}{8\sqrt{2}\pi}\dfrac{R_Q}{R_N\varepsilon}\dfrac{1}{\tau^2}K_1\left(\dfrac{\Delta_s|\tau|}{\hbar}\right) & \text{for } s/d_0 \\[2ex] \dfrac{6}{5\hbar}\dfrac{R_Q\Delta_s\Delta_d}{R_N}K_1\left(\dfrac{\Delta_s|\tau|}{\hbar}\right) & \text{for } s/d_{\pi/4} \end{cases} \quad (4)$$

where R_Q is the resistance quantum, R_N is the normal state resistance, $\Delta_{s(d)}$ is the superconducting gap for $s(d)$-wave superconductors, $\varepsilon = \Delta_d/\Delta_s$, and K_1 is the modified Bessel function. For $|\tau| \gg \hbar/\Delta_s$ the dissipation kernel decays exponentially as a function of the imaginary time τ, i.e.,

$$\alpha(\tau) \approx \begin{cases} \dfrac{3\hbar^{3/2}}{16\sqrt{\pi}} \dfrac{R_Q\sqrt{\Delta_s}}{R_N\Delta_d} \dfrac{1}{|\tau|^{5/2}} e^{-\frac{\Delta_s|\tau|}{\hbar}} & \text{for } s/d_0 \\ \dfrac{6\sqrt{\pi}}{5\sqrt{2}\hbar} \dfrac{R_Q\sqrt{\Delta_s}\Delta_d}{R_N} \dfrac{1}{\sqrt{|\tau|}} e^{-\frac{\Delta_s|\tau|}{\hbar}} & \text{for } s/d_{\pi/4} \end{cases} . \quad (5)$$

If the phase varies slowly with the time scale given by $\hbar/\Delta_s$, then we can expand $\phi(\tau) - \phi(\tau')$ in Eq. (1) about $\tau = \tau'$. This gives

$$S_\alpha[\phi] \approx \frac{\delta C}{2} \int_0^{\hbar\beta} d\tau \left[\frac{\hbar}{2e} \frac{\partial\phi(\tau)}{\partial\tau} \right]^2 . \quad (6)$$

Hence, the dissipation action $\mathcal{S}_\alpha$ acts as a kinetic term so that the effect of the quasiparticles results in an increase of the capacitance, $C \to C + \delta C \equiv C_{ren}$. This indicates that the quasiparticle dissipation in s/d junctions is qualitatively weaker than that in in-plane d/d-wave junctions in which the super-ohmic ($\alpha(\tau) \sim |\tau|^{-3}$)[9,10] or ohmic dissipation ($\alpha(\tau) \sim \tau^{-2}$)[11] appears.

3. MQT

Next, we will investigate MQT in s/d junctions The MQT escape rate from the metastable potential (Fig. 1(b)) at zero temperature is given by[16]

$$\Gamma = \lim_{\beta\to\infty} \frac{2}{\beta} \operatorname{Im} \ln \mathcal{Z}. \quad (7)$$

By using the Caldeira and Leggett theory,[17] the MQT rate is approximated as

$$\Gamma(\eta) = \frac{\omega_p(\eta)}{2\pi} \sqrt{120\pi B(\eta)} \exp[-B(\eta)], \quad (8)$$

where

$$\omega_p(\eta) = \sqrt{\frac{\hbar a_i I_{C_i}}{2eM_{\rm ren}}} (1-\eta^2)^{\frac{1}{4}}, \quad (9)$$

is the Josephson plasma frequency ($i = 1$ for d_0, $i = 2$ for $d_{\pi/4}$, $a_1 = 1$, $a_2 = 2$, and $M_{\rm ren} = (\hbar/2e)^2 C_{\rm ren}$) and $B(\eta) = \mathcal{S}_{\rm eff}[\phi_B]/\hbar$ is the bounce exponent, which is the value of the the action $\mathcal{S}_{\rm eff}$ evaluated along the bounce trajectory $\phi_B(\tau)$. The analytic expression for the bounce exponent is given by

$$B(\eta) = \frac{b_i}{e} \sqrt{\frac{2e}{\hbar} I_{C_i} M_{\rm ren}} \, (1-\eta^2)^{\frac{5}{4}}, \quad (10)$$

where $b_1 = 12/5$ and $b_2 = 3\sqrt{2}/5$. In actual MQT experiments, the switching current distribution $P(\eta)$ is measured, where $P(\eta)$ is related to the MQT rate $\Gamma(\eta)$ as

$$P(\eta) = \frac{1}{v}\Gamma(\eta) \exp\left[-\frac{1}{v} \int_0^\eta \Gamma(\eta') d\eta' \right] . \quad (11)$$

Here $v \equiv |d\eta/dt|$ is the sweep rate of the external bias current. At high temperatures, the thermally activated (TA) decay dominates the escape process. Then the escape rate is given by the Kramers formula[16]

$$\Gamma = \frac{\omega_p}{2\pi} \exp\left(-\frac{U_0}{k_B T} \right), \quad (12)$$

where U_0 is the barrier height. Below the crossover temperature T^*, the escape process is dominated by MQT. Note that T^* is reduced in the presence of dissipation.

In order to see explicitly the effect of the quasiparticle dissipation on MQT, we numerically estimated T^*. We determined T^* from the relation

$$\sigma^{\rm TA}(T^*) = \sigma^{\rm MQT}, \quad (13)$$

where $\sigma^{\rm TA}(T)$ and $\sigma^{\rm MQT}$ are the standard deviation of $P(\eta)$ for the temperature-dependent TA process and the temperature-independent MQT process, respectively. Presently, no experimental data are available for ideal highly under-damped s/d junctions with large McCumber parameters $\beta_M = (2e/\hbar) I_C C R_{\rm sg}^2 \gg 1$ ($R_{\rm sg}$ is the subgap resistance). Thus, we estimate T^* by using the parameters for an actual Nb/Au/YBCO junction[13] ($C = 0.60$ pF, $I_C = 95.2$ μA at $\chi = 0$, and $R_N = 5.68$ Ω) in which $\beta_M \approx 1.5$. We also assume $I_{C_1} = I_{C_2} = 95.2$ μA for simplicity, $\Delta_s = \Delta_{\rm Nb} = 1.55$ meV, $\Delta_d = \Delta_{\rm YBCO} = 20.0$ meV, and $vI_{C_i} = 0.0424$ A/s. In the case of s/d_0 junctions, we obtain $T^* = 336$mK for the dissipationless case

($C_{\rm ren} = C$) and $T^* = 333$mK for the dissipative case ($C_{\rm ren} = C + \delta C$). Thus, the influence of the nodal-quasiparticle is negligibly small. On the other hand, in the case of $s/d_{\pi/4}$ junctions, we obtain $T^* = 601$mK for the dissipationless case and $T^* = 101$mK for the dissipative case. As expected, the ZES have a larger influence on MQT than the nodal-quasiparticles. However, the T^* suppression is small enough to allow experimental observations of MQT.

4. Summary

In conclusion, MQT in the s/d hybrid Josephson junctions with the perfect insulating barrier has been theoretically investigated using the path integral method. The effect of the low energy quasiparticles on MQT is found to be weak. This can be attributed to the quasiparticle-tunneling blockade effect in the s-wave superconductor. We also investigated MQT in a realistic s/d junction and showed that the expected T^* is relatively high in spite of the small β_M. These results strongly indicate that s/d hybrid junctions are expected to be highly applicable to quantum computers.

Acknowledgments

I would like to thank my co-workers A. A. Golubov, Ariando, C. J. M. Verwijs, H. Hilgenkamp, and J. R. Kirtley.

References

1. T. Bauch, F. Lombardi, F. Tafuri, A. Barone, G. Rotoli, P. Delsing, and T. Claeson, Phys. Rev. Lett. **94**, 087003 (2005).
2. T. Bauch, T. Lindström, F. Tafuri, G. Rotoli, P. Delsing, T. Claeson, and F. Lombardi, Science **311**, 57 (2006).
3. K. Inomata, S. Sato, K. Nakajima, A. Tanaka, Y. Takano, H. B. Wang, M. Nagao, H. Hatano, and S. Kawabata, Phys. Rev. Lett. **95**, 107005 (2005).
4. X. Y. Jin, J. Lisenfeld, Y. Koval, A. Lukashenko, A. V. Ustinov, and P. Müller, Phys. Rev. Lett. **96**, 177003 (2006).
5. T. Matsumoto, H. Kashiwaya, H. Shibata, S. Kashiwaya, S. Kawabata, H. Eisaki, Y. Yoshida, and Y. Tanaka, Supercond. Sci. Technol. **20**, S10 (2007).
6. H. Kashiwaya, T. Matsumoto, H. Shibata, S. Kashiwaya, H. Eisaki, Y. Yoshida, S. Kawabata, and Y. Tanaka, J. Phys. Soc. Jpn. **77**, 104708 (2008).
7. S. Kashiwaya and Y. Tanaka, Rep. Prog. Phys. **63**, 1641 (2000).
8. T. Löfwander, V. S. Shumeiko, and G. Wendin, Supercond. Sci. Technol. **14**, R53 (2001).
9. S. Kawabata, S. Kashiwaya, Y. Asano, and Y. Tanaka, Phys. Rev. B **70**, 132505 (2004).
10. T. Yokoyama, S. Kawabata, T. Kato, and Y. Tanaka, Phys. Rev. B **76**, 134501 (2007).
11. S. Kawabata, S. Kashiwaya, Y. Asano, and Y. Tanaka, Phys. Rev. B **72**, 052506 (2005).
12. L. B. Ioffe, V. B. Geshkenbein, M. V. Feigel'man, A. L. Fauchere, and G. Blatter, Nature (London) **398**, 679 (1999).
13. S. Kawabata, S. Kashiwaya, Y. Asano, Y. Tanaka, A. A. Golubov, Phys. Rev. B **74**, 180502(R) (2006).
14. S. Kawabata, A. A. Golubov, Ariando, C. J. M. Verwijs, H. Hilgenkamp, and J. R. Kirtley, Phys. Rev. B **76**, 064505 (2007).
15. U. Eckern, G. Schön, and V. Ambegaokar, Phys. Rev. B **30**, 6419 (1984).
16. U.Weiss, *Quantum Dissipative Systems* (World Scientific, Singapore, 1999).
17. A. O. Caldeira and A. J. Leggett, Ann. Phys. (N.Y.) **149**, 374 (1983).
18. H. J. H. Smilde, A. A. Golubov, Ariando, G. Rijnders, J. M. Dekkers, S. Harkema, D. H. A. Blank, H. Rogalla, H. Hilgenkamp, Phys. Rev. Lett. **95**, 257001 (2005).

SILICON ISOLATED DOUBLE QUANTUM-DOT QUBIT ARCHITECTURES

D. A. WILLIAMS, M. G. TANNER, T. FERRUS, G. PODD, A. ANDREEV and F. CHAPMAN
Hitachi Cambridge Laboratory, Cavendish Laboratory
J.J. Thompson Avenue, Madingley Road, Cambridge, CB3 0HE, United Kingdom

We present both experimental and theoretical results on silicon isolated double quantum-dots (IDQDs) fabricated through trench isolation of silicon-on-insulator, with near-by single electron transistors (SETs) for charge state sensing. The IDQD can be described as an artificial molecule and is therefore capable of exhibiting quantum superposition states suitable for quantum information processing.

Keywords: Silicon isolated double quantum dots; quantum information processing.

A number of new ways of manipulating information, generically known as quantum information processing, have been postulated in the last 15-20 years. Several have been demonstrated experimentally, but there remains a large gap between principle and practice, particularly in quantum computation. The success of solid-state electronics and optoelectronics in classical information processing leads many to the conclusion that condensed matter systems will provide the best way of bridging this gap. We describe one of our approaches to making solid-state structures for quantum information processing, using nanoscale silicon-on-insulator, patterned into conjoined double quantum-dot structures. This silicon system has been chosen as a structure to carry out quantum operations and quantum information processing due to various advantageous characteristics: Low decoherence is expected due to low phonon coupling, the isolation of the active qubit device layer from an electron bath and the lack of direct electrical connection to leads. Essentially, the reasons why silicon technology is useful for classical information processing apply to the quantum equivalent.

The IDQD architecture is also easily scalable to many qubit devices as surface gates are not required to define the structure, as such the optimised architecture can easily be developed in the mature silicon processing industry. Both charge and spin qubit operation can be imagined in the IDQD, currently charge qubit operation is being pursued with the SET for readout. While charge qubit is currently being pursued, spin qubit operation can also be imagined in the IDQD.

This approach recognises that the length scales of commercial silicon manufacture are rapidly approaching the length scales of experimental quantum transport devices, and thus a re-examination of the relationship between industrial systems and experimental investigations is timely.

The devices are manufactured from silicon-on-insulator wafers, with a silicon layer typically 30-50nm thick, and P-doped in the range 10^{18} – 10^{20} cm^{-3}. The layer is then patterned using electron-beam lithography and reactive ion etching, followed by oxidation. The oxidation step serves two purposes; reducing the critical dimensions of the device and passivating surface states which otherwise are responsible for unacceptable levels of random telegraph noise.

Low-noise electrical measurements of IDQD-SET devices have been performed at 4.2K, including microwave stimulation. IDQD and SET features are successfully distinguished through the gate dependence of their DC responses and comparison to SET only devices. Semi-periodic charge polarisation of silicon IDQDs over a large gate range is observed and supported by device simulation, demonstrating device controllability and remote charge state sensing. A complex energy level structure is observed with microwave excitation and is shown to depend on the IDQD polarisation state. The polarisation response of the IDQDs is observed to be controllable with device geometry.

These sets of measurements demonstrate the possibility of a designable, controllable and scalable qubit structure.

Figure 1 shows a key control experiment which also illustrates some of the difficulty in interpretation of the response of these systems – the only output is the current through the electrometer as the environment is changed by altering the fields applied to the device whether statically or dynamically. The change in the environment inevitably alters the condition of the electrometer as well as the device of interest.

Detailed analysis of Coulomb peak gradients with field on an IDQD device using different gates showing suspected IDQD features appearing in different places with different gates has been performed. The multi-electron IDQD polarization response is complex, and a portfolio of simulation techniques has to be employed to understand the detailed electronic behaviour of these systems. To simulate the charge states of the IDQD we employ anisotropic effective mass

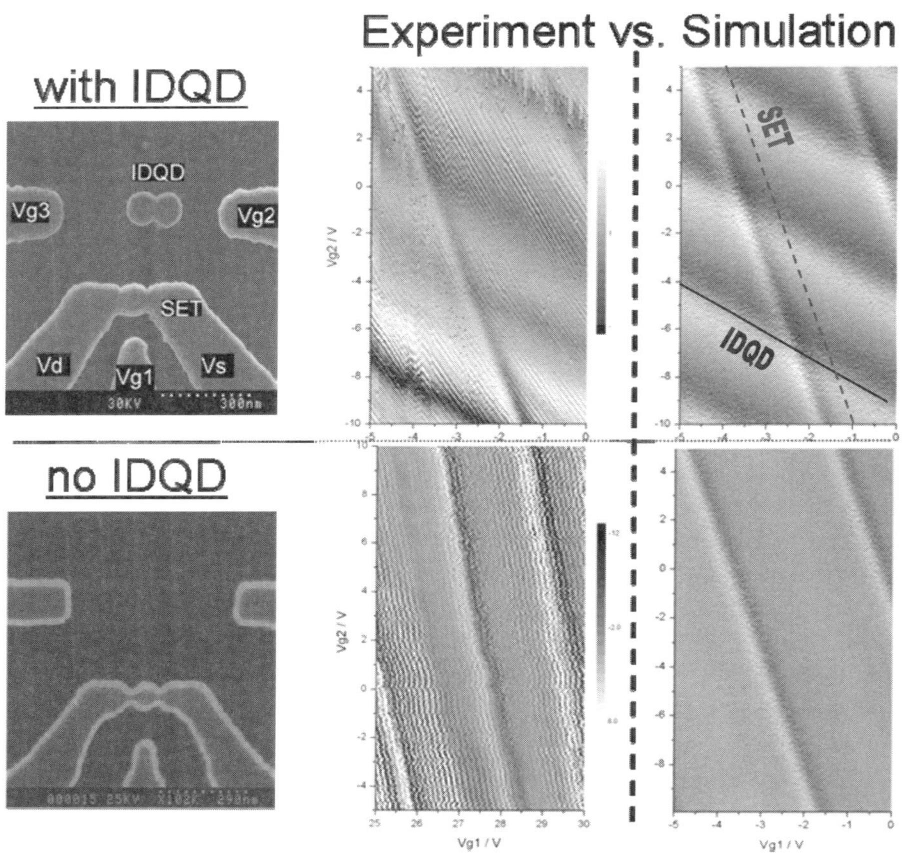

Fig 1 (Above): DC conduction response to varying device gating for IDQD-SET device (top) and SET only control device (bottom). Simulated results are shown for comparison (right). This allows discrimination between the response of the IDQD qubit structure and that of the measurement electrometer, which itself may have a complex response to environmental changes.

density functional theory based on Kohn-Sham equation to take account of realistic device geometry and applied gate voltages. We use the finite difference method to solve self-consistently the Kohn-Sham and Poisson equations for arbitrary potentials on a 3D grid. A series of calculations for different applied gate voltages are performed and compared with experimental SET field measurements. We demonstrate that a careful optimisation of the IDQD' structure is required for effective operation of the charge qubit device.

Figure 2 shows a charge density map of such a double-dot with 20 electrons – electron-electron interactions and structural asymmetries can result in complex shell structures and non-trivial changes in electron polarization as a response to changes in electron number or externally applied fields.

First experiments demonstrating coherence in these systems[2] were performed in a dilution refrigerator with high-bandwidth access to room temperature, and it was believed that the main sources of decoherence were due to noise transmitted down the access waveguide, and random back-action from the SET electrometer on the IDQD qubit. To test this, a custom cryogenic CMOS measurement circuit was developed to increase the measurement bandwidth without direct interaction with room temperature. (See Fig.3) This allows the implementation of the single shot technique to record the time-evolution of a coherent quantum state. Initial tests have been carried out successfully on a single electron transistor without the double dot, and are reported elsewhere in this volume.

As an additional method of exciting the structure, the devices were irradiated with pulsed and continuous microwaves, and selecting a high quality factor resonance in the spectrum. These resonances are shown to be

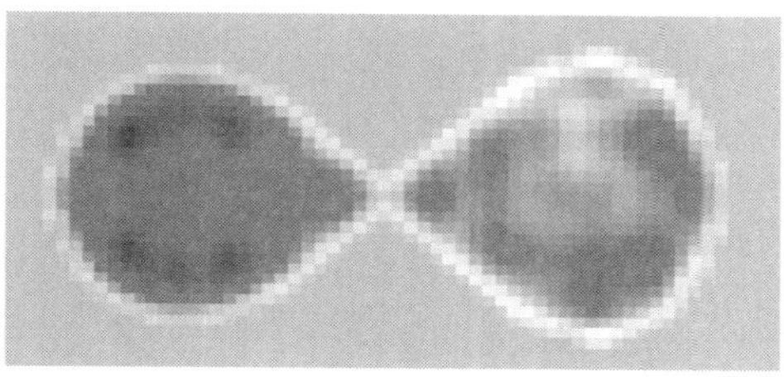

Fig. 2: A map of a slice of the total charge density for a 20 electron case at z=0.

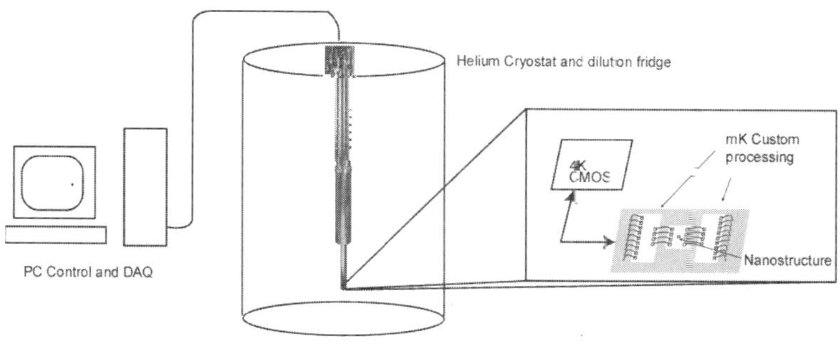

Fig. 3: A schematic view of the custom cryogenic measurement system, built within the UK National project CIQE project in collaboration with the Central Laboratories of the Research Councils, Cambridge University, Oxford University and Oxford Instruments Ltd. The key feature is custom CMOS circuitry at low temperature and custom electronics for control at room temperature.

linked to coherent electron excitations within the SET, and a rich structure of response is observed which is currently being explored.

Dilution refrigerator measurements are also under way at 31mK enabling the fine detail of the effect of the double dot on the SET to be characterised. Fridge measurements are being taken on an Isolated Double Dot (IDD) device; non-variation of a suspected IDD feature with Vsd strongly suggests that the features are indeed from the IDD.Clear signatures of charge redistribution in the double dot is seen in the SET. This consists in either an increase of conduction when the SET is in the blockade regime or a suppression of the Coulomb peak. This happens at specific qubit gate voltages.

Charge qubit device applications are straightforward using this method, with many possible geometries possible, as can be seen in Figure 4. When the IDQD is biased such that it is in a degenerate charging state a quantum superposition may be possible, suitable for use as a charge qubit. Pulsed DC gates and micro-wave excitation can both be used to access the superposition states. Current progress in stimulating coherent evolution between charge states will be reported and potential multiple qubit architectures are beiing examined, and are described in more detail in other papers in this volume[3,4].

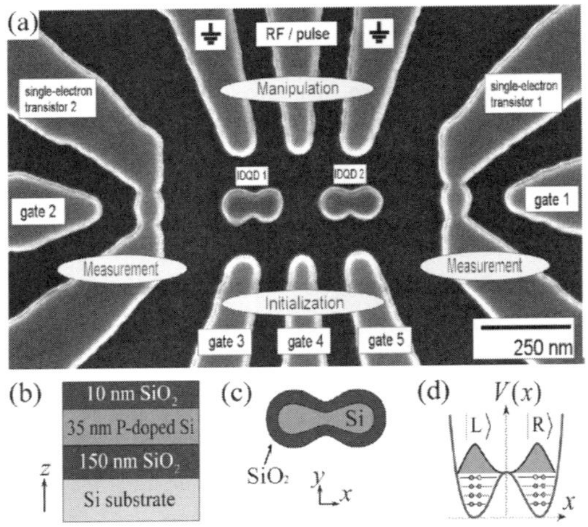

Fig. 4: A double IDQD quantum circuit with two sets of control electrodes and two SET readout electrometers. The response of the circuit is comples, but evidence of strong interaction between the two devices is seen. (a) Scanning electron micrograph of the circuit. (b) The silicon-on-insulator layer structure. (c) A schematic view of the IDQD (d) A schematic view of the electron states.

Acknowledgments

This work was partly supported by Special Coordination Funds for Promoting Science and Technology.

References

1. M. G. Tanner, D. G. Hasko, and D. A. Williams, Microelectronic Engineering **83**, 1818 (2006).
2. J. Gorman, D. G. Hasko, and D. A. Williams, Physical Review Letters **95**, 090502/1 (2005).
3. M.G. Tanner, G. Podd, P. Chapman, and D.A. Williams, Proceedings of ISQM-TOKYO'08 (2009)
4. P. Howard, A.D. Andreev, and D.A. Williams, Proceedings of ISQM-TOKYO'08 (2009)

CONTROLLED POLARISATION OF SILICON ISOLATED DOUBLE QUANTUM DOTS WITH REMOTE CHARGE SENSING FOR QUBIT USE

M. G. TANNER*, G. PODD, P. CHAPMAN and D. A. WILLIAMS
Hitachi Cambridge Laboratory, Cavendish Laboratory ,J. J.Thompson Avenue, Cambridge,CB3 0HE, UK
**E-mail: mgt33@cam.ac.uk*

Nanometre scale quantum dot devices in silicon consisting of a single electron transistor and an isolated double quantum dot with multiple control gates have been fabricated. The polarisation of the double dot when biased with the control gates has been observed indirectly through its effect on the conduction in the single electron transistor. Simulation has confirmed the experimental results and aided in determining the mechanism of this effect. This work demonstrates the possibility of using the isolated double quantum dot as a charge qubit.

Keywords: Silicon; Quantum Information; Quantum dot.

1. Introduction

This work is one part of a project aimed at demonstrating scalable quantum computing architectures in silicon[1, 2]. There have been many recent advances in solid state quantum computing[3, 4]. However, many of the systems successful for demonstration of fundamental principles are not suitable for scaling into practical quantum computers. This architecture has the advantages of industry standard materials, scalable device layouts due to trench isolated design, inherent low phonon coupling of silicon and long coherence times due to the isolation of the qubit from leads. The intended qubit is embodied in an Isolated Double Quantum Dot (IDQD), with read-out through a near by Single Electron Transistor (SET).

2. Device design

The fabrication process of the silicon IDQDs with integrated SET is mature. SOI wafer with a 45nm thick actively layer, highly doped with phosphorus ($2.9\times10^{19}cm^{-3}$), is lithographically patterned both optically and with e-beam for the high resolution elements. The pattern is transferred through reactive ion etching deep into the BOX layer, leaving trench isolated silicon device elements. A thermal oxide is subsequently grown to both decrease the internal dimensions of the quantum dots and passivate silicon surface states. The device is then contacted and mounted for measurement.

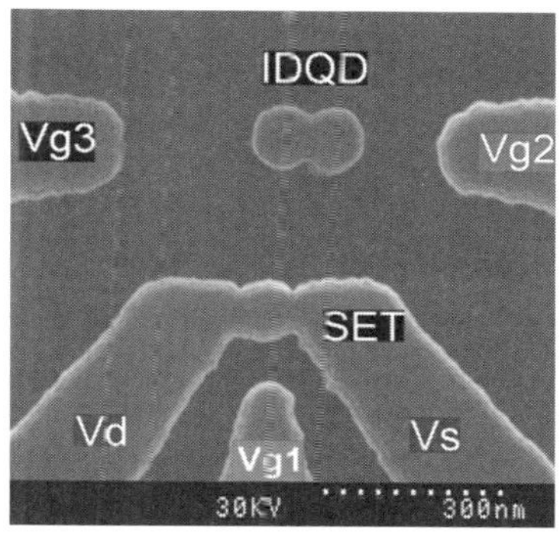

Fig. 1, SEM image of IDQD – SET device. Lighter regions are the active device elements.

The device includes two major elements, the SET and the IDQD. The SET, in the lower part of the micrograph Fig. 1, is a well characterized charge sensor with Coulomb blockade characteristics highly sensitive to the potentials induced by other device elements. This can be used for sensing the charge state of the IDQD.

The IDQD is composed of two strongly coupled quantum dots. If each dot can be considered to be an artificial atom, then the double dot is an artificial molecule. As this element is isolated from the device leads the total number of electrons is fixed, however polarization of electrons across the IDQD is possible.

Control of both the SET and IDQD is performed through biasing of the gates, V_{g1-3}. Each control gate will have an effect both on the SET and IDQD, however the proximity of gates to device elements will change the strength of this effect. It is expected that V_{g1} will have a much stronger coupling to the SET, while $V_{g2,3}$ will be more strongly coupled to the IDQD. It is this difference in gate coupling that allows us to distinguish between effects occurring in the IDQD and the SET.

3. Electrical measurements

Measurements are performed at 4.2K through dipping in liquid helium. Both source-drain bias and gate potential are controlled, while the only measurement is through the drain current of the SET.

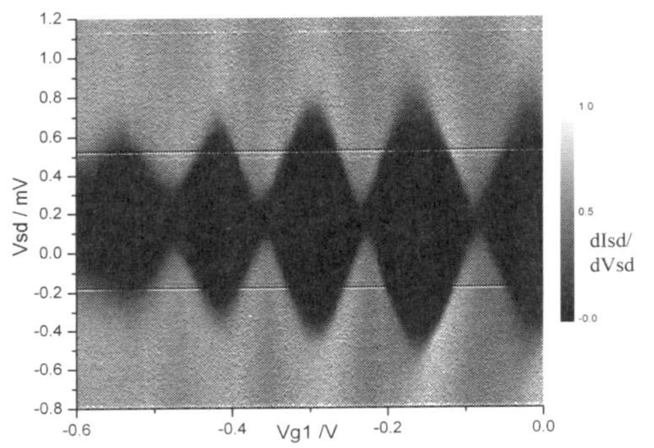

Fig. 2, Clear Coulomb diamonds characterise the SET. Data is differentiated with respect to V_{sd} for clarity.

Well defined Coulomb blockade can be observed in the SET (when measured in a parameter space separate from IDQD effects), as seen in Fig. 2. This demonstrates the use of the SET as a charge sensor suitable for measurement of the IDQD. If the SET is further characterised by sweeping the gates to much more negative potentials a continuing series of equally spaced Coulomb diamonds is observed (Fig. 3). This demonstrates the electron number in the dot is high, as expected for this highly doped system. As such the SET (and the IDQD) can not be operated in the few electron regime. However the accessible electrons in the system are normally just the few closest to the Fermi level while the many lower energy electrons are in a stable state.

In order to observe IDQD polarisation events it is necessary to sweep multiple gates over a large range and observe effects with differing gate dependences as in Fig. 4. In this plot it is possible to see two main sets of features. The more vertical features (also seen in the simulation inset of the figure) are due to the Coulomb blockade of the SET. It should be noted that many more finer pitched Coulomb blockade oscillations relating to those seen in the previous figures are also present in this data. These are not resolvable when looking over the large voltage range of Fig. 4, but the large features observed are also representative of the Coulomb blockade in the SET as discussed in other literature.

In addition to the SET features, many other oscillations in the device conduction are present, seen as darker features coming more horizontally across the plot (Fig. 4). The angle of these features on this plot tells us that they are much more dependent on V_{g2} than the SET oscillations. As has already been noted, the

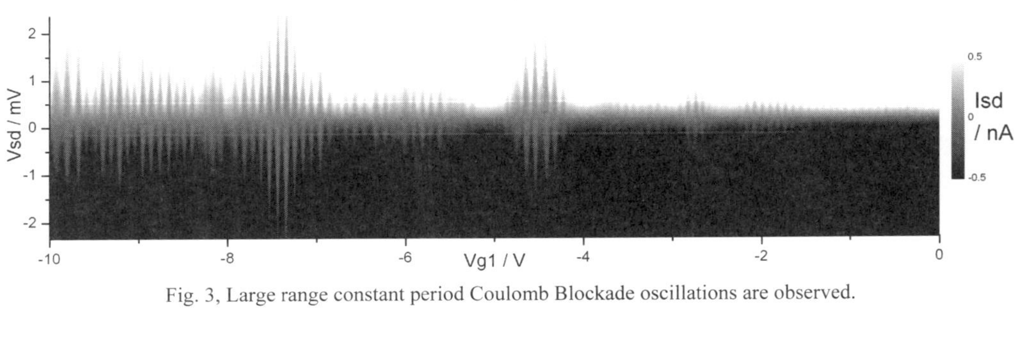

Fig. 3, Large range constant period Coulomb Blockade oscillations are observed.

IDQD is expected to be much more strongly coupled to V_{g2} than other device components. As such it is believed that these additional oscillations are a result of the IDQD.

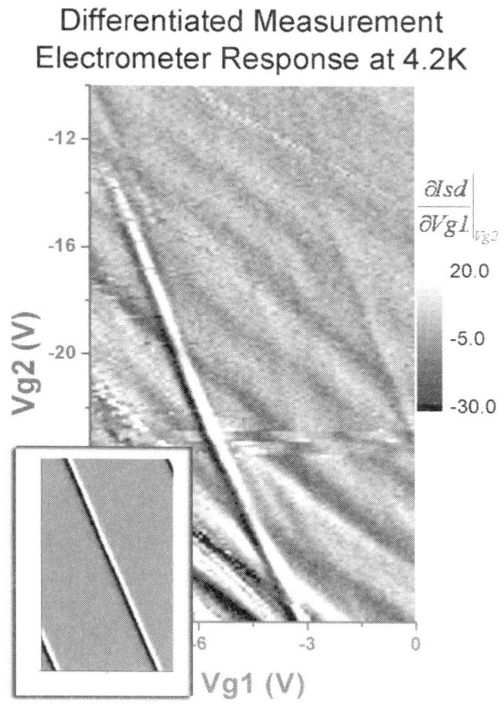

Fig. 4, Device conduction over large gate ranges, showing SET (more vertical) features and IDQD (more horizontal) features. Inset shows SET only simulation.

Through simulation and further measurements the semi periodic features observed here have been confirmed to be a result of IDQD polarization releasing the Coulomb blockade condition on the SET.

4. Simulations

Simulations of the device have been performed through a capacitance model to estimate the coupling between device elements and a single electron circuit simulator[5] to give conduction characteristics. The initially estimated capacitance values were adjusted to match the simulation results to the measured device characteristics.

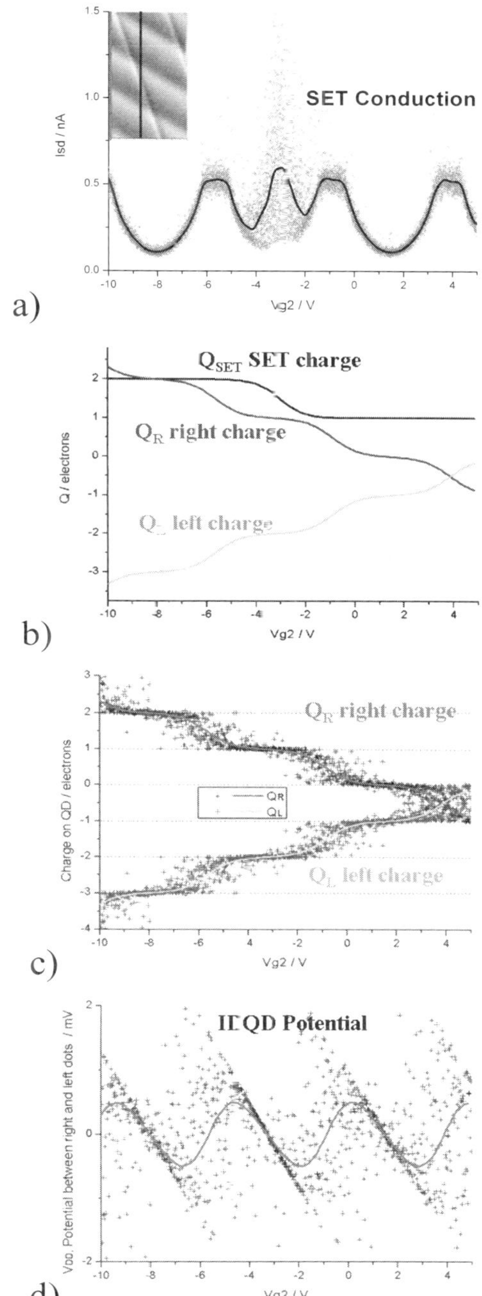

Fig. 5, Simulated SET conduction (a) as IDQD control gate V_{g2} is swept. IDQD potential charging. Simulated IDQD gate sweeps (Vg2). Peaks in conduction are observed when the charge number in the SET (b) or IDQD (b and c) changes Tunnelling stability plot is inset with the sweep marked for reference. One SET and three IDQD peaks are observed. Potential induced across the IDQD dots is shown in (d).

Simulation results demonstrate the SET sensing of charge degeneracy in the IDQD in Fig. 5a. The simulation method described previously yields these results but includes a high level of noise, therefore the current measurement result I_{sd} is plotted with an FFT filtered trend line to show the form of the response to be compared with the experimental data, which itself is smoothed by the 20ms integration time of the multi-meter. Charging state and potential across the dot structures is read directly from the simulation in order to support analysis of the experimental mechanisms.

The data shown in Fig. 5 demonstrates that when the number of electrons on the SET changes (top line, plot b) a corresponding Coulomb blockade peak is observed in the SET current (plot a), as expected from established theory. Additionally when the IDQD charge number changes (lower lines, plot b) a peak in SET conduction is observed. In order to understand the mechanism of this it is useful to consider the nature of the IDQD charging. The switch from an electron occupying the left dot to it occupying the right dot is not the smooth transition shown in plot b, but instead the electron is rapidly switching between the two states in the transition region if considered classically. This can be shown in the simulation by reducing the number of electron tunneling events modeled, plotted in c. When this rapid switching is occurring the potential applied to the SET is also rapidly switching, causing a freedom in the Coulomb blockade of the SET allowing conduction to occur.

The simulation gives the separation of SET charging levels to be approximately 12meV and the polarization energy of the IDQD to be 3meV in this device, in comparison to the thermal energy k_BT of 0.4meV at 4.2K, demonstrating well defined energy levels at this temperature. As the simulation has been fitted to the experimental results these energies can be inferred for the experimental device, demonstrating the value of the simulation in determining experimental parameters.

5. Summary

This paper has demonstrated the successful use of an SET for detection of charge states within a near by possible charge qubit. Simulation has supported the experimental results and given additional insight into the mechanisms involved. This work is one part of a continuing project to demonstrate the use of the IDQD for quantum information processing. Experiments are underway to demonstrate qubit operation, including the use of magnetic fields to access spin states, pulsed gate measurements for coherent state evolution, microwave addressing of excited states and multiple qubit devices to demonstrate a scalable quantum computer architecture with long coherence times. Further details of these experiments will follow.

Acknowledgements

This work was partly supported by an EPSRC PhD studentship, a CASE award from Hitachi Europe Ltd, and Special Coordination Funds for Promoting Science and Technology (Japan).

References

[1] J. Gorman, D. G. Hasko, and D. A. Williams, Phys. Rev. Lett. **95**, 090502/1 (2005).

[2] M. G. Tanner, D. G. Hasko, and D. A. Williams, Microelectron. Eng. **83**, 1818 (2006).

[3] Y. Nakamura, A. Pashkin Yu, and J. S. Tsai, Nature **398**, 786 (1999).

[4] J. R. Petta, A. C. Johnson, J. M. Taylor, E. A. Laird, A. Yacoby, M. D. Lukin, C. M. Marcus, M. P. Hanson, and A. C. Gossard, Science **309**, 2180 (2005).

[5] C. Wasshuber, H. Kosina, and S. Selberherr, IEEE Trans. CA. Design. Int. Circ. Sys, **16**, 937 (1997).

MODELLING OF CHARGE QUBITS BASED ON SI/SIO$_2$ DOUBLE QUANTUM DOTS

P. HOWARD[1], A. D. ANDREEV[1,2] and D. A. WILLIAMS[2]

1. *Advanced Technology Institute, University of Surrey, Guildford, GU2 7XH, United Kingdom*

2. *Hitachi Cambridge Laboratory, Hitachi Europe Ltd., Cambridge, CB3 0HE, United Kingdom*

The finite difference method and anisotropic effective mass density functional theory (DFT) are used to study the dependence of the electronic structure, wavefunctions and charge density on the width of the Silicon neck between the dots in an Si/SiO$_2$ isolated double quantum dot (IDQD) structure. An optimised algorithm for determining the self-consistent DFT potential was used. We demonstrate that the behaviour changes from strong to weak coupling between 2nm and 5nm width of the neck connecting the 30nm diameter dots.

Keywords: charge qubit; silicon; double quantum dot.

1. Introduction

Double Quantum Dot (DQD) structures are a promising candidate for building qubits, the basic unit of quantum computing. The use of DQDs as charge qubits was demonstrated recently[1,2]. In order to efficiently use a DQD for potential quantum computing applications, a detailed knowledge of its electronic structure is required. For quantum dots more than 10nm in diameter, using sophisicated approaches based on ab initio atomistic calculations is too slow and inefficient for modelling the device performance. Therefore, a simpler but physically justified model that takes account of mentioned important factors is required and this paper seeks to use such an approach. Recently[3], we have developed demonstrate an anisotropic effective mass density functional theory (DFT) model, which allows us to model dots sized between 5 and 100 nm.

This paper is focused on modelling a structure similar to that experimentally studied in Ref.[1] and consisting of an Si/SiO$_2$ isolated double quantum dot surrounded by electric gates and a Single Electron Transistor (SET). We studied the dependence of the electronic structure, wavefunctions and charge density on the width of the conducting neck between the dots. We calculate the electronic structure of such a device, including self-consistent consideration of many body effects through DFT for a varying width conducting necks between the dots.

The geometry of the structure determines the quantised states, and modelling these states provides an insight into what is happening inside the quantum dots. This work seeks to model the system, and numerically determine the available electron wavefunctions in a given isolated DQD structure with the aim to investigate what happens when the width of the neck connecting the two dots is varied.

2. DQDs as charge qubits

The two basis states required for use of a DQD as a qubit are related to the electrons existing in one dot or the other (more precisely, to two states of the electron distribution in the DQD). Manipulation of the qubit may be done by varying the electric potential of the gates to change the charge distribution in the DQD system, i.e. to change the state of the qubit between the two basis states. The state of the qubit can be determined by measuring the current through the SET, which is very sensitive to the charge distribution in the DQD. Long decoherence times have been observed in isolated DQDs[1] compared to the structures with dots formed by the gates[2].

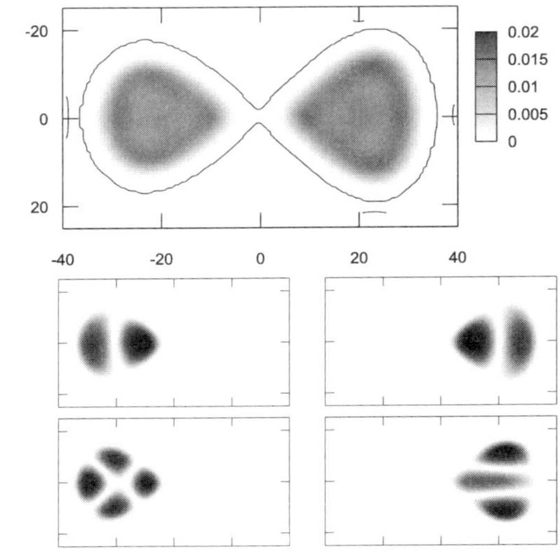

Fig. 1. Upper plot - the total charge density of the electrons localised in the DQD in units of e/nm^3; there are in total 101 electrons in the dot and the same number of electrons are trapped on the surface; the boundary of the dot is outlined. Lower quartet of plots - the examples of the wavefunctions for several levels. The DQD neck is 2nm, the lenght scale on the upper graph is given in nm.

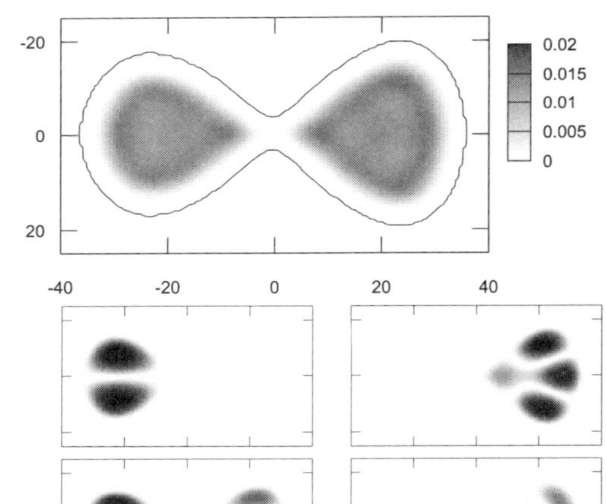

Fig. 2. Same as in Fig.1, but for neck width of 5 nm.

3. Model

We use the Finite Difference Method to solve the Schrodinger and Poisson equations for arbitrary potentials on a 3D grid. A novel method is used to solve the Poisson equation in free space: numerical integration is used to determine the boundary values, which are then applied as Dirichlet boundary conditions to the Poisson equation and solved as a linear algebra problem. The DQD potential cross-section is approximated by a spiric section, with the effect of an applied electric field as a potential difference between the dots. We use full anisotropic effective mass Density Functional Theory, enabling the simulation of 5-100nm Quantum Dots. Surface charges around the DQD at the Si/SiO_2 interface are taken into account as a region of continuous charge near the interface. The outline of the method is given in Ref.[3].

4. Results

We look at the effects of DQD geometry on the charge density of 30nm diameter dots. The shape of the DQD is assumed to be given by the spiric section, which is a type of toric section. It is formed by cutting the torus with a plane perpendicular to the sweeping circle. We found that the spiric section gives a very good fit to experimentally fabricated DQD shapes, giving a simple analytical form of that shape at the same time. The DQD is assumed to be doped[1], therefore some of the electrons may be trapped in the silicon and silicon dioxide interface creating the surface charge; the total DQD structure is electrically neutral[1]. We define the charge per dot as:

$$Q_{pd} = \int_{v_d} [\rho(\mathbf{r}) + \rho_{imp}(\mathbf{r}) + \rho_{sc}(\mathbf{r})]\, d\mathbf{r} \quad (1)$$

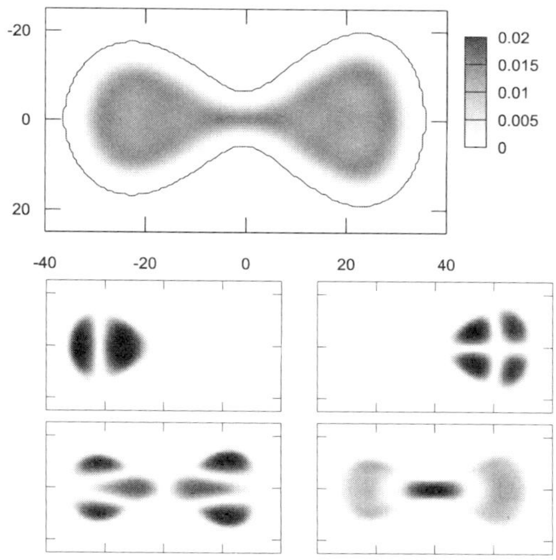

Fig. 3. Same as in Fig.1, but for neck width of 10 nm.

where v_d is the volume region of one dot, delineated by the slice of the $x = 0$ plane; ρ_{imp} is the impurity charge density and ρ_{sc} is the surface charge density. Thus Eq.1 gives a value of the total charge in each dot.

Narrowing the neck between the dots reduces the coupling between the dots, pushing the device towards the weak coupling regime (Fig.1). The charge density of the system with neck width of 2nm and below (top plot in Fig.1) shows no charge within the neck region (the DQD shape is outlined). All the wavefunctions for different DQD energy levels (examples shown in lower quartet in Fig.1) are localised to one dot or the other. In an intermediate stage (Fig.2), some wavefunctions become delocalised across the two dots but do not inhabit the neck region and the charge density remains near zero in the neck region. By contrast, in the strong coupling regime, the wavefunctions are delocalized across the whole DQD, and include electrons entirely localised to the neck, see Fig.3; at a 10 nm neck width, the charge density is high in the neck region. Some wavefunctions are localised to one dot (upper left and right plots in the lower plot quartet in Fig.3), some cover both, and some are localised to the neck region

5. Conclusions

In this paper we have developed a physical and numerical model to calculate the electronic structure of DQD devices for quantum computing applications. We have demonstrated that in the weak coupling regime electrons move wholly from one dot to the other when changing the potential difference between the dots by changing the gate voltage. In the weak coupling regime the wavefunctions for a given state are highly confined to a single dot. In the strong coupling regime, however, the DQD acts more like one large dot, and the electrons flow smoothly from one dot to the other under the variation of the potential difference between the two dots.

We thank Hitachi Cambridge Laboratory and the EPSRC for financial support.

This work was partly supported by Special Coordination Funds for Promoting Science and Technology.

References

1. J. Gorman, D.G. Hasko and D.A. Williams, Phys. Rev.Lett., **95**, 090502 (2005).
2. Hayashi, Fujisawa, Cheong, Jeong and Hiragama, Phys. Rev Lett., **91**, 226804 (2003).
3. P. Howard, A.D. Andreev and D.A.Williams, Physica Status Solidi (C), **5**, 3156 (2008).
4. W. Kohn and L. J. Sham, Phys. Rev., **140**, A1133 (1965).
5. J. Perdew and Y. Wang, Phys. Rev. B, **45**, 13244 (1992).
6. C. Herring, Rev. Mod. Phys. **34**, 631 (1962)
7. S. Kristyan and P. Pulay, Chem. Phys. Lett. **229**, 175 (1994)
8. T.R. Walsh, Phys Chem. Chem. Phys. **7**, 443 (2005)
9. A. Wensauer, O. Steffens, M. Suhrke and U. Rössler, Phys. Rev. B **62**, 2605 (2000)

InAs BASED QUANTUM DOTS FOR QUANTUM INFORMATION PROCESSING: FROM FUNDAMENTAL PHYSICS TO 'PLUG AND PLAY' DEVICES

XIULAI XU*, ALEKSEY ANDREEV†, FREDERIC BROSSARD, KIYOTAKA HAMMURA AND DAVID WILLIAMS

Hitachi Cambridge Laboratory, Hitachi Europe Ltd., JJ Thomson Avenue, Cambridge CB3 0HE, United Kingdom

†*Also at ATI, University of Surrey, GU2 7XH, United Kingdom*

**E-mail: xx757@cam.ac.uk*

Semiconductor quantum dots have attracted much interest in implementing solid-state quantum information processing. Using InAs based quantum dots, we demonstrate quantum coupling between two stacked quantum dot molecules in electroluminescence, controlling Stark shifts of single quantum dots in electroluminescence, and 'plug and play' single photon emission at telecommunication wavelengths.

Keywords: Semiconductor quantum dots; single photon sources

1. Introduction

Semiconductor quantum dots have attracted much interest in implementing solid-state quantum information processing because of their 'artificial atom' nature, such as quantized energy levels, high stability, electroluminescence and long spin coherence time. Single-photon[1,2] and entangled-photon sources[3] have been demonstrated with InAs based quantum dots, showing potential applications in quantum key distribution and quantum teleportation. Also, one qubit or two-qubit gates have been achieved with coherent control methods[4]. However, many issues still remain to be solved, for example, how quantum-dot based qubits can scale up, how to engineer the energy of single quantum dots in a photonic structure to achieve optical quantum circuits, how easy these devices can be used for real applications. In this paper, we discuss our recent work on manipulating the exciton energy of quantum dots, demonstrating quantum coupling between two stacked quantum dots, and implementing single-photon sources at telecommunication wavelength to realize a 'plug and play' device.

2. Stark shifts from quantum dots in electroluminescence

In our previous work, we obtained electrically pumped single photon sources[1]. However, the extracting efficiency was low. Recently, photonic crystals and distributed Bragg reflectors have been used to enhance the collecting efficiency. Coupling quantum dot with photonic crystals, both weakly coupling and strongly coupling regimes have been demonstrated to enhance single photon emission[5] and exciton-photon entanglement[6]. To achieve the coupling between quantum dots and cavities, temperature tuning is normally used to move the peak position of quantum dot emission to match the cavity modes[7,8]. However, an increased temperature reduces the coherence time of the system. One way to shift the excitonic transition in quantum dots is to use the Stark shift, the energy shift due to an electric field applied across the dots. To tune the exciton energy in electroluminescence (EL), it is desirable to have separate electrodes to inject electrons and holes for EL, and to manipulate the electric field for Stark shift. Figure 1(a) shows a typical elec-

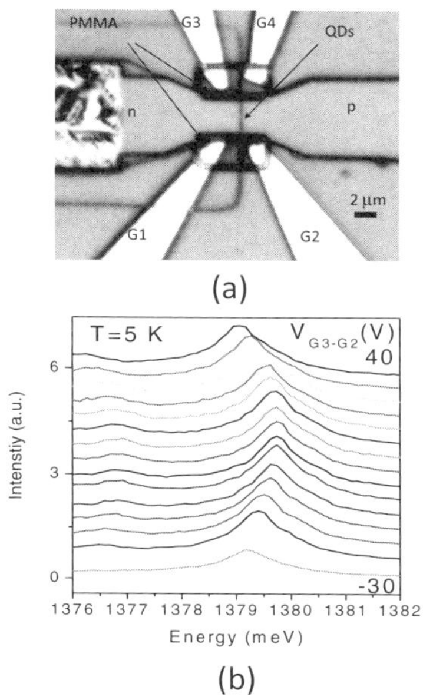

Fig. 1. (a) Lateral p-i-n junction with side gates (b) EL spectra as a function of bias voltage of side gates in a step of 5 V. Forward injection current is 220 μA.

troluminescence device with four side gates. The gates were used to control intensity and direction of the electric field across the quantum dots in the p-i-n junction[9]. Single exciton emission can be observed from single quantum dots, as shown in Fig. 1(b). With a forward bias current at 220 μA, A Stark shift about 500 μeV can be observed with a bias voltage around 40 V on the side gates between G2 and G3. The asymmetric Stark shift on the bias voltage is due to the build-in dipole in the quantum dot structure.

3. Quantum coupling in quantum dot molecules

Double quantum dots (also called quantum dot molecules) are potentially applicable to scale up quantum dot based quantum computation[10,11]. With considering strain, piezoelectric field, and potential, the band structure has been calculated with a plane-wave expansion method in InAs quantum dots[12]. The coupling strength between the two dots is calculated with different dot separations and with an applied electric field, resulting in different wavefuntions in the two dots, as shown in Figure 2(a). When the electron wavefuction occupies two dots, the bonding and antibonding states are formed, resulting in an anticrossing when the electric field is tuned.

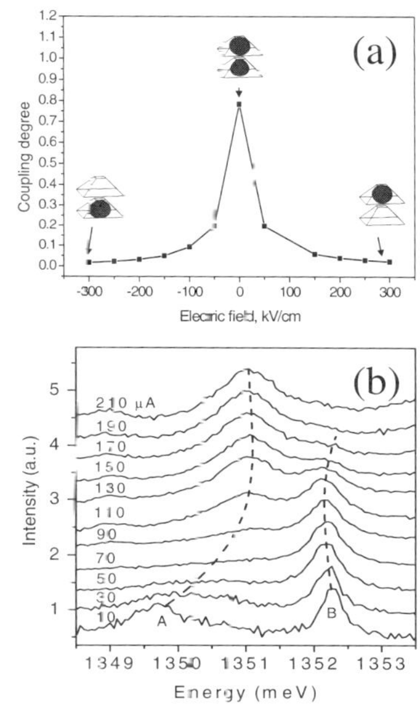

Fig. 2. (a) Coupling strength as a function of electric field. Inset: electron wavefunctions corresponding to the electric fields. (b) Anticrossing of the electroluminescence from the direct exciton and the indrect exciton in a quantum dot molecule.

Figure 2(b) shows the electroluminescence spectra as a function of injection current from a lateral p-i-n device, in which two stacked layers of quantum dots (separated by around 10 nm) are sandwiched[13]. At low-current injection, two peaks around 1.3496 and 1.3523 eV (A and B) are observed. With increasing current, peak A is strongly blue-

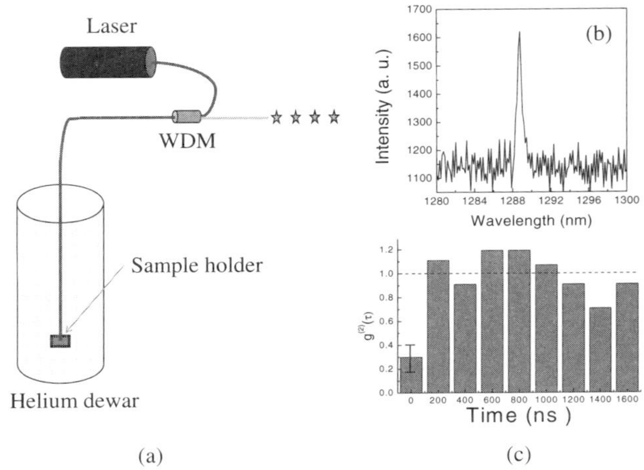

Fig. 3. (a) 'plug and play' single photon system. (b) PL spectrum for a quantum dot emitting at telecommunication wavelength using the fiber system (c) Second order correlation function of the exciton peak in (b).

shifted and its intensity decreases, while peak B is slightly red-shifted with increasing intensity. By increasing the current above 90 A, peak A appears again with increasing intensity while peak B quenches rapidly and shifts to the higher-energy side. The minimum energy splitting between A and B is about 1 meV with a current of 90 A at 2.9 V. In this structure for a dot-dot separation around 10 nm, two kinds of exciton can be observed, direct and indirect. The direct exciton with the electron and hole wavefunctions localized within one dot (small electronic dipole moment) is less sensitive than the indirect exciton with electron and hole in different dots (large electronic dipole moment). At particular dot to dot separations, it is possible to tune the indirect exciton to the resonance of the direct exciton, and to observe bonding and antibonding states in QDMs. With these considerations, the peak A and peak B in Figure 2(b) are ascribed to indirect exciton and direct exciton emission in a coupled QDM. The coupling can be controlled with applied forward bias. Clear anticrossing is a signature of coupling of the two quantum dots in electroluminescence.

4. 'plug and play' single photon sources

From the application point of view, fiber based single photon sources are desirable for quantum key distribution. We designed a 'plug and play' source of single photons, with full integration to a single-mode optical fibre, as shown in Figure 3(a)[14]. One end of the fibre is attached to the top of an InGaAs/GaAs quantum dot wafer. The other end is connected via a wavelength-division multiplexing system to two separate fibres: one for carrying excitation light, and the other for emitted light. The emitted light after an optical tunable filter (OTF) was input into a Hanbury-Brown and Twiss (HBT) interferometer to verify the single photon emission. Great stability of nearly (but not limited) one month has been demonstrated.

Recently, an InAs QD wafer with a 1.3 μm wavelength emission is used in this fiber system[15]. The QD layer is embedded in a 1-lambda cavity between 1-period (top) and 15-period (bottom) GaAs/$Al_{0.9}Ga_{0.1}As$ distributed Bragg reflectors, in order to enhance the extraction efficiency. Figure 3(b) shows a typical photoluminescence spectra from a patterned quantum dot wafer with micro pillars with diameters ranging from 1 to 4 μm, as detected from the fiber output via the tunable optical filter. The average photon counts of the filtered exciton emission are around 3000 cps, detected with an InGaAs APD. The collection efficiency is around 0.094 % for this fiber system. The filtered single exciton emission was fed into the HBT setup. The correlation result is shown in Fig. 3 (c) by subtracting the background noise. The second order correlation function at zero delay $g^{(2)}(0)$ is around 0.3, indicating single photon emission with this fiber system. The following reasons might have prevented us from obtaining a $g^{(2)}(0)$ close to zero: First, the background counts given by a detuned OTF position are not exactly the same as that given by the OTF of the exciton peak. Second, the random electronic noise generated by the long delay cable can be picked up and amplified by the inline signal amplifier.

Because of the large dark counts of InGaAs detectors, the detectors are usually operated in Geiger mode and gated at 5 MHz. With this detector, it is difficult to verify how fast the single photon sources can be operated. To examine the pulse rate of our single-photon gun at 1.3 μm, time resolved photoluminescence (PL) measurements were performed with two separated optical pulses at different wavelengths. Under quasi-resonant pumping, the single photon intensity from first pulse can be well isolated from the second pulse with a separation of 1 ns. This indicates that the single-photon sources can be operated around 1 GHz.

5. Summary

In this paper, we summarized the implementation of quantum information processing using InAs quantum dots. The electroluminescence energy of single excitons can be tuned with side gates independently via the Stark shift. This provides the possibility for electrically tuning the exciton energy to the resonant mode of a photonic cavity. To scale up the quantum dot based quantum computation, quantum coupling has been calculated between quantum dot molecules and demonstrated experimentally in electroluminescence. Integrating the photoluminescence into fiber system, 'plug and play' single photon sources have been achieved at telecommunication wavelengths. These photon sources show a great stability and can be operated around GHz, which are easily integrated into fiber based quantum key distribution system.

Acknowledgments

We wish to acknowledge M. Hopkinson, H. Y. Liu, C Stanley and A. Fiore for growth of quantum dot wafer.

References

1. X. L. Xu, D. A. Williams, and J. R. A. Cleaver, Appl. Phys Lett. 85, 3238 (2004).
2. Z. Yuan, et al., Science 295, 102 (2002).
3. M. R. Stevenson, et al.,*nature* **439**, 179 (2006).
4. X. Q. Li et al.,*Science* **301**, 809 (2003).
5. P. Michler, et al., Science 290, 2282 (2000).
6. K. Hennessy et al., *Nature* **445**, 896 (2007).
7. T. Yoshie, et al., *Nature* **432**, 200 (2004).
8. J. P. Reithmaier et al., *Nature* **432**, 197 (2004).
9. X. L. Xu, et al., *New Journal of Physics* **10**, 053036 (2008).
10. D. Loss and D. P. DiVincenzo, *Phys, Rev. A* **57**, 120 (1998)
11. J. Gorman, D. G. Hasko and D. A. Williams, *Phys. Rev. Lett.* **95**, 090502 (2005)
12. X. L. Xu. A. Andreev and D. A. Williams, in *Self-Assembled Quantum Dots*, ed. Z. Wang (Springer, 2008).
13. X. L. Xu, A. Andreev, D. A. Williams, and J. R. A. Cleaver, *Appl. Phys. Lett.* **89**, 091120 (2006).
14. X. L. Xu, et al., *Appl. Phys. Lett.* **90**, 061103 (2007)
15. X. L. Xu, et al., *Appl. Phys. Lett.* **93**, 021124 (2008)

QUANTUM ASPECTS IN SUPERCONDUCTING QUBIT READOUT WITH JOSEPHSON BIFURCATION AMPLIFIER

HAYATO NAKANO*, SHIRO SAITO and KOUICHI SEMBA
NTT Basic Research Laboratories, Atsugi-shi, Kanagawa 243-0198, Japan
**E-mail: nakano@will.brl.ntt.co.jp*

HIDEAKI TAKAYANAGI
Department of Applied Physics, Tokyo University of Science, Shinjuku-ku, Tokyo 162-8601, Japan
Int. Center for Materials Nanoarchitectonics, NIMS, Tsukuba-shi, Ibaraki 305-0003, Japan

The qubit readout process with a Josephson bifurcation amplifier (JBA) is analyzed quantum-mechanically. We calculate and clarify the dynamics of the density operator of a driven nonlinear oscillator and a qubit coupled system during the measurement. In purely quantum-mechanical JBAs, the bifurcation is impossible. It requires finite decoherence. When the JBA is coupled to a qubit, that is initially in a superposition state, we can observe the entanglement formation between the qubit and the probe (JBA) and the projection into one of two separable states at the moment of the JBA transition begins. To readout the measurement result, however, we must wait until the two JBA states are macroscopically well separated. The waiting time is determined by the strength of the decoherence in the JBA. This would give a trade-off on the strength of the JBA decoherence for a precise readout.

Keywords: Josephson Bifurcation Amplifier; Qubit; Readout Process.

1. Introduction

When we use a Josephson bifurcation amplifier (JBA) as the probe of the superconducting qubit measurement, it provides non-destructive and high visibility readout. JBA technique is now widely and successfully used in actual experiments[1]. The bifurcation phenomenon enhances a small difference in operation conditions[2]. A *classical* nonlinear oscillator becomes bistable under an appropriate driving force. One stable state has a small amplitude (low-amplitude state), and the other has a larger amplitude (high-amplitude state). The critical driving force f_c or the critical detuning δ_c for the transition between these two states is very sensitive to small changes in the operational parameters of the oscillator. Figure 1 shows that the hysteretic behavior of a JBA when we increase or decrease the driving force continuously. When using a JBA as a qubit state readout probe, the JBA detects a small change depending on the qubit state.

A classical analysis of a JBA has been established for long time. However, in order to understand the qubit readout process, a quantum mechanical analysis is indispensable because quantum properties, such as, entanglement formation between the qubit and the probe (JBA) or the decoherence in the composite system, are essential in the readout process.

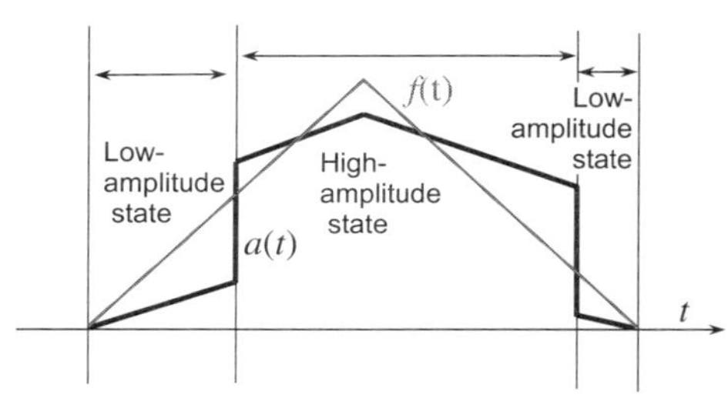

Fig. 1. Hysteretic behavior of the oscillation amplitude $a(t)$ of a JBA when the driving amplitude $f(t)$ is operated (schematic).

In the present work, we analyze the quantum-mechanical time evolution of a JBA, and investigate what happens during the process of the qubit state readout with a JBA by analyzing dynamics of the qubit-JBA composite system. This paper is organized as follows. This section is the introduction. We briefly discuss the impossibility of the JBA bifurcation phenomenon in purely quantum-mechanical situation and that under finite decoherence reproduce bifurcation-like behavior in a quantum-mechanical JBA, in Sec. 2. Section 3 gives the framework of our analysis on the dynamics during the qubit readout process with a JBA. We show the calculation results and analyze the quantum-mechanical important properties in the readout process in Sec. 4. In Sec. 5, we summarize our results.

2. Bifurcation in a Quantum-mechanical Josephson Bifurcation Amplifier

In a highly quantum-mechanical JBA case, tunneling between classically stable states destroys the criticality in a classical oscillator. This type of phenomenon has been precisely discussed[3]. However, actual JBA measurements are made with more classical conditions. Rigo *et al.*[4] investigated such an oscillator.In order to obtain quantum information more directly, here, we analyze the time evolution of a JBA and a qubit during the readout process.

A JBA can be modeled with a Hamiltonian;

$$H_{\mathrm{J}} = (\Omega - \omega)n_a + \alpha {n_a}^2 - \frac{1}{2}f(a^\dagger + a) \quad (1)$$

in a rotating frame approximation, where, $a^\dagger(a)$ is the creation (annihilation) operator of the Josephson plasma oscillation. $n_a = a^\dagger a$, and Ω is the linear resonant frequency of the JBA oscillator. ω is the driving frequency, which is slightly smaller than Ω by the detuning $\delta \equiv \Omega - \omega$. f is the driving strength, and $\alpha(> 0)$ is the nonlinearity.

The quasienergy-states (eigenstates of the Hamiltonian Eq. (1)) are calculated and we find that if the JBA is initially in the ground state without driving, it never moves to the high-amplitude resonant state even if we increase the driving field because the JBA state only moves along the initial quasienergy-state and never jumps to the quasienergy-state which the high-amplitude resonant state belongs.

This suggests that perturbation from outside the system the bifurcation phenomenon is reproduced. Now, we introduce linear loss in the oscillator (JBA). The time evolution of the system (qubit-JBA) is governed by a Liouville equation:

$$\frac{\mathrm{d}\rho}{\mathrm{d}t} = \frac{1}{i}[\rho, H] + \frac{\Gamma}{2}(2a\rho a^\dagger - a^\dagger a\rho - \rho a^\dagger a), \quad (2)$$

where ρ is the density operator of the system, and Γ is the relaxation rate due to the linear loss in the JBA. The Q-value is given by Ω/Γ.

We carried out numerical calculations of JBA dynamics *without a qubit*, and show the trajectory of the JBA in the phase-space (Fig. 2). The parameters $\delta = 0.007\Omega$, $\alpha = 8 \times 10^{-5}\Omega$, $Q = 2500$ are used, and f is operated as $0 \to 0.025\Omega \to 0$. These parameters are similar to those used in actual experiments [1]. However, in order to emphasize quantumness and discuss the influences of decoherence, δ and Γ are a factor of 10^{-2} times smaller than real cases.

In this case, the critical driving f_c is approximately 0.011Ω. Our calculation provides a lot of quantum-mechanical information about the JBA transition. Increasing the driving force f, we found that the purity decreases abruptly ($t = \tau_a$). This corresponds to the beginning of the transition from $|G\rangle_J$ to $|E\rangle_J$ of the JBA. This shows that the transition needs an intense emission/absorption of energy to/from an external energy bath. After the rapid decrease, the purity recovers to some extent and the

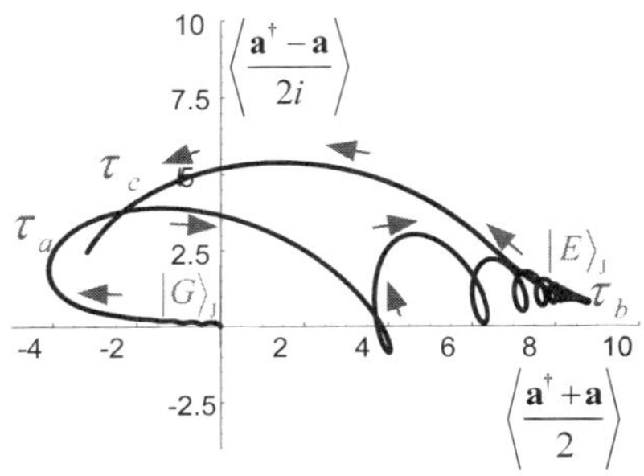

Fig. 2. The trajectory of the quantum expectation value in the phase space of the oscillator when the driving force f is operated as shown in Fig. 1. The starting point is $|G\rangle_J$ without driving, and the right convergence point is $|E\rangle_J$.

JBA approaches to another classically stable state $|E\rangle_J$ $(t = \tau_b)$.

3. Qubit Readout Process with a JBA

What we actually observe is the macroscopic state of the JBA, and merely *postulate* the qubit state. Therefore, the process by which the entanglement is formed and split into separable states due to decoherence ("projection") is very important for understanding the readout process[5].

The qubit-JBA composite system is approximately expressed by the Hamiltonian

$$H = H_J + k\sigma_z n_a + H_q, \quad H_q = \frac{1}{2}(\varepsilon\sigma_z + \Delta\sigma_x) \tag{3}$$

where H_q is the Pauli operator representation of the qubit. k is the interaction constant between the qubit and the JBA. The qubit state (σ_z) slightly changes the effective detuning, resulting in a change in the critical value f_c. By detecting the change in f_c, we can distinguish the qubit state.

The above description about the JBA readout is literally correct for a classical readout. In other words, when the qubit is in a definite state of the measured physical quantity. However, in a quantum-mechanical readout, there is a case that the initial qubit is in a superposition of measured quantity. In this case, the readout process becomes dynamics of the qubit-probe (JBA) composite system, consisting 'entanglement formation' and 'division into separable states'. This dynamics is not so trivial depending on the readout setup.

4. Results and Discussion

We carried out calculations of the dynamics for the qubit-JBA coupled system during the qubit readout process.JBA parameters are the same as for the above example. The initial state is a separable state; $(\frac{1}{\sqrt{2}}|g\rangle_q + \frac{1}{\sqrt{2}}|e\rangle_q) \otimes |G\rangle_J$, that is, the qubit is in a superposition. Here, $|g\rangle_q$ and $|e\rangle_q$ are the ground and excited states of the qubit, respectively. Qubit parameters are $\epsilon = 0.2\Omega$, $\Delta/\epsilon = 1/2$. The coupling between the qubit and the JBA is set at $k = 0.001\Omega$. The driving force f is increased from 0 to 0.012Ω (slightly larger than f_c of the JBA) and maintained. This parameter set gives a typical behavior of successful qubit readout.

We adopt$E \equiv \mathrm{Tr}\left[\rho^2\right] - \mathrm{Tr}\left[(\mathrm{Tr}_q[\rho])^2\right]$, as a measure of the entanglement between the qubit and the JBA, where ρ is the density operator of the qubit-JBA coupled system, and $\mathrm{Tr}_q[\cdots]$ denotes taking partial trace about qubit degrees of freedom.

The time variation of the entanglement measure E is shown in Fig. 3. This process can be schematically expressed as

$$\begin{aligned}
&\rho(0) = |G\rangle_{JJ}\langle G| \otimes \\
&\left(\frac{1}{\sqrt{2}}|g\rangle_q + \frac{1}{\sqrt{2}}|e\rangle_q\right)\left(\frac{1}{\sqrt{2}}{}_q\langle g| + \frac{1}{\sqrt{2}}{}_q\langle e|\right) \\
&\rho(\tau_1) = \frac{1}{2}\left(|G\rangle_J|e\rangle_q + |G'\rangle_J|g\rangle_q\right) \\
&\qquad \left({}_J\langle G|_q\langle e| + {}_J\langle G'|_q\langle g|\right) \\
&\rho(\tau_2) = \frac{1}{2}|G\rangle_J|e\rangle_{qJ}\langle G|_q\langle e| + \frac{1}{2}|G'\rangle_J|g\rangle_{qJ}\langle G'|_q\langle g| \\
&\rho(\tau_3) = \frac{1}{2}|G\rangle_J|e\rangle_{qJ}\langle G|_q\langle e| + \frac{1}{2}|E\rangle_J|g\rangle_{qJ}\langle E|_q\langle g|.
\end{aligned} \tag{4}$$

Entanglement formation and "projection" correspond to the second $(t = \tau_1)$ and third

($t = \tau_2$) lines of Eq. (4), respectively. At this moment (τ_2), however, it is impossible to obtain any information about the qubit from the observed probe (JBA) state because $|G'\rangle$ closely resembles $|G\rangle$ in a classical mechanical sense. When we increase the driving force, one JBA state $|G'\rangle_J$ moves to $|E\rangle_J$. In contrast, the other $|G\rangle_J$ does not move significantly. Then (τ_3), we can easily distinguish $|E\rangle_J$ or $|G\rangle_J$. This is the completion of the readout.

The *entanglement of formation* for each eigenstate of ρ reaches almost unity when it reaches its maximum. This means that the correlation between the qubit and JBA state becomes almost perfect via the interaction between them. As a result an ideal JBA readout exhibits 100% visibility if the qubit relaxation discussed below is negligible.

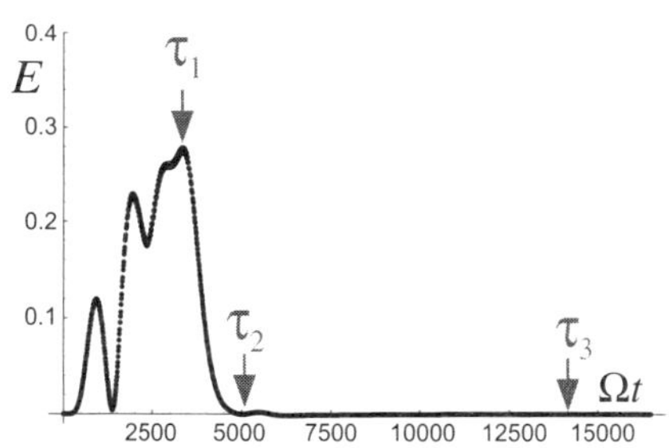

Fig. 3. Time variation of the entanglement between the JBA and the qubit. τ_1: entanglement formation, τ_2: projection, τ_3: end of the readout.

The non-commutation relation between the qubit Hamiltonian and the interaction Hamiltonian gives the readout backaction to the qubit. When the qubit gap Δ is much smaller than other energies, the JBA readout causes only *pure dephasing* on the qubit. This does not pollute the measurement result.However, when Δ is not negligible compared with ε, the measurement simultaneously causes *qubit relaxation*.

5. Conclusion

In conclusion, we analyzed the quantum dynamics of the density operator of a system composed of a qubit and a JBA as the probe of the qubit state readout. From the analysis results, we have succeeded in extracting the essential feature of the JBA readout process of a superconducting qubit. In order to avoid pollution in the readout result, faster readout is feasible. The time we have to wait in order to distinguish two states is mainly determined by the strength of decoherence Γ for the JBA. Stronger decoherence provides rapid readout. However, at the same time, it gives stronger *indirect* relaxation in the qubit. This is a trade-off situation. Optimization of the decoherence in a JBA is a problem in the next work.

Acknowledgments

We thank M. Devoret and H. Mabuchi for fruitful discussions. The authors also thank H. Yamaguchi, K. Kakuyanagi, and Y. Yamada for useful comments from the experimental viewpoints. This work received support from KAKENHI(18001002, 18201018).

References

1. I. Siddiqi *et al.*, Phys. Rev. Lett. **93**, . 207002 (2004); I. Siddiqi *et al.*, *ibid.* **94**, 027005 (2005), A. Lupascu *et al.*, *ibid.* **96**, 127003 (2006); I. Siddiqi *et al.*, Phys. Rev. **B 73**, 054510 (2006); A. Lupascu *et al.*, Nature Physics **3**, 119 (2007); N. Boulant *et al.*, Phys. Rev. **B 76**, 014525 (2007).
2. W. Jordan and P. Smith, 'Nonlinear Ordinary Differential Equations', third ed. (Oxford Univ. Press, 1999).
3. M. I. Dykman and M. V. Fistul, Phys. Rev. **B 71**, 140508 (R) (2005); M. I. Dykman, Phys. Rev. **E 75**, 011101 (2007); V. Peano and M. Thorwart, Chem. Phys. **322**, 135 (2006); V. Peano and M. Thorwart, New J. Phys. **8**, 21 (2006); and references therein.
4. M. Rigo *et al.*, Phys. Rev. **A 55**, 1665 (1997).
5. Y. Makhlin, A. Shnirman and G. Schön, Rev. Mod. Phys., **73**, 357 (2001).

DOUBLE-LOOP JOSEPHSON-JUNCTION FLUX QUBIT WITH CONTROLLABLE ENERGY GAP

Y. SHIMAZU*, Y. SAITO, and Z. WADA

Department of Physics, Yokohama National University, Yokohama 240-8501, Japan
**E-mail: yshimazu@ynu.ac.jp*

A double-loop four-Josephson-junction (4-JJ) flux qubit has an advantage over a three-Josephson-junction flux qubit in that the energy gap Δ at the degeneracy point can be controlled *in situ* by varying the magnetic flux in one of the two loops. We report the results of microwave spectroscopy and ground-state measurements of a 4-JJ qubit. The value of Δ is estimated from the resonances due to one-photon and two-photon processes. The appearance of a qubit step on a two-dimensional flux map is in agreement with the theoretical prediction. The dependence of the maximum slope of the qubit step on the control flux indicates the tunability of the barrier height.

Keywords: Flux qubit; Spectroscopy; Josephson-junction.

1. Introduction

Superconducting qubits are promising candidates for the implementation of scalable quantum information processors.[1] A qubit in the flux regime has an advantage that it can be rendered insensitive to randomly fluctuating charges in the substrate. In particular, three-Josephson-junction (3-JJ) flux qubits[2] have been studied extensively by many researchers. Recently, a tunable coupling scheme[3] and a controlled-NOT quantum gate operation[4] using a qubit pair have been demonstrated.

In a 3-JJ flux qubit, the energy gap Δ of the qubit at the symmetry point, i.e., the point at which the energy splitting is minimum, is determined from sample parameters such as the Josephson energy E_J, charging energy E_C, and junction area ratio α of the Josephson junctions of the qubit. Since the optimal manipulation of a qubit with minimum decoherence can be achieved at the symmetry point using resonant microwave pulses, adequate control over Δ can be very advantageous for the implementation of efficient quantum operation schemes. Mooij *et al.* have introduced the concept of double-loop four-Josephson-junction (4-JJ) flux qubits with an *in situ* controllable energy gap.[2] A double-loop 4-JJ flux qubit is similar to a 3-JJ flux qubit with its smallest junction replaced by a DC-SQUID. It is important to note that the control over two magnetic fluxes in the 4-JJ qubit allows one to perform any unitary transformation of the qubit state.[5] Very recently, efficient coupling schemes for the 4-JJ qubits have been presented.[6] In this paper, the results of spectroscopy and ground-state measurements of a 4-JJ flux qubit have been presented.

2. Samples and Hamiltonian

Figures 1(a) and (b) illustrate schematic diagrams of the samples. The 4-JJ flux qubit is denoted by thick lines. The difference between the two samples was the coupling strength between the control line and the SQUID loop of the qubit. The circulating current in the qubit was measured by using the readout DC-SQUID, which was strongly coupled to the qubit loop for both samples. The samples were fabricated by *e*-beam lithography and the shadow deposition of Al. Two of the junctions had an equivalent area S of approximately 0.02 (μm)^2 and equivalent junction capacitance C, while the other two junctions had a small area of βS, where β was approximately 0.8. The quantum state of the 4-JJ qubit was manipulated through magnetic frustrations f_1 and f_2 in the two loops; magnetic frustration is the amount of magnetic flux in units of the superconducting flux quantum $\Phi_0 = h / 2e$. The areas of the qubit loops enclosing f_1 and f_2 were 75 $(\mu m)^2$ and 33 $(\mu m)^2$, respectively.

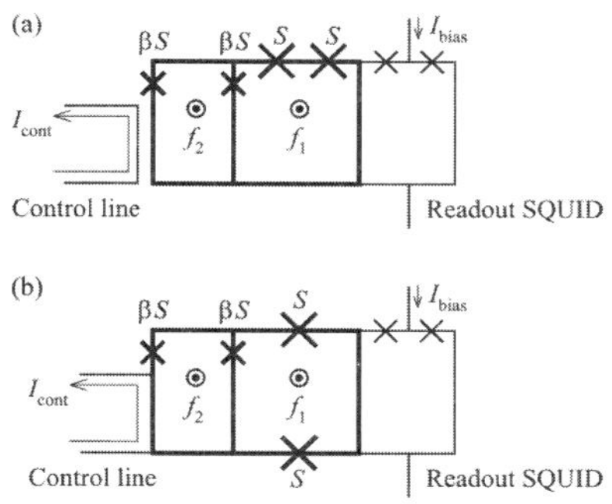

Fig. 1. Schematic diagrams of the 4-JJ qubit samples. The coupling strength between the control line and the DC-SQUID loop of the qubit is different between samples (a) and (b).

It is convenient to use $\varphi_p = (\varphi_1 + \varphi_2) / 2$ and $\varphi_m = (\varphi_1 - \varphi_2) / 2$ as the coordinates of the system, where φ_1 and φ_2 are the phase differences of the two larger junctions. The Hamiltonian of the system can be expressed as

$$H = \frac{P_p^2}{2M_p} + \frac{P_m^2}{2M_m} + E_J\{2 + 2\beta - 2\cos\varphi_p \cos\varphi_m$$
$$-\alpha(f_2)\cos(2\pi(f_1 + f_2/2) + 2\varphi_m)\}, \quad (1)$$

where the momenta P_p and P_m and mass terms M_p and M_m are shown in ref. 7. The term $\alpha(f_2) = 2\beta\cos(\pi f_2)$ plays the role of α in the 3-JJ qubit. By changing f_2, the barrier height of the potential and, in turn, the energy gap can be controlled.

3. Experimental

The samples were mounted in a copper cell, which was thermally anchored on the cold finger of a dilution refrigerator with a base temperature of 20 mK. The switching current I_{SW} of the readout DC-SQUID was measured as a function of the control current I_{cont} and the magnet current I_{mag} flowing in the superconducting magnet that was used to apply a uniform magnetic field. For the measurement of the switching current, the SQUID bias current I_{bias} was swept using a triangular waveform, typically at 1 kHz, and the average switching current was recorded using a sample-and-hold circuit and a low-pass filter. Twisted pairs of electrical leads through which I_{cont} and I_{bias} flowed were filtered by using stainless steel powder low-pass filters. The samples were magnetically shielded by using a superconducting lead shield and a permalloy shield. For spectroscopy measurements, continuous microwaves were applied using a small loop antenna.

4. Results and Discussion

For the sample depicted in Fig. 1 (a) wherein the control line was weakly coupled to the qubit, resonant peaks and dips were observed that were induced by the continuous microwaves. Figure 2 illustrates the switching current as a function of I_{mag} at various microwave frequencies. The linear background was subtracted for these traces. Typically, two peaks and two dips were symmetrically found. Some dips and peaks exhibited a doublet structure, the origin of which has not been understood. These peaks and dips are attributed to one-photon and two-photon resonances. A comparison between the observed dip positions and the theoretical fits is indicated in Fig. 3. The energy gap Δ at the degeneracy point is estimated to be approximately 3.5 GHz and the circulating current I_{circ} to be 120 nA.

The data presented in Figs. 2 and 3 were obtained at $I_{cont} = 0$. The resonant peaks and dips disappeared when I_{cont} was higher than 10 μA due to a heating effect associated with chip resistors in the leads inserted for filtering. The experimental data obtained thus far, however, have indicated that I_{circ} was dependent on I_{cont} in the range of $|I_{cont}| < 5$ μA. This result demonstrates the tunability of this system.[8]

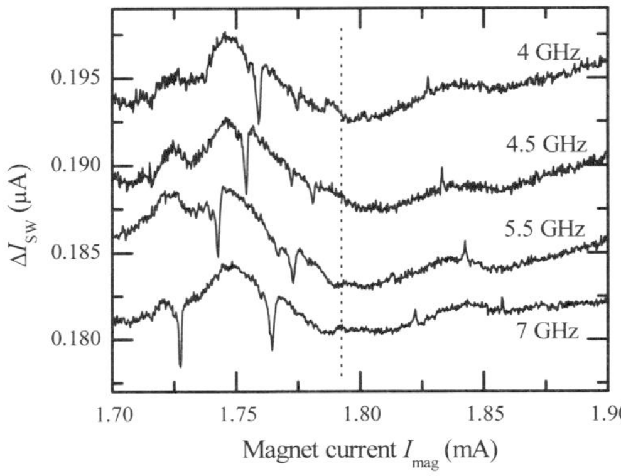

Fig. 2. Switching current of the readout DC-SQUID as a function of I_{mag} at various microwave frequencies applied. The curves are shifted vertically for clarity. Resonant peaks and dips are visible. The vertical dashed line indicates the symmetry point of the resonances.

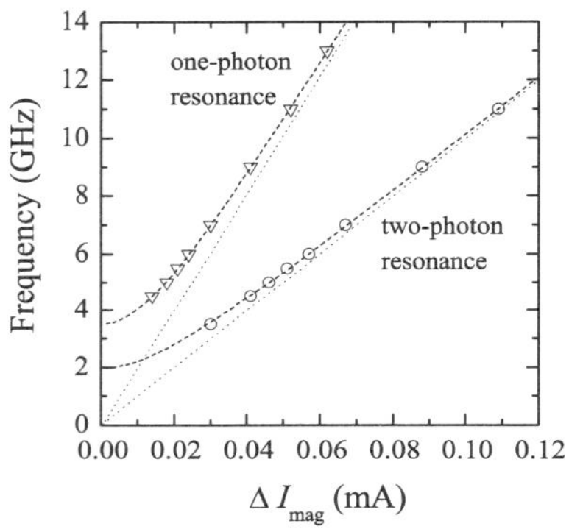

Fig. 3. Resonant dip positions and theoretical fits for microwave resonances. ΔI_{mag} is the offset of the magnet current from the symmetry point. The energy gap and circulating current are estimated from the fitting curves.

For the sample (Fig. 1(b)) wherein the order of the magnitude of the coupling strength between the control line and the qubit was higher than that in the previously mentioned sample, resonant peaks and dips were not observed. This may be explained in terms of the external noise associated with the control current.

Due to the strong coupling, however, the qubit step, which is associated with the reversal of the circulating current, was observed over a very wide range of $f_2\Phi_0$, i.e. over $2\Phi_0$, for this sample. Figure 4 shows the dependence of I_{SW} on I_{mag} for various I_{cont} values. In addition to the usual oscillation of I_{SW} of the DC-SQUID, qubit steps were observed. The qubit steps appeared in limited ranges of I_{cont}, and the positions of the qubit steps changed linearly with I_{cont} in these ranges.

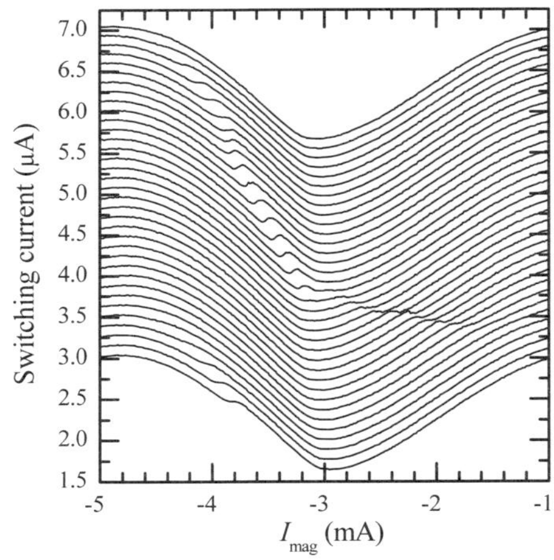

Fig. 4. Switching current of the readout DC-SQUID as a function of I_{mag} for various values of I_{cont}, from –14 to 20 μA. The curves are offset for clarity. The qubit steps are observed.

The positions of the observed qubit step in the I_{cont}-I_{mag} plane are summarized in Fig. 5. The regular pattern of the step position can be elucidated in view of the effective potential indicated in the Hamiltonian (Eq. (1)). It is expected that in the f_1-f_2 plane, the qubit step appears on a line segment satisfying

$$f_1 + f_2 / 2 = N_1 + (N_2 + 1) / 2 \\ = 0, \pm 1/2, \pm 1, \pm 3/2, \cdots \qquad (2)$$

near the lattice point $(f_1, f_2) = (N_1 + 1/2, N_2)$, where N_1 and N_2 are integers. The observed position of the qubit step is in agreement with the theoretical prediction, as shown in Fig. 5. The range of f_2 for the occurrence of the qubit step is approximately $|f_2 - N_2| < 0.33$ for this sample.

We now discuss the change in the shape of the qubit step for variation in f_2. The dependence of the maximum slope of the qubit step on f_2 is shown in Fig. 6, in terms of the current flowing in the qubit. As f_2 decreases below -0.3, the decrease in the maximum slope is exponential. It should be noted that this sample is possibly in the classical regime because microwave resonances were not observed. Irrespective of whether the system is in the classical or quantum regime, the decrease in the maximum slope is a manifestation of the decrease in the potential barrier. The observed dependence of the maximum slope indicates the tunability of the potential barrier height via f_2, which should be closely related to the tunability of Δ.

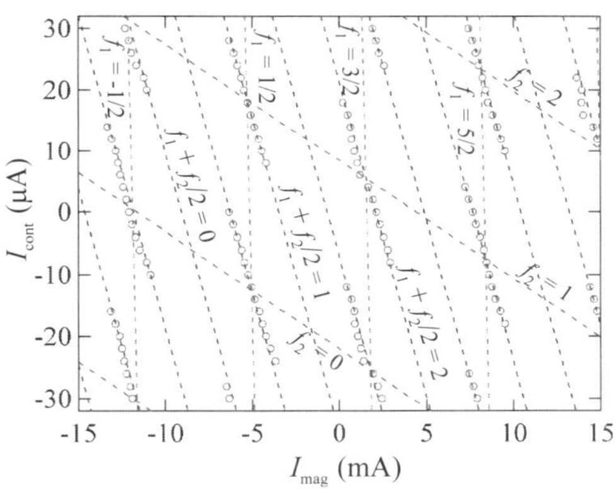

Fig. 5. Circles representing the position of the qubit step in the I_{cont}-I_{mag} plane. The flux maps for f_1 and f_2 are also shown. The position of the qubit step is in agreement with the position obtained by the theory described in the text.

5. Conclusion

In conclusion, we have demonstrated the basic operation of a double-loop 4-JJ flux qubit. In the spectroscopy measurements, the one-photon and two-photon resonances were observed. The energy gap Δ was estimated to be approximately 3.5 GHz at $I_{cont} = 0$. The tunability of the circulating current was observed. The position of the qubit step on the two-dimensional flux map agreed with the theoretical prediction. The change in the maximum slope of the qubit step as a function of the control flux indicates the dependence of the barrier height on the control flux, which should be associated with the tunability of Δ.

Acknowledgements

This work was supported by the CREST-JST project and a Grant-in-Aid for Scientific Research from MEXT, Japan.

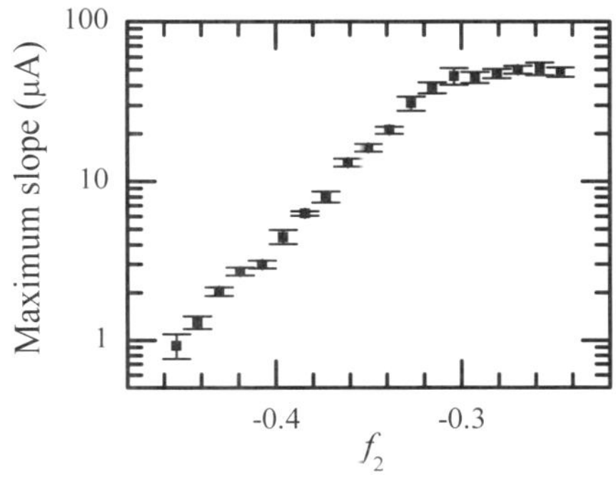

Fig. 6. Dependence of the maximum slope of the qubit step on the control flux f_2, plotted in terms of the circulating current in the qubit. An exponential dependence on f_2 is observed.

References

1. M. A. Nielsen and I. L. Chuang, *Quantum Computation and Quantum Information* (Cambridge University Press, 2000).
2. J. E. Mooij *et al.*, *Science* **285**, 1036 (1999).
3. A. O. Niskanen *et al.*, *Science* **316**, 723 (2007).
4. J. H. Plantenberg *et al.*, *Nature* (London) **447**, 836 (2007).
5. Y. Makhlin *et al.*, *Rev. Mod. Phys.* **73**, 357 (2001).
6. A. J. Kerman and W. D. Oliver, *Phys. Rev. Lett.* **101**, 070501 (2008). Y. D. Wang *et al.*, arXiv: 0807.1584 (2008).
7. T. P. Orlando *et al.*, *Phys. Rev.* **B60,** 15398 (1999).
8. Recently, the tunability of the energy gap was demonstrated spectroscopically. FG Paauw *et al.*, arXiv: 0812.1912 (2008). Y. Shimazu, unpublished.

NOISE CHARACTERISTICS OF THE FANO EFFECT AND FANO-KONDO EFFECT IN TRIPLE QUANTUM DOTS, AIMING AT CHARGE QUBIT DETECTION

TETSUFUMI TANAMOTO, YOSHIFUMI NISHI, AND SHINOBU FUJITA

Advanced LSI laboratory, Corporate R&D Center, Toshiba Corporation, 1, Komukai Toshiba-cho, Saiwai-ku, Kawasaki 212-8582, Japan

We theoretically compare transport properties of Fano-Kondo effect with those of Fano effect. We focus on shot noise characteristics of a triple quantum dot (QD) system in the Fano-Kondo region at zero temperature, and discuss the effect of strong electric correlation in QDs. We found that the modulation of the Fano dip is strongly affected by the on-site Coulomb interaction in QDs.

Keywords: Fano-Kondo effect; quantum dot; charge qubit.

1. Introduction

Quantum dot (QD) systems have attracted a lot of interest over many years because of their variety of controllability of a small number of electrons in order to understand many-body effects in electronic systems. Quantum correlation between localized states in QDs and free electrons in electrodes induces interesting phenomena such as the Fano effect and the Kondo effect. A number of important experiments have been carried out[1,2,3,4,5,6] and many theories have been proposed[7,8,9,10]. The Fano effect occurs as a result of quantum interference between a discrete energy state and a continuum state[1]. The Kondo effect is observed as a result of many-body correlations where internal spin degrees of freedom play an important role[2]. The Fano-Kondo effect, which is a combination of the Fano effect and the Kondo effect, can be observed when on-site Coulomb interaction in a QD is strong[4].

We have theoretically investigated transport properties of the triple QD system depicted in Fig.1, where QDs a and b are connected to electrodes through QD d[10]. When coupling between QD a and b is larger than that between QD b and d ($t_C > t_d$), we can use this setup as apparatus for detecting a two-level system (QD a and QD b) by a QD d with electrodes (we call QD d a detector QD). Moreover, when the number of electrons is controlled, double QD a and b can be regarded as a charge qubit[12,13] with a Fano interference detector QD. In Ref.[10], we have shown that the Fano dip is modulated for a slow detector (small tunneling rate of the detector) with no on-site Coulomb interaction in QD d. We focus on shot noise characteristics at zero temperature. Although $1/f$ noise is shown to be the largest cause of decoherence[14], shot noise induces relaxation and dephasing effects in the detecting process and is an important measure of decoherence.

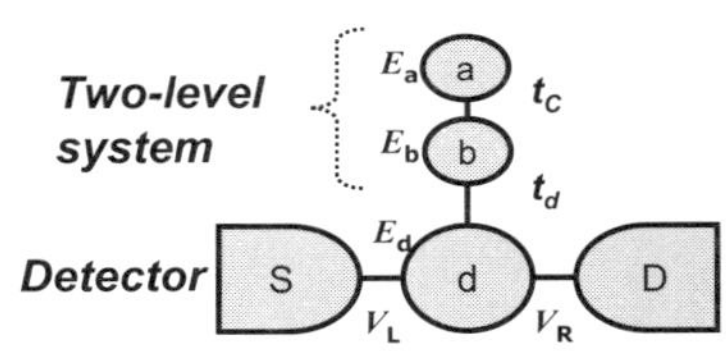

Fig. 1. Schematic plot of triple QD system. QDs a and b constitute a two-level system that is coupled to QD d only which is connected to the electrodes.

Wu *et al.*[8] calculated noise properties of T-shaped QD system and showed that shot noise strongly depends on the coupling strength between a side QD and a detector QD. As tunneling coupling between side QD and detector QD increases, Fano factor γ, which is defined as a ratio of noise power and average current, quickly increases up to the Poisson value ($\gamma = 1$). López *et al.* calculated shot noise of serially and laterally coupled double QD system and showed that γ strongly depends on the coupling strength between QDs[11]. Thus, γ and shot noise reflect the coupling configuration of QD system and provide important information about the electronic structure of the system.

Here, we compare zero temperature shot noise properties of the Fano-Kondo effect with those of the Fano effect, in order to reveal the effect of strong on-site Coulomb interaction on the transport properties. We assume an infinite Coulomb interaction for QD a and b and no Coulomb interaction for QD d ($U_a = U_b = \infty$, $U_d = 0$) for the Fano-Kondo case. For the Fano case, we consider that there is no on-site Coulomb interaction for all QDs ($U_a = U_b = U_d = 0$), which is realized in large QDs[5]. We also assume that there is a single energy level in each QD and that the two energy levels of QD a and QD b coincide and correspond to gate voltages that are applied to those QDs. We use slave-boson mean-field theory (SBMFT) based on the nonequilibrium Keldysh Green's function method.

2. Formulation

Hamiltonian is constructed from electrode parts, QD parts, tunneling parts between an electrode and a QD, and those between QDs. For the Fano-Kondo case, additional constraint is required. The mean-field Hamiltonian for the Fano-Kondo case is described in terms of slave-bosons b_{α_1} ($\alpha_1 = a, b$) as:

$$\begin{aligned} H^{\rm FK} = & \sum_{k_L,k_R,s} E_{k_\alpha} c^\dagger_{k_\alpha s} c_{k_\alpha s} + \sum_{s,\alpha_1=a,b,d} E_{\alpha_1} f^\dagger_{\alpha_1 s} f_{\alpha_1 s} \\ & + \sum_{\alpha_1=a,b} \lambda_{\alpha_1} \Big(\sum_s f^\dagger_{\alpha_1 s} f_{\alpha_1 s} + b^\dagger_{\alpha_1} b_{\alpha_1} - 1\Big) \\ & + \frac{t_C}{N} \sum_s (f^\dagger_{as} b_a b^\dagger_b f_{bs} + f^\dagger_{bs} b_b b^\dagger_a f_{as}) \\ & + \frac{t_d}{N} \sum_s (f^\dagger_{ds} b^\dagger_b f_{bs} + f^\dagger_{bs} b_b f_{ds}) \\ & + \sum_{\alpha=L,R} \frac{V_\alpha}{\sqrt{N}} \sum_{k_\alpha,s} (c^\dagger_{k_\alpha s} f_{ds} + f^\dagger_{ds} c_{k_\alpha s}), \end{aligned} \quad (1)$$

where E_{k_α} is the energy level for source ($\alpha = L$) and drain ($\alpha = R$) electrodes. E_a, E_b and E_d are energy levels for the three QDs, respectively. t_C, t_d and V_α are the tunneling coupling strength between QD a and QD b, that between QD b and QD d, and that between QD d and electrodes, respectively. $c_{k_\alpha s}$ and $f_{\alpha_1 s}$ are annihilation operators of the electrodes, and of the three QDs ($\alpha_1 = a, b, d$), respectively. s is spin degree of freedom with spin degeneracy N; here we apply $N = 2$. λ_{α_1} is a Lagrange multiplier. We take $z_{\alpha_1} \equiv b^\dagger_{\alpha_1} b_{\alpha_1}/2$ and $\tilde{E}_{\alpha_1} \equiv E_{\alpha_1} + \lambda_{\alpha_1}$ as mean-field parameters for QD a and QD b. The Hamiltonian for the Fano case is similar to $H^{\rm FK}$ except that $\lambda_a = \lambda_b = 0$ and $b_a = b_b = 1$ in Eq.(1).

In the Fano-Kondo effect, four self-consistent equations to determine mean-field parameters λ_{α_1} and b_{α_1} are derived by $\partial\langle H^{\rm FK}\rangle/\partial\lambda_{\alpha_1} = 0$ and $\partial\langle H^{\rm FK}\rangle/\partial b_{\alpha_1} = 0$.

Current and noise formula are expressed by Keldysh Green's functions. Keldysh Green's functions are obtained by applying analytic continuation rules to the equations of motion, which are derived from the above Hamiltonian[15,11].

Source current I_L is expressed as $I_L = (2e/h)\int_{-\infty}^{\infty} d\omega T(\omega)(f_L(\omega) - f_R(\omega))$ where transmission probability $T(\omega)$ is given as

$$T(\omega) = \frac{\Gamma_L \Gamma_R |D_{ab}|^2}{[D_{ab}(\omega - \tilde{E}_d) - (\omega - \tilde{E}_a) z_a \tilde{t}_d^2/2]^2 + \Gamma^2 D_{ab}^2/4} \quad (2)$$

and Fermi distribution functions $f_\alpha(\omega) \equiv [\exp((\omega - \mu_\alpha)/T) + 1]^{-1}$ where we set symmetrical bias condition: $\mu_L = E_F - eV/2$ and $\mu_R = E_F + eV/2$. Here, $D_{ab} \equiv (\omega - \tilde{E}_a)(\omega - \tilde{E}_b) - \tilde{t}_C^2$, with $\tilde{t}_C = t_C b_a b^\dagger_b/N$ and $\tilde{t}_d = t_d b^\dagger_b/N$. $\Gamma_\alpha \equiv 2\pi\rho_\alpha(\mu_\alpha)|V_\alpha|^2$ is the tunneling rate between α electrode and QD d with a density of states (DOS), $\rho_\alpha(\mu_\alpha)$, for each electrode at Fermi energy μ_α. $\Gamma \equiv (\Gamma_L + \Gamma_R)/2$ and we assume $\Gamma_L = \Gamma_R$. Note that in the present case we can check that I_L and I_R are symmetric and satisfy a current conservation. Conductance is given as $G = -\frac{2e}{h}\int d\omega T(\omega)\frac{\partial f_L(\omega)}{\partial\omega}$. The transmission probability is related to a DOS of the detector QD $\rho_d(\omega) = -\mathrm{Im}G^r_{dd}(\omega)/\pi$ such as $T(\omega) = \frac{2\Gamma_L\Gamma_R}{\Gamma_L+\Gamma_R}\pi\rho_d(\omega)$, which means that we can discuss characteristics of DOS similar to a transmission probability.

Noise formula at $T = 0$ is derived similarly to that in Ref.[11] and we have

$$S(V) = \frac{4e^2}{h}\int_{-eV/2}^{eV/2} d\omega T(\omega)(1 - T(\omega)). \quad (3)$$

The Fano factor γ at zero bias $V = 0$ is obtained by $\gamma = 1 - T(E_F)$, which indicates that shot noise is in the sub-Poissonian region ($\gamma \leq 1$). Similar to Ref.[10], we classify our triple QD system by magnitude of t_C/t_d and Γ/t_d. The ratio t_C/t_d compares the internal coupling strength in a two-level system with that between the two-level system and the detector, and we regard the case where $t_C/t_d = 5$ as a strongly coupled two-level system and the case where $t_C/t_d = 1$ as a weakly coupled two-level system. If Γ/t_d is large, the electron that flows through QD d is so fast that it cannot detect the oscillation of an electron in the coupled QDs a and b. If Γ/t_d is small, the electron that flows through QD d can observe the evidence of bonding and antibonding states. We call a detector with large $\Gamma/t_d = 2$ a fast detector, and one with smaller $\Gamma/t_d = 0.4$ a slow detector. We assume that $D = 20t_d$, $|E_d| < 0.4t_d$, $\Gamma > 0.4t_d$ and $E_F = 0$ (D is a bandwidth). Then, we have the Kondo temperature $T_K \sim De^{-\pi|\tilde{E}_d - E_F|/\Gamma} \sim 1.6t_d$.

3. Numerical calculations

Here, we show numerical results of our triple QD system in terms of the Fano-Kondo effect and the Fano effect.

Figure 2 shows current-voltage (I-V) characteristics at $T = 0$. We can see that all currents look similar for a fast detector ((a)) in both the Fano effect and the Fano-Kondo effect. This indicates that a fast detector is less sensitive to quantum states of QD system than a slow detector. Current of the Fano-Kondo effect is always less than that of the Fano effect. This indicates that current path from the source to the drain via the two-level system is suppressed by the Kondo effect due to the on-site Coulomb interaction.

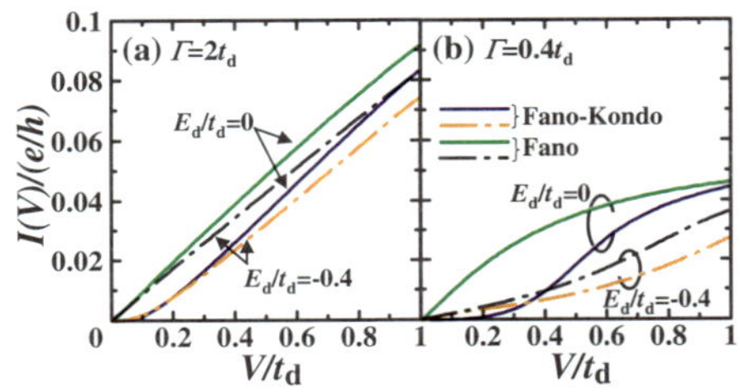

Fig. 2. I-V characteristics for a strong coupling ($t_C/t_d = 5$) at $T = 0$. (a) Fast detector ($\Gamma/t_d = 2$) and (b) slow detector ($\Gamma/t_d = 0.4$). $E_a = E_b = 0 = E_F$.

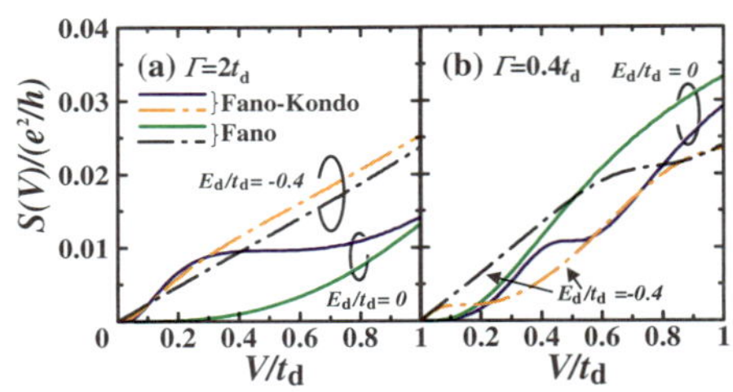

Fig. 3. Shot noise as a function of bias voltage for a strong coupling ($t_C/t_d = 5$). (a) Fast detector($\Gamma/t_d = 2$). (b) Slow detector($\Gamma/t_d = 0.4$). $T = 0$. $E_a = E_b = 0 = E_F$.

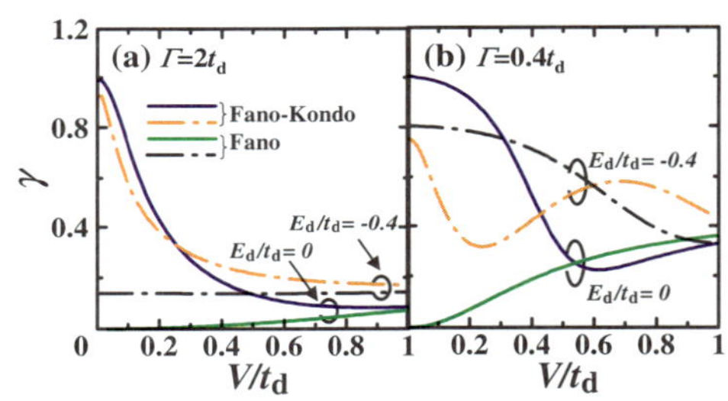

Fig. 4. Fano factor γ as a function of bias voltage. $E_a = E_b = 0 = E_F$ for a strong coupling ($t_C/t_d = 5$). (a) Fast detector($\Gamma/t_d = 2$) and (b) slow detector($\Gamma/t_d = 0.4$).

Figure 3 shows shot noise characteristics as a function of bias voltage across the detector QD. E_d dependence is simpler for a fast detector ((a)). This is because a fast detector is more sensitive to energy level of a detector QD d than to energy levels in two-level system. Shot noise of a slow detector reflects internal states of two-level system.

Although magnitude of current for a fast detector is larger than that for a slow detector (Fig. 2), magnitude of shot noise for a fast detector is of the same order as that of a slow detector (Fig. 3). Thus, γ for a slow detector is relatively larger than that for a fast detector as shown in Fig. 4. This is because of the stronger coupling of flowing electrons with two-level states in a slow detector. Strong nonlinearity around $V = 0$ is considered to reflect the resonant state of $E_d = 0$ $(= E_a = E_b = \mu_L = \mu_R)$ and the dip of $E_d/t_d = -0.4$ in Fig.4(b) comes from the shift of resonant energy level. In both a fast detector and a slow detector, γ for the Fano-Kondo case is larger than that for the Fano case. This indicates that stronger electronic correlation induces more noise.

4. Discussion

Our numerical results show that, as coupling in a two-level state (qubit) becomes stronger, noise increases. In addition, a slower detector, which is found to be more desirable for reading out a two-level state, induces more noise than a fast detector. Thus, there is a trade-off in that reading out more detailed information induces more noise or larger Fano factor. Appropriate parameters (t_C/t_d, Γ/t_d etc.) should be determined depending on sensitivity of the external circuit connected to this triple QD system. As noted in the introduction, in order to use the two-level state as a qubit, only one excess electron should stay in the two-level system (charge qubit condition). This would be realizable, for example, by forming smaller and closely coupled QDs, such that two electrons are not permitted into the QDs because of their repulsive Coulomb interaction. As shown in the numerical results, the Fano factor for stronger correlation (Fano-Kondo case) is relatively larger than that for weaker correlation (Fano case). Thus, it is possible that we will have to accept larger noise when introducing the charge qubit condition. More elaborate control of the measurement setup would be required for a charge qubit system.

5. Conclusion

Focusing on the Fano effect and the Fano-Kondo effect in a two-level state, we studied noise properties of a triple QD system. We showed that, depending on the coupling strength among the triple QDs, noise and the Fano factor are greatly modulated for a slow detector. In particular, we found that detailed reading of a two-level state is inclined to increase noise characteristics of the system.

References

1. U. Fano Phys. Rev. **124**, 1866 (1961).
2. W. G. van der Wiel *et al.*, Science **289** 2105 (2000).
3. J. Göres *et al.*, Phys. Rev. B**62**, 2188 (2000).
4. M. Sato *et al.*,Phys. Rev. Lett. **95**, 066801 (2005).
5. T. Otsuka *et al.*,J. Phys. Soc. Jpn. **76** 084706 (2007).
6. A. W. Rushforth *et al.*, Phys. Rev. B **73**, 081305(R) (2006).
7. K. Kang *et al.*, Phys. Rev. B **63**, 113304 (2001).
8. B. H. Wu *et al.*,Phys. Rev. B**72**, 165313 (2005).
9. C. A. Büsser *et al.*, Phys. Rev. **70**, 035402 (2004).
10. T. Tanamoto and Y. Nishi,Phys. Rev. B**76**, 155319 (2007).
11. R. López *et al.*, Phys. Rev. B **69**, 235305 (2004).
12. T. Tanamoto and X. Hu, Phys. Rev. B**69**, 115301 (2004).
13. H. S. Goan *et al.*, Phys. Rev. B **63**, 125326 (2001).
14. O. Astafiev *et al.*,Phys. Rev. Lett. **96**, 137001 (2006).
15. D M. Newns and N. Read, Adv. Phys. **36**, 799 (1987).

GEOMETRIC UNIVERSAL SINGLE QUBIT OPERATION OF COLD TWO-LEVEL ATOMS

HIROMITSU IMAI* and ATSUO MORINAGA

Department of Physics, Tokyo University of Science, Chiba, Japan

**E-mail: hero@makizou.com*

A universal single qubit operation on an ensemble of cold sodium atoms was accomplished by pure geometric rotations of the Bloch vector around axes 2 and 3 using carefully prepared resonant laser-controlled pulses. The phase shift between two states was measured using geometric atom interferometer and it was shifted double the phase difference of control with an uncertainty of 1%. The fidelity of the universal single qubit operation was evaluated to be 90%.

Keywords: geometric phase, resonant laser-controlled pulses, universal single qubit operation, cold two-level atom, atom interferometry, fidelity of operation.

1. Introduction

Quantum computation is one of the prospective subjects in the scientific field. There are two blocks to realize the quantum computation. The one is a universal single qubit operation for one qubit such as atoms or spins [1,2] and the other is a controlled-NOT gate for two qubits [3]. Especially, the single qubit operation is important because the operation is a basis of the controlled-NOT gate. These operations have been induced through the dynamical evolution of interaction Hamiltonian from external sources such as laser pulses and magnetic field, but can also be induced through a geometric effect. The geometric manipulations are more robust than dynamic ones since a geometric phase solely depends on the amount of the solid angle enclosed by the evolution, not on the details of the path, the time spent, the driving Hamiltonian, or the initial and final states of the evolution [4]. Up to now, experimental demonstrations using partial geometric operation have been presented in nuclear magnetic resonance [5] and ion trap system [6].

In 2004, Tian et al. proposed a scheme for geometric rotations of Bloch vector around axes 3 and 2 composed purely by resonant laser-controlled pulses [7]. They demonstrated a rotation around axis 3 using heterodyned-photon echoes applied to inhomogeneous detuned two-level Tm^{3+} doped in a YAG crystal. Recently, we developed geometric manipulations of rotations around axes 3 [8] and 2[9] with an ensemble of cold atoms and have accomplished a universal single qubit operation with pure geometric phase changes [10]. However, it was found that careful preparations for resonant laser-controlled pulses are necessary, in order to accomplish the universal single qubit operation with a high fidelity. In this paper, we report the fidelity of the present pure geometric universal single qubit operation on the ground-state wave function.

2. Principle

2.1. Universal single qubit operation

We consider a two-level atom with a wave function A described as $\Psi_A = a|0\rangle + b|1\rangle$, where $|0\rangle$ represents the ground state ($\mathbf{r}_{-3}$) and $|1\rangle$ the excited state ($\mathbf{r}_3$). When a universal single qubit operation is applied to the wave function A, it can be transformed to an arbitrary wave function B described as $\Psi_B = a'|0\rangle + b'|1\rangle$. The universal single qubit operation is decomposed of successive unitary rotations of the Bloch vector around axes 3, 2 and 3 on the Bloch sphere [1], as $U = U_3(\gamma)U_2(\beta)U_3(\alpha)$, where U_3 and U_2 are represented as rotations around axes 3 and 2 respectively. It is noted that U_3 changes a phase difference between $|0\rangle$ and $|1\rangle$, and U_2 changes population probabilities for them. In the present case, as shown in Fig. 1, we consider the initial condition where the Bloch vector is $|0\rangle$. Then the universal single qubit operation is simplified as

$$U = U_3(\gamma)U_2(\beta). \tag{1}$$

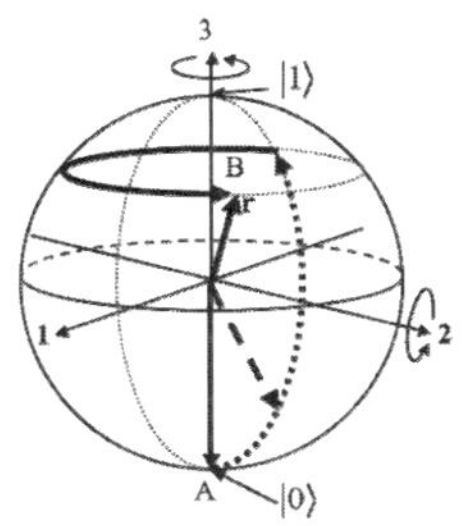

Fig. 1. Universal single qubit operation when the Bloch vector starts from a ground state.

2.2. *Geometric manipulation for rotations around axes 3 and 2*

According to the proposal by Tian et al. [7], the U_3 and U_2 can be accomplished geometrically by plural resonant laser-controlled pulses with a relative phase between them. For U_3, two resonant π–pulses whose second pulse has a relative phase of δ/2 are applied to the wave function. Then the Bloch vector of a ground state $\mathbf{r}_{-3}$ rotates on the Bloch sphere, tracing an enclosed circuit with a solid angle of 2(π-δ/2) and gains a phase shift of -(π-δ/2) because Berry's phase is defined as –Ω/2 for SU(2). Similarly, by the same pulses the excited state gains a phase shift of (π-δ/2). These produce a matrix

$$U_3(\delta) = e^{-i\pi}\begin{pmatrix} e^{i\delta/2} & 0 \\ 0 & e^{-i\delta/2} \end{pmatrix}, \tag{2}$$

where $e^{-i\pi}$ is called a global phase and can be ignored as it has no effect on the observations. Therefore, this manipulation is equivalent to a rotation around axis 3.

For U_2, three resonant pulses with pulse areas of π/2, π and π/2, whose second pulse has a relative phase of π+δ/2, are applied to the Bloch vector $\mathbf{r}_2$ (Fig. 2(b)). Then the vector rotates tracing a closed circuit with a solid angle of δ. Therefore, the wave function gains a phase shift of -δ/2. Similarly, an initial vector of $\mathbf{r}_{-2}$ gains a phase of δ/2. If we convert those basis vectors to $|0\rangle$ and $|1\rangle$, these produce a matrix

$$U_2(\delta) = \begin{pmatrix} \cos\left(\delta/2\right) & \sin\left(\delta/2\right) \\ -\sin\left(\delta/2\right) & \cos\left(\delta/2\right) \end{pmatrix}, \tag{3}$$

which is equivalent to a rotation around axis 2.

With these resonant laser-controlled pulses, we can rotate geometrically the wave function of two-level atoms around axes 3 and 2 on the Bloch sphere.

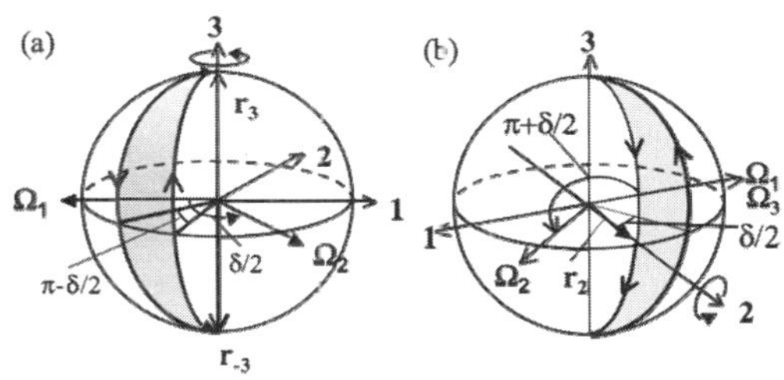

Fig. 2. (a) Traces of the Bloch vector for a manipulation of a rotation around axis 3 and (b) around axis 2.

3. Experiments

3.1. *Setup*

We used an ensemble of cold sodium atoms as a qubit. The energy diagram of sodium atom is shown in Fig. 3, together with related transitions. The ground hyperfine levels of $3S_{1/2}$ F=1, m_F=0 and F=2, m_F=0 were used as the two states of atom. The transition between the two states was generated by the two photon stimulated Raman transition via an intermediate state of $3P_{3/2}$, F'=1 or 2 with detuning below about 500 MHz [11,12]. The two photons were produced through an electro-optic modulator (EOM) driven at 1.772 GHz by a synthesizer. The frequency is equivalent to an frequency difference between the F=1 and F=2 states. The synthesizer was

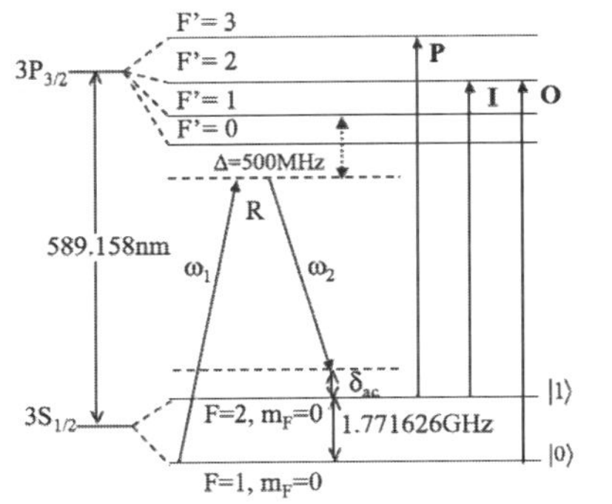

Fig. 3. Partial energy diagram of sodium atom and related transitions. P: probe beam, I: initializing beam, O: optical pumping beam, and R: Raman beam.

used to sweep the relative phase or the frequency difference between two photons.

The present experimental procedure is briefly summarized as follows; (i) Sodium atoms were trapped by Magneto-optical Trap (MOT) at 1mK. After release from the trap, those atoms were cooled down to 200 μK by polarization gradient cooling (PGC). (ii) After being initialized to the F=1 state by an initializing beam (F=2→F=1), those atoms were manipulated by laser-controlled pulses. (iii) The population probability of the excited state was measured from the transmittances of a probe beam (F=2→F'=3) before and after the geometric manipulations. The resonant laser-controlled pulses were prepared carefully so as to be free from the ac Stark shift, to have a precise pulse area and to have a little distortion and decoherence.

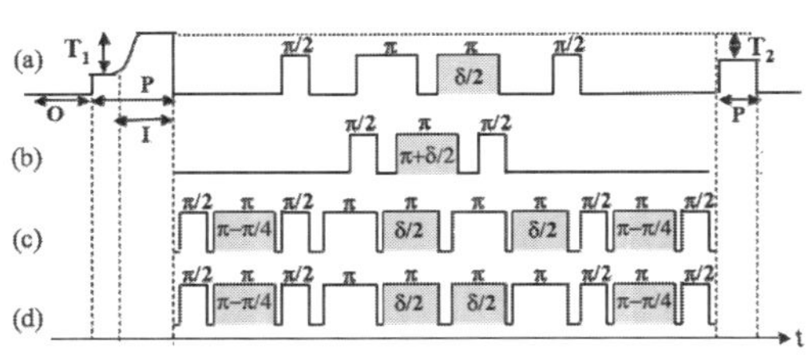

Fig. 4. Time sequence. (a) Rotation around axis 3 with atom interferometry. (b) Rotation around axis 2. (c) Geometrical universal single qubit operations. The phase shift of a couple of rotation around axis 3 by δ/2 is measured by geometric atom interferometer. (d): Subtraction of rotations around axis 3 is used for evaluation of the fidelity of universal single qubit operation.

The time sequences of typical operations with resonant laser-controlled pulses are shown in Fig. 4(a) to (d).

3.2. Results

Figure 5 (a) and (b) show typical results of geometric rotations around axes 3 and 2, observed using time sequences of Fig. 4(a) and (b). The phase shift between the two states caused by the rotation around axis 3 was measured by an atom interferometer adding two π/2 pulses to both sides of the laser-controlled pulses of U_3. The phase shift was found to be (2.01±0.02)×δ/2 or just double the phase difference of control. On the other hand, the geometric rotation around axis 2 was observed as the variation of the population probability of the excited state. It varies according to a sinusoidal function of δ with an uncertainty of 1%, that is, (1.98±0.02)×δ/2. The rotation around axis 2 is equivalent to the Rabi oscillation with an excitation pulse, but it should be noted that the oscillation due to geometric rotation is more robust than dynamic one over 2π.

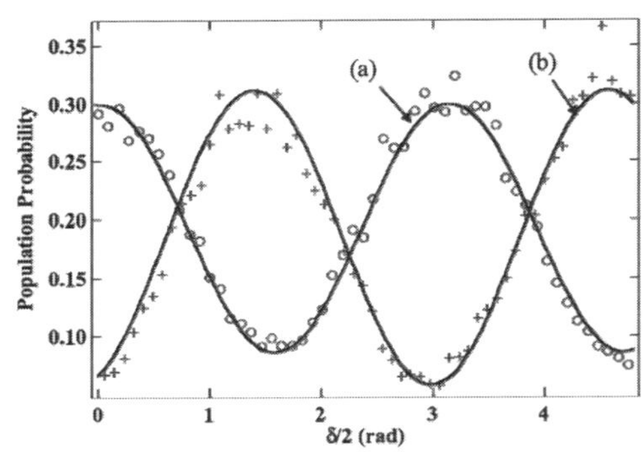

Fig. 5. (a) Phase shift of a geometric rotation around axis 3. (b) Variation of a population probability of a geometric rotation around axis 2.

Next, we examined a geometric universal single qubit operation when the initial state is the ground state, as shown in Fig. 1. The Bloch vector was first rotated around axis 2 by δ_2 and next rotated around axis 3 by δ_3. In the present experiment, δ_2 was fixed to $\pi/2$ and δ_3 was changed continuously. Then we can observe the phase difference of δ_3 with an atom interferometry formed by adding the geometric rotation around axis 2 by $\pi/2$ after the resonant laser-controlled pulses. Figure 4(c) shows a time sequence of atom interferometer for twice rotations around axis 3 by $\delta/2$. Figure 6(a) shows the observed phase shift of $(4.01\pm0.03)\times\delta/2$. The phase shift is suitable, except for the amplitude of fringes was reduced so that the total amount of pulse area was 8π.

Last, we evaluated the fidelity of such a geometric universal single qubit operation. To evaluate the fidelity, the sense of the second rotation around axis 3 was reversed, as shown in Fig. 4(b). Then we should observe a straight line not depending on $\delta/2$ at the maximum of fringes. The observed signal is shown in Fig. 6(b). The signal is near the maximum of fringes of Fig. 6(a) and the sinusoidal component of δ is cancelled out, but there is a small sinusoidal component of $\delta/2$, together with scattering of the data. We defined the fidelity of 1-b/a, where a is a peak-to-peak amplitude of fringes and b is an amplitude of the component of $\delta/2$ and the scattering of the data. Then we obtained the fidelity of about 90% with an uncertainty of 10%. If we could remove such a component by appropriate adjustment of the pulse areas, the fidelity would be improved more.

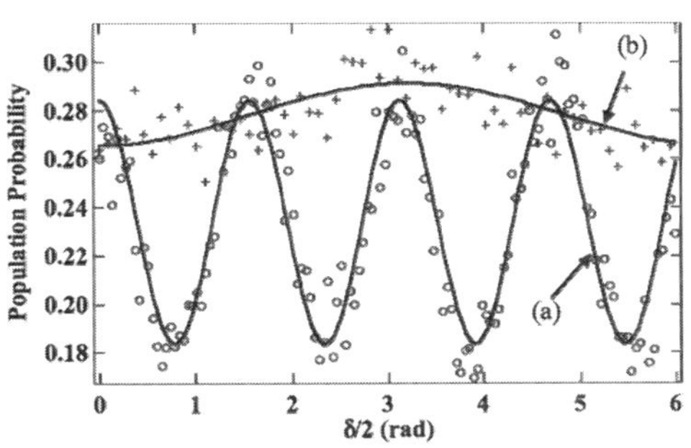

Fig. 6. (a) Phase shift for a couple of rotations around axis 3 by $\delta/2$. (b) Fidelity of the universal single qubit operation.

4. Conclusions

A universal single qubit operation on an ensemble of cold sodium atoms was accomplished by pure geometric rotations of the Bloch vector around axes 2 and 3 using carefully prepared resonant laser-controlled pulses. The phase shift between two states was measured using geometric atom interferometer and it was shifted double the phase difference of control with an uncertainty of 1%. The fidelity of the universal single qubit operation was evaluated to be 90%. At present, we are planning to realize the controlled-NOT gate with geometric manipulations using two-level atoms in optical lattices entangled with a vibrational state, like ion trap [13].

References

1. M. A. Nielsen and I. L. Chuang, Quantum Computation and Quantum Information (Cambridge University Press, Cambridge, 2000).
2. T. Calarco, U. Dorner, P. S. Julienne, C. J. Williams and P. Zoller, Phys. Rev. A **70**, 012306 (2004).
3. Barenco, et. al. Phys. Rev. A **52**, 3457 (1995).
4. M. V. Berry, Proc. R. Soc. London, Ser. A **392**, 45 (1984).
5. J. A. Jones, V. Vedral, A. Ekert and G. Castagnoli, Nature(London) **403**, 869 (2000).
6. D. Leibfried, et.al, Nature (London) **422**, 412 (2003).
7. M. Tian, Z. W. Barber, J. A. Fischer and Wm. Randall Babbitt, Phys. Rev. A **69**, 050301(R), (2004).
8. H. Imai, Y. Otsubo and A. Morinaga, Phys. Rev. A **76**, 012116 (2007).
9. H. Imai, and A. Morinaga, Phys. Rev. A **76**, 062111 (2007).
10. H. Imai and A. Morinaga, Phys. Rev. A, **78**, 010302(R) (2008).
11. M. Kasevich and S. Chu, Appl. Phys. B **54**, 321 (1992).
12. T. Aoki, K. Shinohara and A. Morinaga, Phys. Rev. A **63**, 063611 (2001).
13. C. Monroe, D. M. Meekhof, B. E. King, W. M. Itano, and D. J. Wineland, Phys. Rev. Lett, **75**, 4714 (1995).

ENTANGLEMENT DYNAMICS IN QUANTUM BROWNIAN MOTION

K. SHIOKAWA*

Institute of Physics, Academia Sinica,
128, Sec. 2, Academia Road, Nankang, Taipei 115, Taiwan
**E-mail: kshiok@phys.sinica.edu.tw*

Dynamical aspects of quantum Brownian motion at low temperatures are investigated. Exact calculations of quantum entanglement among two Brownian oscillators are given without invoking the Born-Markov or Born approximation widely used for the study of open systems. Our approach is applicable for arbitrary time scale, in particular, suitable to probe short time regime at cold temperatures where many experiments on quantum information processing are performed. We study separability criteria based on the uncertainty relation, negativity, and entanglement of formation. We found a crossover behavior in a disentanglement process between quantum and thermal fluctuation dominated regimes. The fluctuation-dissipation relation which holds for one-particle quantum Brownian particle in the long time limit does not hold for two particles interacting with common environment. The deviation from the relation is originated in the interaction mediated by environment which drives two particles into a steady oscillatory state preventing the system to thermalize. Consequently there is a residual entanglement at low temperatures for arbitrary coupling strength. This entanglement is generated from environment nonperturbatively even when two modes are not entangled initially. When a distance of two oscillators are varied, competition between entanglement induced from environment and modified decoherence due to finite separation leads to the characteristic distance where entanglement is minimized. We also discuss the influence of Unruh effect on entanglement dynamics.

Keywords: Quantum entanglement; Decoherence; Relativity

1. Introduction

Implementing quantum information processing devices requires precise knowledge of quantum open systems. The detailed study of quantum open systems are often prevented by a limitation of analytical or numerical resources to probe systems with large degrees of freedom. The first principle calculation of open quantum systems yields reduced dynamics that follows non-Markovian evolution carrying the memory of environment. Solving the equations of motion obtained by these method directly is generally difficult to deal with as they are given by integro-differential equations.

Various approximation schemes have been used to simplify the exact evolution to study open system evolution. Most commonly used approximation is Born approximation in which the system-bath interaction is treated perturbatively. However, naive Born approximation applied to quantum open systems does not guarantee the positivity of the density matrix and is not suitable when evaluating quantities sensitive to positivity such as quantum entanglement. In a commonly used Born-Markov approximation, the bath correlation is ignored. At sufficiently low temperature, however, the bath memory can persist for a long time and this approximation is expected to break down. Regardless of the temperature, the noise with long range correlation such as $1/f$ noise cannot be treated under this approximation. Further approximation such as rotating-wave approximation is not suitable for describing short time dynamics in a strong coupling regime.

Many works have been devoted to clarify the rigorous criteria for entanglement in quantum information science. For continuous variables, the necessary and sufficient criteria for entanglement can be given in terms of Peres-Horodecki criteria[1], negativity[2], and entanglement of formation[3]. Most of these works are devoted to study the static properties of general pure and mixed states. In light of information processing, it is also desirable to study the dynamical properties based on the realistic models in an open system setting.

In this work, we study the precise mechanism of open system entanglement dynamics based on a Brownian oscillator model[4][5][6], where both system and environment consist of harmonic oscillators. Realization of quantum protocols based on continuous variables such as quantum teleportation[7] and quantum key distribution[8] show that they also play the similar crucial role in quantum information science.

So far most of the studies of continuous variables for open systems are based on the master equation under Born-Markov rotating wave approximation such as Lindblad equations or the phenomenological model with high temperature Markovian bath[9]. Applications of these methods are limited to the high temperature regime or slow dynamics with Ohmic noise, while actual solid state implementation operates at low temperatures manipulated with fast pulses and often suffers from 1/f noise.

Since our model is formally exactly solvable, it makes us possible to probe the precise dynamics of decoherence effects on entanglement without conventional approximations[10][11]. Previous works for the study of a two-level system revealed that non-Markovian dynamics that plays the major role at low temperatures is highly nontrivial[12].

2. General Formulation

We consider the system composed of $\mathcal{N}$ particles. Its Hamiltonian is given by

$$H_S = \sum_{j=1}^{\mathcal{N}} \frac{P_j^2}{2M} + \sum_{j=1}^{\mathcal{N}} \frac{M\Omega^2}{2} R_j^2, \qquad (1)$$

Each oscillator variable R_j located at x_j couples linearly with the scalar field ϕ_j as

$$H_I = \lambda \sum_{j=1}^{\mathcal{N}} R_j(\tau)\phi(x_j(\tau), t_j(\tau)). \qquad (2)$$

For Lorentz scalars R_j viewed as internal coordinates as a function of a proper time τ, our total Hamiltonian is Lorentz invariant. The field Hamiltonian for ϕ is

$$H_F = \frac{1}{2}\int dx \left[\Pi_\phi^2 + (\partial_x\phi)^2 + m^2\phi^2\right] \qquad (3)$$

Then Heisenberg equations that R_j satisfy is

$$\begin{aligned} M_j\ddot{R}_j(\tau) + M_j\Omega^2 R_j \qquad (4) \\ + 2\sum_l \int_0^\tau ds\alpha_{Ijl}(\tau, s)R_l(s) = 0, \end{aligned}$$

where

$$\begin{aligned} \alpha_{Ijl} = -\sum_k \frac{\lambda^2}{2\omega_k} \cos\left[k(x_j(\tau) - x_l(s))\right] \\ \sin\left[\omega_k(t_j(\tau) - t_l(s))\right] \end{aligned}$$

is an imaginary part of the response function.

First we consider the case when there are two static observers. In this case, $t_1 = t_2 = \tau$. Let us assume that the distance between two observable is negligibly small. Then the Heisenberg equations of motion for this system can be written in a form of two coupled damped oscillators with the damping constant γ.

The necessary and sufficient condition for separability of the density matrix for a bipartite two-level-system is to have only nonnegative eigenvalues after partial transpose of one of its subsystem. The extension of this criteria to continuous Gaussian variables was also given [?]. For Gaussian variables, the partial transpose of a density matrix in a coordinate representation for one oscillator component is equivalent to a mirror reflection of that component in the Wigner distribution function. The necessary and sufficient condition for a continuous variable quantum state to be separable is that the partial mirror reflected state is still a physical quantum state that satisfies the uncertainty principle. In the phase space representation, the partial mirror reflection on the second variable can be stated as $(R_1, P_1, R_2, P_2) \rightarrow (R_1, P_1, R_2, -P_2)$.

Thus entanglement property of our system can be studied by considering the symmetric correlation:

$$(\Delta\mathcal{X})^2 \equiv \langle\{X_i, X_j\}\rangle \tag{5}$$

where $\{A, B\} \equiv (AB + BA)/2$ is an anti-commutator. We defined $X_{2j-1} \equiv \sqrt{\Omega}R_j$, $X_{2j} \equiv P_j/\sqrt{\Omega}$ for $j = 1, 2, ..., \mathcal{N}$ and also write, for $\mathcal{N} = 2$ case,

$$(\Delta\mathcal{X})^2 = \begin{pmatrix} D & A \\ A^T & D \end{pmatrix} \tag{6}$$

One can write uncertainty relations explicitly by using the symplectic invariants constructed from the determinants of covariances $|A|$, $|D|$, $|(\Delta\mathcal{X})^2|$ as

$$\zeta_\pm^2 = |D| + |A| \pm \sqrt{(|D| + |A|)^2 - |(\Delta\mathcal{X})^2|} \geq \frac{1}{4}.$$

Let us consider a two-mode squeezed state with a squeezing parameter r as an initial state. Its correlation matrix is

$$(\Delta\mathcal{X})^2(0) = \frac{1}{2}\begin{pmatrix} \cosh(2r)\mathbf{1} & -\sinh(2r)\sigma_3 \\ -\sinh(2r)\sigma_3 & \cosh(2r)\mathbf{1} \end{pmatrix}.$$

In the Wigner representation, the same state can be expressed as

$$W\ (R_1, R_2, P_1, P_2) = \frac{4}{\pi^2} \tag{7}$$
$$\exp\left\{-e^{2r}\left[\Omega(R_1 - R_2)^2 + (P_1 + P_2)^2/\Omega\right] -e^{-2r}\left[\Omega(R_1 + R_2)^2 + (P_1 - P_2)^2/\Omega\right]\right\}.$$

This state can be obtained by acting a squeezing operator $e^{ir(R_1P_2 - P_1R_2)}$ on the vacuum. In Fig. 1, the temporal behavior of the uncertainty function ζ_- is plotted. The initial state is a pure two mode squeezed state introduced above that satisfies the uncertainty relation with the minimum uncertainty 1/4. As the state becomes mixed, the uncertainty increases monotonically in time even for a zero temperature case. At higher temperature, the rate of increase is faster.

Similarly the separability conditions are

$$\lambda_\pm^2 = |D| - |A| \pm \sqrt{(|D| - |A|)^2 - |(\Delta\mathcal{X})^2|} \geq \frac{1}{4}.$$

Note that the inequalities for ζ_+ and λ_+ follow automatically from those for ζ_- and λ_-. Thus λ_- carries the essential information on the separability of quantum states. In Fig. 2, time evolution of the λ_- is plotted. For an initial coherent state, the uncertainty relation for the partial transposed state is always satisfied throughout the whole evolution indicating that there is no entanglement. For an initial squeezed state ($r = 0.1$), the uncertainty relation is violated initially but eventually satisfied indicating that there is a crossover from an entangled to a separable state. The asymptotic value of separability seen in λ_- appears to be independent of the degree of initial squeezing.

Next we consider the case when the first

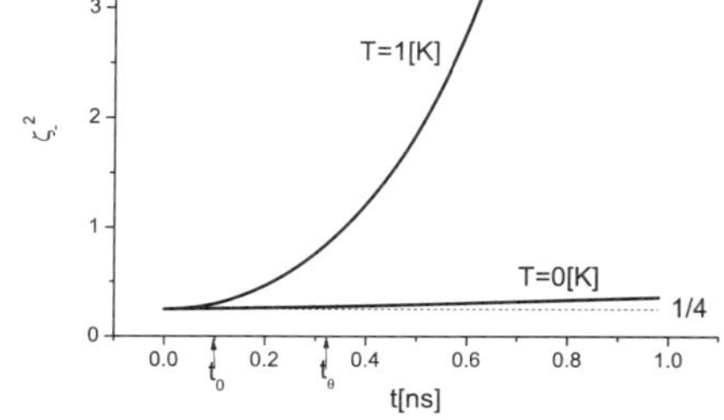

Fig. 1. The temporal evolution of the uncertainty function ζ_- is plotted. The initial condition is a two mode squeezed state with $r = 0$. Other parameters are $\gamma = 0.01[GHz]$. We take $\Omega = 1.0[GHz]$ for all the figures.

observer is under constant acceleration a. Its coordinate is given by $t_1(\tau) = \sinh(a\tau)/a$ and $x_1(\tau) = \cosh(a\tau)/a$. The second observer is fixed at the location $x_2 = 0$. The Heisenberg equations of motion for this case can be written in the same form as in (4). The world lines of two observers are shown in Fig.3. Due to accelerated motion, there is an event horizon for the first observer such that no signal from the second observer reaches the first after he crosses the horizon.

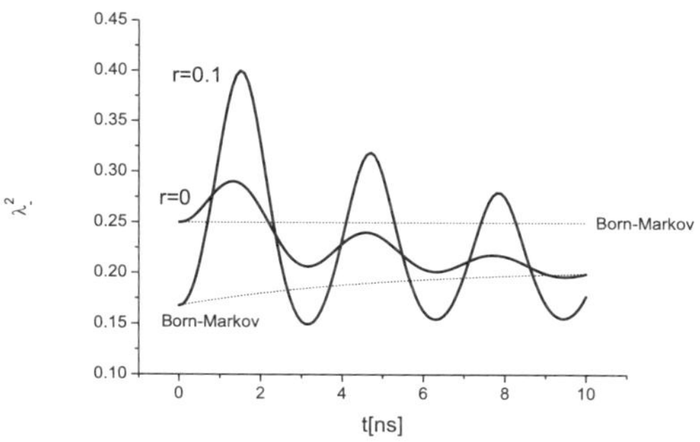

Fig. 2. The temporal evolution of the uncertainty function λ_- after partial transpose is plotted. The initial condition is a two mode squeezed state with $r = 0.1$ for the thick solid curve and a coherent state ($r = 0$) for the thin solid curve. Other parameters are $\gamma = 0.1[GHz]$, $\Omega = 1.0[GHz]$, $T = 0$.

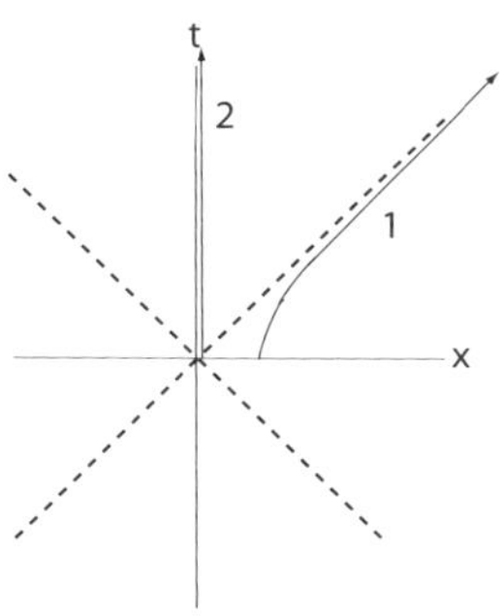

Fig. 3. The world line of each observer.

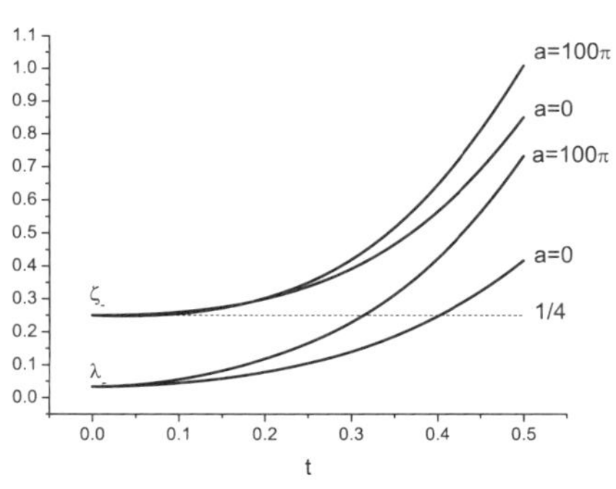

Fig. 4. The temporal evolution of the uncertainty function ζ_- and λ_- before and after partial transpose is plotted. The initial condition is a two mode squeezed state with $r = 0.5$. Other parameters are $\gamma = 0.05$, $\Omega = 1.0$.

3. Conclusion

We showed the exact entanglement dynamics of quantum states of static and accelerated observers. Due to interaction with environment, quantum states of the observer suffer decoherence and disentanglement. As a result of Unruh effect[13], accelerated observer sees vacuum as a thermal state that causes rapid disentanglement of accelerated observer's state faster for the larger acceleration.

References

1. A. Peres, *Phys. Rev. Lett.* **77**, 1413 (1996); P. Horodecki, *Phys. Lett.* **A232**, 333 (1997); R. Simon, *Phys. Rev. Lett.* **63**, 2726 (2000).
2. G. Vidal and R. F. Werner, *Phys. Rev.* **A65**, 032314-1 (2002).
3. G. Giedke, M. M. Wolf, O. Krüger, R. F. Werner, and J. I. Cirac, *Phys. Rev. Lett.* **91**, 107901-1 (2003).
4. A. O. Caldeira and A. J. Leggett, *Physica* **A121**, 587 (1983)
5. F. Haake and R. Reibold, *Phys. Rev.* **A32**, 2462 (1985).
6. H. Grabert, P.Schramn and G. L. Ingold, *Phys. Rep* **168**, 115 (1988).
7. A. Furusawa, et. al, *Science* **282**, 282 (1998).
8. F. Grosshans, G. Van Assche, J. Wenger, R. Brouri, N. J. Cerf, and P. Grangier, *Nature* **421**, 238 (2003).
9. A. K. Rajagopal and R. W. Rendell, *Phys. Rev.* **A63**, 022116 (2001).
10. K. Shiokawa, *Jour. of Math. Chem.* , online version available (2008).
11. K. Shiokawa, to appear in Phys. Rev. A.
12. K. Shiokawa and B. L. Hu, *Phys. Rev.* **A70**, 062106 (2004).
13. W.G. Unruh, *Phys. Rev.* **D14**, 870 (1976).

COUPLING SUPERCONDUCTING FLUX QUBITS USING AC MAGNETIC FLXUES

YU-XI LIU[1,2] and FRANCO NORI[1,2,3]

[1] *Advanced Science Institute, The Institute of Physical and Chemical Research (RIKEN), Wako-shi 351-0198, Japan*

[2] *CREST, Japan Science and Technology Agency (JST), Kawaguchi, Saitama 332-0012, Japan*

[3] *Center for Theoretical Physics, Physics Department, Center for the Study of Complex Systems, The University of Michigan, Ann Arbor, Michigan 48109-1120*

Exploiting the intrinsic nonlinearity of superconducting Josephson junctions, we propose a scalable circuit with superconducting (SC) qubits which is very similar to the successful one now being used for trapped ions. The SC qubits are coupled to the "vibrational" mode provided by a superconducting LC circuit or its equivalent (e.g., a SQUID). Both single-qubit rotations and qubit-LC-circuit couplings/decouplings can be controlled by varying frequencies of the applied ac magnetic fluxes. The circuit is scalable since the qubit-qubit interactions, mediated by the LC circuit, can be selectively performed, and the information transfer can be realized in a controllable way. We also discuss how to control the couplings between the qubit and the data bus via "dressed" qubit states. This approach can also be reduced to the one used for trapped ions.

Keywords: superconducting flux qubits; controllable qubit-data-bus coupling; dressed states.

1. Introduction

Superconducting quantum circuits with Josephson junctions are currently studied for their potential applications in quantum information processing [1]. A major challenge for superconducting (SC) qubits is how to design an experimentally realizable circuit where the couplings for different qubits can be selectively switched on and off, and then "scaled up" to many qubits. Although two-qubit gates can be generated with always-on interbit couplings, it is still very difficult to scale up experimental circuits. Theoretically, the circuits can be scaled up via a common data bus (DB). The DB modes are either virtually excited or excited. In the former case, the effective qubit couplings are controlled by the magnetic flux within nanoseconds through the circuit, which is a challenge for current experiments. In the later case, the qubit and the DB are coupled or decoupled when they have the same (resonant) or different (non-resonant) frequencies, realized via a sudden non-adiabatic change of either the qubit or the DB eigenfrequency. This introduces additional errors for the qubits.

Combining the "ac-signal controlled coupling approach" proposed by us [2] with the experimental versatility of the quantum LC circuit [3], we have studied two different approaches to realize controllable coupling: one uses the nonlinearity of the Josephson junction, another uses "dressed" qubit states.

In our approaches, the individual properties (e.g., eigenfrequencies) of the DB and SC qubits are *always fixed*, but the SC qubit-DB *couplings* can be conveniently *controlled by changing the frequencies* of the applied ac signal. The quantum LC circuit or its equivalent has to be excited when the information is transferred from one qubit to another; therefore it is an active element, not a passive one which is just virtually excited.

Our proposal [4,7] can be essentially reduced to the one used for trapped ions [5]. Both single-qubit rotations and qubit-LC-circuit couplings/decouplings can be controlled by the *frequencies* of the applied ac magnetic fluxes. It means that these qubits

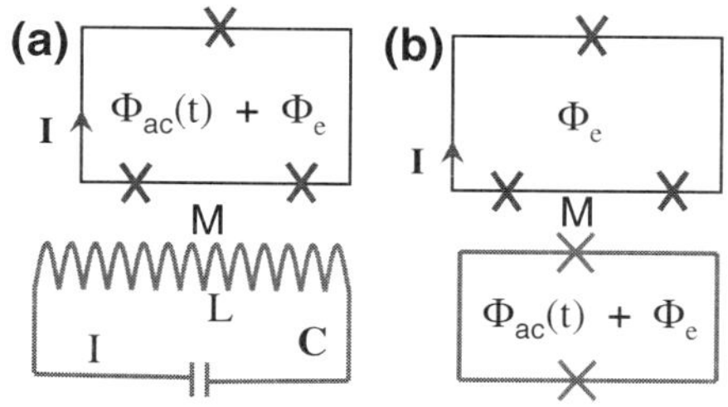

Fig. 1. Schematic diagram for the coupling between a superconducting flux qubit with three Josephson junctions and an LC circuit through mutual inductance M in (a), here the ac signal is applied to the qubit loop. In (b), we replace the LC circuit by a superconducting loop with two Josephson junctions, in this case, the ac signal is applied to the loop with the two Josephson junctions.

can be coupled and separately addressed similarly to trapped ions.

2. Switchable coupling using the nonlinearity of the junctions

We now study how to control the coupling between a superconducting flux qubit and an LC circuit via an ac magnetic flux $\Phi_{\rm ac}(t) \propto \cos(\omega_c t)$ with frequency ω_c, as shown in Fig. 1(a) (or its equivalent in Fig. 1(b)). The system Hamiltonian for both Figs. 1(a) and 1(b) can be written [4] as

$$\begin{aligned} H = \hbar\omega a^\dagger a + \frac{\hbar}{2}\omega_q\sigma_z + (\lambda\sigma_- + \text{h.c.})\cos(\omega_c t) \\ - (a^\dagger + a)(\Omega\sigma_- + \text{h.c.})\cos(\omega_c t) \\ + \hbar\,(\chi\sigma_+ a + \text{h.c.}) \qquad (1) \end{aligned}$$

with qubit frequency ω_q and LC circuit frequency ω. Here, the LC circuit is modeled as a harmonic oscillator with creation (annihilation) operator $a^\dagger$ (a). The always-on coupling constant between the qubit and the LC circuit is χ. The coupling between the qubit and the ac magnetic flux is λ. The nonlinear coupling between the qubit, the LC circuit, and the ac magnetic flux is Ω; this coupling Ω is from the nonlinearity of the Josephson junctions. We assume that the difference Δ' between ω_q and ω is much larger than χ, that is,

$$\Delta' = \omega_q - \omega \gg |\chi|, \qquad (2)$$

which can be satisfied when the circuit is initially fabricated. The condition in Eq. (2) implies that the always-on coupling χ between the LC circuit and the qubit can be neglected, then the Hamiltonian in Eq. (1) has the same form as the one used for trapped ions, in the standard Lamb-Dicke limit (see, e.g., Ref. [5]).

Analogous to the case of trapped ions, in our proposed devices, three-types of dynamical evolutions (*carrier process; red sideband excitation; and blue sideband excitation*) can be produced by the ac magnetic flux using the *frequency-matching* (resonant) condition and neglecting all fast oscillating terms. These three dynamical evolutions can be described as follows.

(i) If $\omega_c = \omega_q$, the qubit and the DB evolve independently in the large detuning condition. The external flux $\Phi_{\rm ac}(t)$ is only used to separately address the lth qubit rotations. These rotations are described by

$$H_c = \lambda\sigma_- + \text{H.c.}, \qquad (3)$$

in the interaction picture and using the rotating-wave approximation (RWA) (also for the H_r and H_b shown below). This is the so-called *carrier process* in the trapped ions approach.

(ii) If the frequencies satisfy the condition $\omega_c = \omega_q - \omega$, then the $\Phi_{\rm ac}(t)$ assists the qubit to couple resonantly with the DB. This is the *red sideband* excitation, described by

$$H_r = \Omega\, a^\dagger\, \sigma_- + \text{H.c.}. \qquad (4)$$

(iii) In the *blue sideband* excitation, the frequencies satisfy the condition $\omega_c = \omega_q + \omega$, with the Hamiltonian

$$H_b = \Omega a \sigma_- + \text{H.c.}. \qquad (5)$$

It can be easily proved that our derived Hamiltonian in Eq. (1) can naturally *explain experimental results on the sideband excitations*, e.g., in Ref. [3]. The qubit-LC couplings/decouplings can be *controlled* by appropriately selecting the ω_c of $\Phi_{ac}(t)$ to match/mismatch the above *frequency* conditions of the sideband excitations. We point out that when the ac signal is applied to the qubit as shown in Fig 1(a), the qubit cannot work at the optimal point because of the symmetry of the qubit potential energy [6]. This problem can be solved by either replacing the LC circuit by a superconducting loop with Josephson junctions [as in Fig. 1(b)] or by using dressed states as discussed below.

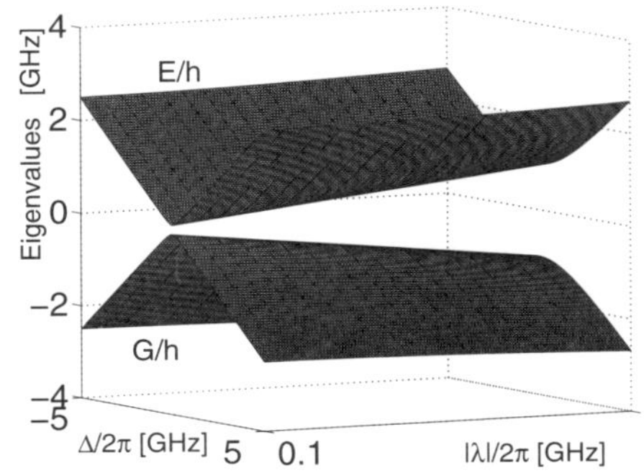

Fig. 2. The dependence of the eigenvalues E and G on the detuning $\Delta/2\pi = (\omega_q - \omega_c)/2\pi$. Here, the eigenvalues have been rescaled as E/h and G/h. The gap $[E(\Delta,|\lambda|) - G(\Delta,|\lambda|)]/h$ between two surfaces denotes the transition frequency $\Omega_{\rm dressed}$ of the dressed qubit (i.e., the qubit plus the applied magnetic flux).

3. Switchable coupling using dressed qubit states

Let us now consider another approach to achieve controllable coupling between qubits and the data bus by using dressed qubit states [7]. In this approach, the data bus is a simple LC circuit, as shown in Fig 1(a). We assume that the ac magnetic flux is still applied to the qubit loop but *the qubit now works at the optimal point.* Thus the nonlinear coupling term Ω, i.e., the fourth term of Eq. (1) is zero due to the symmetry of the potential energy [6] at the optimal point. In this case, the Hamiltonian of the system shown in Fig 1(a), with the always-on coupling, is written as

$$H = \hbar\omega a^\dagger a + \frac{\hbar}{2}\omega_q\sigma_z + \hbar\left(\chi\sigma_+ a + \text{h.c.}\right) + \hbar\left(\lambda e^{-i\omega_c t}\sigma_+ + \text{H.c.}\right). \qquad (6)$$

In the rotating reference frame via the unitary transform $U_R = \exp(-i\omega_c\sigma_z t/2)$, the Hamiltonian in Eq. (6) is transformed to an effective Hamiltonian

$$H_e = U_R^\dagger H\, U_R - iU_R^\dagger\left(\frac{dU_R}{dt}\right). \qquad (7)$$

We can divide the Hamiltonian H_e into two parts, one part is the interaction Hamiltonian $H_e^{(1)}$ between the quantized field the qubit, and the ac magnetic flux, another part is the driven qubit Hamiltonian $H_e^{(2)}$ by the ac magnetic flux, i.e., $H_e = H_e^{(1)} + H_e^{(2)}$ with

$$H_e^{(1)} = \hbar\omega a^\dagger a + \hbar\left(\chi a\sigma_+ e^{i\omega_c t} + \text{H.c.}\right), \qquad (8a)$$

$$H_e^{(2)} = \frac{\hbar}{2}\Delta\sigma_z + \hbar\left(\lambda\sigma_+ + \text{H.c.}\right), \qquad (8b)$$

where $\Delta = \omega_q - \omega_c$. The Hamiltonian $H_e^{(2)}$ can be diagonalized as

$$H_e^{(2)} = \frac{\hbar}{2}\Omega_{\rm dressed}\,\rho_z \qquad (9)$$

with the transition frequency

$$\Omega_{\rm dressed} = \sqrt{\Delta^2 + 4|\lambda|^2}.$$

Here, ρ_z is given as $\rho_z = |E\rangle\langle E| - |G\rangle\langle G|$ in the new basis states (dressed states)

$$|E\rangle = \cos\frac{\eta}{2}|e\rangle + e^{i\phi}\sin\frac{\eta}{2}|g\rangle, \qquad (10a)$$

$$|G\rangle = -\sin\frac{\eta}{2}|e\rangle + e^{i\phi}\cos\frac{\eta}{2}|g\rangle, \qquad (10b)$$

with $\eta = \tan^{-1}(2|\lambda|/\Delta)$. The eigenvalues E and G, corresponding to the eigenstates $|E\rangle$ and $|G\rangle$, are denoted by

$$E = -G = \frac{\hbar}{2}\sqrt{\Delta^2 + 4|\lambda|^2}. \qquad (11)$$

The phase ϕ is related to the coupling constant λ, i.e., $\lambda = |\lambda|e^{-i\phi}$, it can be controlled by the applied ac magnetic flux. In Fig. 2, the dependence of the eigenvalues E/h and

G/h on the detuning Δ have been plotted for a given coupling constant, e.g., $|\lambda|/2\pi \sim 200$ MHz. It clearly shows that the transition frequency $\Omega_{\rm dressed}$ can be changed by changing ω_c when $|\lambda|$ and ω_q are given.

If $\omega > \omega_c$, then the effective Hamiltonian H_e in the dressed state basis of Eq. (10) can be rewritten as

$$H_e = \hbar\,\omega a^\dagger a + \frac{\hbar}{2}\Omega_{\rm dressed}\rho_z + \hbar(\kappa\,\rho_- a^\dagger e^{i\omega_c t} + \text{H.c.}) \quad (12)$$

with the coupling constant $\kappa = \chi\cos^2(\eta/2)$. The Hamiltonian in Eq. (12) also has the same form as the one for trapped ions, as in Ref. [5]. The ladder operator ρ_- is defined as $\rho_- = |G\rangle\langle E|$. Here, the terms $[\chi\sin(\eta)/2]\rho_z a e^{i\omega_c t}$, $\sin^2(\eta/2)\rho_- a e^{-i\omega_c t}$, and their complex conjugates have been neglected for the following reason: there is no way to conserve the energy in these terms, and thus these terms can be neglected by using the usual rotating wave approximation.

To easily understand the coupling mechanism, in the interaction picture, we can rewrite the Hamiltonian in Eq. (12) as

$$H_{e,\rm int} = \hbar\kappa\,\rho_- a^\dagger e^{i[(\omega-\omega_c)-\Omega_{\rm dressed}]t} + \text{H.c.}. \quad (13)$$

If $\Omega_{\rm dressed} = \omega - \omega_c$, then the fast-oscillating factor $e^{i[(\omega-\omega_c)-\Omega_{\rm dressed}]t}$ and its complex conjugate are always equal to one. In this case, the Hamiltonian (13) becomes

$$H_{e,\rm int} = \hbar\kappa\,\rho_- a^\dagger + \text{H.c.}. \quad (14)$$

The condition $\Omega_{\rm dressed} = \omega - \omega_c$ can always be satisfied by choosing the appropriate frequency ω_c of the ac magnetic flux and the Rabi frequency λ. Therefore, the dressed qubit can resonantly interact with the LC circuit, and then the coupling between the qubit and the LC circuit is implemented with the assistance of the ac magnetic flux.

4. Conclusions

In summary, we proposed two approaches to achieve the controllable coupling between the qubit and a data bus (e g., the LC circuit or its equivalent, a cavity field etc.). Using our approaches [4,7], a one qubit circuit can be scaled up to many ones by using a data bus, and then the information can be transferred from one qubit to other distant ones. We should also mention that the experimentalists have demonstrated the dressed qubit states [8] and the side-band excitations [9] after our two proposals.

Acknowledgments

We thank our collaborators in Refs. [4,7]. F. Nori gratefully acknowledges partial support from the NSA, LPS, ARO, NSF grant No. EIA-0130383, JSPS-RFBR 06-02-91200.

References

1. J.Q. You and F. Nori, Phys. Today **58** (11), 42 (2005).
2. Y. X. Liu, L. F Wei, J. S. Tsai, and F. Nori, Phys. Rev. Lett. **96**, 067003 (2006).
3. I. Chiorescu *et al.*, Nature **431**, 159 (2004); K. Semba *et al.*, Proc. of 24th Inter. Conf. on Low Temp. Phy., Florida, 939 (2005); J. Johansson *et al* , Phys. Rev. Lett. **96**, 127006 (2006).
4. Y. X. Liu, L. F. Wei, J. R. Johansson, J. S. Tsai, and F. Nori Phys. Rev. B **76**, 144518 (2007).
5. M. Sasura and V. Buzek J. Mod. Opt. **49**, 1593 (2002).
6. Y. X. Liu, J. Q. You, L. F. Wei, C. P. Sun, and F. Nori Phys. Rev. Lett. **95**, 087001 (2005).
7. Y. X. Liu, C. P. Sun, and F. Nori, Phys. Rev. A **74**, 052321 (2006).
8. C. M. Wilson *et al.*, Phys. Rev. Lett. **98**, 257003 (2007).
9. A. Wallraff *et al* , Phys. Rev. Lett. **99**, 050501 (2007).

ENTANGLEMENT PURIFICATION USING NATURAL SPIN CHAIN DYNAMICS AND SINGLE SPIN MEASUREMENTS

KOJI MARUYAMA[1,*] and FRANCO NORI[1,2,3]
[1] *Advanced Science Institute, The Institute of Physical and Chemical Research (RIKEN), Wako-shi 351-0198, Japan*
[2] *CREST, Japan Science and Technology Agency, Kawaguchi, Saitama 3320012, Japan*
[3] *Center for Theoretical Physics, Physics Department, Center for the Study of Complex Systems, University of Michigan, Ann Arbor, MI 48109-1040, USA*
[*] *E-mail: kmaruyama@riken.jp*

We present a simple protocol to purify bipartite entanglement in spin-1/2 particles by utilising only natural spin-spin interactions, i.e. those that can commonly be realised in realistic physical systems, and S_z-measurements on single spins. Even the standard isotropic Heisenberg interaction is shown to be sufficient to purify mixed state entanglement if there are at least three pairs of spins. This approach could be useful for quantum information processing in solid-state-based systems.

Keywords: Entanglement purification; Spin chains.

1. Introduction

Entanglement purification is one of the most important tasks in quantum information processing[1,2]. It is a process to extract strongly entangled pairs out of initially weakly entangled ones using local operations and classical communication. Most well-known entanglement purification protocols for bipartite two dimensional systems were proposed by Bennett et al.[1] (the *BBPSSW* protocol) and Deutsch et al.[2] (the *DEJMPS* protocol). Both operate on two pairs shared between two separated parties, Alice and Bob, to enhance entanglement in one of the pairs. An operation commonly used in these protocols is a controlled-NOT (CNOT) gate applied at both ends of the channel. Yet, in general, carrying out a CNOT operation is one of the toughest challenges in experiments. Here, we will consider spin-1/2 particles as a representative of two-level systems because our main interest is in the behaviour of entanglement in spin chains with standard spin-spin interactions.

2. Physical motivation of this work

Our aim is to show that entanglement purification is possible only with unmodulated natural interactions between spins and simple S_z-measurements on single spins, as long as the fidelity of the initial state to a Bell state is larger than 1/2. The physical intuition behind this idea is that measuring a subset of spins would project the remaining pairs onto a more strongly entangled state since there are constructive and destructive interference of magnons in a spin chain during its dynamical evolution[3,4]. Here, there is *no* need of mapping quantum information to different physical systems, e.g. photons. Also, there is *no* need of artificial multi-qubit gate operations, such as CNOT. With an isotropic Heisenberg interaction, the implementation of a CNOT has to involve at least using two two-spin operations[5], that is, the accumulation of errors is very likely to occur. It does not entail even single spin operations after the first round of the protocol.

3. Entanglement purification by spin chain dynamics

A general scenario of entanglement purification by spin chain dynamics is depicted in Fig. 1. Alice and Bob share n pairs of weakly entangled spins forming two local spin chains. Each chain will evolve due to the exchange interaction between nearest-neighbouring spins. We will consider the isotropic Heisenberg interaction,

$$H = -J \sum_{\langle i,j \rangle} \vec{S}_i \cdot \vec{S}_j. \tag{1}$$

After a certain time lapse they turn off the interaction and then measure $n-m$ spins each in the $\{|\uparrow\rangle, |\downarrow\rangle\}$ basis to enhance or reduce entanglement in the remaining m pairs compared with that of the initial state. All active controls we consider for a single run of the protocol are one switching-on and one switching-off of the exchange interaction, and single-spin measurements in the $\{|\uparrow\rangle, |\downarrow\rangle\}$ basis. Single spin operations can be added to the list of available controls, however, the fewer controls, the better. In order to avoid complications due to residual correlations between unmeasured pairs, we consider cases in which only one pair is left unmeasured after a run of the protocol. We will denote vectors $|\uparrow\rangle$ and $|\downarrow\rangle$ by $|1\rangle$ and $|0\rangle$, respectively, and take $\hbar = 1$.

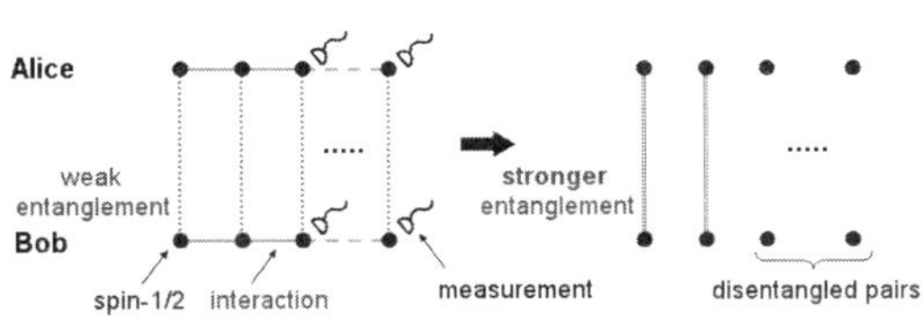

Fig. 1. Entanglement purification by spin chains.

If the initial state of the pair is a partially entangled pure state, then a maximally entangled pair can be attained in a simple fashion. One of the parties, say, Alice, attaches one spin in $|0\rangle$ to her spin to form a

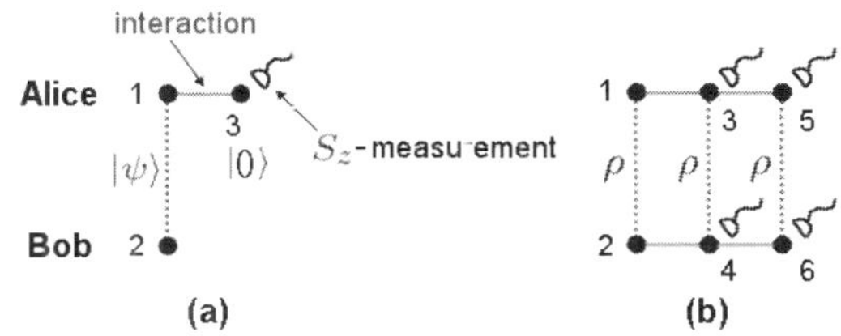

Fig. 2. Configurations of spin chains for entanglement purification.

two-spin chain locally (Fig. 2(a)). Then she lets the two spins evolve under the isotropic Heisenberg interaction. The calculation is quite straightforward[6], so let us move on to the case in which the initial state of the pairs is mixed. In order to simplify the discussion, we shall assume that each pair is in a Werner state, that is

$$\rho_W = F|\Phi^+\rangle\langle\Phi^+| + \frac{1-F}{3}(|\Phi^-\rangle\langle\Phi^-| + |\Psi^+\rangle\langle\Psi^+| + |\Psi^-\rangle\langle\Psi^-|), \tag{2}$$

where $|\Phi^\pm\rangle = (|00\rangle \pm |11\rangle)/\sqrt{2}$ and $|\Psi^\pm\rangle = (|01\rangle \pm |10\rangle)/\sqrt{2}$ are the maximally entangled Bell states, and F is the fidelity between ρ_W and $|\Phi^+\rangle$.

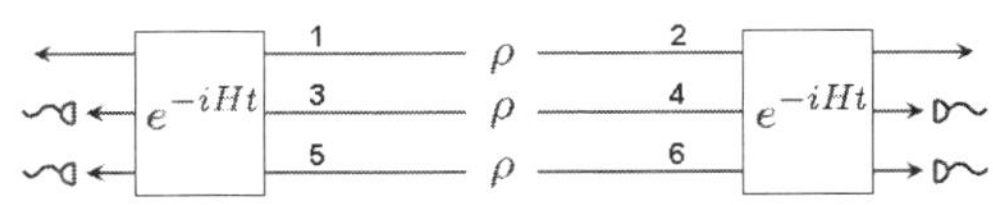

Fig. 3. Diagram representation of the protocol.

It turns out that higher entanglement can never be attained by starting with two pairs of spins, $\rho_W^{\otimes 2}$. This situation does not change even if we introduce anisotropy in the Hamiltonian. Nevertheless, entanglement in a pair can be enhanced if there are two extra pairs, i.e. three pairs in total (Fig. 2(b)). With the initial state $\rho_W^{\otimes 3}$, the measurement outcomes $(0,0,1,1)$ or $(1,1,0,0)$ at sites $\langle 3,4,5,6\rangle$ or $\langle 1,2,5,6\rangle$ lead to a successful fidelity increase (see Fig.3 for the circuit representation and the numbering of spins). The maximum fidelity is achieved at time T such that $JT = (2n+1)2\pi\,(n = 0, 1, 2, ...)$ and it can be calculated as

$$F'_{\max} = \langle\Phi^+|e^{-iHT}\rho_W e^{iHT}|\Phi^+\rangle = \frac{16 - 53F + 118F^2}{59 - 106F + 128F^2}, \quad (3)$$

which is always larger than F if $F > 1/2$. Figure 4 shows the comparison between the maximum reachable fidelity by our protocol and that by the BBPSSW protocol. The fidelity increase in the spin-chain-based protocol is approximately *twice* as large as that by BBPSSW.

Another nice feature of this protocol is that the purified pair stays in the Werner state with the new fidelity $F'_{\max}$. This means that further twirling operations are *not* necessary when iterating the protocol with three new pairs in ρ', unlike the BBPSSW.

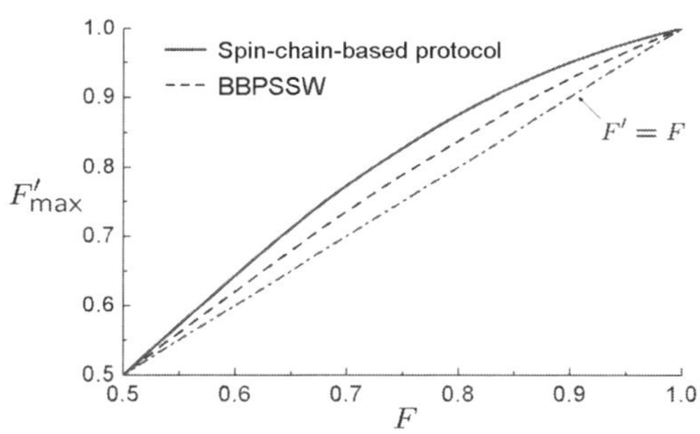

Fig. 4. Maximum attainable fidelity between the resulting state and a Bell state.

4. Comparison between different protocols

Why does the protocol not work with two pairs and why does it work with three pairs? Physically, this difference can be attributed to the behaviour of spinless fermions that appear after the Jordan-Wigner transformation. When there are only two spins in a chain and if there is only one fermion, the fermion propagates (or hops) freely between sites. If there are zero or two fermions, nothing happens (because fermions cannot move to the other site). Hence nothing useful happens from the point of view of entanglement purification. On the contrary, in a three-spin chain, the fermions interact with each other and it gives rise to some nontrivial 'interferences' between them. Then, depending on the measurement results from the subset of the chain, the concentration of entanglement occurs (more details are given in ref [6]).

A price we have to pay for the simplicity and the larger fidelity increase is the probability of success. An appropriate figure of merit for efficiency comparison is the number l of initial pairs to obtain a final pair achieving a desired fidelity increase. Comparing these numbers for BBPSSW (l_B) and for our spin-chain-based protocol (l_{SC}), we find that $l_{\mathrm{SC}}/l_B \simeq 1.16^{r_{\mathrm{SC}}}$ holds for any fidelity increase, where r_{SC} is the number of rounds (or the depth of iteration tree) with our protocol (similarly, r_B below is the number of rounds in the BBPSSW case). This might make our protocol appear less attractive than the BBPSSW. However, more important is the number of multi-spin operations the resulting pair experiences. In our scheme this number is r_{SC}, which is only 23% of $2r_B$, the operations needed for the standard CNOT-based BBPSSW protocol. In addition, the difference in l is rather small. Even if $r_{\mathrm{SC}} = 5$, which corresponds to a fidelity increase from 0.75 to 0.99, l_{SC}/l_B is only 2.07. We therefore wish to state that the efficiency of our protocol is still notable, considering that *only* natural evolutions and initial one-time twirling are involved.

5. Concluding remarks

We have shown that entanglement purification is possible with only the free dynamics of spin chains and simple S_z-measurements on single spins. Arbitrarily high entanglement can be obtained by repeating the process. The amount of operations is minimal, in the sense that single spin operations (twirling) are needed only once for the entire process and only one activation of the inter-spin interaction is required for each run. Hence the error accumulations due to artificial controls should be much less likely compared with conventional schemes.

The proposed scheme might also be experimentally realisable in the near future as the number of spins in this protocol is not large. Candidate physical systems would be, for example, semiconductor quantum dots, superconducting qubits, atoms trapped in optical lattices, etc., where the inter-qubit couplings are often described by Heisenberg-type interactions.

Acknowledgments

Some calculations were performed with the RSCC system of RIKEN. This work was supported in part by the Incentive Research Grant of RIKEN, the National Security Agency (NSA), the Army Research Office (ARO), the Laboratory for Physical Sciences (LPS), the National Science Foundation (NSF) grant No. EIA-0130383, the CREST program of the Japan Science and Technology Agency (JST), the Japan Society for the Promotion of Science Core-to-Core (JSPS-CTC) program, and the JSPS-RFBR grant 06-02-91200.

References

1. C. H. Bennett, G. Brassard, S. Popescu, B. Schumacher, J. A. Smolin, and W. K. Wootters, Phys. Rev. Lett. **76,** 722 (1996); C. H. Bennett, D. P. DiVincenzo, J. A. Smolin, and W. K. Wootters, Phys. Rev. A **54,** 3824 (1996).
2. D. Deutsch, A. Ekert, R. Jozsa, C. Macchiavello, S. Popescu, and A. Sanpera, Phys. Rev. Lett. **77,** 2818 (1996).
3. S. Bose, Phys Rev. Lett. **91,** 207901 (2003).
4. K. Maruyama, T. Iitaka, and F. Nori, Phys. Rev. A **75,** 012325 (2007).
5. D. Loss and D. P. DiVincenzo, Phys. Rev. A **57,** 120 (1998).
6. K. Maruyama and F. Nori, Phys. Rev. A **78,** 022312 (2008).

EXPERIMENTAL ANALYSIS OF SPATIAL QUTRIT ENTANGLEMENT OF DOWN-CONVERTED PHOTON PAIRS

GEN TAGUCHI, TATSUO DOUGAKIUCHI, NOBUAKI YOSHIMOTO, KATSUYA KASAI, MASATAKA IINUMA, HOLGER F. HOFMANN, and YUTAKA KADOYA

Graduate School of Advanced Sciences of Matter, Hiroshima University, Higashihiroshima, 739-8530, Japan

We experimentally demonstrate remote state preparation and 9×9 density matrix reconstruction for spatial qutrits generated by passing down-converted photons through triple slits.

1. Introduction

As was first demonstrated by Neves and co-workers, entangled spatial qudits can be obtained by passing photons generated in spontaneous parametric down-conversion (SPDC) through multi-slits [1]. Recently, we have investigated the possibility of accessing the complete Hilbert space of these multi-dimensional quantum systems by measurements between the focal and image planes of a lens [2]. It then becomes possible to perform quantum state tomography in a single scan in one plane.

In our initial work, we demonstrated this single scan tomography for the case of a double-slit qubit. However, our method can easily be extended to general multi-slit qudits. In this paper, we report the first experimental results using entangled triple-slit qutrits. As in the previous work on qubits, several qutrit states were prepared in one of two arms, A, through the measurements in the other arm, B, and the 3×3 density matrices were reconstructed from the scans in arm A. The complete 9×9 density matrix of the entangled state can then be obtained from a sufficiently large number of settings in arm B.

2. Measurement between the Image and Focal Planes

To analyze the spatial qutrit, we place a lens of focal length f at a distance of L from the triple-slit and set the detector at a distance of z between the focal and image planes, as shown in Fig. 1. The transverse wave-function $\varphi(z;x)$ in the plane of the detector can then be calculated using the Fresnel-Kirchhoff diffraction integral for an effective distance of $R = (Lf + zf - Lz)/(z - f)$ from the multi-slit. If the detection plane is far enough from the image plane, so that the slit width a is sufficiently smaller than $\sqrt{R\lambda}$, the wavefunction takes the conventional slit diffraction pattern given by the sinc function $\mathrm{sinc}(x) = \sin x/x$ as

$$\varphi_n(z;x) = \sqrt{\frac{K}{\pi}} e^{-i\frac{2r_n}{a}Kx} \times \mathrm{sinc}\left[K\left(x + \frac{z-f}{f} r_n\right)\right], \quad (1)$$

where $K = \pi a f/\lambda R(z - f)$ and r_n is the distance between the optical axis and the nth slit. This expression is valid for arbitrary numbers of slits N.

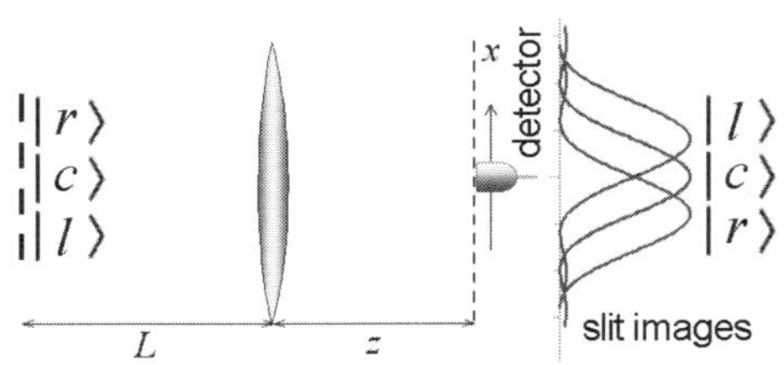

Fig. 1. Measurement between the focal and image planes.

The wavefunctions $\varphi_n(z;x)$ define the N-dimensional Hilbert space of the qudit in the measurement plane. In this Hilbert space, the measurement of x can be given by a projection onto a non-normalized superposition of slits states,

$$|m^{(N)}(x)\rangle = \sum_{n=1}^{N} \varphi_n^*(z;x)|n\rangle. \quad (2)$$

For our triple-slit system, the projection states can then be expressed as

$$|m(x)\rangle = \varphi_l^*(z;x)|l\rangle + \varphi_c^*(z;x)|c\rangle + \varphi_r^*(z;x)|r\rangle, \quad (3)$$

where $|l\rangle$, $|c\rangle$, and $|r\rangle$ correspond to the photon passing through the left, center, and right slit, respectively, as indicated in Fig. 1, with the slit positions $r_l = -d$,$r_c = 0$, and $r_r = d$. The positive operator valued measure describing this projective qudit measurement is then given by [3]

$$\hat{M}(x) \equiv |m^{(N)}(x)\rangle\langle m^{(N)}(x)|. \quad (4)$$

Thus the probability of photon detection at a position x for a qudit state given by a density operator $\hat{\rho}$ is

$$P(x) = \mathrm{Tr}[\hat{M}(x)\hat{\rho}]. \quad (5)$$

The normalization of the wavefunctions $\varphi_n(z;x)$ ensures that the set of operators $\{\hat{M}(x)\}$ satisfies the completeness relation

$$\int_{-\infty}^{\infty} \hat{M}(x)dx = 1. \quad (6)$$

3. Remote State Preparation and Single Scan Tomography

The entangled pair of spatial qutrits generated by using SPDC is ideally given by

$$|\Psi_{\mathrm{pair}}\rangle_{AB} = \frac{1}{\sqrt{3}}\big(|l\rangle_A|r\rangle_B + |c\rangle_A|c\rangle_B + |r\rangle_A|l\rangle_B\big), \quad (7)$$

where the suffixes A and B denote two photons of the entangled pair in the different paths. We can use this entanglement for an arbitrary state preparation of one qutrit by defining a target state through the selection of a specific measurement outcome for the other qutrit. The non-normalized state prepared in arm A by selecting a measurement result of x_B in arm B is

$$|\Psi_{\mathrm{prep}}\rangle_A = {}_B\langle m(x_B)|\Psi_{\mathrm{pair}}\rangle_{AB} = \frac{1}{\sqrt{3}}\big[\varphi_r(z;x_B)|l\rangle_A + \varphi_c(z;x_B)|c\rangle_A + \varphi_l(z;x_B)|r\rangle_A\big]. \quad (8)$$

The 3×3 density matrix of the conditional qutrit state can be reconstructed from the scan data in arm A. In the slit basis, the conditional density matrix of the spatial qutrit is

$$\hat{\rho}_A = \begin{pmatrix} \rho_{ll} & \rho_{lc} & \rho_{lr} \\ \rho_{cl} & \rho_{cc} & \rho_{cr} \\ \rho_{rl} & \rho_{rc} & \rho_{rr} \end{pmatrix}. \quad (9)$$

Using Eq. (5), the probability of detection at a point x is then

$$\begin{aligned} P_A(x) =& \rho_{ll}\,|\varphi_l(z;x)|^2 + \rho_{lc}\,\varphi_l^*(z;x)\varphi_c(z;x) \\ &+\rho_{lr}\,\varphi_l^*(z;x)\varphi_r(z;x) + \rho_{cl}\,\varphi_c^*(z;x)\varphi_l(z;x) \\ &+\rho_{cc}\,|\varphi_c(z;x)|^2 + \rho_{cr}\,\varphi_c^*(z;x)\varphi_r(z;x) \\ &+\rho_{rl}\,\varphi_r^*(z;x)\varphi_l(z;x) + \rho_{rc}\,\varphi_r^*(z;x)\varphi_c(z;x) \\ &+\rho_{rr}\,|\varphi_r(z;x)|^2. \end{aligned} \quad (10)$$

Each density matrix element is thus connected to a unique pattern defined by $\varphi_l(z;x)$, $\varphi_c(z;x)$ and $\varphi_r(z;x)$. In general, the detection probability $P_A(x)$ is a linear function of the nine density matrix elements. Therefore, it is in principle always possible to invert and solve the equation for the elements, so that the density matrix elements can be determined from specific integrals of $P_A(x)$. Thus the density matrix of the spatial qudits can be directly identified with the corresponding spatial patterns observed in the measurement.

4. Experiments

Figure 2 shows our experimental setup. The entangled photons were generated in type II SPDC by a 405 nm CW laser incident on a β-barium borate (BBO) crystal in the collinear condition. The output beam was split into two arms by a polarizing beam splitter (PBS). In each arm, a triple-slit (slit width 40 μm, slit distance 100 μm) was placed in the focal plane of a lens in order to select a pair of transverse momenta of the photons. A lens of focal length $f = 50$ mm was placed at a distance of $L = 2f$ from the triple-slit. A detector system including an 810 nm band pass filter was placed at a distance of $z = 1.8f$ from each lens. The coincidence count rates between the detectors were obtained by scanning the detector position x in arm A, while fixing the other detector position in arm B.

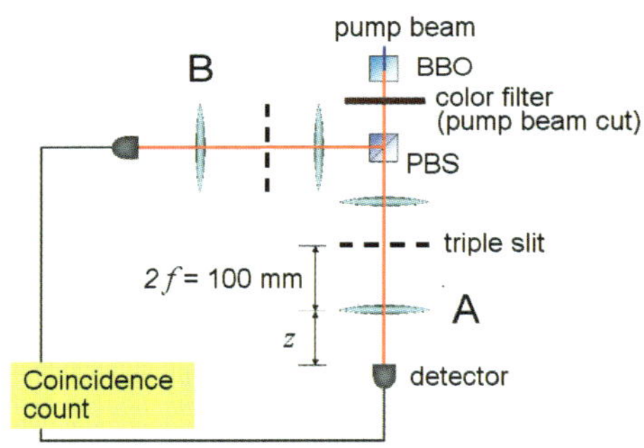

Fig. 2. Experimental setup.

5. Results

Conditional measurements of the pattern $P_A(x)$ in arm A were performed for 27 different detector positions in arm B. The experimental results for nine of them are shown in Fig. 3. From each interference pattern, we can determine the conditional density matrix of the states in arm A by inverting Eq. (10). The density matrices so obtained are shown on the right side of each scan. The fidelity $F = {}_A\langle\Psi_{\text{prep}}|\hat{\rho}_A|\Psi_{\text{prep}}\rangle_A$, of the experimentally determined state is given above each interference pattern in Fig. 3. The results are between $F = 0.741$ and $F = 0.938$, confirming the successful performance of remote state preparation.

The complete 9×9 density matrix can be reconstructed from scanning data in arm A obtained for a sufficient number of different detector positions in arm B. The non-normalized conditional density matrices $\hat{\rho}_A(x_i)$ obtained by setting the detector in arm B to x_i depend on the density matrix $\hat{\rho}_{AB}$ of the entangled qutrit as $\hat{\rho}_A(x_i) = {}_B\langle m(x_i)|\hat{\rho}_{AB}|m(x_i)\rangle_B$. The reconstruction of the density matrix $\hat{\rho}_B$ in arm B from the probabilities $P_B(x_i)$ and the reconstruction of $\hat{\rho}_{AB}$ from $\hat{\rho}_A(x_i)$ are related by a set of reconstruction operators $\hat{\Lambda}_i$ with

$$\hat{\rho}_B = \sum_i P_B(x_i)\hat{\Lambda}_i \tag{11}$$

$$\hat{\rho}_{AB} = \sum_i \hat{\rho}_A(x_i) \otimes \hat{\Lambda}_i. \tag{12}$$

In principle, nine different detection points in arm B are enough to reconstruct the complete two qutrit density matrix $\hat{\rho}_{AB}$. Therefore, nine scans in arm A would be sufficient to reconstruct the density matrix. Figure 4 shows the result obtained from all 27 scans in the $\{|ll\rangle, |lc\rangle, |lr\rangle, |cl\rangle, |cc\rangle, |cr\rangle, |rl\rangle, |rc\rangle, |rr\rangle\}$ basis and the density matrix of the maximally entangled pure state for comparison. The fidelity of the ideal state given in Eq. (7) is $F = {}_{AB}\langle\Psi_{\text{pair}}|\hat{\rho}_{AB}|\Psi_{\text{pair}}\rangle_{AB} = 0.703$. At present, this fidelity seems to be limited by experimental errors such as the precise distances between optical elements. It should, therefore, be possible to improve the experiment by identifying and correcting these errors without changing the setup for the entangled qutrit generation.

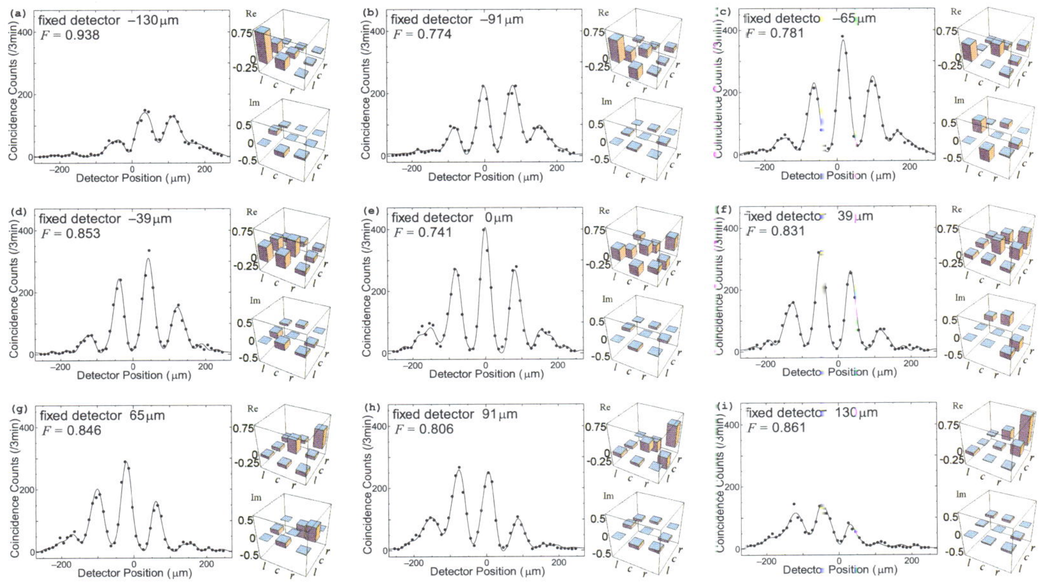

Fig. 3. Coincidence counts data. Each measurement was obtained by placing the fixed detector in a different position. The solid lines are fitted to the data points. The corresponding conditional density matrix is shown on the right side of each interference pattern. The fidelity F of the theoretically expected state is shown above each interference pattern.

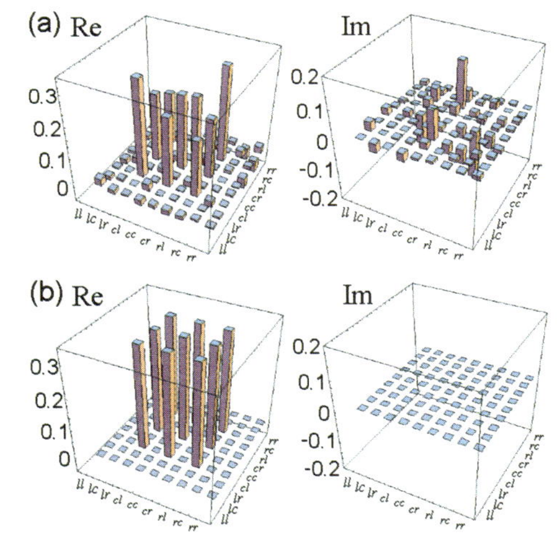

Fig. 4. Real and imaginary part of the density matrix. (a) shows the density matrix reconstructed from the experimental data and (b) shows the maximally entangled pure state for comparison.

6. Conclusion

We have experimentally realized the generation and characterization of spatial qutrits generated by passing down-converted photon pairs through triple-slits. The possibilities of remote state preparation and of accessing the complete Hilbert space of the qutrit have been demonstrated. This method may thus offer a particularly simple and straightforward way of generating and manipulating spatial qutrits for quantum information.

References

1. L. Neves, G. Lima, J.G. AguirreGómez, C. H. Monken, C. Saavedra, and S. Pádua, Phys. Rev. Lett., **94**, 100501, (2005).
2. G. Taguchi, T. Dougakiuchi, N. Yoshimoto, K. Kasai, M. Iinuma, H. F. Hofmann, and Y. Kadoya, Phys. Rev. A, **78**, 012307, (2008).
3. M. A. Nielsen and I. L. Chuang, Quantum Computation and Quantum Information, (Cambridge University Press, Cambridge, UK, 2000).

ON THE PHASE SENSITIVITY OF TWO PATH INTERFEROMETRY USING PATH-SYMMETRIC N-PHOTON STATES

HOLGER F. HOFMANN

Graduate School of Advanced Sciences of Matter, Hiroshima University,
Kagamiyama 1-3-1, Higashi Hiroshima 739-8530, Japan
**E-mail: hofmann@hiroshima-u.ac.jp*

It is well known that the estimation of a small phase shift ϕ from the associated changes in the photon distribution in the output of a two-path interferometer can maximally achieve a sensitivity $1/\delta\phi$ equal to the uncertainty $\Delta(n_1 - n_2)$ of the photon number difference between the two paths. In the following, it is shown that the condition for achieving this quantum Cramer-Rao bound at any phase shift Φ_0 is that the state is symmetric with respect to an (unphysical) exchange of the two paths in the interferometer. If this condition is met, the phase sensitivity should be independent of the initial phase shift Φ_0, implying that experimentally observed phase dependences of sensitivity indicate either a non-optimal estimation strategy, or an effect of decoherence.

Keywords: quantum metrology, two-path interferometers, uncertainty relations

1. Introduction

The increasing precision with which quantum states of light can be controlled have intensified the interest in exploring the ultimate limits of precision of measurements using quantum systems to obtain estimates of small classical parameters, a procedure also referred to as quantum metrology [1]. Specifically, the generation of non-classical N-photon states of light should make it possible to estimate small optical phase shifts between the two arms of a two-path interferometer from the small changes in the photon detection statistics at the output of the interferometer. Recently, there have been several experiments demonstrating the possibility of enhancing the phase sensitivity by using non-classical states such as path-entangled states (also known as NOON states)[2] and pair states [3,4,5]. In these experiments, the interference properties of the non-classical N-photon states are evaluated in terms of the complete phase dependence of the probabilities of obtaining a given photon number distribution in the output. Initially, the focus was on the achievement of N-photon interference fringes, and the question of phase sensitivity was largely ignored. Proper evaluations of the phase sensitivity corresponding to the experimental results have only been performed recently [5,4]. Interestingly, these evaluations show a highly non-trivial phase dependence of the sensitivity. In principle, the sensitivity for a given quantum state depends on the choice of the measurement basis. Since the phase of the interferometer defines a specific output basis, it would be possible that the phase dependence originates from this choice of basis. It is therefore of interest to consider the possibility of performing an optimized phase estimate using only the output photon number at a given phase shift Φ_0.

As will be shown below, the condition for achieving the highest possible phase sensitivity of a given N-photon state (its quantum Cramer-Rao bound) at any given phase shift Φ_0 is simply that the quantum state should be symmetric under an (unphysical) exchange of the two paths in the interferometer. Since this condition applies to the states used in [5,4], it can be concluded that the phase dependences observed come from (a) not measuring all possible output distributions and (b) decoherence effects.

2. Generator-estimator uncertainty relation

For the following arguments, it is convenient to express the photon number statistics of N-photon two-path interference in terms of the Schwinger representation, which corresponds to the operator algebra of a spin-$N/2$ system. In terms of the annihilation operators $\hat{a}_1$ and $\hat{a}_2$ of the two paths inside the interferometer, the three components of the coherence vector $\hat{\mathbf{J}}$ can be defined as

$$\begin{aligned}\hat{J}_1 &= \frac{1}{2}(\hat{a}_1^\dagger \hat{a}_2 + \hat{a}_2^\dagger \hat{a}_1)\\ \hat{J}_2 &= \frac{-i}{2}(\hat{a}_1^\dagger \hat{a}_2 - \hat{a}_2^\dagger \hat{a}_1)\\ \hat{J}_3 &= \frac{1}{2}(\hat{a}_1^\dagger \hat{a}_1 - \hat{a}_2^\dagger \hat{a}_2). \qquad (1)\end{aligned}$$

In this representation, a phase shift between the two arms of the interferometer corresponds to a rotation of the $\hat{\mathbf{J}}$ vector around the J_3-axis, while the interference effect at the output beam splitter can be described by a $\pi/2$ rotation around the J_2-axis, converting $\hat{J}_1$ into $\hat{J}_3$ and vice versa. Effectively, the photon number difference measured at the output ports of the two-path interferometer is therefore given by $2\hat{J}_1$.

The effects of the phase shift of Φ between the two arms of the interferometer on an arbitrary two mode N-photon state $\mid \psi\rangle(0)$ is given by the unitary transformation

$$\mid \psi(\Phi)\rangle = \exp(-i\Phi\hat{J}_3) \mid \psi(0)\rangle. \qquad (2)$$

Thus the operator $\hat{J}_3$ is the generator of the phase shift Φ, in the same sense that the Hamiltonian is the generator of an evolution in time t. It is therefore possible to derive a generator-phase uncertainty corresponding to the Mandelstam-Tamm uncertainty of energy and time [6,7]. Specifically, a small change of phase ϕ around a known phase shift of Φ_0 can be determined measuring an estimator observable $\hat{A}$. The average change of the estimator $\hat{A}$ caused by small phase shifts ϕ is given by

$$\frac{\partial}{\partial\Phi}\langle\hat{A}\rangle = -i\langle[\hat{A}, \hat{J}_3]\rangle, \qquad (3)$$

that is, the rate of change in the expectation value of $\hat{A}$ is given by the expectation value of the commutation relation of $\hat{A}$ and the generator $\hat{J}_3$. The same commutation relation also defines the lower limit of the uncertainty product $\Delta J_3 \Delta A$ for any quantum state $\mid \psi\rangle$. The uncertainty relations therefore limit the rate of change of any expectation value $\langle\hat{A}\rangle$ to

$$\frac{\partial}{\partial\Phi}\langle\hat{A}\rangle \leq 2\Delta A \Delta J_3. \qquad (4)$$

This generator-estimator uncertainty relates the statistical fluctuations in a measurement of the estimator $\hat{A}$ to the average rate of change in $\hat{A}$ caused by ϕ. In an experimental estimate of ϕ, these fluctuations result in a statistical error $\delta\phi^2$ given by

$$\delta\phi^2 = \left(\frac{\Delta A}{\frac{\partial}{\partial\phi}\langle\hat{A}\rangle}\right)^2 \geq \frac{1}{(2\Delta J_3)^2}. \qquad (5)$$

Thus the optimal phase sensitivity that can be achieved by a quantum state $\mid \psi\rangle$ is given by its path uncertainty $2\Delta J_3$ [1,6,7]. In terms of parameter estimation theory, eq.(5) defines the quantum Cramer-Rao bound in terms of a maximal Fischer information of $F = 4\Delta J_3^2$.

In an actual experiment, the possibility of achieving the quantum Cramer-Rao bound depends on the kind of measurements that are performed. In the case of conventional N-photon interferometry, this means that the optimal estimator $\hat{A}$ should be a function of the output photon number distribution observed at a phase bias of Φ_0. In the next section, we therefore consider the explicit construction of an optimal estimator $\hat{A}$ for the output measurement of $\hat{J}_1$.

3. Path symmetry condition

To achieve the quantum Cramer-Rao bound, the estimator $\hat{A}$ must be chosen so that $\mid \psi\rangle$ is a minimal uncertainty state of the generator-estimator uncertainty relation (4). For pure states, this uncertainty relation corresponds to the Cauchy-Schwarz inequality of the Hilbert space vectors $\hat{A} \mid \psi\rangle$ and $-i\hat{J}_3 \mid \psi\rangle$, that is, the state $\mid \psi\rangle$ is a minimal uncertainty state if

$$\lambda\hat{A} \mid \psi\rangle = -i\hat{J}_3 \mid \psi\rangle, \tag{6}$$

where λ is an arbitrary factor used to normalize $\hat{A}$.

If $\hat{A}$ is a function of the measurement outcome $\hat{J}_1$, its eigenstates must be the eigenstates $\mid m_1\rangle$ of $\hat{J}_1$. The eigenvalues A_m of $\hat{A}$ can then be found by multiplying eq. (6) with $\langle m_1 \mid$. The result reads

$$\lambda A_m = \frac{-i\langle m_1 \mid \hat{J}_3 \mid \psi\rangle}{\langle m_1 \mid \psi\rangle}. \tag{7}$$

Since the eigenvalues of the estimator observable must be real, the condition for achieving the quantum Cramer-Rao bound reduces to the requirement that the right hand side of eq.(7) should be real [6]. We can now make use of the symmetry properties of the spin algebra in eq.(1) to further simplify this requirement. Specifically, it is straightforward to confirm that the matrix elements of $\hat{J}_3$ in the $\mid m_1\rangle$ basis are all imaginary. Real eigenvalues of A_m can therefore be obtained if the amplitudes $\langle m_1 \mid \psi\rangle$ are all real. A sufficient condition for achieving the quantum Cramer-Rao bound by conventional two-path interferometry is therefore given by

$$\langle m_1 \mid \psi\rangle = \langle m_1 \mid \psi\rangle^*. \tag{8}$$

To visualize the meaning of this relation, it is useful to consider the non-physical transformation associated with complex conjugation in the $\mid m_1\rangle$ basis. Since the matrix elements of both $\hat{J}_1$ and $\hat{J}_2$ are all real in this basis, the complex conjugation acts only on the all imaginary matrix elements of $\hat{J}_3$, effectively inverting $\hat{J}_3$,

$$\begin{aligned} \hat{J}_1 &\rightarrow \hat{J}_1 \\ \hat{J}_2 &\rightarrow \hat{J}_2 \\ \hat{J}_3 &\rightarrow -\hat{J}_3. \end{aligned} \tag{9}$$

Thus the complex conjugation in the $\mid m_1\rangle$ basis corresponds to an exchange of the two paths in the interferometer that leaves the coherence between the paths unchanged. A state that is invariant under this operation (and hence fulfills eq.(8)) is therefore a path-symmetric state.

Obviously, path symmetry is preserved by phase rotations around the J_3-axis. We can therefore conclude that path-symmetric states achieve their quantum Cramer-Rao bound of $F = 4\Delta J_3^2$ at any phase shift Φ_0, if the corresponding optimal estimator $A_m(\Phi_0)$ for the output photon number differences of $2m$ at Φ_0 is used.

4. Experimental limitations and decoherence effects

Since the two paths in the interferometer are themselves symmetric, it is natural to use path-symmetric states in N-photon two-path interferometry. In particular, maximally path entangled states [2], pair states [3,4,5], and the states generated by interference between coherent state and down-conversion or squeezing [8,9,10] are all path-symmetric and should therefore achieve a phase independent sensitivity at their respective quantum Cramer-Rao bounds. This result is consistent with the work of Pezze et al., that indicates a phase independent sensitivity for interferences between coherent light and squeezed light [10]. However, it seems to be at odds with the very strong phase dependences reported in the two recent works that evaluated the phase sensitivity of four photon pair states [4,5].

As mentioned in the introduction, two reasons contribute to the phase dependence of sensitivity in the case of actual experiments such as the ones reported in [4,5]. Firstly, the experiments reported up to now selected only a particular set of outcomes and did not distinguish all possible photon number distributions in the output. This is the reason why the methods of analysis used in [4,5] even predict a phase dependent sensitivity in the absence of decoherence. On the other hand, another important aspect of the work in [4,5] is the inclusion of decoherence effects. The results suggest that the robustness against decoherence effects is strongly phase dependent, with well defined regions of very low robustness showing up as dips in the phase sensitivity around specific values of Φ_0.

Based on the above analysis, it is possible to explain the robustness against errors by considering the effect of each measurement result m on the phase estimate. As eq.(7) shows, the estimator eigenvalues A_m can be very large for small measurement probabilities $p(m) = |\langle m_1 \mid \psi \rangle|^2$. This means that very rare results still have an appreciable effect on the estimate, since their weight is considerably increased by the high value of A_m. However, these contributions are very sensitive to errors in the probability. In particular, A_m goes to infinity for $p(m) \to 0$, but in a real experiment, the probability of measuring m will not be zero. Consequently, even the smallest background noise will reduce the sensitivity at a phase Φ_0 where $p(m)$ is zero. In an experiment with very small decoherence effects, we can therefore expect that the quantum Cramer-Rao bound will be achieved for all phases except those around an output probability of zero. In the vicinity of those phases, the sensitivity will show a Lorentz-shaped dip to the Fisher information obtained by setting the value of A_m to zero for the results with nearly zero probability.

5. Conclusions

It has been shown that all path-symmetric pure states should achieve their quantum Cramer-Rao bound of $\delta\phi = 1/(2\Delta J_3)$ in a phase estimate based on conventional photon counting statistics in the output of a two-path interferometer. This means that the phase sensitivity at all phases Φ is given only by the uncertainty $2\Delta J_3$ in the photon number difference between the two paths of the interferometer. Phase dependences of the sensitivity indicate either incomplete measurement data or decoherence effects, where the latter should show up as sharp dips around phases where one of the output probabilities goes to zero. In the near future, it should be possible to obtain the complete sets of data to experimentally confirm this fundamental and universal property of highly phase sensitive N-photon states of light.

References

1. V. Giovannetti, S. Lloyd and L. Maccone, *Phys. Rev. Lett.* **96**, 010401 (2006)
2. M.W. Mitchell, J.S. Lundeen, and A.M. Steinberg, *Nature (London)* **429**, 161 (2004)
3. T. Nagata, R. Okamoto, J.L. O'Brien, K. Sasaki and S. Takeuchi, *Science* **316**, 726 (2007).
4. F.W. Sun, B.H. Liu, Y.X. Gong, Y.F. Huang, Z.Y. Ou, and G.C. Guo, *EPL* **82**, 24001 (2008).
5. R. Okamoto, H.F. Hofmann, T. Nagata, J.L. O'Brien, K. Sasaki and S. Takeuchi, *New J. Phys.* **10**, 073033 (2008).
6. S.L. Braunstein and C.M. Caves, *Phys. Rev. Lett.* **72**, 3439 (1994).
7. S. L. Braunstein, C.M. Caves and G.J. Milburn, *Ann. Phys.* **247**, 135 (1996).
8. H.F. Hofmann and T. Ono, *Phys. Rev.* **A76**, 031806R (2007).
9. T. Ono and H.F. Hofmann, *J. Phys. B: At. Mol. Opt. Phys.* **41**, 095502 (2008).
10. L. Pezze, A. Smerzi, G. Khoury, J. F. Hodelin, and D. Bouwmeester, *Phys. Rev. Lett.* **99**, 223602 (2007).

CONTROL OF MULTI-PHOTON COHERENCE USING THE MIXING RATIO OF DOWN-CONVERTED PHOTONS AND WEAK COHERENT LIGHT

TAKAFUMI ONO* and HOLGER F. HOFMANN

Graduate School of Advanced Sciences of Matter, Hiroshima University, Kagamiyama 1-3-1, Higashi Hiroshima 739-8530, Japan
** E-mail:d075972@hiroshima-u.ac.jp*

We show that a class of non-classical superposition states can be generated by interference between down-converted photons and a weak coherent laser light. These interferometric states are characterized by a single parameter η which is determined by the ratio of down-converted amplitude to squared coherent amplitude. For $\eta > 2$, the state is a superposition of two states, where all N-photons are either in one or the other of two non-orthogonal modes.

Keywords: quantum superposition, N-photon state, interferometry

1. Introduction

One of the most interesting tasks in quantum physics is the generation and analysis of non-classical superposition states. In optics, such a state can be realized in the form of a maximally path entangled N-photon NOON state which is a superposition state where all photons are either in one arm or in the other arm of an interferometer. The NOON state is known to be the state which has the maximum phase sensitivity at the Heisenberg limit.

Recently, we have investigated the possibility of generating non-classical N-photon states by interferometric addition of down-converted photons with weak coherent light. We found that it is possible to generate a NOON state with high probability when the mixing ratio $\eta = N\gamma/\alpha^2$ is equal to 2, where η is defined as the ratio of the down-conversion amplitude γ to the squared coherent amplitude α^2 scaled by photon number N [1]. In this paper, we show that by increasing the mixing ratio η to values greater than 2 we can obtain a class of non-classical superposition states where all photons are either in one or the other of two non-orthogonal modes. The overlap between the modes is zero at $\eta = 2$ and increases with increasing η. We can thus control the separation of the quantum state components in the non-classical superposition.

2. Characterization of the interferometric N-photon state

Figure 1 shows an illustration of the generation and measurement of the N-photon interferometric state. Weak coherent laser light $\mid \alpha\rangle$ in input mode $\hat{a}$ interfers with down-converted photons $\mid \gamma\rangle$ in input mode $\hat{b}$ at the input beam splitter of a conventional Mach-Zehnder interferometer. The N-photon interferometric state is discribed by the N-

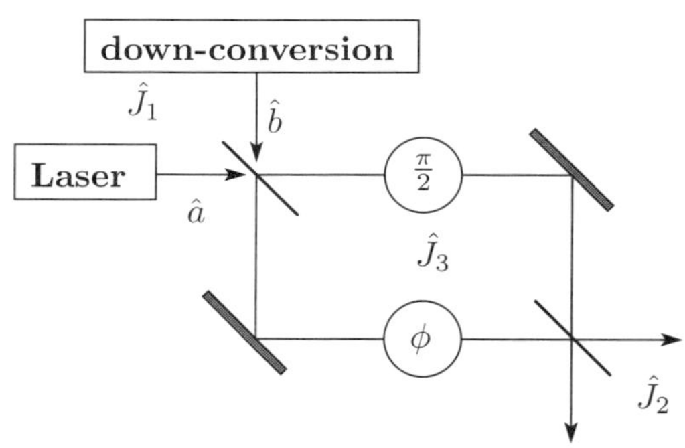

Fig. 1. Illustration of the generation and measurement of the N-photon interferometric state.

photon components of this two mode state. It can be expressed by a superposition of states with $2k$ photons from down-conversion and $N-2k$ photons from the laser, where the coherences are given by a combination of the coherences from $\mid \alpha\rangle$ and from $\mid \gamma\rangle$ [1,2,3]. However, we found that a more compact representation of the coherences can be obtained by using the operator relations that characterize the coherent state in mode $\hat{a}$ and the single mode down-converted state in mode $\hat{b}$,

$$\begin{aligned} \hat{a} \mid \alpha\rangle &= \alpha \mid \alpha\rangle \\ \hat{b} \mid \gamma\rangle &= -\gamma \mid \gamma\rangle. \end{aligned} \tag{1}$$

The quantum coherence of the N-photon component $\mid \eta\rangle$ of the two mode state can then be characterized by photon number preserving combinations of creation and annihilation operators for the two modes. The most compact relation reads[13]

$$\begin{aligned} \hat{a}^\dagger \hat{b} \mid \eta_N\rangle &= \frac{\eta}{N} \hat{a}^\dagger \hat{a} \hat{b}^\dagger \hat{a}, \\ \text{where} \quad \eta &= \frac{N\gamma}{\alpha^2}. \end{aligned} \tag{2}$$

Thus the interferometric N-photon state $\mid \eta\rangle$ is defined by a single parameter η.

To illustrate the properties of the interferometric N-photon state, it is convenient to use the Schwinger representation, which is based on the equivalence of all N+1 dimensional Hilbert spaces. Specifically, the two mode Fock states $\mid N-m; m\rangle$ are equivalent to the Hilbert space of a spin-N/2 system. The corresponding spin components can then be expressed in terms of the input modes $\hat{a}$ and $\hat{b}$ as follows,

$$\begin{aligned} \hat{J}_1 &= \frac{1}{2}\left(\hat{a}^\dagger \hat{a} - \hat{b}^\dagger \hat{b}\right) \\ \hat{J}_2 &= \frac{1}{2}\left(\hat{a}^\dagger \hat{b} + \hat{b}^\dagger \hat{a}\right) \\ \hat{J}_3 &= -\frac{i}{2}\left(\hat{a}^\dagger \hat{b} - \hat{b}^\dagger \hat{a}\right). \end{aligned} \tag{3}$$

As indicated in Fig.1, the three orthogonal components of this vector can be interpreted as follows: $\hat{J}_1$ is half of the photon number differences between the input modes, $\hat{J}_2$ is half of the photon number differences between the output modes at $\phi = 0$, and $\hat{J}_3$ is half of the photon number differences between the paths of the interferometer. Using the Schwinger representation, eq. (2) can be transformed into a non-linear squeezing relation,

$$\begin{aligned} &\left(1 + \left(\frac{1}{2} + \frac{\hat{J}_1}{N}\right)\eta\right) \hat{J}_2 \mid \eta\rangle \\ &\quad = -i\left(1 - \left(\frac{1}{2} + \frac{\hat{J}_1}{N}\right)\eta\right) \hat{J}_3 \mid \eta\rangle. \end{aligned} \tag{4}$$

This relation defines the properties of the η-states for all N

3. Solutions in the classical limit

In the classical limit, states correspond to points on the J-sphere. The points defined by eq. (4) can be found by regarding the operators $\hat{J}_i$ as real numbers. Equation (4) then gives the relation

$$\begin{aligned} &\left(1 + \left(\frac{1}{2} + \frac{J_1}{N}\right)\eta\right) J_2 \\ &\quad + i\left(1 - \left(\frac{1}{2} + \frac{J_1}{N}\right)\eta\right) J_3 = 0. \end{aligned} \tag{5}$$

The classical value of J_2 must always be 0 and the value of J_3 must be 0 if $\eta < 1$. However, for η greater than 1, J_3 can be non-zero if $(1 - (1/2 + J_1/N)) = 0$. The points on the J-sphere given by eq. (5) are therefore given

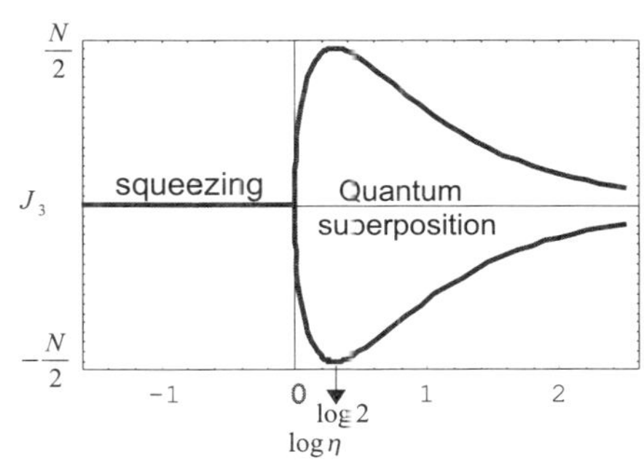

Fig. 2. η-dependence of the classical values of J_3.

by

$$\mathbf{J} = \begin{pmatrix} N/2 \\ 0 \\ 0 \end{pmatrix} \qquad \text{for} \quad \eta < 1$$

$$\mathbf{J} = \begin{pmatrix} \frac{N}{2}\left(\frac{2}{\eta} - 1\right) \\ 0 \\ \pm\frac{N}{\eta}\sqrt{\eta - 1} \end{pmatrix} \qquad \text{for} \quad \eta > 1 \qquad (6)$$

Thus the classical value of J_3 for $\eta > 1$ has two solutions as shown in Fig.2.

For $\eta < 1$, the η-state points in the J_1 direction of the J-sphere. As explained in Ref.3, eq. (4) then represents a linear squeezing relation between J_2 and J_3, where the ratio of uncertainties is given by a squeezing factor of $\Delta J_3/\Delta J_2 \approx (1+\eta)/(1-\eta)$. When η is larger than 1, the η-state splits into a superposition of two J-vectors moving away from the J_1-axis toward the two poles on the J_3-axis. For $\eta = 2$, the J-vectors of the two components point to the opposite poles of the J_3 axis, corresponding to a NOON state superposition of all photons in one arm or in the other arm of an interferometer. For η greater than 2, the J-vectors move towards the negative direction of J_1, approaching the J_1-axis for $\eta \to \infty$. Thus the classical limit of eq. (4) allows us to characterize the η-state on the J-sphere.

4. Quantum superposition for η>2

Once we have determined the classical J-vectors of the quantum state components, the amount of squeezing can be determined by linearizing the squeezing relation given by eq. (4) around the corresponding points in the $J_1 - J_3$ plane. The approximate J_2-squeezing is then described by $\Delta J_2^2 \approx N/4 \times (\eta - 1)/\eta$, indicating that the J_2-squeezing continuously drops back towards the shot noise limit of N/4 when η increases. For $\eta > 2$, the squeezing effects quickly become negligible and the quantum states can be described directly by a superposition of two states where all N photons are in one of two non-orthogonal modes,

$$\hat{c}(\theta_\pm) = \sin\left(\frac{\theta}{2}\right)\hat{a} \pm i\cos\left(\frac{\theta}{2}\right)\hat{b}, \qquad (7)$$

where θ is the angle between the J-vectors and the J_1-axis, as shown in Fig.3. According to eq. (6), this angle is given by $\cos\theta = (1 - 2/\eta)$. Therefore the mode overlap can be controlled by adjusting the mixing ratio η. The approximate η-state for a given value of θ can then be written as

$$\mid \eta\rangle \approx \frac{1}{\sqrt{2 + 2\mathrm{Re}[\langle\theta_+ \mid \theta_-\rangle]}}\left(\mid \theta_+\rangle + \mid \theta_-\rangle\right), \qquad (8)$$

where $\mid \theta_\pm\rangle$ are the states of all N photons in one of the two non-orthogonal modes $\hat{c}(\theta_\pm)$ given by eq. (7). The overlap between the two states is

$$|\langle\theta_+ \mid \theta_-\rangle|^2 = \left(\cos^2\theta\right)^N \approx \mathrm{e}^{-\frac{4N}{\eta}}. \qquad (9)$$

Thus η should be sufficiently smaller than 4N for the two states to be clearly distinguishable.

The effects of the superposition of the two states can be seen in the output photon number distribution J_2. According to eq. (8), the J_2 amplitudes of the η-state are the sums of the amplitudes of the two components $\mid \theta_\pm\rangle$. In the J_2-basis, the absolute values of the amplitudes of both components $\mid \theta_\pm\rangle$ are the same, and only the phases are different. As a result, an interference pattern

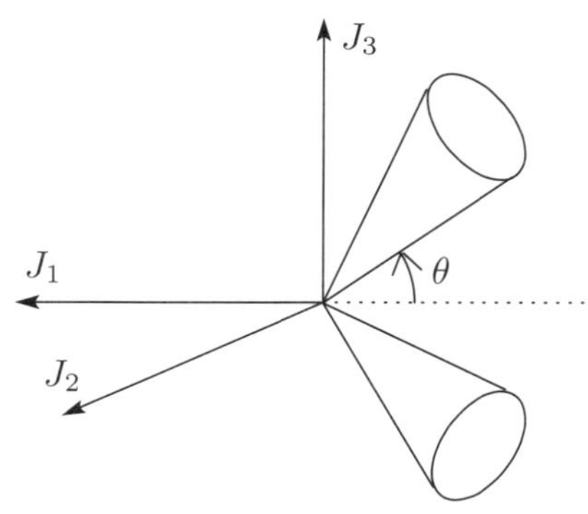

Fig. 3. Illustration of the superposition of the two states on the J-sphere.

appears in the J_2-distribution, where the periodicity of the interference fringes is equal to π/θ, that is, the fringes become narrower as θ increases. The envelope of the interference fringes is determined by the binomial distribution given by the absolute values of the J_2 amplitudes of the components $\mid \theta_{\pm}\rangle$.

The interference fringes of the output can mostly be seen within the width of the envelope, which is approximately given by two times the fluctuations of the photon number distribution of $\mid \theta_{\pm}\rangle$, $2\sqrt{N}$. The number of the interference fringes within the width of the envelope can therefore be estimated from the ratio of the width of the envelope to the periodicity of the interference fringes, $2\sqrt{N}/(\pi/\theta)$. For θ sufficiently smaller than 1, this number is approximately $4/\pi \times \sqrt{N/\eta}$, indicating that there is only a single fringe left around $\eta = N$. In this region, the interference of the two components then appears as a squeezing effect in J_2, where the J_2 fluctuation is determined by the width of the interference fringe instead of the width of the envelope[4].

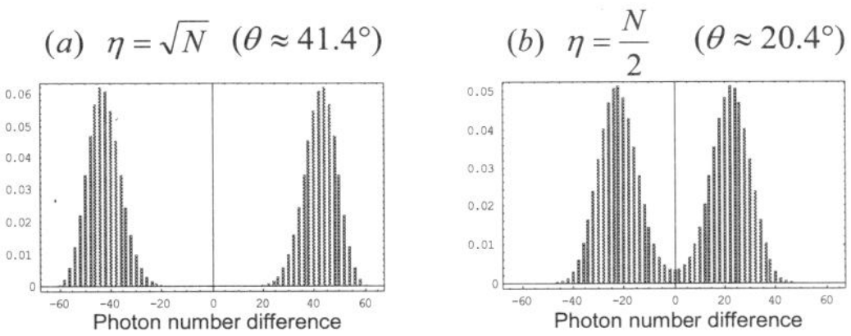

Fig. 4. Photon number distribution inside the interferometer(J_3) of η-state for $N = 64$.

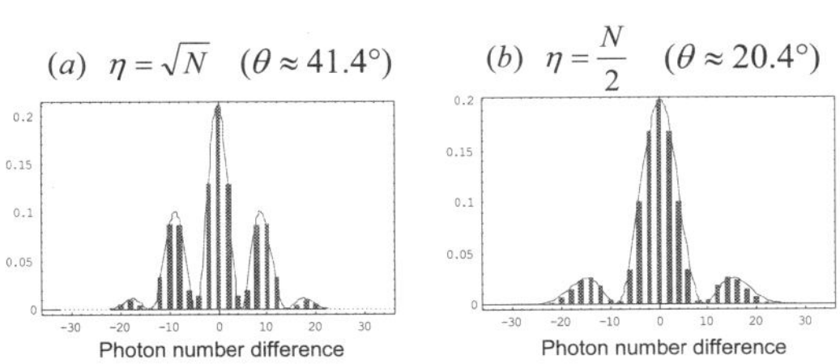

Fig. 5. Output photon number distribution(J_2) of η-state for $N = 64$

Figures 4 and 5 show the photon number distributions of J_2 and J_3 for the precise solution of eq. (4) at $N = 64$. The graphs (a) and (b) show the case of $\eta = \sqrt{N} = 8(\theta \approx 41.4°)$ and $\eta = N/2 = 32(\theta \approx 20.4°)$, respectively. As shown in Fig. 4, the J_3 distributions have two peaks corresponding to the two J-vectors of the superposition. With the increases of η, the two peaks of the J_3 distribution move closer to each other. Figure 5 shows the interference patterns between the two components of the superposition state observed in the output photon number distribution J_2. With the increases of η, the interference fringes grow wider. The numerical results thus correspond well with the theoretical prediction of the simple superposition model of eq. (8).

5. Conclusions

We have shown that a superposition of two states where all N photons are in one of two non-orthogonal modes can be generated by interference between down-converted photons and weak coherent laser light. The separation between the components of the superposition can be controlled by varying the mixing parameter η. It may therefore be possible to experimentally realize the transition between microscopic (kitten) superpositions and macroscpic (cat) superpositions in this system.

References

1. H.F. Hofmann and T. Ono, *Phys. Rev. A* **76**, 031806(R) (2007).
2. F. Shafiei, P. Srinivasan, and Z. Y. Ou, B. Liu and Z. Y. Ou, *Phys. Rev. A* **74**, 035802 (2006)
3. T. Ono, H.F. Hofmann *J. Phys. B: At. Mol. Opt. Phys.* **41**, 095502 (2008).
4. V. Buzek, A. Vidiella-Barranco, and P.L. Knight, *Phys. Rev. A* **45**, 6570 (1992).

RATTLING AS A NOVEL ANHARMONIC VIBRATION IN A SOLID

Z. HIROI* and J. YAMAURA

Institute for Solid State Physics, University of Tokyo, Kashiwa, Chiba 277-8581, Japan
**E-mail: hiroi@issp.u-tokyo.ac.jp*

We study the superconducting properties of the β-pyrochlore oxides AOs_2O_6, where A is Cs, Rb and K. The variations in the superconducting transition temperature T_c from 3.3 K for Cs to 6.3 K for Rb and to 9.6 K for K and also in the superconducting character from weak- to strong-coupling with increasing T_c are ascribed to the enhancement in the rattling of the A ions. The rattling is essentially a local, anharmonic mode with low energy and serves as an effective media for the Cooper pairing in the β-pyrochlore oxides.

Keywords: rattling; superconductivity; pyrochlore oxides.

1. Introduction

Pyrochlore oxides exhibit various intriguing phenomena such as magnetic frustration, metal-insulator transition, colossal magnetoresistance and ferroelectricity.[1] In 2001, superconductivity with T_c = 1.0 K was found in cadmium rhenium oxide $Cd_2Re_2O_7$ for the first time in the family of pyrochlore oxides.[2] Later, alkali metal osmium oxides AOs_2O_6 (A = Cs, Rb and K) were found to show superconductivity below T_c = 3.3, 6.3, and 9.6 K for A = Cs,[3] Rb,[4-6] and K,[7] respectively. This family of compounds are now called the β-pyrochlore oxides, in distinction from the conventional pyrochlores with the general formula $A_2B_2O_7$ that are called the α-pyrochlore oxides. Extensive study has been carried out on the three members of the β-pyrochlore oxides in order to elucidate the nature of their superconductivity. As for the superconducting gap symmetry, most experimental results seem to draw the conclusion that it is of an s-wave BCS type,[8] suggesting conventional phonon-mediated superconductivity: no evidence of contributions from other "glue" such as magnetic fluctuations as sometimes found in an exotic superconductor with a nodal gap. Interesting enough, however, is the observation of pronounced changes of various properties over the series.[9] It is obvious, at least, that a simple BCS picture cannot explain the systematic variation of T_c from Cs to K. There must be a special ingredient for the superconductivity of the β-pyrochlore oxides.

An important structural feature to be focused on the β pyrochlores is the rattling of the A atoms. The β-pyrochlore oxides crystallize in a cubic structure of the space group Fd-3m, where a relatively small A cation lies at the 8b site in the center of an oversized atomic cage made of OsO_6 octahedra, as depicted in Fig. 1.[10, 11] Due to the high symmetry of the site as well as the large size mismatch, the A atoms can "rattle" intensively inside the cage.[10, 12] The rattling has been recognized as a fascinating phenomenon for a class of cage compounds like filled skutterudites[13] and Ge/Si clathrates[14] and attracted many researchers, because it may suppress thermal conductivity leading to an enhanced thermoelectric efficiency. On the

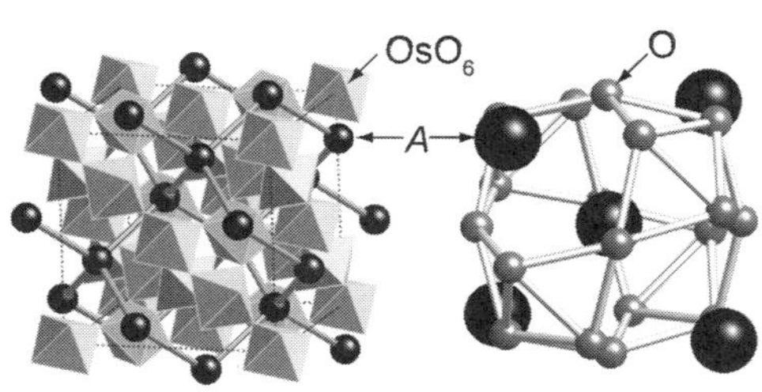

Fig. 1. Crystal structure of the β-pyrochlore oxide AOs_2O_6. The A ion (big ball) is located in an oversized atomic cage made of OsO_6 octahedra and can move along the 4 <111> directions pointing to the neighboring A ions in adjacent cages.

other hand, the rattling itself is also intriguing from the viewpoint of lattice dynamics: it gives an almost localized mode even in a crystalline material and often exhibits unusual anharmonicity. Hence, it is considered that the rattling is a novel type of low-lying excitations that potentially influence various properties in a crystal at low temperature.

The evidence for the rattling has been obtained through various methods; structural refinements,[10, 11] specific heat measurements,[15, 16] NMR relaxation rates from the A nuclei,[17] and more recently spectroscopic methods such as Raman[18, 19] and inelastic neutron scattering experiments.[20, 21] These experiments revealed large atomic displacements for the A cations and found low-energy excitations as well as their softening. For example, the specific heat measurements indicated that the contributions of the rattling were described approximately by the Einstein model and determined the Einstein temperature Θ_E to be 70 K, 60 K and 61 / 22 K for A = Cs, Rb, and K, respectively.[8, 9, 15, 16] Thus, the characteristic energy decreases with decreasing the cation size or mass. This tendency illustrates the uniqueness of the rattling, because, to the contrary, one expects a higher energy for a lighter atom in the case of conventional phonons. Moreover, it was demonstrated that the specific heat shows an unusual T^5 dependence at low temperature below 7 K for A = Cs and Rb, instead of a usual T^3 dependence expected from a normal Debye-type phonon.[15]

2. Superconductivity

Figure 2 shows the superconducting transition in specific heat for all the members of the β-pyrochlores and α-$Cd_2Re_2O_7$. It is apparent that the jump at T_c increases enormously with increasing T_c and also that the Sommerfeld coefficient γ, which is proportional to the density-of-states (DOS) at the Fermi level, increases accordingly. These chemical trends of the superconducting properties over the series are summarized in Fig. 3, which reveal that the character of superconductivity changes systematically from a weak-coupling BCS type to an extremely strong-coupling with increasing T_c. Since the superconducting coherence length ξ deduced from the upper critical field H_{c2} decreases in the same direction, the size of Cooper pairs becomes effectively small with increasing T_c, implying an enhanced pair potential.

According to the first-principles study, the basic electronic structures of the β pyrochlores are identical over the series.[9, 12, 22, 23] The band structure around the Fermi level is given by a manifold of 12 bands arising from Os 5d and O 2p states with a total bandwidth of ~3 eV. The formal valence of metals is A^+ or $Os^{5.5+}$. The guest alkali metals are always purely ionic, having the related states far above the Fermi level, that is, donate one electron to the d-p bands on the cage to induce metallicity. Since the DOS at the Fermi level is nearly the same over the series, the large enhancement in γ, as well as T_c, from Cs to K cannot be understood based on the calculations.

We discuss the mass enhancement in the normal state. The mass enhancement has sometimes a common origin as the Cooper pairing in superconductors. As shown in Fig. 3, the mass enhancement factor increases with

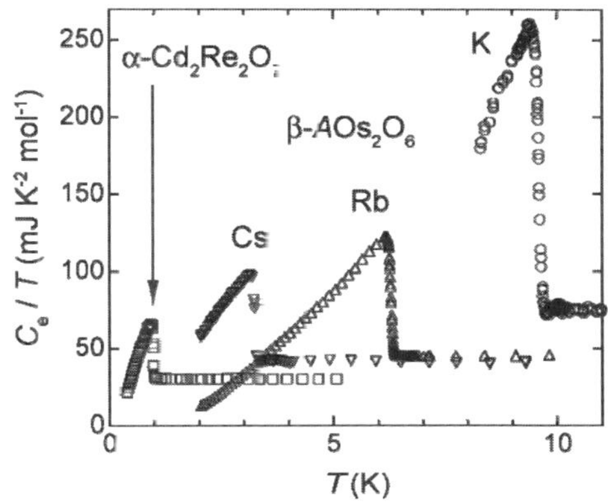

Fig. 2. Superconducting transitions in specific heat for α-$Cd_2Re_2O_7$ and the three members of the β-pyrochlore oxides.

increasing T_c: $\gamma_{exp} / \gamma_{band}$ = 2.6 (Re), 3.6 (Cs), 3.9 (Rb), and 7.4 (K). These give large renormalization factors λ in $\gamma_{exp} / \gamma_{band} = 1 + \lambda$. To be stressed here is the exceptionally large enhancement for K. This is also the case for other parameters, $\Delta C / \gamma T_c$ and H_{c2}, suggesting that some additional effect must be present for K.

The large value of 7.3 for K may not be explained by a single electron-phonon (*e-ph*) interaction, because λ cannot become larger than 2 in such a metallic system owing to the screening effects.[24] Thus, it is required to assume at least two independent renormalization processes; $1 + \lambda = (1 + \lambda_{sc})(1 + \lambda_n)$, where λ_{sc} is effective for superconductivity and

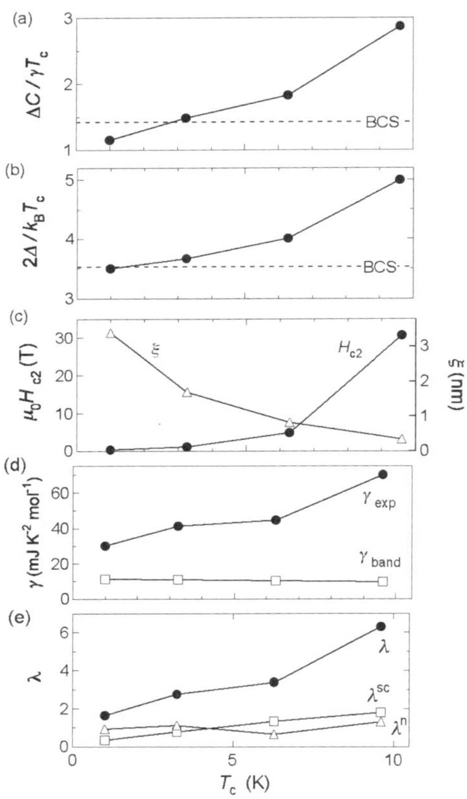

Fig. 3. Evolution of superconducting and normal-state parameters as a function of T_c for α-$Cd_2Re_2O_7$ and β-AOs_2O_6. (a) jump in specific heat at T_c divided by γ, (b) superconducting gap, (c) upper critical field and coherence length, (d) Sommerfeld coefficients from the experiments and the calculations and (e) mass enhancement factor.

λ_n comes from the other effects. The λ_{sc} has been estimated on the basis of strong-coupling analyses on the thermodynamic parameters.[25] Interestingly, the remaining λ_n is always close to 1, as shown in Fig. 3. This means that there is already a mass enhancement of factor 2 commonly to the α and β pyrochlores, which is not related to the superconductivity. This additional enhancement may be attributed to a conventional *e-ph* interaction on the cage, not to an electron-electron (*e-e*) interaction, because there are no evidence for a predominant *e-e* scattering in the β pyrochlores. In fact, Saniz and Freeman calculated the *e-ph* coupling constant assuming a Debye phonon of 285 K from the cage and obtained $\lambda_{ep} \sim 0.8$ for all the members,[23] which is comparable to our λ_n. In this context, it is plausible to ascribe λ_{sc} to the electron-rattler interaction, that is, $\lambda_{sc} = \lambda_{er}$. The observed large increase of $\lambda_{sc} = \lambda_{er}$ toward K is definitely related to the corresponding enhancement in the rattling. Thus, the electron-rattler interaction plays a crucial role governing all the electronic properties of the β pyrochlores.

3. Rattling of the *A* cations

One of the important findings from the previous structural analyses is the unusually large isotropic atomic displacement parameter U_{iso} for the *A* atoms, which equals to U_{11} in the present case. It is decided that U_{iso} is 2.5 × 10^{-4}, 4.3 × 10^{-4} and 7.4 × 10^{-4} nm^2 for Cs, Rb and K, respectively, which are plotted against their ionic radii in Fig. 4. The largest value for K is approximately 17- and 6-times greater than that of the Os and O atoms at 300 K.[10] To gain an intuitive insight into the systematic variation in the ADP over the series, we have estimated the 'size' of the free space inside the $Os_{12}O_{18}$ cage shown in Fig. 1. A simple measure is the clearance $d - r_i(A^+)$, where $r_i(A^+)$ is the ionic radius of A^+ ions and d is the distance obtained by extracting the ionic radius

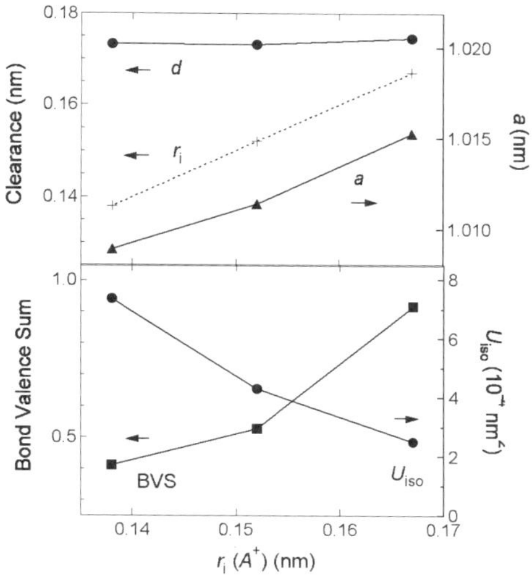

Fig. 4. Clearance for the rattling, lattice constant a, bond valence sum and atomic displacement parameter U_{iso} as a function of the ionic radius of the A cations.

of oxygen (0.14 nm) from the experimentally determined A–O bond length. The clearance should be negligibly small compared with $r_i(A^+)$ in a normal crystal structure. In β-pyrochlores, as shown in Fig. 4, the d is almost insensitive to the changes of $r_i(A^+)$, indicating that the Os–O framework is rigid. Thus, the clearance increases enormously with decreasing $r_i(A^+)$ from Cs to K. It is obvious that U_{iso} increases with increasing the clearance, which determines the space available for the A atom. Therefore, the K atom is the most weakly bound to the surrounding atoms and can rattle heavily.

The bond valence sum (BVS) of the A cations, which is a measure for the charge of a cation surrounded by a set of anions, is calculated using the structural parameters and plotted in Fig. 4. It is close to unity for Cs, as expected, while decreases even below 0.5 for K. This large deviation is very unusual for such ionic elements as alkali metals in oxides.

Kunes *et al.* found an anharmonic potential for the A cation along the <111> direction in AOs_2O_6 and larger anharmonicity for a smaller A cation.[12] In fact, the electron distribution of the K atom in the Fourier map is considerably elongated along the <111> directions, suggesting that there are electron lobes of the tetrahedral shape pointing to the four open channels of the $Os_{12}O_{18}$ cage. It is not possible to explain this elongation at the 8b site within the harmonic treatment of thermal vibrations because of the high symmetry of the site in the expression of the anisotropic displacement parameters ($U_{11} = U_{22} = U_{33}$ and $U_{12} = U_{13} = U_{23} = 0$). This is the clear evidence for the existence of an anharmonic potential for the K atom at 300 K, which seems to remain even at low temperature.

4. Concluding remarks

We have studied the superconducting and lattice properties of the β-pyrochlore oxides, $CsOs_2O_6$, $RbOs_2O_6$ and KOs_2O_6. From the chemical trends of various parameters over the series including α-$Cd_2Re_2O_7$, it becomes evident that the rattling of the A atoms, which is essentially a local, anharmonic mode with low energy, serves as an effective media for the Cooper pairing in the β-pyrochlore oxides, as schematically illustrated in Fig. 5 The enhancement of T_c and the characteristic change from weak- to strong-coupling superconductivity from Cs to K through Rb are successfully attributed to the corresponding enlargement of the electron-rattler interactions.

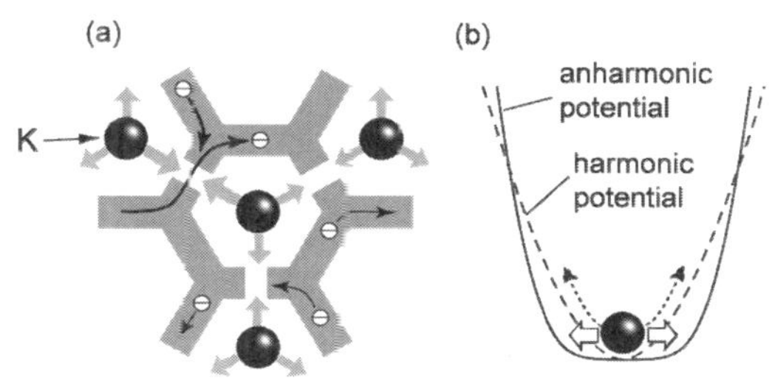

Fig. 5. (a) schematic representation of the rattling-induced superconductivity in the β-pyrochlore oxides. A pair of electrons on the cage form a Cooper pair through the rattling motion of the K ions. (b) The K ion is moving in an anharmonic potential with a flat bottom, which enables the K ion to move with a large excursion even at very low temperature.

References

1. M. A. Subramanian *et al.*, *Prog. Solid State Chem.* **15** 55 (1983).
2. M. Hanawa *et al.*, *Phys. Rev. Lett.* **87** 187001 (2001).
3. S. Yonezawa *et al.*, *J. Phys. Soc. Jpn.* **73** 1655 (2004).
4. S. Yonezawa *et al.*, *J. Phys. Soc. Jpn* **73** 819 (2004).
5. S. M. Kazakov *et al.*, *Supercond. Sci. Technol.* **17** 1169 (2004).
6. M. Brühwiler *et al.*, *Phys. Rev. B* **70** 020503(R) (2004).
7. S. Yonezawa et al., J. Phys.: Condens. Matter **16** L9 (2004).
8. Z. Hiroi *et al.*, *Phys. Rev. B* **76** 014523 (2007).
9. Z. Hiroi *et al.*, *Physica C* **460-462** 20 (2007).
10. J. Yamaura *et al.*, *J. Solid State Chem.* **179** 336 (2006).
11. R. Galati *et al.*, *J. Mater. Chem.* **17** 160 (2007).
12. J. Kunes *et al.*, *Phys. Rev. B* **70** 174510 (2004).
13. V. Keppens *et al.*, *Nature* **395** 876 (1998).
14. G. S. Nolas *et al.*, *Appl. Phys. Lett.* **73** 178 (1998).
15. Z. Hiroi *et al.*, *J. Phys. Soc. Jpn.* **74** 1255 (2005).
16. M. Brühwiler *et al.*, *Phys. Rev. B* **73** 094518 (2006).
17. M. Yoshida *et al.*, *Phys. Rev. Lett.* **98** 197002 (2007).
18. T. Hasegawa *et al.*, *Phys. Rev. B* **77** 064303 (2008).
19. J. Schoenes *et al.*, *Phys. Rev. B* **77** 134515 (2008).
20. K. Sasai *et al.*, *J. Phys. Soc. Jpn.* **76** 104603 (2007).
21. H. Mutka *et al.*, *Phys. Rev. B* **78** 104307 (2008).
22. R. Saniz *et al.*, *Phys. Rev. B* **70** 100505(R) (2004).
23. R. Saniz, A. J. Freeman, *Phys. Rev. B* **72** 024522 (2005).
24. H. Fröhlich, *Adv. Phys.* **3** 325 (1954).
25. Y. Nagao *et al.*, in preparation.

MICRO/NANOMECHANICAL SYSTEMS FOR INFORMATION PROCESSING

H. YAMAGUCHI* and I. MAHBOOB

NTT Basic Research Laboratories, NTT Corporation, Atsugi-shi, Kanagawa, 243-0198 Japan
**E-mail: hiroshi@will.brl.ntt.co.jp*

A parametric electromechanical resonator was fabricated using a GaAs/AlGaAs modulation-doped heterostructure. Three on-chip operations - actuation, detection and mechanical resonance frequency control - were all demonstrated using piezoelectric based strain-voltage transduction. We clearly confirmed that the device can perform bit storage operation in the same way a *Parametron* computer does. Finally, methods for constructing an electromechanical shift register and AND/OR gates are also proposed.

Keywords: NEMS; Mechanical Logic; Piezoelectricity.

1. Introduction

More than 50 years ago, a graduate student at the University of Tokyo, Eiichi Goto, invented the "Parametron," an electrical circuit that formed the basis for digital computers [1]. The Parametron utilizes a parametrically excited resonance of a harmonic oscillator, whose oscillation has two stable phases separated by π radians, as the basis for logic operations. The computational architecture based on this principle was developed using an LC oscillator and was used for many practical applications. It was, however, rendered obsolete by the transistor due to its comparatively low operation speed, high power consumption, and the difficulties encountered in integrating it into dense system architechtures. Here, we propose to implement mechanical logic in the mould of the Parametron with micro/nanoelectro-mechanical systems (MEMS/NEMS).

MEMS/NEMS have recently attracted much attention as they enable the development of new applications for lithographically defined fine structures [2,3]. In the field of fundamental research, they are especially promising in applications to ultimate sensors, such as single-charge [4], single-spin [5], and single-atom detectors [6]. Recently a new application for MEMS/NEMS has been proposed by several groups [7-12]. The aim is to construct novel logic systems with qualities that can not be realized by using present CMOS-based technologies, such as high energy efficiency and robustness against diverse environments. Mechanical memory operations using lithographically fabricated strained beams [7,9,11] and carbon nanotubes [8,12] have been demonstrated. The XOR operation using piezoelectric cantilevers has also been realized [10]. However, a detailed computational architecture for these proposals has not been developed. In this paper, we report the application of electromechanical parametric resonators to the "Parametron" computer scheme [13]. Because the computational architecture for the Parametron has already been established [1], the construction of a binary logic system with electromechanical systems is straightforward. As a first step, we demonstrate that our parametric resonator can perform electromechanical bit storage and bit reset operations in the same way as the LC-based Parametron does. Finally, we also propose an electromechanical architecture to construct a shift register and AND/OR logic gates.

2. Device Fabrications

The fabricated device [13-15] is a micro-electromechanical parametric resonator based on the piezoelectricity in a GaAs/AlGaAs modulation-doped heterostructure, which was grown by solid source molecular beam epitaxy and is schematically shown in Fig. 1. The two-dimensional electron system (2DES) is formed 90 nm below the surface. After a shallow etching to laterally define the 2DES mesa, Au deposition and lift-off process formed Schottky gates. The beam structure was defined by deep mesa etching, and the $Al_{0.65}Ga_{0.35}As$ sacrificial layer was then selectively etched to release the beam from the substrate (Fig. 2).

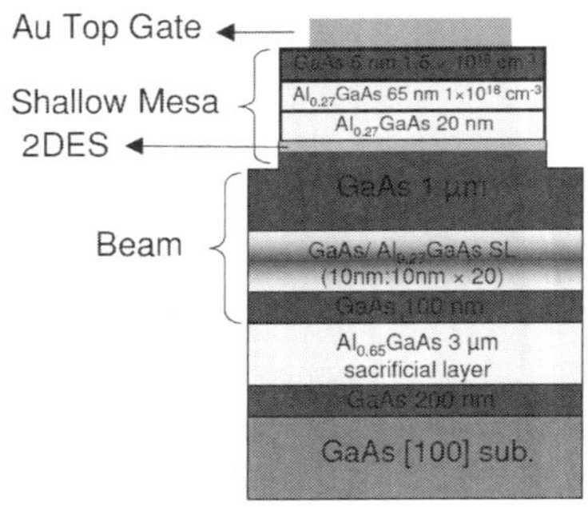

Fig. 1. The modulation-doped heterostructure used to fabricate a piezoelectric parametric resonator [13].

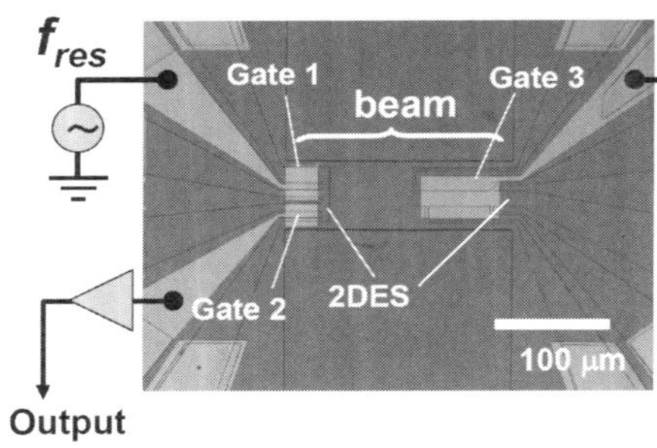

Fig. 2. The top-view microscope image of the fabricated device. The three gates are used for actuation (Gate 1), and detection (Gate 2), and frequency modulation (Gate 3).

3. Experimental Results

3.1. *Basic device operations*

Applying a voltage between the Schottky gate and 2DES induces a strain along the beam direction. Because the strain is concentrated only in the near-surface region, it generates a bending moment that is used to actuate the out of plane beam motion. Reversibly, the beam motion can also be detected by measuring the generated voltage. In addition, the resonance frequency can be modulated by voltage-induced stress in the beam. Therefore, we can realize actuation, detection, and frequency modulation all on chip by using the piezoelectric effect.

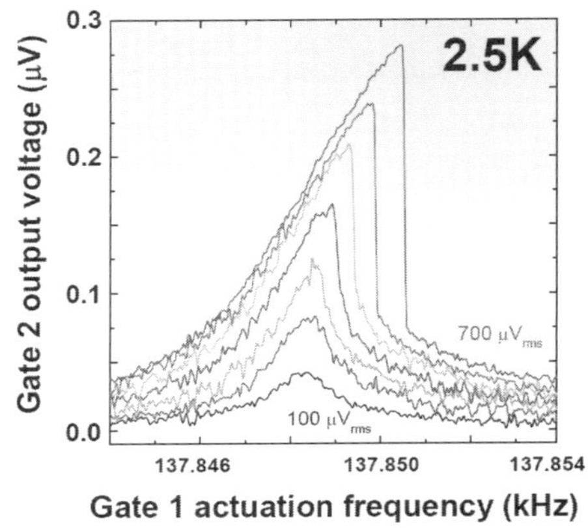

Fig. 3. Rms voltage measured at Gate 2 when actuation voltage with various amplitudes (100–700 μV_{rms}) was applied on Gate 1. The frequency was swept from low to high values.

Figure 3 shows the detected mechanical resonance characteristics when we actuated the beam motion by applying an alternate voltage on Gate 1 and detected it by measuring the generated voltage on Gate 2 (see Fig. 2). An amplitude jump due to the nonlinear response of the resonator was clearly seen when the actuation voltage was higher than 500 μV_{rms}.

Figure 4 shows the resonance frequency and the Q-factor when DC voltage was applied to Gate 3. We clearly confirmed the linear dependence of resonance frequency on Gate 3 voltage. No shift was observed when electrons were depleted (V<-0.5V), strongly supporting that the frequency shift was induced by the voltage between Gate 3 and the 2DES. The Q-factor was nearly constant but slightly increased in the region with low carrier concentration, suggesting that the electrons

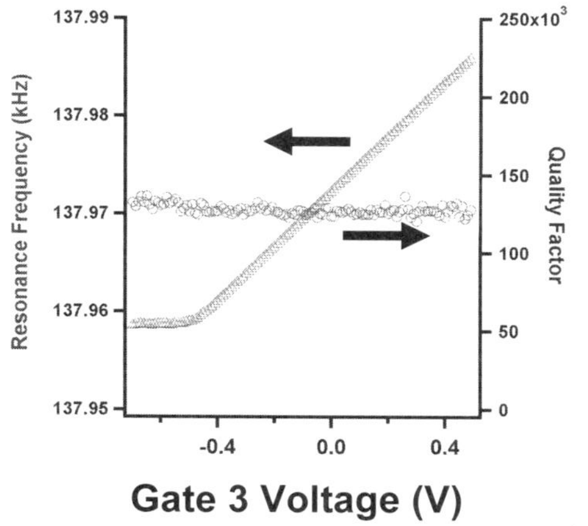

Fig. 4. Resonance frequency and Q-factor measured as a function of Gate 3 voltage.

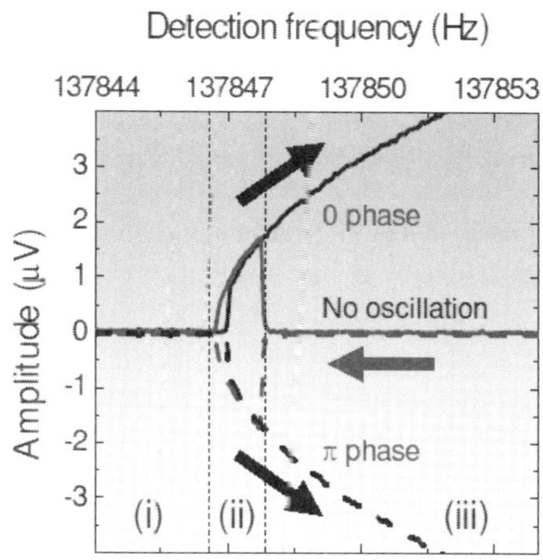

Fig. 5. Frequency response of mechanical oscillation amplitude of the parametrically actuated beam.

also induce an internal friction on the beam motion.

3.2. *Parametric actuation*

We then actuated the beam motion by modulating the spring constant at twice the resonance frequency. This phenomenon is known as parametric actuation and has been studied in many classical systems [15]. Figure 5 shows the experimentally obtained frequency response. The parametric resonance line shape is highly asymmetric and it does not become damped as the actuation frequency is swept forward across the resonance; conversely, the resonance does become damped when the actuation frequency is swept backwards.

The frequency response can be summarized as follows. When the excitation frequency is below the resonance, only one stable state with zero amplitude exists [region (i) in Fig. 5]. For excitations around the resonance frequency [region (ii)], the parametric resonance is bi-stable; that is, this oscillation has a choice between two stable phases, 0 and π. In contrast, when the excitation is above the natural frequency [region (iii)], the parametric resonance is tri-stable, because the oscillation has three stable states: no oscillation, 0 phase, and π phase. This difference appears as a hysteresis in the parametric resonance frequency response.

The parametrically actuated beam motion is described by the following non-linear Mathieu's equation:

$$m\frac{d^2x(t)}{dt^2} + m\omega_0 Q^{-1}\frac{dx(t)}{dt} + m\omega_0^2[1+\beta x(t)^2 - 2\Gamma\cos(2\omega t)]x(t) = 0, \quad (1)$$

where m is the beam mass, $x(t)$ the effective beam displacement, ω_0 the resonance frequency, β is mechanical nonlinearity, Γ the parametric actuation amplitude, and ω half the parametric actuation frequency. When $\Gamma \geq Q^{-1}$, the beam can be parametrically excited. The frequency response of the beam can be reproduced by numerically solving this equation of motion and thus demonstrating that our device is an ideal parametric resonator.

3.3. *Bit storage and bit flip operations*

The bi-stability in the parametric resonance response (region (ii) in fig. 5) was used to demonstrate binary data storage, in the same way as the LC-based Parametron computer. The oscillation modes with 0-phases and π-phase can be assigned to "0" and "1" states, respectively. Here we simply show the result when we store these two states alternatively into the phase of the mechanical oscillation. The details are described in Ref. 13. Figure 6 shows the amplitude of the output voltage when the two states are alternatively stored as the phase states. Periodic switching between the two phases of the oscillation is observed with an output voltage of 2 μV. Although this value is the half of minimum input voltage (4 μV) that ensures 100 % switching fidelity, we expect to obtain a sufficiently large fan-out when we directly connect the output electrode to the input one.

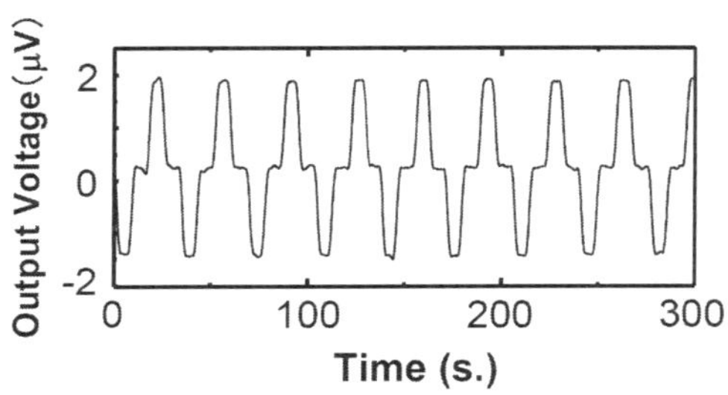

Fig. 6. Output voltage when "0" and "1" information is alternately stored as phase states in the parametrically actuated mechanical resonance.

4. Logic Operations

Now we describe how to construct a shift resister and AND/OR logic gates using our mechanical parametric resonators. Figure 7 illustrates the electromechanical parametron shift-register that can be constructed by connecting six electromechanical resonators. The output electrode of one mechanical bit is electrically connected to the input electrode of the next mechanical bit. The binary information, i.e. the phase state, can be transferred from one bit to the next when the parametric actuation of the first mechanical resonator is followed by that of the next. The parametric actuation is excited in series between adjacent bits. Following the formalism used in the LC-based Parametron, a series of mechanical resonators are formed into three groups. For example, resonators 1 and 4 are grouped into A, resonators 2 and 5 are into B, and 3 and 6 are into C. After actuating the group A resonators, the resonators in B are parametrically excited. The phase state stored in a group A resonator is transferred to the next one in B. Then the actuation of the group A resonators is terminated and that of group C is excited. The information stored in group B is now transferred to group C. By repeating this sequence, the bit information can be transferred from the one end to the other.

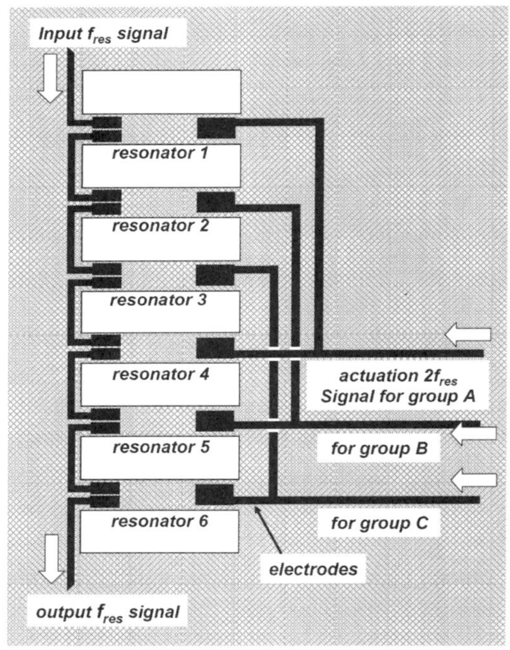

Fig. 7. A schematic illustration of a Parametron shift register comprising of six electromechanical resonators.

To achieve AND/OR operations, we can use the concept of majority voter. The controlled AND/OR device is illustrated in Fig. 8. The output electrodes from the three resonators, A-C, are all connected to the input of the forth resonator D. After the resonance for A-C has been parametrically activated, the three output signals are superimposed and the major phase is transferred to D. When A is actuated in the "0" state, the input of D becomes “1” only if both B and C are “1”. However, when A is actuated in “1”, the input of D becomes “0” only when both B and C are “0”. Therefore, the output from D becomes “B OR C” and “B AND C” when A is “1” and “0”, respectively. In this way, we can construct any kind of logic gate by interconnecting the electromechanical resonators.

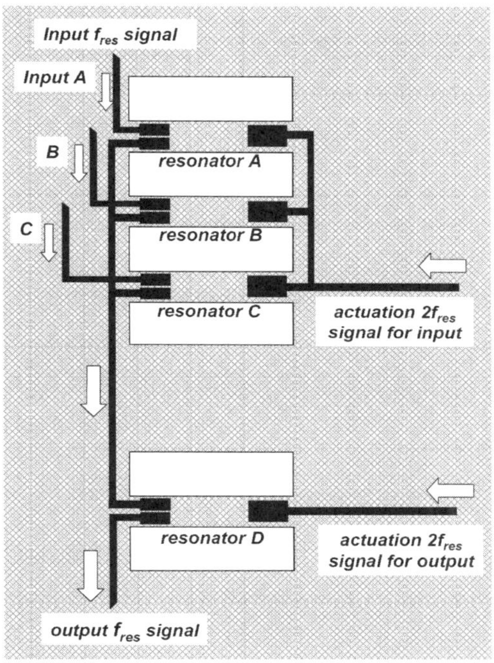

Fig. 8. A schematic illustration of controlled AND/OR logic gates constructed by four electromechanical resonators.

5. Conclusion

We have demonstrated mechanical bit storage operation using piezoelectric parametric resonators in the same way as in the LC-based Parametron computer. The natural frequency of the mechanical resonator can be increased to more than 1 GHz by minimizing the beam size and much faster logic operation can be expected. The energy consumption in such mechanical resonators is extremely small and electromechanical logic might be used for a highly energy-efficient computer system. The dynamic bistable state is also promising for detecting quantum phenomena in macroscopic objects [16-18] because the barrier between the two stable phases of oscillation can be precisely controlled.

Acknowledgments

We thank J. Yumoto for continuous encouragement. This work was partly supported by JSPS KAKENHI (16206003 and 1920601).

References

1. E. Goto, *Proc. IRE* **47**, 1304 (1959).
2. see for example, T.R. Hsu, *MEMS and Microsystems* (John Wiley & Sons, New Jersey, 2008)
3. A. N. Cleland, *Foundations of Nanomechanics* (Springer, Berlin, 2003).
4. A. N. Cleland and M. L. Roukes, *Nature* **392**, 160 (1998)
5. D. Rugar, R. Budakian, H. J. Mamin, B. W. Chui, *Nature* **430**, 329 (2004)
6. K. Jensen, K. Kim, and A. Zettl, *Nature* Nanotechnology 3, 533 (2008).
7. B. Halg, Proc. of 1990 IEEE-MEMS Conference pp172 (1990).
8. T. Rueckes, K. Kim, E. Joselevich, G. Y. Tseng, C. L. Cheung, and C. M. Lieber, Science, **289**, 94 (2000).
9. R. L. Badzey, G. Zolfagharkhani, A. Gaidarzhy, P. Mohanty, *Appl. Phys. Lett.* **85**, 3587 (2004).
10. S.C. Masmanidis, R. B. Karabalin, I. De Vlaminck, G. Borghs, M. R. Freeman, and M. L. Roukes, *Science* **317**, 780 (2007).
11. B. Charlot, W. Sun, K. Yamashita, H. Fujita, and H. Toshiyoshi, *J. Micromech. Microeng.* **18**, 045005 (2008).
12. J.E. Jang, S. N. Cha, Y. J. Choi, D. J. Kang, T. P. Butler, D. G. Hasko, J. E. Jung, J. M. Kim and G. A. J. Amaratunga, *Nature Nanotechnol.* 3, 26 (2008).
13. I. Mahboob and H. Yamaguchi, *Nature Nanotech.* **3**, 275 (2008).
14. I. Mahboob and H. Yamaguchi, *Appl. Phys. Lett.* **92**, 173109 (2008).

15. I. Mahboob and H. Yamaguchi, *Appl. Phys. Lett.* **92**, 253109 (2008).
16. B. Wielinga and G. J. Milburn, *Phys. Rev. A* **48**, 2494 (1993).
17. M. Marthaler and M. I. Dykman, *Phys. Rev. A* **73**, 042108 (2006).
18. M. Marthaler and M. I. Dykman, *Phys. Rev. A* **76**, 010102(R) (2007).

ELECTRON PHASE MICROSCOPY FOR OBSERVING SUPERCONDUCTIVITY AND MAGNETISM

A. TONOMURA
Initial Research Project, Okinawa Institute of Science and Technology, Kunigami, Okinawa 904-0411, Japan
Advanced Research Laboratory, Hitachi Ltd., Hatoyama, Saitama 350-0395, Japan
Advanced Science Institute, RIKEN, Wako, Saitama 351-0198, Japan
E-mail: akira.tonomura.zk@hitachi.com

Electron phase microscopy based on the Aharonov-Bohm (AB) effect principle has been used to illuminate fundamental phenomena concerning superconductivity and magnetism by revealing the microscopic distributions of magnetic flux. These phenomena include the unconventional behaviors of interlayer Josephson vortices in anisotropic layered high-Tc superconducting $YBa_2Cu_3O_{7-\delta}$ (YBCO) thin films, which show various behaviors, especially when the applied magnetic field is greatly tilted to the layer plane and the magnetization of tiny magnetic heads for perpendicular recording.

Keywords: Electrons; Phase; Vortex; Superconductor; YBCO; Magnetic head.

1. Introduction

Although electrons are detected as particles, they also show wave properties. Thanks to the development of brighter electron beams, we can now practically use the wave nature of electrons to directly image the quantum world by obtaining the phase information of electrons [1]. Projected magnetic lines of force can be quantitatively observed as phase contours of transmitted electrons by using holographic interference microscopy [2] since a phase shift of 2π is produced between two electron beams enclosing a magnetic flux of h/e due to the Aharonov-Bohm (AB) effect [3], and quantized vortices in superconductors can be dynamically observed by Lorentz microscopy [4].

This paper describes the challenges we faced in our efforts to investigate the phenomena related to Josephson vortices in layered high-Tc superconductors and in magnetic heads used for perpendicular recording.

2. Experimental Method

We carried out experiments on superconductors and magnetic heads using 1-MV and 300-kV field-emission transmission electron microscopes, respectively. We observed vortices in superconducting thin films as Lorentz micrographs by defocusing electron microscopic images (Fig. 1). When vortices penetrated the thin films parallel to the film surfaces, they could be observed even without tilting the film (Fig. 1(a)). However, when vortices, or parts of vortices, were perpendicular to the thin film, the films had to be tilted so that an incident electron beam could interact with the vortex magnetic fields (Fig. 1(b)) Projected magnetic lines of force of magnetic heads were directly observed as contour lines in the interference micrographs obtained from off-axis electron holograms formed with electron biprisms [1].

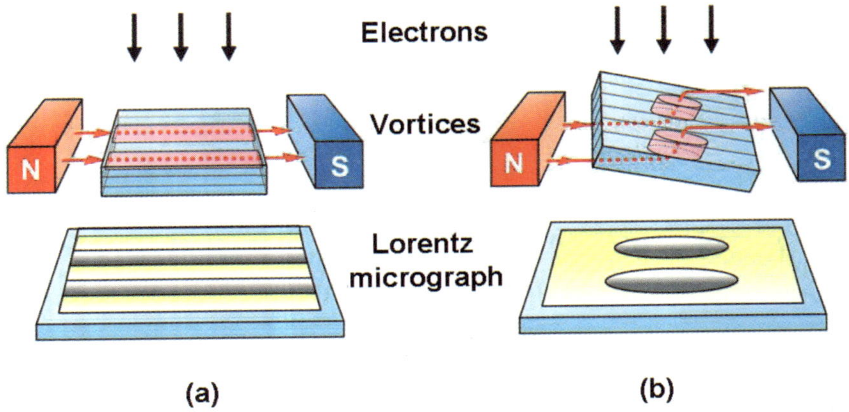

Fig. 1 Setups for observing Josephson vortices in layered superconductors by Lorentz microscopy: (a) vortices in *ab*-plane film; (b) vortices in *ac*-plane film.
Electrons are phase-shifted by magnetic fields of vortices, and the phase changes of the transmitted electrons are transformed into intensity changes by image defocusing. When vortices penetrate the film perpendicular to it, the film is tilted.

3. Experimental Results

3.1. Josephson Vortices in YBCO Thin Films

Josephson vortices in a highly anisotropic $Bi_2Sr_2CaCu_2O_{8+\delta}$ (Bi-2212) superconductor play an important role in the magnetic response of the superconductor and are expected to be used practically, for example, as Terahertz radiation sources. However, they had never been directly observed since they were hidden inside the crystals. Using Lorentz microscopy and our 1-MV microscope (Fig. 1 (a)), we were the first to observe the static and dynamic behaviors of Josephson vortices in *ab*-plane films of moderately isotropic YBCO crystals as shown in Fig. 2 [5].

Each line in the Lorentz micrograph in Fig. 2 corresponds to an individual Josephson vortex. We confirmed this by measuring a relative

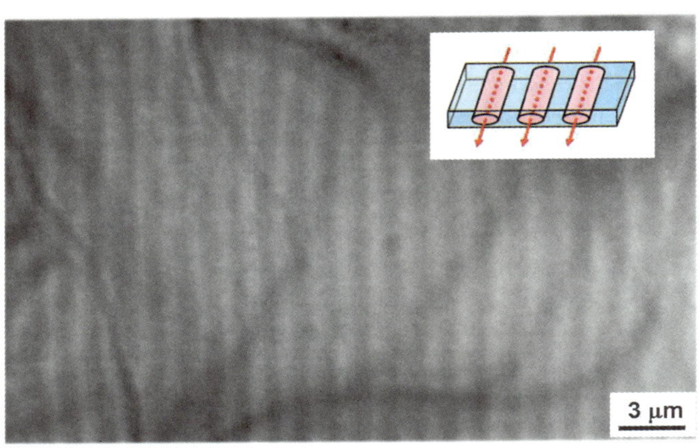

Fig. 2 Josephson vortices in YBCO thin film (T = 68 K).

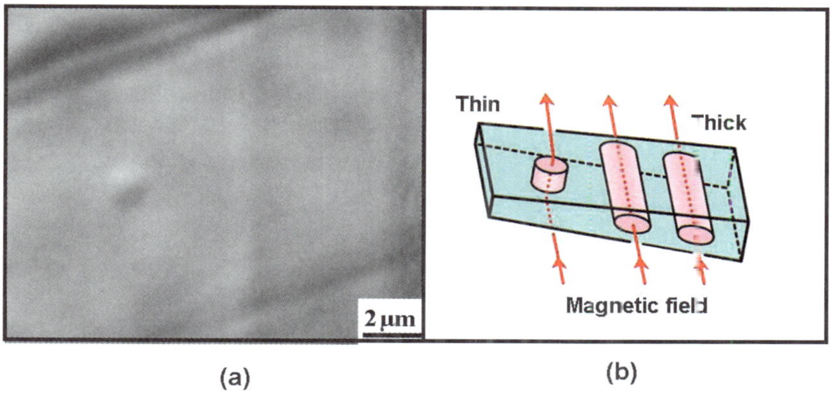

Fig. 3 Vortices in YBCO thin film: (a) Lorentz micrograph; (b) schematic. Josephson vortices penetrate film thicker than 0.4 μm.

phase-shift of π between two electron beams passing through on both sides of a single line, which indicates that the magnetic flux of h/2e was enclosed. When the applied magnetic field strength was increased, the density of vortex lines increased proportionally with the magnetic field.

The vortex lines, however, were not produced in samples thinner than 0.4 μm when the applied magnetic field was less than 100 Oe. An example Lorentz micrograph of a tapered film is shown Fig. 3. Two Josephson vortices are evident in the thick region on the right, while in the thin region, only the spot of a vertical vortex is evident. The contrast of the vertical vortex was due to the film being tilted around the axis determined by the direction of the applied magnetic field. It can be concluded from the bright-and-dark contrast features that the magnetic field of this vortex was upward.

Josephson vortices were produced only when the magnetic field was parallel ($\pm 0.5^\circ$) to the layer plane under field-cooled conditions. Once produced, they behaved as if they were firmly trapped between layers. However, when the magnetic field was tilted 3°, the images of the vortex lines suddenly terminated at spots near the film edge. One end of each Josephson vortex line must have stood up near the film edge due to the tilting.

This state of the vortices was not static. The spots immediately began to move along the vortex lines, and the lines shortened and disappeared. This movement could not be stopped. This behavior of Josephson vortices indicates that they are formed by circulating elliptical vortex currents and that they cannot be easily transformed into tilted vortices consisting of disc-like vortex currents within layers.

We also attempted to observe elliptical cross sections of Josephson vortices. We fabricated *ac*-plane film samples (Fig. 1(b)) instead of *ab*-plane film samples (Fig. 1(a)) and applied a magnetic field almost perpendicular to the film plane. An example of a Lorentz micrograph of Josephson vortices penetrating *ac*-plane film is shown in Fig. 4.

The images of the vortices are elongated along the *a*-axis or the *c*-plane, revealing the elliptical cross sections of the vortices. Since we are observing the vortices themselves inside

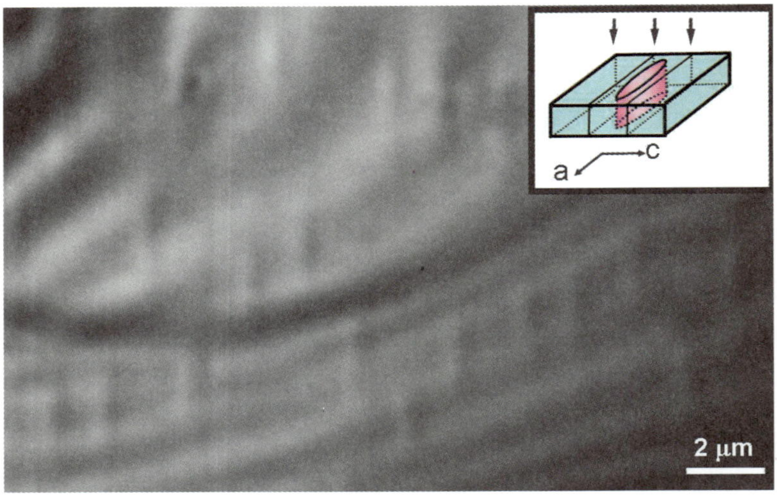

Fig. 4 Lorentz micrograph of elliptical Josephson vortices in YBCO thin film

superconductors, we can observe their dynamics as well. In fact, when the magnetic field was reduced, these vortices began to move only in the direction of the *a*-axis, never in different directions. This indicates strong intrinsic pinning due to the layer structure. The velocity of the vortices was 300 nm/sec on average at 5 K.

3.2. Magnetic heads

Magnetic lines of force radiating outward from a tiny magnetic head were observed as holographic interference micrographs [6] (Fig. 5 (a)). This head is used for high-density perpendicular recording, so it must produce a high yet localized magnetic field. A strong magnetic field is rather simple to produce. However, this is not sufficient. We also need to be able to control magnetic fields in any way we want. For example, we need to get a zero field outside a head, which poses difficulties since the density of magnetic lines inside ferromagnets is a constant determined by the strength of the magnetization.

In the case of our FeNi head, straight magnetic lines inside the head were changed to "U-turn" lines [6] (Fig. 5 (b)) by reducing the applied magnetic field to zero. At the pole tip, which is used to localize magnetic fields, the magnetic lines formed a tiny closed circuit. The two distributions in (a) and (b) of magnetic lines are both symmetrical.

The question is how to change straight lines into U-turn lines so that the outside magnetic lines disappear. The series of Lorentz micrographs in Fig. 6 show how this is done. When the applied magnetic field was reduced from 120 to 80 Oe, the straight magnetic lines inside the head ((a)) begin to bend ((b)). This "blue" bent region moved upward ((c)) when the magnetic field was further reduced. The reason for this movement is evident from the pictures. The "red" U-turn lines appeared from the lower left in (c), pushing the blue lines up..

The U-turn lines gradually moved upwards ((d) and (e)) and finally covered the whole head region ((f)). The "blue" bent magnetic lines finally changed into closed circles at the tip.

The leakage magnetic field gradually decreased from (a) to (f) and almost vanished in the remanent state (f), where all the magnetic lines inside the head were closed, restoring the pattern symmetry.

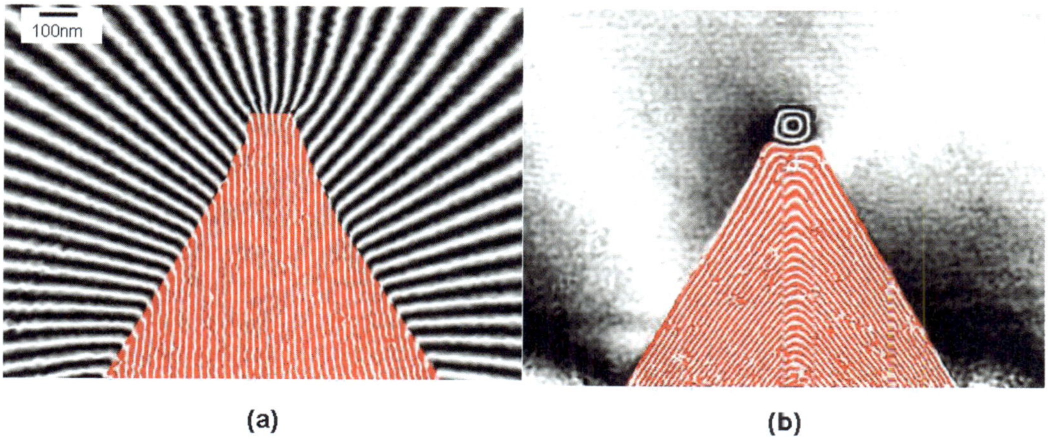

Fig. 5 Interference micrographs showing magnetic lines of force of a FeNi magnetic head for perpendicular recording: (a) ON; (b) OFF.
Magnetic lines are produced radially from the magnetic head when it is magnetized in the upward direction by applying an external magnetic field ((a)). When the magnetic field is removed, magnetic lines outside the head almost disappear, since magnetic lines inside it form a closed circuit ((b)). These pattern are both symmetrical.

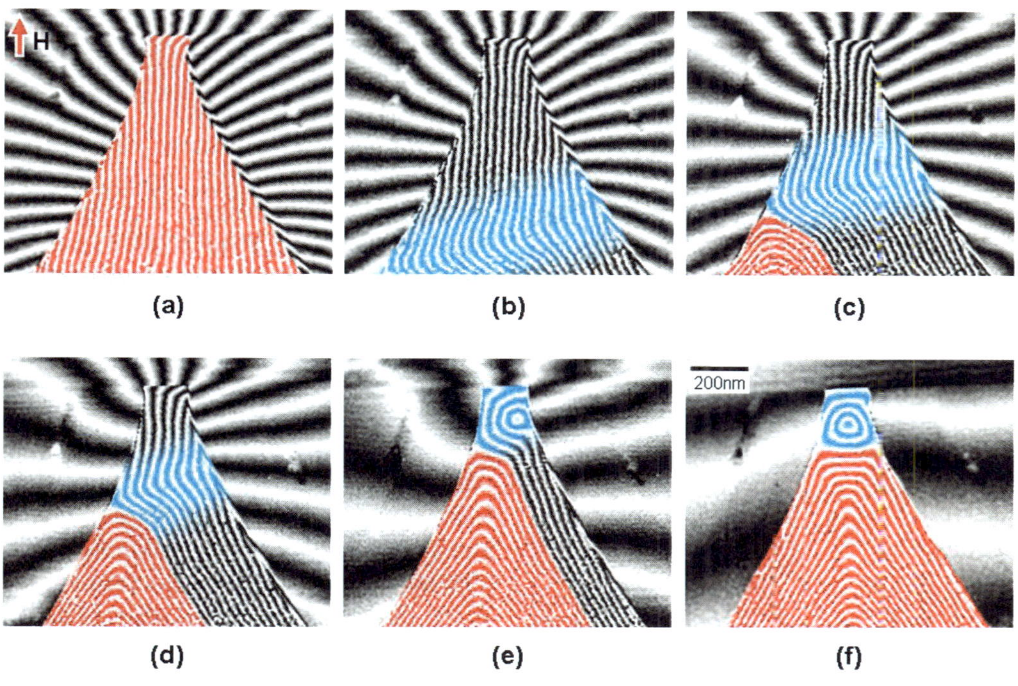

Fig. 6 Interference micrographs showing how magnetic lines inside the head change from straight lines to U-turned lines when the applied magnetic field decreases: (a) H=120 Oe; (b) H=80 Oe; (c) H=60 Oe; (d) H=35 Oe; (e) H=15 Oe; (f) H=0 Oe. "Red" U-turn lines appear from the lower left ((c)) and move upwards pushing the "blue" region of bent magnetic lines ((d), (e)). The patterns of magnetic lines are no longer symmetrical. The bent magnetic lines form circular closed circuit in the remanent state in (f).

4. Conclusions

Thanks to recent developments in advanced technologies such as coherent yet bright electron beams, the wave nature of electrons can now be used to observe microscopic objects previously unobservable by conventional electron microscopy. Specific examples include the quantitative observation of the microscopic distribution of magnetic lines in h/e units by interference microscopy and the dynamic observation of quantized vortices in superconductors by Lorentz microscopy.

Electron phase microscopy in a field-emission electron microscope is expected to play a more and more important role in future research and development not only in nanotechnology but also in biology, since microscopic information about materials reaching down even to atoms and molecules has become more pivotal.

References

1. A. Tonomura, *Proc. Natl. Acad. Sci. USA* **102**, No. 42, 14952 (2005).
2. A. Tonomura, T. Matsuda, J. Endo, T. Arii and K. Mihama, *Phys. Rev. Lett.* **44**, No. 21, 1430 (1980).
3. Y. Aharonov and D. Bohm, *Phys. Rev.* **115**, 485 (1959).
4. K. Harada, T. Matsuda, J. Bonevich, M. Igarashi, S. Kondo, G. Pozzi, U. Kawabe and A. Tonomura, *Nature* **360,** 51 (5 November 1992).
5. A. Tonomura, T. Akashi, H. Kasai, S. Mamishin, T. Yoshida, T. Matsuda, S. Tajima, H. Tadatomo, T. Masui and A. Buzdin, (2009) to be submitted.
6. J. J. Kim, K. Hirata, Y. Ishida, D. Shindo, M. Takahashi and A. Tonomura, *Appl. Phys. Lett.* **92**, 162501 (2008).

RATIO OF THE AL+ AND HG+ OPTICAL CLOCK FREQUENCIES TO 17 DECIMAL PLACES

W. M. ITANO,* T. ROSENBAND, D. B. HUME, P. O. SCHMIDT,† C. W. CHOU, A. BRUSCH,‡ L. LORINI,§ W. H. OSKAY,¶ R. E. DRULLINGER, S. BICKMAN, T. M. FORTIER, J. E. STALNAKER,|| S. A. DIDDAMS, W. C. SWANN, N. R. NEWBURY, D. J. WINELAND, AND J. C. BERGQUIST

National Institute of Standards and Technology, 325 Broadway, Boulder CO 80305, USA
**E-mail: itano@boulder.nist.gov*

Frequency standards (atomic clocks) based on narrow optical transitions in $^{27}Al^+$ and $^{199}Hg^+$ have been developed over the past several years at NIST. Both types of standards are based on single ions confined in Paul traps, but differ in the methods used to prepare and detect the internal atomic states. Al^+ lacks a strong, laser-accessible transition for laser-cooling and for state preparation and detection. Coupling with a Be^+ ion, trapped simultaneously with the Al^+ ion, enables state manipulation, detection, and cooling of the Al^+ ion. Both standards have achieved absolute reproducibilities of a few parts in 10^{17}. Development of femtosecond laser frequency combs makes it possible to directly compare optical frequencies. The present determination of $f_{\rm Al}/f_{\rm Hg}$ is 1.052 871 833 148 990 438 (55), where the uncertainty is expressed in units of the least significant digit. Measurements of $f_{\rm Al}/f_{\rm Hg}$ made over about one year show a drift rate consistent with zero. This result can be used to place limits on time variations of fundamental constants such as the fine structure constant α.

Keywords: atomic clocks, fundamental constants, optical frequency standards.

1. Introduction

Time, or its inverse, frequency, is the physical quantity that can be measured the most accurately and reproducibly. Bohr frequencies ν_{ij} corresponding to differences between two atomic energy eigenvalues E_i and E_j according to $\nu_{ij} = (E_i - E_j)/h$, where h is the Planck constant, are the basis for the most reproducible frequency standards. As opposed to frequency standards based on astronomical cycles or on vibrations of macroscopic objects, such atomic frequency standards have the advantage that the frequency-determining objects, e.g., atoms of a particular type, are perfectly identical, so that two atomic frequency standards of the same type should have the same frequency.

In practice, frequency standards based on the same atomic resonance frequency do not agree exactly because of errors introduced for example by motion (Doppler shifts), by external fields that shift the frequencies, or by errors in the apparatuses used to determine the frequencies. The reproducibilities of atomic frequency standards have steadily improved. In 1967, the 13th Conférence Générale des Poids et Mesures redefined the SI (Système International) unit of time, previously based on astronomical motions, in terms of an atomic transition frequency.[1] Specifically, the SI second was defined as 9 192 631 770 periods of the radiation corresponding to the transition between the two hyperfine levels of the ground state of ^{133}Cs. At that time, frequency standards

†Present address: Institut für Experimentalphysik, Universität Innsbruck, Austria.
‡Present address: Niels Bohr Institute, University of Copenhagen, Denmark.
§Present address: Istituto Nazionale di Ricerca Metrologica (INRIM), Torino, Italy.
¶Present address: Stanford Research Systems, Sunnyvale, CA 94089, USA.
||Present address: Department of Physics and Astronomy, Oberlin College, Oberlin, OH 44074, USA.

based on cesium were accurate to about one part in 10^{12}. Currently the accuracy of cesium standards is about 3.3 parts in 10^{16}.[2]

There is nothing in principle to prevent an atomic frequency standard from being more reproducible than the best cesium standards. However, it is impossible to express a frequency in SI units (Hz) any better than the accuracy of cesium standards, since they define the SI unit of time. This limit does not apply to the measurement of a frequency *ratio* between two different frequency standards.

Frequency standards based on optical transitions in the mercury ion and, more recently, the aluminum ion have been developed at the National Institute of Standards and Technology (NIST). Both are based on narrow optical transitions in single trapped ions. According to estimates of their systematic uncertainties, both the mercury and aluminum standards are more reproducible than the primary cesium frequency standards. By means of a relatively new device, the self-referenced femtosecond laser frequency comb,[3–5] the ratio of their frequencies can be measured without being limited by the accuracy of the cesium frequency standards. Measurements of this ratio as a function of time place limits on the variation of the fundamental constants, such as the fine-structure constant α.

2. Optical Frequency Measurements

2.1. *Narrow Optical Sources*

A prerequisite for a high-quality optical atomic frequency standard is a light source with extremely narrow linewidth and high frequency stability that can be tuned into resonance with the atomic transition. Lasers are now available with linewidths much less than 1 Hz.[6] These lasers are frequency-stabilized in the short-term to optical resonators based on mirrors with extremely high reflectivity. Since the mode frequencies of the resonator depend on its physical dimensions, they drift with time.

2.2. *Optical Frequency Standards and Clocks*

In an optical atomic frequency standard, the average frequency of a highly frequency-stable optical source is servo-locked to an atomic resonance frequency. In most cases this is done by stepping the frequency of the optical source slightly above and then slightly below the atomic resonance and detecting the difference in the probability of making a transition between the atomic levels. The apparatus uses this difference in signals to keep the average frequency of the optical source close to that of the atomic resonance.

An atomic frequency standard is not automatically usable as an atomic clock. Some means of counting the oscillations of the servo-locked electromagnetic field is required. With frequency standards based on atomic transitions at microwave frequencies, this is relatively straightforward, by use of conventional frequency multiplication, division, and mixing techniques.

Until relatively recently, the counting of optical frequencies was difficult, involving complex chains of frequency-doubled and frequency-mixed lasers. Now, with the self-referenced femtosecond laser frequency comb, the linkage of optical to microwave frequencies, required for the generation of time from a stable optical frequency, is relatively simple.

2.3. *The Femtosecond Laser Frequency Comb*

A self-referenced femtosecond laser frequency comb generates a series of discrete, equally frequency-spaced modes, called a "comb." The mode spacing, f_{rep}, is given by the repetition rate of a mode-locked laser, typically

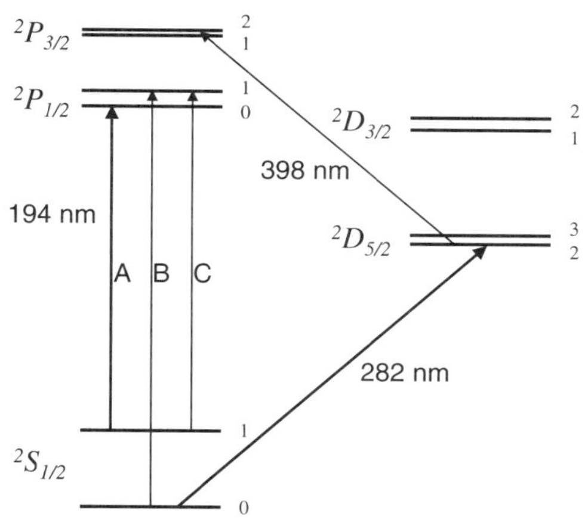

Fig. 1. The energy levels of $^{199}Hg^+$ relevant to the operation of the frequency standard. The numbers to the right of the hyperfine energy levels are F, the total atomic angular momentum. The transitions induced by the lasers are indicated by arrows. For clarity, the energy differences between the hyperfine levels are expanded relative to the electronic energy differences.

around 100 MHz to 1 GHz. The frequency of an individual mode can be expressed as $f(m) = f_0 + m \times f_{\text{rep}}$, where m is an integer. The repetition frequency is measured by directing some of the comb light to a photodetector. The offset frequency f_0 is measured by a method called self-referencing.[4,5] If the heterodyne beat frequency f_{beat} between an optical frequency standard and the nearest tooth of the optical comb is measured, then the frequency of the standard is known in terms of f_0, f_{rep}, and f_{beat}, provided the integer m is known. This provides a link between the optical frequency and the frequencies, f_0, f_{rep}, and f_{beat}, which can be referenced to a microwave frequency standard.

3. The Mercury Ion Frequency Standard

3.1. *Atomic Level Structure*

The mercury ion frequency standard is based on the transition from the ground $5d^{10}6s$ $^2S_{1/2}$ state to the metastable $5d^9 6s^2$ $^2D_{5/2}$ state. The metastable state has a natural lifetime of around 90 ms, giving this transition a Q of around 6×10^{14}. Doppler broadening is eliminated for single-photon absorption by confinement of the ion to less than the wavelength of the radiation.[7] Line broadening due to the magnetic field fluctuations is reduced by using the $(F = 0)$ to $(F = 2, m_F = 0)$ hyperfine-Zeeman component in $^{199}Hg^+$, which has only a quadratic Zeeman shift. The 282 nm resonance has been observed with a linewidth as low as 6.7 Hz.[8]

3.2. *Operation of the Frequency Standard*

The basic methods used for laser cooling, state preparation, and detection of the clock transition have been described previously.[8,9] The energy levels of $^{199}Hg^+$ that are relevant to the operation of the frequency standard are shown in Fig. 1. The 194 nm $5d^{10}6s$ $^2S_{1/2}$ to $5d^{10}6p$ $^2P_{1/2}$ transition is used for Doppler laser cooling and laser-induced-fluorescence detection. The laser cooling and detection beam is tuned to component A in Fig. 1. To prevent trapping in a nonabsorbing dark state, the laser cooling and detection laser is polarization-modulated.[10] A laser beam tuned to the B component prevents trapping in the $(F = 0)$ ground state. To determine whether the 282 nm clock transition has been driven, radiation at the A and B frequencies is applied. If 194 nm fluorescence is observed, then the transition out of the $(F = 0)$ ground state did *not* occur. A laser tuned to the 398 nm transition empties the $^2D_{5/2}$ state. Prior to driving the 282 nm clock transition, the ion is prepared in the $^2S_{1/2}$ $(F = 0)$ hyperfine state by introducing a laser tuned to the C component to drive the ion from the $^2S_{1/2}$ $(F = 1)$ state to the $^2S_{1/2}$ $(F = 0)$ state.

With the frequency of the laser servo-locked to the atomic resonance, the apparatus functions as a frequency standard. With the aid of a self-referenced femtosecond laser frequency comb, the frequency of the Hg^+-stabilized laser has been compared to microwave or other optical frequency standards.[11–17]

3.3. *Limitations and Current Performance*

The various contributions to the systematic uncertainty of the mercury ion frequency have been evaluated or estimated.[14,17] The second-order Doppler shift due to thermal motion is reduced by laser cooling to near the Doppler cooling limit. The second-order Doppler shift due to rf micromotion is minimized by compensation of the stray electric fields.[18]. The static magnetic field is periodically measured by observing the resonance line of a first-order magnetic-field dependent Zeeman component of the $^2S_{1/2}(F = 0)$ to $^2D_{5/2}(F = 2)$ line. The frequency shift due to the interaction of the atomic quadrupole moment with a static electric field gradient is canceled by switching the magnetic field direction between three mutually perpendicular directions.[14,19] The blackbody radiation shift is negligible because the trap is operated at liquid helium temperature. An AC Zeeman shift due to unbalanced rf currents in the trap electrodes has been estimated and found to be small.[17]

It has been estimated that none of these individual sources of uncertainty is greater, fractionally, than 1×10^{-17}. The total fractional systematic uncertainty, given by the quadrature sum of the individual terms, is 1.9×10^{-17}.[17]

4. The Aluminum Ion Frequency Standard

4.1. *Atomic Level Structure*

The energy levels of $^{27}Al^+$ relevant to the operation of the frequency standard are shown in Fig. 2. The lifetime of the $^3P_0$ state has been measured to be 20.6 ± 1.4 s,[20] so the

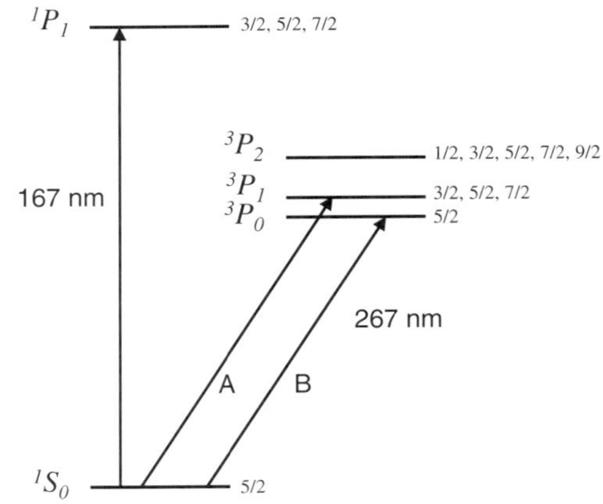

Fig. 2. The energy levels of $^{27}Al^+$ relevant to the operation of the frequency standard. The numbers to the right of the energy levels are the values of F, the total atomic angular momentum. The energy separations between hyperfine energy levels are not shown. The strong transition at 167 nm is not directly driven, for lack of a tunable laser at that wavelength. Transitions near 267 nm induced by the lasers are are labeled A and B. The energy differences between the ^3P_J fine-structure levels are expanded for clarity.

transition has a Q of 1.45×10^{17}, and is well suited as the basis for a frequency standard. However, there is no strongly-allowed transition that is available for laser cooling and fluorescence detection. The $3s3p\ ^3P_1$ state has a lifetime of $305 \pm 10\ \mu$s,[21] and the $3s3p\ ^3P_2$ state has an even longer lifetime. The strongly-allowed 167 nm $3s^2\ ^1S_0$ to $3s3p\ ^1P_1$ transition would be useful for laser cooling and state detection, but narrowband, tunable lasers are not available at that wavelength. Since the nuclear spin of ^{27}Al has spin $I = 5/2$, both the $^1S_0$ and the $^3P_0$ state have total angular momentum $F = 5/2$. Therefore, each m_F sublevel of the $^1S_0$ and $^3P_0$ states has a linear Zeeman shift.

4.2. *Operation of the Frequency Standard*

To get around the cooling and detection problems, Wineland proposed to simultaneously trap an auxiliary ion, which could be laser cooled and optically detected at a more convenient wavelength.[22] Since the two ions are coupled through the Coulomb interaction, the $^{27}Al^+$ "clock" ion is also cooled. Further, the superposition state of the clock ion can be transferred to the auxiliary ion, making use of the fact that they share a vibrational degree of freedom. That is, if the state of the clock ion is $(\alpha|S\rangle + \beta|P\rangle)$, the state of the auxiliary ion becomes $(\alpha|1\rangle + \beta|2\rangle)$, where $|1\rangle$ and $|2\rangle$ are two of the hyperfine ground state sublevels. This makes it possible to detect whether the clock ion has been driven to the metastable state by observing the fluorescence of the auxiliary ion.

The basic methods were demonstrated at NIST with a $^{27}Al^+$ clock ion and a $^9Be^+$ auxiliary ion.[23] The $^{27}Al^+$ $^1S_0$ to $^3P_1$ transition (A in Fig. 2) was used for this demonstration. More recently, the $^{27}Al^+$ $^1S_0$ to $^3P_0$ transition (B in Fig. 2) has been observed.[20] This transition is the basis for the $^{27}Al^+$ optical frequency standard. Detection of the $^1S_0$ to $^3P_0$ transition depends on the mapping of the $^1S_0$ to $^3P_1$ superposition state to a superposition state of the $^9Be^+$ ion. Details of the methods used to frequency-lock a laser to the $^9Be^+$ $^1S_0$ to $^3P_0$ resonance have been published.[20]

4.3. *Limitations and Current Performance*

The total fractional systematic uncertainty of the Al^+ clock frequency is 2.3×10^{-17}. The various contributions are discussed in detail elsewhere.[17,20] The greatest part of the systematic uncertainty is due to the second-order Doppler shift, from both the thermal motion and the micromotion. The linear Zeeman shift is cancelled by observing Zeeman components with reversed m_F and averaging the frequencies. The blackbody radiation shift is unusually small for an optical frequency standard because of a fortuitous cancelation between the $^1S_0$ and $^3P_0$ quadratic

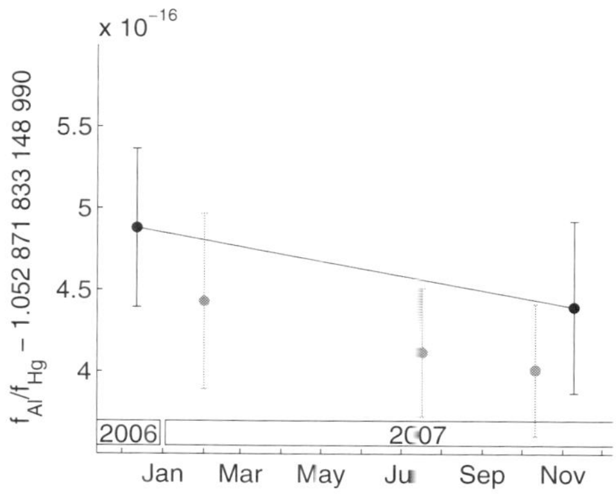

Fig. 3. Measurements of the ratio of the Al^+ and Hg^+ clock frequencies as a function of the date the measurements were made. The error bars are a combination of the statistical measurement uncertainties and the systematic uncertainties. The line connects the first point to the last one with a slope of $(-5.3 \pm 7.9) \times 10^{-17}$/year. The three points plotted in gray are not used in the calculation of the slope because not all of the systematic uncertainties were evaluated at the time of the measurements.

Stark shifts.[24] The fractional frequency shift at 300K is $(-8 \pm 5) \times 10^{-18}$. At the normal operating temperature of the frequency standard, it is $(-12 \pm 5) \times 10^{-18}$.

5. Hg^+ - Al^+ Frequency Comparisons

The Hg^+ and Al^+ frequency standards have been operated simultaneously. If one tooth of the femtosecond comb is phase-locked to one of the standards, and the heterodyne beat frequency between the other standard and the nearest tooth is measured, the frequency ratio of the two standards can be obtained in a way that does not depend on the accuracy of any microwave frequency standards used as references. The instability of the frequency ratio is described by an Allan deviation $\sigma_y(\tau)$ (fractional error for a measurement duration τ) of $3.9 \times 10^{-15}\ \tau^{-1/2}$ for $\tau > 100$, where τ is expressed in seconds.[17] Both standards are thought to contribute approximately equally, in an uncorrelated manner, to the instability of the ratio, so the fractional frequency instabilities of both standards are given by $\sigma_y(\tau) = 2.8 \times 10^{-15}$ $\tau^{-1/2}$. Figure 3 is a plot of measurements of f_{Al}/f_{Hg}, the ratio of the frequency of the Al^+ standard to that of the Hg^+ standard. The reproducibility of the ratio is seen to be better than 1×10^{-16}, which is better than the accuracy of the primary Cs frequency standards. The weighted mean of the measurements of f_{Al}/f_{Hg} is 1.052 871 833 148 990 438 (55), where the uncertainty is expressed in units of the least significant digit.[17]

6. Limits on Time Variation of Fundamental Constants

It is of interest to see whether the ratio f_{Al}/f_{Hg} remains constant over time. If a systematic change in this quantity were observed, this might be due to a time variation of fundamental constants, such as the fine structure constant α, the strong interaction coupling constant α_s, or the ratios of the masses of elementary particles.[25] At present, no such change has been observed, which places stringent limits on variations of the fundamental constants. The slope of the fitted line in Fig. 3 corresponds to a fractional rate of change in α of $(1/\alpha)(d\alpha/dt) = (-1.6 \pm 2.3) \times 10^{-17}$/year, consistent with zero.

Acknowledgments

This work was supported by the Office of Naval Research, IARPA, and NIST. P. O. S. acknowledges support from the Alexander von Humboldt Foundation. This work was performed by an agency of the U. S. government and is not subject to U. S. copyright.

References

1. J. Terrien, *Metrologia* **4**, 41 (1968).
2. S. R. Jefferts *et al.*, *Proc. SPIE* **6673**, 667309 (2007).
3. S. A. Diddams, *et al.*, *Phys. Rev. Lett.* **84**, 5102 (2000).
4. D. J. Jones, *et al.*, *Science* **288**, 635 (2000).
5. R. Holzwarth, *et al.*, *Phys. Rev. Lett.* **85**, 2264 (2000).
6. B. C. Young, F. C. Cruz, W. M. Itano, and J. C. Bergquist, *Phys. Rev. Lett.* **82**, 3799 (1999).
7. R. H. Dicke, *Phys. Rev.* **89**, 472 (1953).
8. R. J. Rafac, *et al.*, *Phys. Rev. Lett.* **85**, 2462 (2000).
9. W. M. Itano, *et al.*, *Proc. SPIE* **6673**, 667303 (2007).
10. D. J. Berkeland and M. G. Boshier, *Phys. Rev. A* **65**, 033413 (2002).
11. S. A. Diddams, *et al.*, *Science* **293**, 825 (2001).
12. T. Udem, *et al.*, *Phys. Rev. Lett.* **86**, 4996 (2001).
13. S. Bize, *et al.*, *Phys. Rev. Lett.* **90**, 150802 (2003).
14. W. H. Oskay, *et al.*, *Phys. Rev. Lett.* **97**, 020801 (2006).
15. T. M. Fortier, *et al.*, *Phys. Rev. Lett.* **98**, 070801 (2007).
16. J. E. Stalnaker, *et al.*, *Appl. Phys. B* **89**, 167 (2007).
17. T. Rosenband, *et al.*, *Science* **319**, 1808 (2008).
18. D. J. Berkeland, *et al.*, *J. Appl. Phys.* **83**, 5025 (1998).
19. W. M. Itano, *J. Research National Institute of Standards and Technology* **105**, 829 (2000).
20. T. Rosenband, *et al.*, *Phys. Rev. Lett.* **98**, 220801 (2007).
21. E. Träbert, *et al.*, *J. Phys. B* **32**, 537 (1999).
22. D. J. Wineland, *et al.* in *Proceedings of the 6th Symposium on Frequency Standards and Metrology*, ed. P. Gill, (World Scientific, Singapore, 2002).
23. P. O. Schmidt, *et al.*, *Science* **309**, 749 (2005).
24. T. Rosenband, *et al.* in *Proc. 20th European Time and Frequency Forum* (2006); preprint: arXiv:physics/0611125v2.
25. S. N. Lea, *Rep. Prog. Phys.* **70**, 1473 (2007).

STM AND STS OBSERVATION ON TITANIUM-CARBIDE METALLOFULLERENES: $Ti_2C_2@C_{78}$

N. FUKUI*
World Premier International (WPI) Research Center, Advanced Institute for Material Research, Tohoku University, Sendai, 980-8577, Japan
Department of Chemistry, Nagoya University, Nagoya, 464-8602
**E-mail: n-fukui@wpi-aimr.tohoku.ac.jp*

H. MORIBE, H. UMEMOTO and H. SHINOHARA
Department of Chemistry, Nagoya University, Nagoya, 464-8602, Japan

Y. SUWA, S. HEIKE, M. FUJIMORI
ARL, Hitachi Ltd., Hatoyama, Saitama, 350-0395, Japan

T. HASHIZUME
ARL, Hitachi Ltd., Hatoyama, Saitama, 350-0395, Japan
World Premier International (WPI) Research Center, Advanced Institute for Material Research, Tohoku University, Sendai, 980-8577, Japan
Department of Physics, Tokyo Institute of Technology, Oh-okayama, Tokyo, 152-8551, Japan

A metallofullerene $Ti_2C_2@C_{78}$, in which two titanium atoms and C_2 cluster are encapsulated, is studied by scanning tunneling microscopy and spectroscopy. Measurements of $Ti_2C_2@C_{78}$ on Cu(111) surface reveal that their cage symmetry is C_{78}-D_{3h}. There is a preferential orientation of $Ti_2C_2@C_{78}$ resulting from electrostatic interaction same as other metallofullerene.

Keywords: Scanning Tunneling Microscopy, Fullerene, Metallofullerene

1. Introduction

Fullerenes have hollow spaces inside their carbon cage large enough to encapsulate small clusters such as metal atoms[1, 2] (metallofullerenes), tri-metal nitride[3, 4], metal-carbon cluster[5] (metal-carbide-fullerene), and small molecules like hydrogen[6]. The effects of encapsulation of metal atom/cluster are well-known in that the electronic structure of the "cages" are modified by charge transfer from "encapsulated metal/cluster" to "cages"[1]. There are lots of combinations of encapsulated metal and cage, and such a variety is the origin of difference in each fullerene series.

Scanning tunneling microscopy/spectroscopy (STM/STS) measurements are available to visualize the local density of states near the Fermi level and are one of the most powerful tools to clarify physics of fullerene series such as the interaction between fullerenes[7, 8], the interaction between the fullerene and substrate[9], and the interaction between the fullerene cage and encapsulated metal atoms[10, 11].

$Ti_2C_2@C_{78}$ is a new class of fullerenes, which encapsulates two titanium atoms and two carbon atoms. Their electronic structure have attracted much attention[12]. Their cage symmetries are D_{5h} and I_h-C_{80} cages[13]. However, their cage structure is turned out C_{78}-D_{3h} symmetry based on ^{13}C-NMR[12], TEM[14], X-ray[12] and UPS[15, 16] experiments using highly precise sample and theoretical calculations[17]. Encapsulated C_2 cluster are expected to rotate on the center of molecules (inside the titanium position). Additionally, there is large electron density distribution between the cages and encapsulated titanium atoms.

In this study, we show the results of STM/STS observations of $Ti_2C_2@C_{78}$ and discuss their cage structures and electronic structures.

2. Experimental

The $Ti_2C_2@C_{78}$ mettallofullerene was prepared by the DC arc discharge method with Ti/C composite rod, followed by separation, purification and isolation with high performance liquid chromatography (HPLC)[12, 13]. The purity and isolation of $Ti_2C_2@C_{78}$ was checked by LD-TOF mass, ^{13}C-NMR measurement and UV-Vis NIR spectrum[12]. The purified sample was put on a tantalum crucible and degassed in UHV conditions to prevent contamination.

The Cu (111) substrate was prepared by repetitive argon ion sputter and annealing processes. The $Ti_2C_2@C_{78}$ was sublimed in UHV on Cu(111), via a crucible kept at 673-873 K. Typical deposition rate was normally below 0.05 ML/min.

The STM/STS measurements were performed at room temperature and 100K using home-build FI-STM with a W-tip[9].

3. Result and Discussion

Figure 1 shows bias-dependent STM images of a monolayer of $Ti_2C_2@C_{78}$ on Cu(111) surfaces at 100 K. Two different orientations found in Fig. 1 indicate that $Ti_2C_2@C_{78}$ has an anisotropic cage symmetry. The apparent height at the step between $Ti_2C_2@C_{78}$ layers is about 0.2 nm, which corresponds to the Cu mono-atomic steps. The intramolecular patterns are observed on the surface of $Ti_2C_2@C_{78}$ and depend on the bias voltage. Such intramolecular patterns are also found at room temperature, indicating that $Ti_2C_2@C_{78}$ molecules are trapped on Cu(111) surface at room temperature as other fullerenes[9]. The intramolecular pattern having a banana shape and a 6-fold symmetry are observed in layers and in other places such as steps and kinks(Fig.1c).

To analyze these intramolecular patterns and orientations, first-principle calculations using generalized gradient approximation with ultra-soft pseudopotentials were performed[18, 19]. The structure models for calculations have been based on MEM-Reitvelt analysis results[20, 21]. Figure 2(a) shows the calculated wave functions of $Ti_2C_2@C_{78}$ whose cage-structure has C_{78}-D_{3h} symmetry. At the LUMO+1 level, a hybridized state of *d*-orbital of encapsulated titanium atom and π-electron of carbon cage appears consistent with a previous report[17].

Figure 2(b) shows the simulated STM images, according to Tersoff–Hamann calculations of the tunneling current[22].

Banana-shaped patterns can be seen in both of two side views of $Ti_2C_2@C_{78}$. This pattern is parallel to the encapsulated Ti–Ti direction in filled-state images, but perpendicular to it in empty-state images. These shapes are sufficient to identify the orientation of $Ti_2C_2@C_{78}$ on Cu(111) surface.

The long axis of each $Ti_2C_2@C_{78}$ forming a monolayer takes random direction inside the

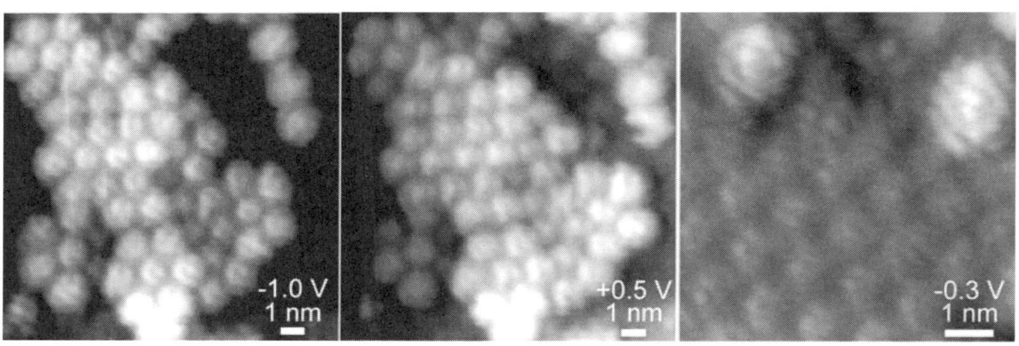

Fig. 1: STM images of $Ti_2C_2@C_{78}$ on Cu(111)
(a),(b) bias dependent image in the same island; banana shape (c) on the island; 6-fold symmetry.

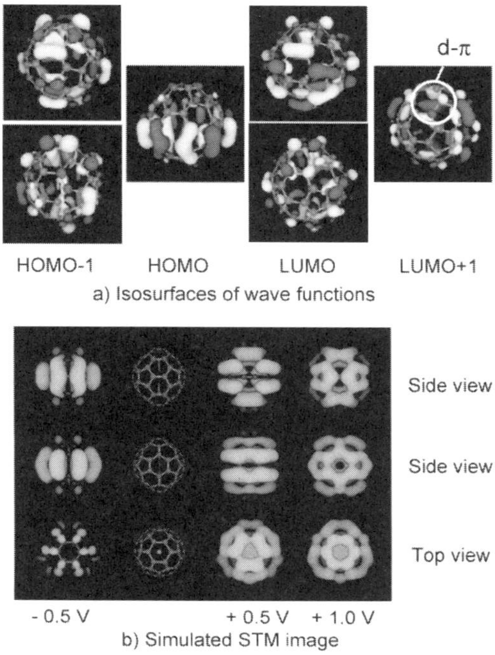

Fig.2 : (a) Isosurfaces of the wave function of each state. HOMO-1 and LUMO state are degenerate orbital. At the LUMO+1 level, d-π hybridized orbital has emerged.
(b) Simulated STM image by Tersoff-Hamann method.

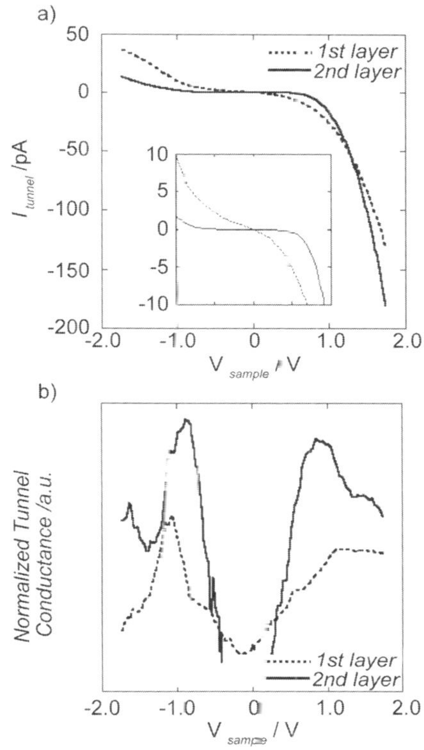

Fig.3 : STS results of $Ti_2C_2@C_{78}$ on Cu(111)
(a) I-V curve (b) Normalized tunnel conductance
(b)The first layer of $Ti_2C_2@C_{78}$ is metallic and the second layer shows band gap (0.61eV).

parallel plane to the substrate, while that of $Ti_2C_2@C_{78}$ placed on such a layer points the direction perpendicular to the substrate. This fact can be explained by a charge transfer from the inner metal to the cage in which the positional distribution of inner metal strongly affects the electronic state of the cage[23, 24].

According to the present calculation, the contribution of hybridized orbital of titanium atoms and carbon cages to the LDOS observable in the STM measurement is very small. It is thus not easy to obtain the direct image of the hybridized state. However, a direct imaging of hybridized state is not totally impracticable because the position of the metal atom is successfully identified in metallofullerenes, such as in $Gd@C_{82}$[10] and $Dy@C_{82}$[11].

The STS results of $Ti_2C_2@C_{78}$ on Cu are shown in Figure 3. The I-V curve of the first layer on Cu(111) clearly shows metallic behavior. This is a result of charge transfer from the substrate to the cage[9]. In contrast, $Ti_2C_2@C_{78}$ at the second layer has a small energy-gap as a semiconductor. This indicates that the second layer of $Ti_2C_2@C_{78}$ is less affected by charge transfer than the first layer.

Figure 3(b) shows the normalized tunneling conductance ((dI/dV)/(I/V)) curves derived from Fig. 3(a). The peak of the conductance curve for the first layer was slightly shifted outward; the charge transfer occurred from Cu(111) to the surface (0.3 eV). This perturbation is small but appears in the second layer of $Ti_2C_2@C_{78}$ on Cu(111). The estimated values of the band gap for the second layer on Cu(111) is 0.61 ± 0.20 eV and those of a theoretical calculation is 0.76 eV. The experimental peak position of each level is consistent with the DFT calculation. These features indicate that $Ti_2C_2@C_{78}$ has D_{3h} symmetry.

Experimentally obtained value of the band gap also agrees with the DFT calculation, in which the D_{3h} symmetry is the most appropriate candidate among the possible isomers[17]. Based on experimental data and calculations, it is reasonable to suppose that there is covalent-like bond between carbon atoms at the cage and encapsulated titanium atoms and presence of innermost two carbon atoms.

4. Conclusions

The STM images of $Ti_2C_2@C_{78}$ with D_{3h} symmetry are observed and compared with the simulated images. A preferential orientation of $Ti_2C_2@C_{78}$ on Cu(111) depends on the configuration of encapsulated titanium atoms as other examples of simple metallofullerenes. The electronic properties of the cage of $Ti_2C_2@C_{78}$ at the first layer have been strongly affected by both encapsulated titanium and Cu(111) substrate.

Acknowledgments

We are grateful to Profs. E. Nishibori and M. Takada for providing MEM data for DFT calculation. This work was supported by World Premier International Research Center Initiative (WPI Initiative), MEXT, Japan.

References

[1]H. Shinohara, Rep. Prog. Phys., 843 (2000).
[2]M. Takata, *et al.*, Phys. Rev. Lett. **83**, 2214 (1999).
[3]S. Stevenson, *et al.*, Nature **401**, 55 (1999).
[4]D. F. Leigh, *et al.*, Surf. Sci. **601**, 2750 (2007).
[5]T. Inoue, *et al.*, J. Phys. Chem. B **108**, 7573 (2004).
[6]K. Komatsu, M. Murata, and Y. Murata, Science **307**, 238 (2005).
[7]Y. Nakamura, *et al.*, Appl. Phys. Lett **85**, 5242 (2004).
[8]S. Fujiki, *et al.*, Chem. Phys. Lett **420**, 82 (2006).
[9]T. Hashizume, *et al.*, Phys. Rev. Lett. **71**, 2959 (1993).
[10]M. Grobis, *et al.*, Phys. Rev. Lett. **94**, 136802 (2005).
[11]K. Wang, *et al.*, Phys. Rev. Lett. **91**, 185504 (2003).
[12]H.Moribe, *et al.*, in *Abstracts of 27th Fullerene.Nanotubes General Symposium* Fullerene Nanotubes Research Association, Nagoya, 2004, p. 170.
[13]B. Cao, *et al.*, J. Am. Chem. Soc. **123**, 9679 (2001).
[14]Y. Sato, *et al.*, Phys. Rev. B **73**, 193401 (2006).
[15]K. Iwasaki, *et al.*, Chem. Phys. Lett **397**, 169 (2004).
[16]S. Hino, *et al.*, Phys. Rev. B **75**, 125418 (2007).
[17]M. Otani, S. Okada, and A. Oshiyama, Chem. Phys. Lett **438**, 274 (2007).
[18]D. Vanderbilt, Phys. Rev. B **41**, 7892 (1990).
[19]J. P. Perdew, K. Burke, and Y. Wang, Phys. Rev. B **54**, 16533 (1996).
[20]E. Nishibori, *et al.*, in *Abstracts of 28th Fullerene.Nanotubes General Symposium* Fullerene Nanotubes Research Association, Nagoya, 2005.
[21]E.Nishibori, unpublished
[22]J. Tersoff and D. R. Hamann, Phys. Rev. B **31**, 805 (1985).
[23]C. Ton-That, *et al.*, Phys. Rev. B **68**, 045424 (2003).
[24]C. Ton-That, *et al.*, Phys. Rev. B **71**, 045419 (2005)

SINGLE SHOT MEASUREMENT OF A SILICON SINGLE ELECTRON TRANSISTOR

T. FERRUS[1],D. G. HASKO[2],Q. R. MORRISSEY[3],S. R. BURGE[3],E. J. FREEMAN[3], M. J. FRENCH[3],A. LAM[2],L. CRESWELL[2],R. J. COLLIER[2],D. A. WILLIAMS[1], G. A. D. BRIGGS[4]

[1] *Hitachi Cambridge Laboratory, J. J. Thomson Avenue, CB3 0HE, Cambridge, United Kingdom*

[2] *Cavendish Laboratory, University of Cambridge, J. J. Thomson Avenue, CB3 0HE, Cambridge, United Kingdom*

[3] *Rutherford Appleton Laboratory, Chilton, Didcot, OX11 0QX, Oxon, United Kingdom*

[4] *Department of Materials, University of Oxford, Parks Road, Oxford OX1 3PH, United Kingdom*

We describe measurements on a silicon single electron transistor (SET) carried out using a custom cryogenic CMOS measurement circuit (LTCMOS) in close proximity to the device. Quantum mechanical states in the SET were mapped by continuous microwave spectroscopy. The real time evolution of a particularly long lived quantum mechanical state was observed in a single shot measurement, made possible by the much faster measurement rate (50kHz bandwidth). This technique is intended to be applied to the measurement of coherent states in a charge qubit device made of a silicon double dot.

Keywords: single electron transistor; silicon; microwaves; single shot

1. Introduction.

The observation of quantum mechanical effects in small electronic devices usually requires their measurement to be carried out at cryogenic temperatures. However, direct measurements made using room temperature electronics are subject to severe bandwidth limitations. These are due to the low pass filtering effect of the output impedance of the device under test and the capacitance of the connecting doped semiconductor leads and of the sensing amplifier input. In a typical Coulomb blockade device giving currents of tens of pA, this effect limits the practical measurement rate to <10Hz. This rate is too small to observe the evolution of a quantum state in real time. Alternative methods based on radio frequency techniques offer a much higher bandwidth, but can lead to unwanted changes in the quantum mechanical state [1]. Here we describe a measurement on a silicon single electron transistor (SET) [2] carried out using a low temperature custom CMOS measurement integrated circuit (LTCMOS) in close proximity to the device under test. This arrangement enables a significant improvement in measurement speed compared to conventional measurements using room temperature source measure units. Quantum mechanical states in the SET were mapped by continuous microwave spectroscopy and single-shot operation was demonstrated using a high quality factor resonance in the silicon single electron transistor.

2. Device and experimental setup.

A silicon-on-insulator substrate was used to fabricate the SET. The 30 nm-thick silicon layer was doped with phosphorous at a density of $10^{20}\,\mathrm{cm}^{-3}$. A single dot of diameter ~60 nm and 30-nm width tunnel barriers was then patterned by e-beam lithography and reactive ion etching. The dot size and tunnel barrier widths were subsenquently reduced by thermal oxidation. The devices were controlled by an in-plane gate that was formed using the same SOI layer (Fig. 1a and b) and mounted in a 20 pin leadless header package at the end of the custom probe. The microwave signal was coupled to the SET by using an open ended semi-rigid waveguide that terminates ~0.5 mm away from the silicon surface (Fig. 1c). The LTCMOS was used to provide the voltages sources as well as to measure the SET current. Electrical lines to the source, drain and gate leads were filtered by single stage *RLC* filters with a cut-off of about 2 MHz to suppress electrical noise from the CMOS circuit. Both the device and the LTCMOS were kept at 4.2 K by immersing the probe into liquid helium.

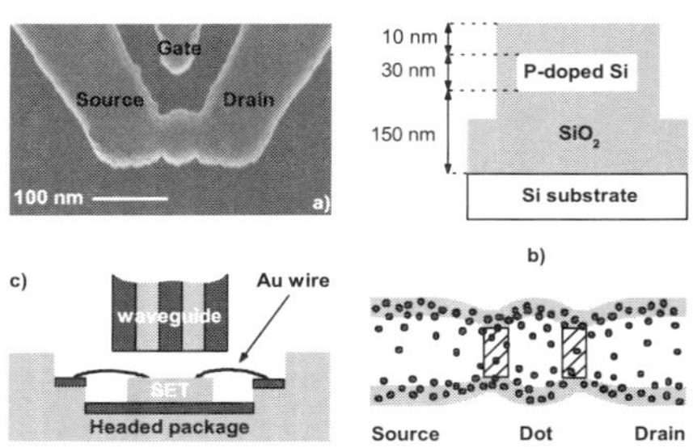

Fig. 1. a) SEM image of the SET. b) Cross-section and plane view of the device with the side depletion (grey), the source and drain potential barriers (dashed) and localising centres (black dots). c) Mounting of the device on the headed package with the open ended waveguide.

3. SET response to microwaves.

3.1. *Continuous wave measurement.*

At a microwave power $P \geq -20\,\text{dBm}$, the SET response to continuous microwave (CW) excitation consists of a large number of resonances (Fig. 2) [3] that exhibit a wide range of amplitudes and quality factors with the largest amplitudes corresponding to the lowest Q values.

Low Q resonances, typically below 10^4, may be explained in terms of cavity resonances. Indeed these modifies the frequency dependence of the coupled microwave power and may contribute to electron heating in the SET island.[4] Indeed, a rise in electron temperature would lead cotunneling and sequential tunneling effects.[5] and increase the conductance of the SET. However some of these resonances appear as an increase or a decrease in the SET current and in a few cases, the SET current at the resonance is even seen to change its polarity respectively to the source-drain potential difference. This also suggests that some of these low Q resonances are actually due to current pumping.

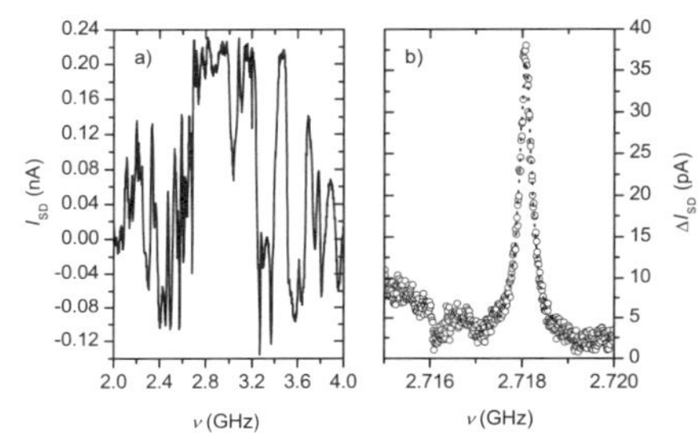

Fig. 2. a) Source to drain current dependence on microwave frequency. I_{SD} is measured relatively to its value for $P = -20\,\text{dBm}$. b) High Q resonance used for the single shot measurement ($Q \sim 48,000$). The Lorentzian fit is shown by the dotted line. Measurements were taken with $P = 9\,\text{dBm}$, a gate voltage of -8.3 mV and a source voltage of 4 mV.

However the large Q values suggest a mechanism that involves weak interaction with the electromagnetic environment as this is the case for tunneling or trap-mediated transport. These phenomenons are relevant here because of the presence of phosphorous donors in the SET island and tunneling barriers. Also a significant number of localised states are present especially on the sidewalls of the SET.[6] This effect is enhanced by the large surface state density found on the non-(100) surfaces forming the sidewalls. Also, the resonance shapes are found to be

Lorentzian with a width giving an energy scale two or three orders of magnitude lower than k_BT. This is clearly inconsistent with phonon-mediated hopping for which the resonance linewidth is proportional to the temperature. Indeed, according to Landauer theory, the conductance of a single channel wire with a barrier is given by the transmission probability [7], and, for a sufficiently high barrier relatively to the Fermi energy, this probability is energy-independent, giving a temperature independent width in the resonance spectrum.[8] The previous observation also implies a weak coupling to the phonon bath and non-adiabatic processes (e.g. the timescale for the transport mechanism is shorter than the thermalization time). In this case, the resonance widths $\Delta\nu_0$ are related to the energy loss mechanism in the system and $Q = \pi\nu_0/\Delta\nu_0 \sim \pi T_0\nu_0$, where T_0 is the energy relaxation time of the system [9] and ν_0 the resonance frequency. In the absence of observations of photon-assisted tunneling effects in the gate dependence of the source-drain current [10], trap-assisted tunneling in the tunnel barrier or potential shifts due to reversible dipoles [11] in the SET island are most probably responsible for the highest Q resonances. In the latter case, electron hops most likely occur between localised states surrounding the SET island although non-uniformity in the doping may result in localised states being present inside the SET island. The electron displacement modifies the electrostatic environment inside the SET island and leads to a change in the potential close to the tunnel barrier. This changes the magnitude of the barrier and is responsible for an increase or a decrease in the measured current depending on the location of the active traps. Localised states in the direct transport pathway would lead to low Q resonances with large amplitude due to the short residence time of the electron whereas remote traps could be responsible for the high Q but low amplitude resonances.

3.2. *Single shot measurement.*

From the quality factor for the high Q peaks the expected lifetime for these excitations is found to be as long as $100\,\mu$s. Due to the measurement bandwidth of the system, it was thus possible to implement single-shot operations and get access to the real time evolution of the SET current. This consisted in sending a single and short on-resonance microwave excitation pulse of duration whilst carrying out current measurements at regular intervals. To this purpose, we first selected a well-isolated resonance at a frequency $\nu_0 \sim 2.718$ GHz from a CW measurement at a power of +9 dBm. The charge integration time was set to $200\,\mu$s and results averaged over 200 measurements to reduce the signal-to-noise ratio of about 100:1 (Fig. 2). The width of the chosen resonance was about 177 kHz, for which the corresponding relaxation time is $T_0 \sim 6\,\mu$s. A single-shot measurement was then performed by reducing the charge integration time to $20\,\mu$s and using a microwave pulse of duration of $2\,\mu$s. The microwave power was increased to about +25 dBm using a power amplifier to compensate for the decrease in the signal amplitude. These conditions also greatly reduced the signal to noise ratio compared to the CW measurement, so that the averaging had to be increased to 1000 measurements per point. The delay in the trigger signal causes the microwave pulse to arrive at $t \sim 0.8$ ms (Fig. 3). The signal measured in single-shot mode contains some very large damped oscillations that are present even in the absence of the microwave pulse. They are even present in the absence of the microwave pulse and are measurement artefacts due to cable capacitance charging effects. By tuning the microwave frequency to on or off resonance values the effects of these background oscillations may be removed. The difference between these is an exponential decay with a time constant $\tau \sim 90\,\mu$s. This value is more than one order of magnitude larger than that found from the CW linewidth measurement. As the time-averaged power in the pulsed measurement is a factor of 1000 less than that used in the CW measurement, this difference in relaxation time may result from microwave-induced electron heating in the CW measurement. The lifetime τ corresponds to the time for the microwave-excited electron to recombine with its original ionised trap. Its large value suggests glassy behaviour [12], a characteristic manifestation of Coulomb interaction in systems with localized states. The disorder is coming from the random placement of phosphorous donors and the glassy dynamics would result from the competition between the disorder and the Coulomb interaction.

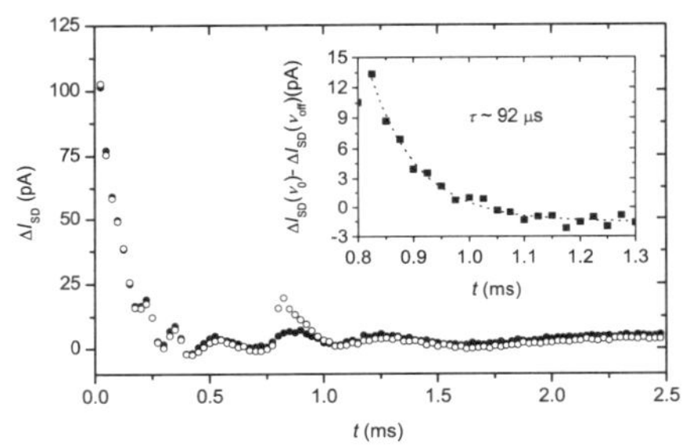

Fig. 3. Relative variation of the source to drain SET current with time t for $\nu_{\rm off} = 2.716\,{\rm GHz}$ (off resonance, •) and $\nu_0 = 2.718\,{\rm GHz}$ (on resonance, ∘). The inset shows the resulting exponential decay in the current.

4. Conclusion.

The use of a low temperature custom CMOS measurement circuit has enabled use to observe for the first time the real-time decay of a quantum state by implementing the single-shot technique in a single electron transistor. The especially long lifetime observed suggest the present experimental setup is suitable for the study of decoherence in charge qubit device made of a silicon double dot.

5. Acknowledgments.

This work was partly supported by Special Coordination Funds for Promoting Science and Technology in Japan.

References

1. W. Lu, Z. Ji, L. Pfeiffer, K. W. West and A. J. Rimberg, Nature 423, 422-425 (2003)
2. T. A. Fulton and G. J. Dolan, Phys. Rev. Lett. 59, 109 (1987); P. Lafarge, H. Pothier, E. R. Williams, D. Esteve, C. Urbina and M. H. Devoret, Z. Phys. B 85, 327 (1991)
3. L. A. Creswell, D. G. Hasko and D. A. Williams, Microelectronic Engineering 84, 1614 (2007)
4. M. H. Manscher, M. T. Savolainen, and J. Mygind, IEEE Trans. Appl. Supercond. 13, 2 (2003)
5. D. V. Averin and A. A. Odintsov, Phys. Rev. A 140, 251 (1989)
6. J. Eldridge and D. Kerr, J. Electrochem. Soc. 118 986 (1971); P. Balk and J. Eldridge, Proc. IEEE 57, 1558 (1969)
7. R. Landauer, IBM J. Res. Develop. 1, 233 (1957); *ibid* Phil. Mag. 21, 863 (1970)
8. P. J. Price, Phys. Rev. B 38, 1994 (1988)
9. C. N. Banwell and E. M. McCash, Fundamentals of molecular spectroscopy, 4th ed., McGraw-Hill (1994)
10. L. P. Kouwenhoven, S. Jauhar, K. McCormick, D. Dixon, P. L. McEuen, Yu. V. Nazarov, N. C. van der Vaart and C. T. Foxon, Phys. Rev. B 50,2019 (1994); D. Dovinos and D. Williams, Phys. Rev. B 72, 085313 (2005)
11. H. O. Müller, M. Furlan, T. Heinzel, and K. Ensslin, Europhys. Lett. 55, 253 (2000)
12. B. I. Shklovskii and A. L. Efros, Electronic Properties of Doped Semiconductors (Springer-Verlag Berlin 1984)

DERIVATION OF SENSITIVITY OF A GEIGER MODE APDS DETECTOR FROM A GIVEN EFFICIENCY TO ESTIMATE TOTAL PHOTON COUNTS

KIYOTAKA HAMMURA* and DAVID ARFON WILLIAMS

Hitachi Cambridge Laboratory, Cavendish Laboratory, University of Cambridge, J. J. Thomson Avenue, Cambridge CB3 0HE, United Kingdom

**E-mail: kh401@cam.ac.uk*

The detection sensitivity (DS) of the commercial single-photon-receiver based on InGaAs gate-mode avalanche photodiode is estimated. Instalment of a digital-blanking-system (DBS) to reduce dark current makes the difference between DS and the detection efficiency (DE). By numerical simulations, it is found that the blanked number of light-pulses by DBS is as many as a quarter of all incident pulses for a specific operation condition. DS is estimated at 0.27, which is 35% larger than a given DE.

Keywords: Quantum key distribution, single-photon detector, afterpulsing, InGaAs avalanche photodiode.

1. Introduction

The emergence of a commercial single-photon-receiver (SPR) would help reduce development time to integrate quantum key distribution (QKD) system. Since QKD[1] was firstly experimentally demonstrated[2], tremendous efforts have continued towards QKD's further performance improvements[3]. Many efforts have been dedicated to the development of two essential devices which constitutes QKD, i.e., single-photon-emitters [4,5,6] and SPRs[7,8,9,10,11,12,13,14,15,16,17]. The commercial SPR of Princeton Lightwave[18], for example, is a highly integrated one. This SPR features not only high detection efficiencies, DEs, at telecom wavelengths and low dark count probabilities but also a digital-blanking system (DBS) to dramatically reduce the effect of afterpulsing on dark count.

It should be noted, however, that the given DE does not clarify about some intrinsic figure-of-merit of the SPR concerning its sensitivity to incident single-photons. We should have a figure-of-merit of how sensitive the SPR in its open-gate state is (which will be designated as a detection sensitivity, DS, hereafter) while the given DE is simply a ratio of the number of registration to that of all incident light-pulses (see Eq.(1) and Eq.(2)).

$$DE \equiv \frac{\textit{How many single-photons are registered}}{\textit{How many single-photons hit the detector}}, \quad (1)$$

$$DS \equiv \frac{\textit{How many single-photons are registered}}{\textit{How many single-photons hit the detector in open-gate status}}. \quad (2)$$

Thanks to the DBS, a designated number of bias pulses for gating the detector are neglected once a registration occurs. In this paper, DS of the commercial SPR[18] is estimated as DS is helpful to estimate the registration number for various operation conditions while DE is helpful only for a set of conditions to give the DE.

DS should be defined as a ratio of the number of registrations to that of light-pules which hit the detector in open-gate status (Eq.(1)). Such number of light-pulses is to be estimated by subtracting the number of light-pulses blanked by DBS, N_B, from all incident ones. The N_B will be numerically estimated as below.

2. Method

DS can be derived using one experimental data and the number of light-pulses blanked by DBS. Eq.(1) is expressed as,

$$DS = \frac{N_G}{N_A + N_G} \quad (3)$$

$$= \frac{N_G}{(N_A + N_G + N_B) - N_B}, \quad (4)$$

where N_G and N_A designate the numbers per second of how many light-pulses are registered, how many light-pulses hit the detector in gate-open status but are not registered due to detector's poor sensitiveness as a whole, respectively. The relations among N_G,N_A, and N_B are illustrated in FIG 1. We use an experimental data given by the manufacturer[19] for N_G in the numerator in Eq.(4).

N_B is calculated under the following experimental conditions: the light-pulse arriving rate, $N_T = 250 \times 10^3$ s^{-1}, the mean number of photons per pulse, $\mu = 0.1$, the triggering rate of the detector, $N_{TR} = 500 \times 10^3$ s^{-1}, the number of gatings blanked by DBS, $B_L = 6$. The simulation is conducted according to the following 3 steps. All figures below are those for one second.

(1) Distribute 25000 ($= N_T \cdot \mu \times 1$ s $\equiv$ N$_0$) incident single-photons randomly over 250000 ($= N_T \times 1$ s) time-bins. The specific procedure is to take a random number between 1 to 250000, for each of 25000 figures. Then, arrange the 25000 figures in ascending order of their associated random numbers. The random numbers mean the occupied time-bins.
(2) Perform blanking. For a time-bin which is occupied, the next 3 adjacent time-bins from the occupied one are called the blanked bins. The reason why the number of blanked bins is not 6 ($= B_L$) but 3 comes from the fact that the triggering rate, N_{TR}, is twice faster than the light-pulse arriving rate, N_T. Blanking 6 time-bins only means blanking 3 pulses.
(3) Count up how many light-pulses are found sitting in the blanked bins, resulting in N_B.

3. Result

Numerical simulations of N_B results in 6250 s^{-1}. With Eq.(4), DS is estimated at 0.27 as follows:

$$DS = \frac{N_G}{(N_A + N_B + N_G) - N_B} \quad (5)$$

$$= \frac{5033}{25000 - 6250} = 0.27\,. \quad (6)$$

Here, $N_G = 5033$ s^{-1} is a given value[19] and $N_A + N_B + N_G \equiv N_0$.

4. Discussions

By estimating the effect of blanking on the number of photon detection registration, DS of the detector is derived under the condition that dark count events are all neglected.

We established a method to derive DS, which is more suitable figure-of-merit to characterise device parameters of SPADs.

The estimation shows that as many as a quarter of incident light-pulses (=6250/25000) is in fact blanked by DBS. While it has been well known that measures for gate suspension is quite helpful to prevent the effect of aftepulsing from adding to dark count rate[7,8,9], the measures also causes nonnegligible amount of decrease in detector's sensitivity to incident light-pulses. DS, which shows an intrinsic sensitivity, is 35 % larger than DE under DBS, which is 0.20 (=5033/25000).

DS value enables us to evaluate the registration number for the experiment in which operation conditions are different from those listed in spec sheet. Applying $DS = 0.27$ to our experimental configurations: $N_{TR} =$

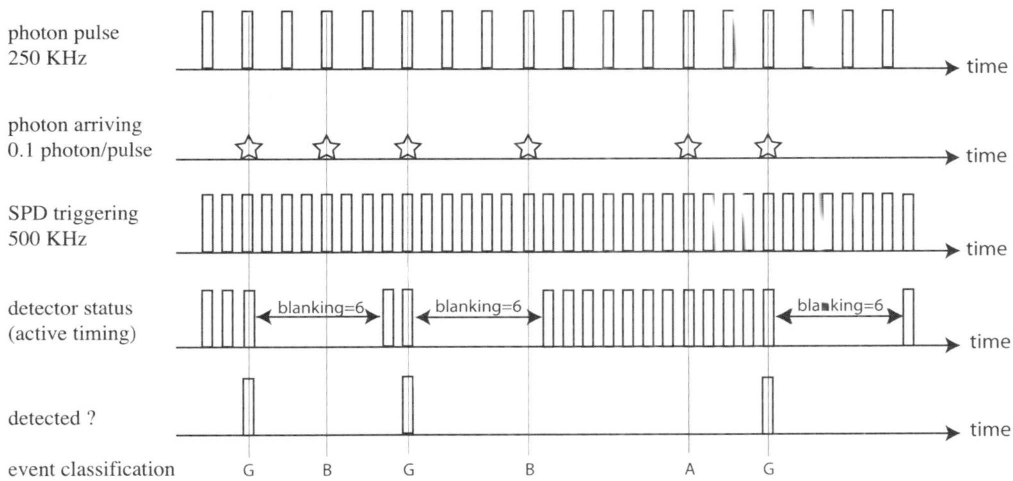

Fig. 1. N_A, N_B, and N_G in the text refer to the numbers per second of incidents designated by A, B, and G.

$5 \times 10^6\ s^{-1}$ and $N_T = 100 \times 10^6\ s^{-1}$, the total photon count for a single photon stream is estimated at $81000\ s^{-1}$. This value is requisite for QKD system integration.

We will optimise the performance of the QKD system under development more efficiently with the detailed knowledge of detector's performance.

DEs at different frequencies help unveiling intrinsic characteristics of the detector. For example, DE at 5 MHz gating is quantified to be 0.162 (=81000/500000) which is less than 0.201 (=5033/25000) at 500 kHz gating despite the same blanking of 6 for both. It suggests that there is an intrinsic factor in the detector to suppress its performance by 20 %.

Acknowledgments

Authors thank Dr. D. S. Bethune of IBM, Dr. B. Nyman of Princeton Lightwave, and Dr. A. Wonfor of Cambridge University for many useful comments and valuable discussions.

References

1. C. H. Bennett, and G. Brassard, *Quantum cryptography: public key distribution and coin tossing*, Proceedings of the International Conference on Computer Systems and Signal Processing, Institute of Electrical and Electronics Engineers, New York, 175 (1984).
2. Charles H. Bennett, Francois Bessette, Gilles Brassard, Louis Salvali, and John Smolin, *Experimental Quantum Cryptography*, Journal of Cryptography **5**, 3 (1992).
3. N. Gisin, G. Ribordy, W. Titel, and H. Zbinden, *Quantum Cryptography*, Rev. Mod. Phys. **74**, 145 (2002).
4. Xiulai Xu, D. A. Williams, and J. R. A. Cleaver, *Electrically pumped single-photon sources in lateral p-i-n junctions*, Appl. Phys. Lett. **85**, 3238 (2004).
5. Xiulai Xu, Ian Toft, Richard T. Phillips, Jonathan Mar, Kiyotaka Hammura, and David A. Williams, *"Plug and play" single-photon sources*, Appl. Phys. Lett. **90**, 061103 (2007).
6. Xiulai Xu, Frederic Brossard, Kiyotaka Hammura, David A Williams, B. Alloing, L. H. Li, and Andrea Fiore, *"Plug and Play" single photons at 1.3 μm approaching gigahertz operation*, Appl. Phys. Lett. **93**, 021124 (2008).
7. D. Bethune, W. Risk, and G Pabst, *A high-performance integrated single-photon detector for telecom wavelengths*, J. Mod. Opt. **15**, 1359 (2004).
8. D. Stucki, N. Gisin, O. Guinnard, G. Ribordy, and H. Zbinden, *Quantum key distri-*

bution over 67 km with a plug&play system, New J. Phys, **4**, 41.1 (2002).
9. A. Yoshizawa, R. Kaji, and H. Tuschida, *After-pulse-discarding in single-photon detection to reduce bit errors in quantum key distribution*, Opt. Express, **11**, 13 (2003).
10. Y. Kang, H. X. Lu, and Y. -H. Lo, D. S. Bethune, and W. P. Risk, *Dark count probability and quantum efficiency of avalanche photodiodes for single-photon detection*, Appl. Phys. Lett. **83**, 2955 (2003).
11. A. Lacaita, F. Zappa, S. Cova, and P. Lovati, *Single-photon detection beyond 1μm: performance of commercially available InGaAs/InP detectors*, Appl. Opt. **35**, 2986 (1996).
12. F. Zappa, A. Lacaita, and C. Cova, *Nanosecond single-photon timing with InGaAs/InP photodiodes*, Opt. Lett. **19**, 846 (1994).
13. S. Cova, M Ghioni, A. Lacaita. C. Samori, and F. Zappa, *Avalanche photodiodes and quenching circuits for single-photon detection*, Appl. Opt. **35**, 1956 (1996).
14. H. Henri, P. Deschamps, B. Dion, A. D. MacGregor, D. MacSween, R. J. Mclntyre, C. Trottier, and P. P. Webb, *Photon counting techniques with silicon avalanche photodiodes*, Appl. Opt. **32**, 3894 (1993).
15. V. Golovin, and V. Saveliev, *Novel type of avalanche photodetector with Geiger mode operation*, Nucl. Instr. and Meth. A **518**, 560 (2004).
16. S. Vasile, P. Gothoskar, R. Farrell, and D. Sdrulla, *Photon Detection with High Gain Avalanche Photodiode Arrays*, IEEE Trans. Nucl. Sci. **45**, 720 (1998).
17. J. Kim, S. Takeuchi, and Y. Yamamoto, *Multiphoton detection using visible light photon counter*, Appl. Phys. Lett. **74**, 902 (1999).
18. Princeton Lightwave Inc., Single Photon Benchtop Receiver PLI-AGD-SC-Rx.
19. Figures 5033 is given through private communications with the manufacture of the detector (Princeton Lightwave Inc.).

FIRST PRINCIPLES STUDY OF ELECTROLUMINESCENCE IN ULTRA-THIN SILICON FILM

YUJI SUWA*

Advanced Research Laboratory, Hitachi, Ltd., Kokubunji, Tokyo 185-8601, Japan
**E-mail: yuji.suwa.uq@hitachi.com*

SHIN-ICHI SAITO†

Central Research Laboratory, Hitachi, Ltd., Kokubunji, Tokyo 185-8601, Japan
† SORST, JST

Optical gains in electroluminescence of ultra-thin Si films sandwiched between SiO_2s are calculated from first principles. The gain of the most efficient film is comparable to that of the bulk GaAs if one-order-of-magnitude higher density of carriers is assumed. The importance of the surface structure of the Si film is also investigated and the interface with quartz crystal is found to be favorable for efficient light-emission.

Keywords: optical gain; silicon; first principle calculation

Observations of light emission from silicon (Si) nanostructures[1] such as porous Si[2], Si nanocrystals[3] and super lattice of Si crystal[4] have attracted much attention because they apparently contradict the fact that Si is an indirect-bandgap semiconductor. They are also important in their potential applications. If an efficient light-emission from Si is possible, that would have a significant impact on electronics industry because they can be easily incorporated into current silicon technology.

Recently, Saito *et al.*[5,6] have succeeded to observe electroluminescence from a ultra-thin silicon film with (001) surface. Because the structure of the device is simple and fabrication process such as oxidation is similar to that of usual silicon devices, industrial applications are within reaching distance.

A fundamental mechanism of this light-emission is known; the confinement of carriers into a finite region (thin film) modified the band structure and enabled direct optical transition. However, details of this mechanism, how the band structure is modified and how a finite transition matrix element is created, are not so clear. Such an knowledge is necessary for further enhancement of efficiency. So we have performed first principles calculations and studied optical direct transitions in this system.

Calculations in this letter were based on density functional theory and used generalized gradient approximation (GGA)[7] with plane-wave-based ultrasoft pseudopotentials[8,9]. The energy cutoff was set at 20.25 Rydberg. The convergence criterion for geometry optimization was that all forces acting on the atoms be less than 1×10^{-3} Hartree/a.u. The optical matrix elements are calculated with the aid of core-repair terms[10].

Figure 1(a) shows the (001)-1×1 unit cell of the bulk silicon and the two-dimensional band structure at $k_z = 0$ calculated using this unit cell. Because this is twice the volume of the primitive unit cell, bands are folded back and appear to have nearly direct bandgap at Γ-point. This is an important feature to have a true direct bandgap when it is processed into a thin film.

To be precise, Fig. 1(a) has still indi-

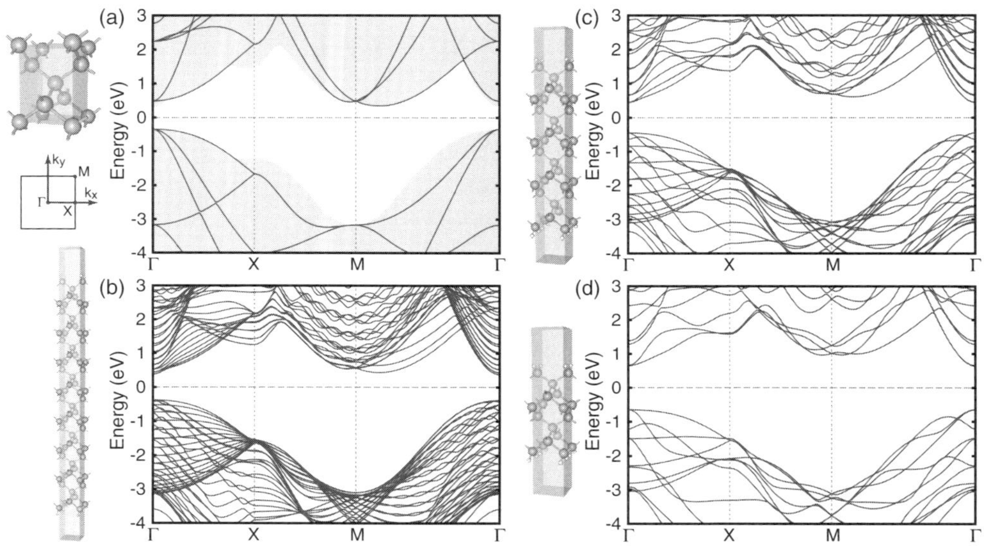

Fig. 1. (a) Two dimensional band structure of the bulk silicon calculated in (001)-1×1 unit cell, $k_z = 0$. Other slices in different k_z will appear within the shaded region. (b-d) Band structures of H-terminated Si(001) film with 32, 16, and 8 atomic layers, respectively. The bottom of the conduction band near the M point rises relatively to that at the Γ point when the silicon film become thin.

rect bandgap since the bottom of the conduction band is the position near M-point between M and Γ. The position near Γ with a finite k_z is also bottom of the conduction band. Because the bulk silicon has a dispersion along the k_z direction, two-dimensional band structures at different k_z have different shapes. All those bands will appear within the shaded region in Fig. 1(a) if they are superposed.

Figures 1(b)–1(d) show the band structures of silicon films with the thicknesses of 32, 16, and 8 atomic layers, respectively. Their surface structures are as that of the bulk silicon (not relaxed) and dangling bonds are terminated by hydrogen atoms. Due to the existence of the (001) surfaces, they have no k_z dependence. Here we assumed rather simple surface-structure in order to focus on the finite-size effect due to film-thinning. The effect of the surface structure will be discussed later. The result shows that even the film with 32 atomic layers have a direct bandgap, i.e., the bottom of the conduction band at Γ is slightly lower than that near M-point in Fig. 1(b). The difference of the energy levels between these two minima grows when the film is thinned down, as shown in Figs. 1(c) and 1(d). This is because the energy of the conduction band is raised by a finite-size effect near the M-point more than at the Γ-point due to different effective masses along the vertical direction[11].

This result shows that an ultra-thin silicon film with (001) surfaces has a direct bandgap, which is a necessary condition to emit lights efficiently. We also calculated thin films with (111) surfaces, but found no direct bandgap. This is also as expected because no pocket of the conduction band appears near the Γ-point when the two-dimensional band structure is calculated using (111)-1×1 unit cell. This disadvantage is insuperable even if the bottom of the conduction band far from Γ is raised by the finite-size effect.

For an efficient electroluminescence, having a direct bandgap is not sufficient, but also

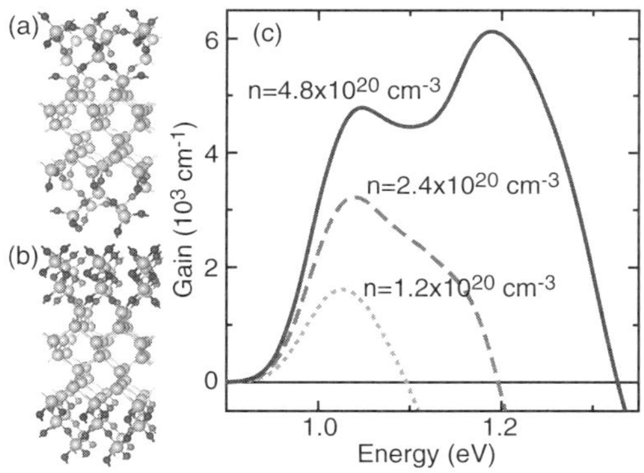

Fig. 2. Structure of the silicon film with seven atomic layers sandwiched (a) between quartzs and (b) between cristobalites. (c) Calculated gain spectrum of the structure (a) at three kinds of carrier densities.

a not small optical matrix element between the gap is necessary. The surface structure of the silicon film is very important for such a matrix because it has a large influence on the wave function of a film with an atomic-scale thickness. We have calculated the matrix elements for three kinds of surface structures varying film thickness. One of the three structure is the hydrogen terminated surface as the above band calculation, and the other two are the interfaces with a quartz and a cristobalite as shown in Figs. 2(a) and 2(b), respectively. The interface with quartz is constructed by Si(001) and $SiO_2(\bar{1}100)$ faces using Si(001)-$\sqrt{2} \times \sqrt{2}$ unit cell as Ref. [12]. The interface with cristobalite is constructed by inserting oxygen atoms between the Si-Si bond of the bulk silicon crystal using Si(001)-1×1 unit cell, where SiO_2 tetrahedrons are twisted around their c-axis as a result of geometry optimization. For both structures with SiO_2s, cell-sizes along the c-axis (vertical direction) are optimized allowing relaxation of atomic positions, while the cell-sizes along the horizontal direction are fixed to those of Si crystal.

Because the silicon film is thinned down by thermal oxidation in the experiment, ultra-thin silicon layer is sandwiched between SiO_2. Although the structure of the bulk part of SiO_2 is expected to be amorphous, its atomic structure at the interface including a few atomic layers of SiO_2 is not known exactly. We assume here that the interface region of SiO_2 with the thickness of a few atomic layers is crystalline, according to an experimental result that an ordered SiO_2 structure exist at the interface[13]. This assumption is consistent with the fact that the surface of the Si layer is atomically flat.

For these structures, we have calculated matrix elements $\langle\psi_{c,\boldsymbol{k}}|\hat{e}\cdot\boldsymbol{p}|\psi_{v,\boldsymbol{k}}\rangle$ where $\psi_{c,\boldsymbol{k}}$, $\psi_{v,\boldsymbol{k}}$, $\hat{e}\cdot\boldsymbol{p}$ are wave functions of the conduction band and the valence band at $\boldsymbol{k}$, and a projection of the momentum operator along the direction $\hat{e}$, respectively. Then, gain spectra have been calculated using following equations[14],

$$g(E) = \frac{he^2}{2n_r c\varepsilon_0 m_0^2 E}\frac{2}{V}\sum_{\boldsymbol{k}} |\langle\psi_{c,\boldsymbol{k}}|\hat{e}\cdot\boldsymbol{p}|\psi_{v,\boldsymbol{k}}\rangle|^2 \times\delta(E_c(\boldsymbol{k}) - E_v(\boldsymbol{k}) - E),(f_c - f_v) \quad (1)$$

$$f_c = \frac{1}{1+e^{(E_c(\boldsymbol{k})-F_c)/k_B T}}, \quad (2)$$

$$f_v = \frac{1}{1+e^{(E_v(\boldsymbol{k})-F_v)/k_B T}}. \quad (3)$$

Here, n_r and m_0 are the refractive index of silicon and mass of electron, respectively. $E_c(\boldsymbol{k})$, $E_v(\boldsymbol{k})$, f_c, f_v, F_c, and F_v are energy levels, distribution functions and pseudo-Fermi levels of the conduction band and the valence band, respectively.

Obtained gain spectra parallel to the surface are shown in Fig. 2(c). Here, optical transitions crossing energy gap are only taken into account, and contributions from k-mesh points which are discrete in energy are smoothed by using gaussian function.

Figure 3 shows the peak values of the gain spectrum for various thickness of the silicon film with three kinds of surface structures. Here, the densities of injected electrons and holes in the films are assumed to be 4.8×10^{20} cm^{-3}. The peak values are obtained by calculating gain spectra as Fig. 2(c)

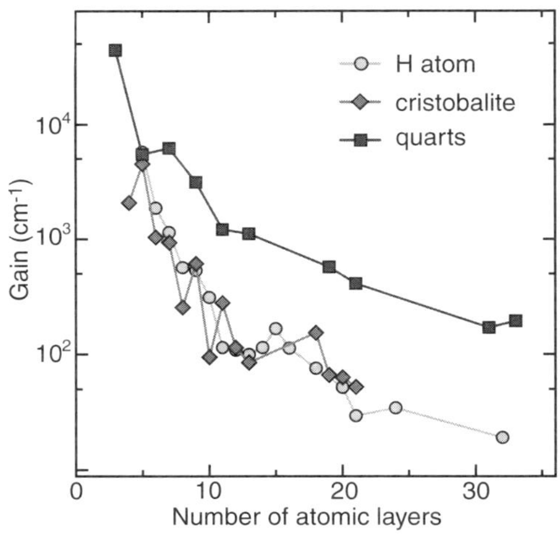

Fig. 3. Peak values of the gain spectrum for various thickness of the silicon film sandwiched between quartz, cristobalites, or H atoms.

for each structure. The overall tendency in Figure 2 is similar to that of oscillator strength given in Ref. [15] calculated by the tight-binding method.

It can be seen that the films sandwiched between quartz give 5-to-10 times higher gains than other films in the wide range of film thickness. This result shows that the surface structure of the silicon film is important to obtain efficient electroluminescence as expected. The magnitudes of the gains are comparable to that of bulk GaAs calculated in the same manner, if ten times larger carrier density is given to the silicon film. Actually, higher density of carriers is achievable in silicon devices than in GaAs.

In summary, we calculated optical gains in electroluminescence of ultra-thin silicon films from first principles. As a result of calculations for three kind of surface structures, we found that the the interface with quartz crystal most enhances light-emission. Because the gain of this film is comparable to that of the bulk GaAs, if the carrier density is large enough, we can expect lasing by such a film, when the cavity for induced emission is prepared.

Acknowledgments

We would like to thank Dr. H. Kageshima, Dr. J. Yamauchi and Dr. Y. Yoshimoto for valuable discussions on the methods of calculations. The first-principles calculations were performed with TAPP (the Tokyo Ab-initio Program Package).

References

1. S. Ossicini, L. Pavesi, and F. Priolo, "Light Emitting Silicon for Microphotonics", Springer, (2004).
2. L. T. Canham, Appl. Phys. Lett. **57**, 1045 (1990).
3. L. Pavesi *et al.*, Nature **408**, 440 (2000).
4. D. J. Lockwood, Z. H. Lu, and J.-M. Baribeau, Phys. Rev. Lett. **76**, 539 (1996).
5. S. Saito *et al.*, Jpn. J. Appl. Phys. **45**, L679 (2006).
6. S. Saito *et al.*, Appl. Phys. Lett. **89**, 163504 (2006).
7. J. P. Perdew, K. Bruke, and A. Wang, Phys. Rev. B **54**, 16533 (1996).
8. D. Vanderbilt, Phys. Rev. B **41**, 7892 (1990).
9. K. Laasonen *et al.*, Phys. Rev. B **47**, 10142 (1993).
10. H. Kageshima and K. Shiraishi, Phys. Rev. B **56**, 14985 (1997).
11. T. Ohno, K. Shiraishi, and T. Ogawa, Phys. Rev. Lett. **69**, 2400 (1992).
12. M. Watarai, J. Nakamura, and A. Natori, Phys. Rev. B **69**, 035312 (2004).
13. E.-C. Cho *et al.*, J. App. Phys. **96**, 3211 (2004).
14. S. L. Chuang, "Physics of Optelectronic Devices", Wiley-Interscience (1995).
15. M. Nishida, Phys. Rev. B **58**, 7103 (1998).

FIRST PRINCIPLES NONLINEAR OPTICAL SPECTROSCOPY

T. HAMADA*

Advanced Research Laboratory, Hitachi, Ltd., 1-280 Higashi-Koigakubo, Kokubunji-shi, Tokyo 185-8601, Japan

**E-mail: tomoyuki.hamada.tk@hitachi.com*

T. OHNO

National Institute of Materials Research, 1-2-1 Sengen, Tsukuba, Ibaraki 305-0047, Japan

First principles nonlinear optical spectroscopy, which is a theoretical alternative of the conventional nonlinear optical spectroscopy, has been realized by our new software calculating the linear and nonlinear optical spectra of materials on the basis of the density functional pseudopotential method. The new software can calculate second- and third-order nonlinear optical susceptibilities of insulating solids, accurately, taking into account the frequency dependence of the susceptibilities. The software is assumed to be a promising theoretical tool of design of new nonlinear optical materials, which is needed in opto-electronics and laser optics.

Keywords: First Principles Calculation; Nonlinear Optics.

1. Introduction

Spectroscopy was the cradle of quantum mechanics when Bohr and Sommerfeld.[1,2] introduced their theory of atoms to explain atomic spectra. Quantum mechanics is well established today, however the situation is the same as that of the age of Bohr and Sommerfeld; spectroscopy is an important subject of quantum mechanics and the source of inspirations of theoretical studies.

We have been working on developing first principles spectroscopy, a new field of computational physics, which aims at studying optical spectra of materials at the atomistic level only by using the quantum mechanical theory of materials. First principles spectroscopy is possible because optical spectra of materials originate from the quantum mechanical structure of the materials and the first principles method enables to calculate the structures of materials at desired accuracy, suitably treating many body effects in the materials. If it works well, first principles spectroscopy would become an attractive theoretical alternative of the conventional spectroscopy, being free from the physical limitation of optical systems such as the frequency scope of light source and detectors and experimental errors.

In this study, we have developed first principles spectroscopy software capable of calculating the linear and nonlinear optical spectra of materials by using the density functional pseudopotential (DFT-PP) method. The developed software called UVSOR[3,4] (Universal Virtual Spectroscopy of Optoelectronics Research) can calculate second harmonic generation (SHG) and third harmonic generation (THG) susceptibilities of solids, reproducing the experimental susceptibilities fairly well. UVSOR appears to be a promising tool of first principles spectroscopy and quantum mechanical studies of material optical spectra, which have been of interest since the age of Bohr and Sommerfeld.

2. Method of First Principles Spectroscopy

2.1 *Linear optical Spectroscopy*

UVSOR is a first principle spectrometer, which calculates the electronic structure of materials from first principles and the optical spectra of the materials from calculated electronic structure. UVSOR uses the DFT PP method for the electronic structure calculation and the time dependent perturbation (TDPT) method for the optical spectra calculation. Kohn-Sham pseudo-electron orbitals are obtained by solving the KS equation in the LDA and GGA approximation, and optical spectra are obtained by investigating direct electronic transitions between KS orbitals caused by an external optical field. Dielectric function $\varepsilon(\omega)$ of a material is calculated from KS orbitals by using the second-order TDPT as[5],

$$\mathrm{Im}[\varepsilon(\omega)] = \frac{2e^2}{V\varepsilon_0}\sum_{\mathbf{k},c,v}\left|\left\langle \varphi_c^{\mathbf{k}}\left|\mathbf{u}\cdot\mathbf{r}\right|\varphi_v^{\mathbf{k}}\right\rangle\right|\delta(E_c^{\mathbf{k}} - E_c^{\mathbf{k}} - \hbar\omega), \qquad (1)$$

where φ is KS orbital, E is energy of KS orbital, $\mathbf{k}$ is electron wave vector index, and c and v are conduction and valence orbital indices, respectively. $\mathbf{r}$ is positional operator, u and ω are, respectively, polarization vector and frequency of an external optical field, V is crystal volume, ε_0 is permittivity of a vacuum, and e is electron charge. The sum is taken over all possible $\mathbf{k}$, c and v. The linear tetrahedron method[6] is used for the summation over $\mathbf{k}$. Real part of $\varepsilon(\omega)$ is obtained by making the Kramers-Kronig transformation (KKT) of Im[$\varepsilon(\omega)$]. Linear optical spectra including refractive index, reflection coefficient and absorption coefficient are directly calculated from $\varepsilon(\omega)$. Transition moment in the $\mathbf{r}$ representation (the length formula) [7] is used in Eq.(1), because UVSOR uses the non-local PP scheme, in which the length formula gives correct results the same as all electron (AE) calculations. UVSOR evaluates the transition moment by using a method by Read and Needs (RN)[8] for norm-conserving (NS) PPs and a method by Kageshima and Shiraishi (KS)[9] for NS and ultra-soft (US) PPs. Use of these methods makes UVSOR possible to calculate optical spectra the same as those by AE calculations, using the PP approximation. The band gap underestimation in the DFT method is corrected by the scissors operator approach, which introduces a rigid shift of conduction bands in energy so as to reproduce experimental band gap.

2.2 *Nonlinear Optical Spectroscopy*

Nonlinear optical (NLO) effects are phenomena accompanying generation of frequency change of photons and are conspicuously observed in NLO materials, which show large nonlinear electronic polarizations. The phenomena are of interest from both academic and practical viewpoints, and are a subject of substantial subject of theoretical calculations. The present UVSOR can calculate second harmonic generation (SHG) and third harmonic generation (THG) NLO susceptibilities of materials from the material electronic structure by using higher-order TDPT methods. SHG susceptibilities are calculated by using Eqs (2) and (3) by third-order TDPT.[10]

$$\begin{aligned}
&\mathrm{Im}[\chi^{(2),el}_{\alpha\beta\gamma}(-2\omega;\omega,\omega)] = \\
&-\frac{\pi}{2}\left|\frac{e\hbar}{m}\right|\sum_{i,j,l}\int_{BZ}\frac{d\mathbf{k}}{4\pi^3}\left(\frac{\mathrm{Im}[\mathbf{p}^{cc}_{jl,\mathbf{k},\alpha}\{\mathbf{p}^{cv}_{li,\mathbf{k}}\mathbf{p}^{vc}_{ij,\mathbf{k}}\}_{\beta\gamma}]}{E^3_{li,\mathbf{k}}(E_{li,\mathbf{k}}+E_{ji,\mathbf{k}})}\delta(E_{li,\mathbf{k}}-\hbar\omega)\right. \\
&\qquad -\frac{\mathrm{Im}[\mathbf{p}^{vc}_{ij,\mathbf{k},\alpha}\{\mathbf{p}^{cc}_{jl,\mathbf{k}}\mathbf{p}^{cv}_{li,\mathbf{k}}\}_{\beta\gamma}]}{E^3_{li,\mathbf{k}}(2E_{li,\mathbf{k}}-E_{ji,\mathbf{k}})}\delta(E_{li,\mathbf{k}}-\hbar\omega) \\
&\qquad \left.+\frac{16\,\mathrm{Im}[\mathbf{p}^{vc}_{ij,\mathbf{k},\alpha}\{\mathbf{p}^{cc}_{jl,\mathbf{k}}\mathbf{p}^{cv}_{li,\mathbf{k}}\}_{\beta\gamma}]}{E^3_{ji,\mathbf{k}}(2E_{li,\mathbf{k}}-E_{ji,\mathbf{k}})}\delta(E_{ji,\mathbf{k}}-2\hbar\omega)\right)
\end{aligned} \qquad (2)$$

$$\begin{aligned}
&\mathrm{Im}[\chi^{(2),hl}_{\alpha\beta\gamma}(-2\omega;\omega,\omega)] = \\
&\frac{\pi}{2}\left|\frac{e\hbar}{m}\right|\sum_{i,j,l}\int_{BZ}\frac{d\mathbf{k}}{4\pi^3}\left(\frac{\mathrm{Im}[\mathbf{p}^{vv}_{li,\mathbf{k},\alpha}\{\mathbf{p}^{vc}_{ij,\mathbf{k}}\mathbf{p}^{cv}_{jl,\mathbf{k}}\}_{\beta\gamma}]}{E^3_{jl,\mathbf{k}}(E_{jl,\mathbf{k}}+E_{ji,\mathbf{k}})}\delta(E_{jl,\mathbf{k}}-\hbar\omega)\right. \\
&\qquad -\frac{\mathrm{Im}[\mathbf{p}^{vc}_{ij,\mathbf{k},\alpha}\{\mathbf{p}^{cv}_{jl,\mathbf{k}}\mathbf{p}^{vv}_{li,\mathbf{k}}\}_{\beta\gamma}]}{E^3_{jl,\mathbf{k}}(2E_{jl,\mathbf{k}}-E_{ji,\mathbf{k}})}\delta(E_{jl,\mathbf{k}}-\hbar\omega) \\
&\qquad \left.+\frac{16\,\mathrm{Im}[\mathbf{p}^{vc}_{ij,\mathbf{k},\alpha}\{\mathbf{p}^{cv}_{jl,\mathbf{k}}\mathbf{p}^{vv}_{li,\mathbf{k}}\}_{\beta\gamma}]}{E^3_{ji}(2E_{jl,\mathbf{k}}-E_{ji,\mathbf{k}})}\delta(E_{ji,\mathbf{k}}-2\hbar\omega)\right)
\end{aligned} \qquad (3)$$

$\mathbf{p}^{cv}_{ij,\mathbf{k}}$ is transition moment in the velocity formula between i-th conduction and j-th valence KS orbitals with wave vector $\mathbf{k}$, and $E_{ij,\mathbf{k}}$ is the energy difference between these orbitals. α, β, and γ are Cartesian index xyz and $\{\mathbf{p}, \mathbf{p}'\}_{\beta\gamma}$ is the index symmetrized product $(\mathbf{p}_\beta\mathbf{p}_\gamma'+\mathbf{p}_\gamma\mathbf{p}_\beta')/2$

Eqs.(2) and (3) give imaginary part of electron component of SHG susceptibilities, $\chi^{(2),el}(-2\omega;\omega,\omega)$ and that of hole component of the susceptibilities, $\chi^{(2),hl}(-2\omega;\omega,\omega)$, respectively. Imaginary part of total SHG susceptibilities, $\chi^{(2)}(-2\omega;\omega,\omega)$ is given as the sum of Im[$\chi^{(2),el}$] and Im[$\chi^{(2),hl}$], and the KKT of the imaginary part gives real part of $\chi^{(2)}$. **p** is calculated from the corresponding transition moment in the length formula that evaluated by using the RN or KS methods. The Brillouin zone (BZ) integration of Eqs. (2) and (3) is done by a linear tetrahedron method for $\chi^{(2)}$, the method of which perform the integration in a way similar to that of the original linear tetrahedron method for $\varepsilon(\omega)$ calculations. UVSOR similarly calculates THG susceptibilities, $\chi^{(3)}(-3\omega;\omega,\omega,\omega)$ of materials by using the fourth order TDPT method.[11] The calculation formulas are lengthy and are not shown here (see reference [11] for details). The scissors operator band gap correction is applied also in SHG and THG calculations.

3 Results and Discussion

3.1 *SHG Calculation Results*

UVSOR was applied to the SHG calculation of semiconductors. Electronic structure of zinc-blende (ZB) and wruzite (W) compound semiconductors was calculated by using the DFT PP method in GGA approximation, and $\chi(2)(-2\omega;\omega,\omega)$ of the semiconductors were calculated by using the second-order TDPT scheme with the scissors operator band gap correction. Table 1 shows calculated static $\chi^{(2)}(\chi^{(2)}(0;0,0))$ of several ZB semiconductors in comparison with experimental values and theoretical values of previous FLAPW studies. $\chi^{(2)}(0)$ of the ZB semiconductor has $\chi^{(2)}{}_{xyz}(0)$ as unique non-zero independent component, and the table shows $\chi^{(2)}{}_{xyz}(0)$ values. Value of other non-zero components can be generated from $\chi^{(2)}{}_{xyz}(0)$ value by using the Kleinmann symmetry condition that value of $\chi^{(2)}{}_{ijk}$ is invariant under permutation of indices i, j, and k. Calculated $\chi^{(2)}{}_{xyz}(0)$ values for GaN and GaP are slightly smaller than the FLAPW and LMTO values of previous studies, but the difference is small and is acceptable. In the case of GaAs, the present calculation well-reproduced the FLAPW results, though the calculation remarkably underestimated the experimental value. The reason for the underestimation could be due to surface and excitonic effects neglected in the present calculation. Table 2 shows calculated static $\chi^{(2)}$ of W GaN and AlN, in comparison with LMTO values of a previous study and experimental values. The W semiconductors have $\chi^{(2)}{}_{zzz}$ and $\chi^{(2)}{}_{zxx}$ as non-zero independent components, and values of these components are shown. A good agreement is seen between the present calculated and LMTO values, both in the cases of GaN and AlN. In the GaN case, calculated $\chi^{(2)}{}_{zzz}$ and $\chi^{(2)}{}_{zxx}$ values are slightly smaller than experimental values in magnitude, while in the AlN case, calculated $\chi^{(2)}{}_{zzz}(0)$ values fall in the experimental range, showing the adequateness of the present calculation. It is likely that calculated values of the other

Table 1 Static $\chi^{(2)}{}_{xyz}$ of zinc-blende semiconductors(10^{-8} esu)(GGA).

	UVSOR	Previous Work	Exp.
GaN	1.60	2.8[a]	—
GaP	14.50	17.9[b]	17.7
GaAs	22.46	23.0[b] 25.0[c]	38.6

[a] GGA-FLAPW [b] LDA-FLAPW [c] LDA-LMTO.

Table 2 Static $\chi^{(2)}(0)$ of GaN and AlN in the wurzite structure(10^{-8} esu).

	GaN		AlN	
	χ_{zzz}	χ_{zxx}	χ_{zzz}	χ_{zxx}
UVSOR	1.83	-1.05	-1.14	0.08
Other Work[a]	1.34	-1.03	-1.03	0.06
Exp.	2.56	-1.27	-1.3~-4.7	—

[a] LDA-LMTO.

components of $\chi^{(2)}$ also fall in the experimental range.

3.2 *THG Calculation Results*

The theoretical calculation of THG susceptibilities $\chi^{(3)}(-3\omega;\omega,\omega,\omega)$ is a difficult task. Previous attempts to evaluate $\chi^{(3)}$ of materials from their electronic structure failed to accurately reproduce the experimental values at desired accuracy. In this study, we calculated $\chi^{(3)}(-3\omega;\omega,\omega,\omega)$ of Si, C(diamond), and α-quartz by using UVSOR. Calculated static $\chi^{(3)}(0)_{xxxx}$ are shown in Table 3. The UVSOR calculation successfully reproduced the experimental static $\chi^{(3)}(0)_{xxxx}$ values of these materials. Calculated $\chi^{(3)}{}_{xxxx}(0)$ values for Si and diamond are in the experimental ranges, and that for α-quartz is quite close to an accurately determined experimental $\chi^{(3)}(0)$ of α-quartz. The excellent agreement between calculated and experimental $\chi^{(3)}$ values indicates that UVSOR is effective also for THG susceptibilities calculations.

Table 3 Static $\chi^{(3)}(0)$ of several crystals.

Crystal	UVSOR	Other Work	Exp.
Si	25.629	48[a]	24±14
C(Diamond)	0.222		0.18±0.03
SiO_2(α-quartz)	0.0109	—	0.0095

[a]Empirical tight binding

4. Conclusions

First principles nonlinear optical spectroscopy has been made possible by using our newly developed software UVSOR, which calculates linear optical spectra, SHG, and THG susceptibilities by using the first principles density functional method. A good agreement between theory and experiment is seen, especially in THG susceptibility. The results indicate that UVSOR is a promising first principles non-linear optics spectrometer enabling theoretical studies of optical spectra of materials, which are a subject of substantial interest in quantum mechanics from the age of Bohr and Sommerfeld.

Acknowledgments

This work is done in Revolutionary Simulation Software for Industrial Science (RSS21) Project supported by MEXT, Ministry of Education, Culture, Sports, Science and Technology, Japan.

References

1. N. Bohr, *Phil. Mag.* **26**, 1 (1913).
2. A.Sommerfeld, *J. Opt. Soc. Am.* **7**, 509 (1923).
3. UVSOR version 3.18 (Institute of Industrial Science, University of Tokyo, Tokyo, 2008);
4. http://www.ciss.iis.u-tokyo.ac.jp/rss21/.
5. Harbeke, , in Optical properties of solids, ed F. Abelés(North-Holland Publishing Co., Amsterdam, 1972)
6. G. Lehmann and T. Taut, *phys. stat. solidi* (b) **54**, 469 (1972).
7. A. F. Starace, *Phys. Rev. A*, **3**, 1242 (1971).
8. A. J. Read and R. J. Needs, *Phy. Rev. B*, **44**, 13701 (1991).
9. H. Kageshima and K. Shiraishi,, *Phys. Rev. B*, **56**, 14985 (1997).
10. E. Ghahramani, D. J. Moss, and J. E. Sipe, *Phys. Rev. B*, **43**,8990 (1991).
11. D. J. Moss, E. Ghahramani, J. E. Sipe, and H. M. van Driel, *Phys. Rev. B*, **41**, 1542 (1990).

FIELD-INDUCED DISORDER AND CARRIER LOCALIZATION IN MOLECULAR ORGANIC TRANSISTORS

M. ANDO*

Hitachi Cambridge Laboratory, Cavendish Laboratory, J.J. Thomson Avenue, Cambridge CB3 0HE, UK
**E-mail: masahiko.ando@hitachi-eu.com*

T. MINAKATA

Asahi-Kasei corporation, 2-1 Samejima, Fuji, Shizuoka, 4168501 Japan

C. DUFFY, H. SIRRINGHAUS

Cavendish Laboratory, J.J. Thomson Avenue, Cambridge CB3 0HE, UK

We propose a "field-induced polymorphous disorder" model to explain bias-stress instability in molecular organic thin-film transistors, based on the experimental results showing the strong correlation between the micro-structural change in semiconductor layer composed of penrtacene molecules and the threshold voltage (Vth) shift due to electron trapping in a reversible manner under the successive bias-stress, thermal annealing, and light irradiation.

Keywords: Molecular semiconductor, Pentacene, Thin-film transistor, Charge localization

Introduction

Organic field-effect transistors (OFET) are presently attracting significant research and development interests as they potentially offer unique advantages over their inorganic counterparts in terms of cost-reduction, compatibility with low-temperature and printing-based manufacturing[1]. However, carrier transport mechanism at the interface between organic semiconductor and gate insulator is still not clear because it is difficult to clarify the relationship between complex structures and transport at the channel interface which are composed of self-assembled molecules and/or polymers via weak Van der Waals force. Recently, however, it has been revealed that dynamic and static disorders at the interface play important roles in both charge-transport and localization of these systems based on the quantum-mechanical principles[2),3)].

Here, an experimental evidence of the disorder-induced carrier localization in molecular semiconductor will be shown, in which prolonged gate-bias stress were applied at the interface between pentacene semiconductor and polymer insulator. Pentacene is the most popular molecule applied to OFET applications and has at least four different crystallographic morphologies (polymorphs) sensitively depending on the temperature, thickness, surface energy, and other parameters[4)]. We observed a coincidence of electron trapping at the interface and formation of the structural-disorder by polymorphic mixture in pentacene film by using micro-Raman spectroscopy after continuous gate-bias stress for 15 hours. The trapped electrons and the disorders were reversibly recovered by thermal annealing and it was accelerated by direct photo-excitation of the pentacene films, which clearly indicated that electrons were trapped in the disordered molecular matrix.

1. Experimental

For this purpose, we have developed solution-processed unsubstituted pentacene TFTs. By combining zone-cast process of pentacene and spin-coated benzocyclobutene as a gate insulator, platelet crystalline films with reduced grain boundaries and higher crystallinity were obtained and the transistor performance such as field-effect mobility, on/off ratio, and threshold voltage stability to the negative-bias stress, becomes almost the same as that of commercially available amorphous silicon TFTs.

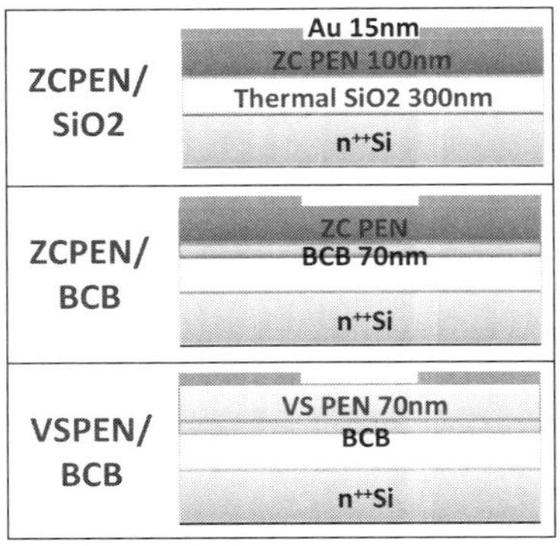

Fig.1. Cross sections of pentacene TFTs.

In this work, in order to distinguish the microscopic origins of positive bias-stress instability between semiconductor and gate insulator, we prepared three kinds of TFTs by combining zone-cast and vacuum sublimed pentacene films with thermal silicon dioxide and benzocyclobutene (BCB) as gate insulators as shown in **Fig.1**. Especially, high temperature zone-cast process of pentacene films and the post-annealing process should exclude extrinsic impurities such as H_2, O_2, and H_2O which are known to become causes for charge trapping states in pentacene films[5),6)].

All samples were made and characterized in N_2 atmosphere. Gate bias stress voltage (Vst) at +40 V shifted the transfer curve to the positive direction with a little degradation shown by an increase of the slope. When the bias-stress was removed, the threshold voltage (Vth) recovered spontaneously, and light irradiation accelerated the recovery. Post annealing at 130 C for 1h in N_2 completely recover the initial transfer curve.

Microstructural changes in pentacene films were monitored by using micro Raman spectroscopy which has been known to be a powerful tool to characterize pentracene microstructure[7)]. In our present set-up, these measurement were done in air and without bias application.

In order to release electrons trapped in pentacene semiconductor, Halogen lamp light was irradiated through transparent Indium Tin oxide (ITO) coated glass filter. The ITO/glass filter cut both infrared and UV light with photon energy above the band gap of pentacene. Therefore the direct photo-excitation of pentacene is expected to accelerate the detrapping of electron and recovery of Vth shift.

2. Results and Discussion

In contrast to the relatively small Vth shift induced by the negative bias stress, large Vth shift were induced by positive bias stress for all three kinds of TFT configurations. **Figure 2** indicates the typical cases of transfer-curves of TFTs with zone-casted pentacene on BCB insulators.

Vth shifts under negative stress tend to be dominated by gate insulator materials (**Fig.3(a)**) because the negative stress is applied only to gate insulator because of the complete accumulation of holes at semiconductor/insulator interface.

In contrast, Vth shifts under positive stress tend to be dominated by semiconductor (**Fig.3(b)**) because the positive stress is also applied to pentacene semiconductor because of the incomplete accumulation of electrons at the interface. After an incubation period, the positive Vth shift become increasing and vacuum sublimed pentacene TFTs showed Vth

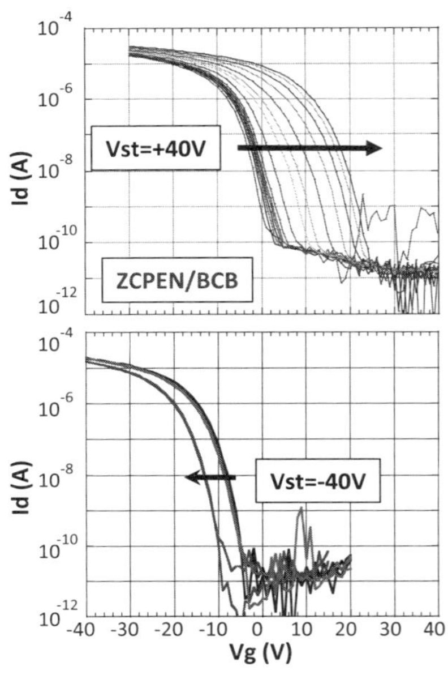

Fig.2. Transfer -curve shift under gate bias stresses.

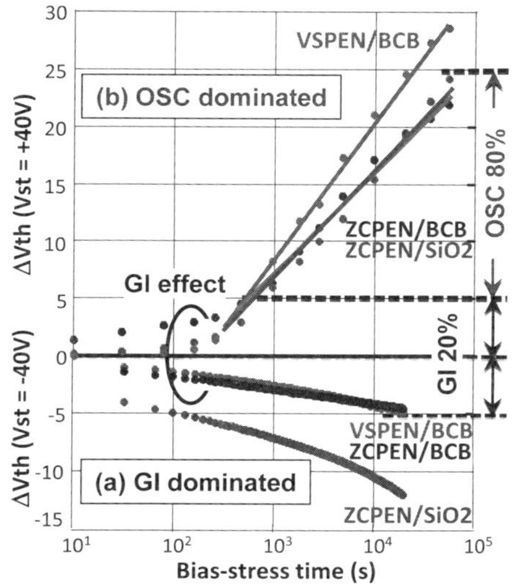

Fig.3. Relative Vth shifts vs. Bias-stress time.

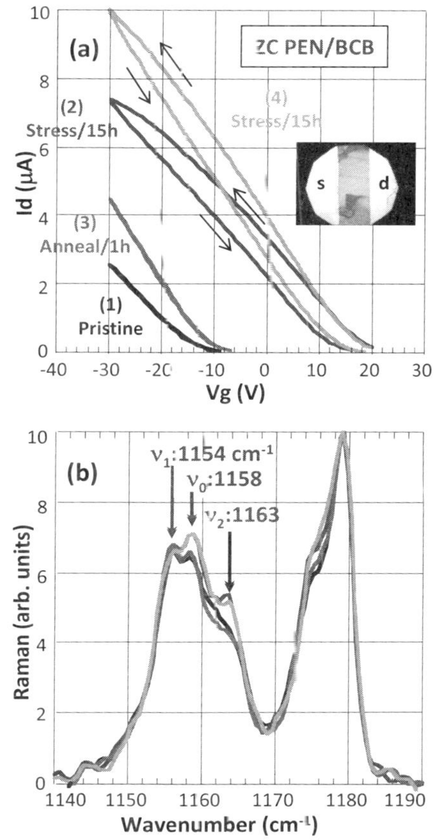

Fig.4. Bias-stress and annealing effect on transfer curves and Raman spectra.

shift larger than zone cast pentacene TFTs. In each sequence of bias stress protocol as shown in **Fig.4(a)**, microstructure of pentacene films was monitored by using micro-Raman spectroscopy. The pristine pentacene film shows typical three Raman vibrational peaks at 1154, 1158, and 1163 cm^{-1} (**Fig.4(b)**). Heights of these peaks changed reversibly by the first bias stress, anneal, and the second bias stress. These results indicate that the microstructures of pentacene films were reversibly changed by electrons trapped and released at the organic semiconductor/insulator interface.

Vth recovery speed after the prolonged positive bias stress in the dark were accelerated by direct photoxcitation of pentacene films from the back surface (**Fig.5(a)**), which indicates that the electron were trapped in pentacene films and assisted to be released from pentacene films by the excitation light.

At the same time, the direct photo-excitation also accelerated the microstructure of pentacene films as shown in the micro Raman peaks. Especially 1163 cm^{-1} peak height normalized by 1158 cm^{-1} peak height ($\nu2/\nu0$), which was increased by bias stress, were recovered by the excitation light (**Fig.5(b)**). These results indicate that structural changes were induced by electrons trapped in pentacene films and not in the gate insulator.

It was known that pentacene thin films have four kinds of polymorph with different d spaces from 14.1 to 15.4 angstroms[4)]. These polymorphs were generated depending on process temperature, substrate, film thickness, and so on. Each single morph is known to have its own Raman peaks with different relative intensities at 1154, 1158, and 1163 cm^{-1}. 1154 and 1158 cm^{-1} peaks are Ag modes due to the motion of hydrogen at the end of pentacene molecule and 1163cm^{-1} is a B1g mode due to the motion of hydrogen at the side of the molecule. Compared with these d spaces and Raman peaks, zone-cast pentacene films on BCB insulator have d space at 14.2 A which is between 14.1A and 14.4A and Raman peaks which superposes those of 14.1A and 14.4A polymorphs. Therefore, it was concluded that this pentacene thin film is a mosaic of these

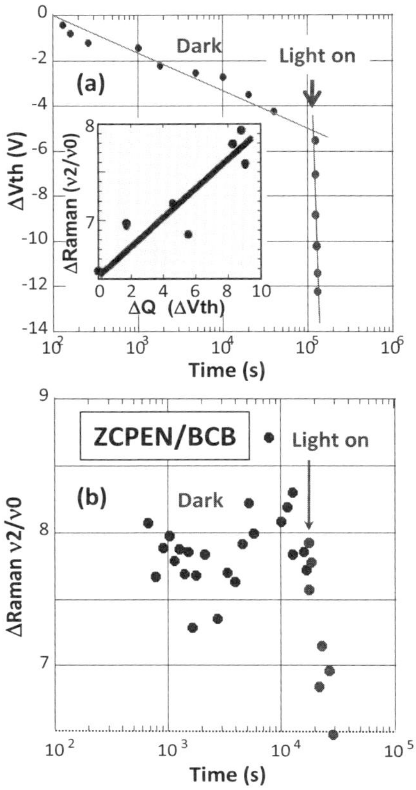

Fig.5. Effect of light irradiation on recovery of Vth and Raman spectra after bias-stress.

two polymorphs which should increase B1g Raman peak by positive bias stress.

Recently, these polymorphs were found to coexist in powder to make a mixture in the wide temperature range and can be reversibly transformed by thermal annealing[8] because small slide between inter-layers can easily change these polymorphs as shown in upper part of **Fig.6**. We conjecture these polymorph mixtures can be also induced by bias-stress applied to the pentacene films in TFT configuration and at some interfaces between the two polymorphic domains as shown in the lower part of **Fig.6**, distances between the molecules are becoming larger to increase the free vibration of hydrogen at the side of the molecule to increase the peak of B1g mode.

It was also known that only a small increase of intermolecular distance can induce a large fluctuation of transfer integral and if the polymorphic mixture is induced by applying bias stress to pentacene thin films, microscopic potential disorders are induced to trap electrons[2),3)].

In summary, electrons accumulated at the hetero interface were trapped mainly in pentacne and not in insulator. Raman spectroscopy indicated that microstructures of pentacene films were reversibly changed by trap and release of electrons. Accelerated recoveries of Vth and Raman peaks by direct photo-excitation of pentacene supported that electrons were trapped at the structural defects within pentacene films as recently reported[9)]. Bias-induced polymorphous disorder model was proposed to generate electron self trapping states in pentacene films.

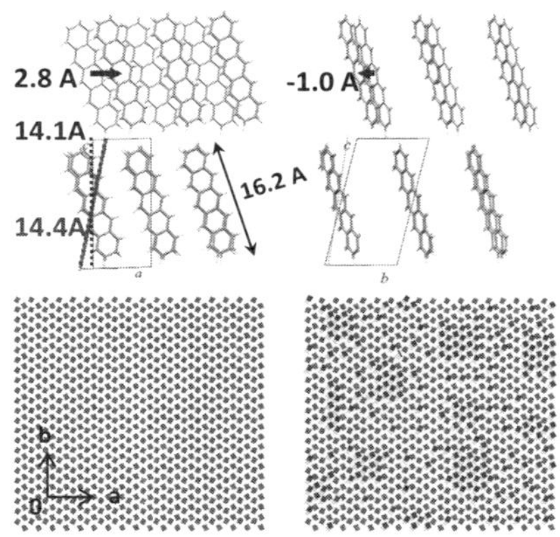

Fig.6. Field-induced polymorphous disorder model.

References

1. H. Sirringhaus & M. Ando, MRS Bulletin **33**, 676 (2008).
2. A. Troisi et al., Phys. Rev. Lett. **96**, 086601 (2006).
3. M. Unge et al., Phys. Rev. **B74**, 235403 (2006).
4. C. C. Mattheus et al., Synth. Metals **138**, 475(2003).
5. D.V. Lang et al., Phys. Rev. Lett. **93**, 076601 (2004).
6. L. Tsetseris et al., Phys. Rev. **B75**, 153202 (2007).
7. H. L. Cheng et al., Adv. Funct. Mater. **18**, 1 (2008).
8. T. Siegrist et al., Adv. Mater. **19**, 2079 (2007).
9. W. L. Kalb et al., Phys. Rev. **B76**, 184112 (2007).

SWITCHING DYNAMICS IN STRONGLY COUPLED JOSEPHSON JUNCTIONS

H. KASHIWAYA, T. MATSUMOTO, H. SHIBATA, S. KAWABATA and S. KASHIWAYA*

National Institute of Advanced Industrial Science and Technology (AIST), Umezono, Ibaraki, Tsukuba 305-8568, Japan

*E-mail: *s.kashiwaya@aist.go.jp*

Recently, quantum dynamics and coherence in Josephson junctions (JJs) made of high Tc superconductors (HTSCs) attract much attention. The HTSCs have relatively large energy gap and higher plasma frequency compared to those of metal superconductors, which suggests their high potential for quantum electron devices. On the other hand, the HTSC JJs have a disadvantage that the pairing symmetry is *d*-wave, and thus the intrinsic damping effect in the dynamic due to the nodal quasiparticles is inevitable. In addition to this, the coupling effect due to the atomic scale stack structure induces complex switching. Therefore, the peculiarities in quantum dynamics of the HTSC JJs must be clarified in order to realize the HTSCs quantum electronics. Here we report our efforts to observe macroscopic quantum tunneling and quantum coherence in the HTSC JJs

Keywords: Intrinsic Josephson junction. MQT. High-Tc. Strong coupling.

1. Introduction

In recent decades, a wide variety of candidates for qubit have been presented. Among them, superconducting qubits are believed to have the advantage that the decoherence due to the low-energy quasiparticles is largely suppressed because of the opening of the superconducting energy gap in the quasiparticle density of states. Actually, various groups have reported the successful observation of quantum coherence in Josephson junctions (JJs) made of metal superconductors such as Al and Nb. High-Tc superconductors (HTSCs) are also fascinating materials for qubit application because of their high critical temperatures (Tcs). Also their layered structures allow us to fabricate high quality intrinsic Josephson junctions (IJJs) only by utilizing c-axis transport [1]. The insulating layers of IJJs are naturally formed during the crystal growth. This feature is quite favorable for the realization of low damping JJ because of the lack of the decoherence due to the defects and impurities in the insulating layer.

In spite of these advantages, experimental trials to build the qubits using HTSCs are limited to few cases. It has long been believed that the nodal quasiparticles of *d*-wave superconductivity caused serious damage on the coherence. Recently, Kawabata et al. have theoretically pointed out that the damping due the *d*-wave was quite limited because of their large superconducting energy gap [2]. Actually observations of conventional dynamics in HTSCs have been reported recently[3-6]. At this stage, it is strongly expected that detailed experimental evaluation will be accomplished on HTSC IJJ.

Another unsolved complexity in the HTSC JJs is due to their stacked structures. Atomic size superconducting layer thickness is smaller than the Debye length and the magnetic penetration depth. Thus the unshielded electromagnetic field couples the dynamics of each junction in the single stack[7,8]. It is expected that the influences of the strong coupling effects induce complex quantum dynamics such as simultaneous switching of multiple junctions. Clarifying of such peculiarities is important issue both in basic physics as well as in application aspects.

Here we report our efforts to figure out the quantum dynamics in the HTSC IJJs made from double layer $Bi_2Sr_2CaCu_2O_{8+\delta}$ (Bi2212, Tc-90K) and $Bi_{1.6}Ln_{0.4}Sr_2CuO_{6+\delta}$ (Bi2201, Tc-30K).

2. Fabrication

We have succeeded in fabricating high quality IJJs both of Bi2212 and Bi-2201. The variety of device selection from among the Bi-based cuprates enables us to tune the rf frequency required to control the quantum states of the qubit to suitable values. The IJJ were milled by a focused ion beam (FIB) from single crystals grown by using a floating zone method. Typical junction areas and the number of junctions contained in a stack were 1μm*2μm and several to several tens, respectively. A current(I)-voltage(V) characteristics obtained on a junction is shown in Fig. 1. The junction exhibits sequential switching to finite voltage states as the bias current is increased.

Recently we have developed a new device process that can fabricate the IJJ's of a size down to 0.5μm without the degradation of the switching properties [9]. The size flexibility is a key technology to realize HTSC flux qubit because the ratio of the charging energy to the Josephson coupling energy is an important parameter.

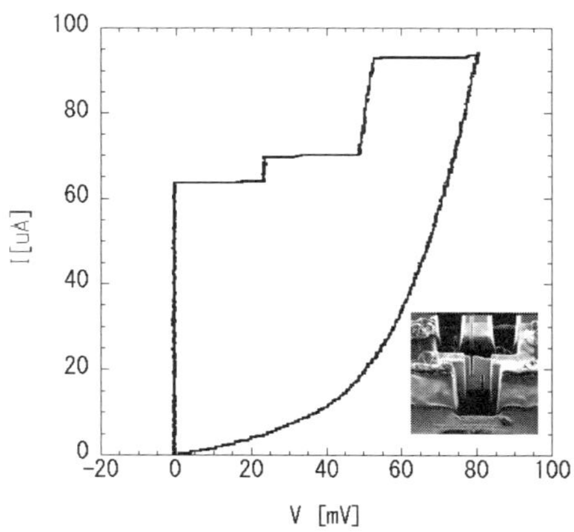

Fig. 1 I-V characteristics of a Bi2212 intrinsic Josephson junction obtained at 4.2K. The curve indicates the sequential switching of three junctions. The lack of the retrapping down to zero voltage indicates low damping of this IJJ.

3. Limit of Q value in *d*-wave Josephson junctions

Since HTSCs are d-wave superconductors, the intrinsic damping due to nodal quasiparticles cannot be avoided. The upper value of quality (Q) factor of HTSC is limited by the intrinsic damping corresponding to the quasiparticle resistance R_{qp} in the energy gap. We have presented a theoretical formulation of the damping effects in the quantum dynamics for IJJs in Ref. [2]. According to this result, the effect of damping on the macroscopic quantum tunneling (MQT) is suggested to be limited contrary to previous anticipation. Here we further evaluate the Q value in more realistic situation. The temperature dependence of R_{qp} can be calculated from I-V characteristics by taking account of thermal smearing using realistic parameters for Bi2212[10]. The Q factors are estimated from the gradient of the I-V curve near the origin. The result shown in Fig. 2 suggests that the Q factor as high as 5,000-20,000 at 4K and about 10^5 at 1K can be realized owing to the relatively large energy gap (Δ) values of Bi2212. These Q values are high enough to observe the phase coherence in actual experiments. However, the realization of

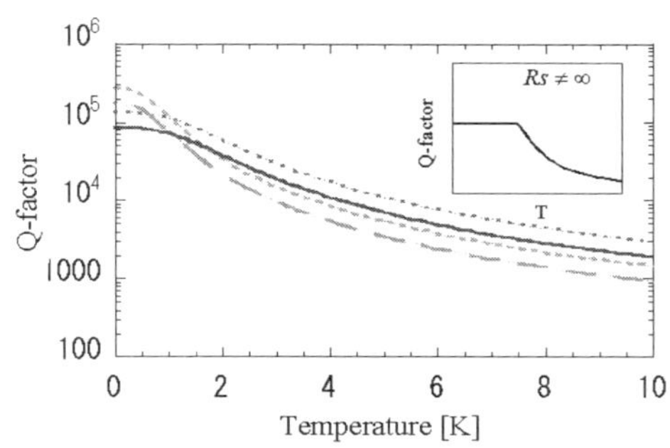

Fig. 2 Temperature dependences of Q factor for *d*-wave IJJs by assuming realistic parameters for Bi2212 (R_n=100 Ω, Δ=40 meV or 50 meV, ω_{pl}=65 GHz or 130 GHz, C=100 fF). The inset shows the schematic of Q factor in a realistic situation.

such high-Q values would not be possible in real situation. The practical limit may be given by the degradation of junction through the fabrication process. Therefore, we conclude the effect of nodal quasiparticle is not fatal for qubit application.

4. MQT and energy level quantization

We have performed two types of switching current distribution (SCD) measurements to make clear the quantum dynamics of the Bi2212 IJJs. One is the temperature dependence of the SCD on a Bi2212 JJ in a temperature range from 4.2K to 20mK[5]. The width of the SCD showed monotonic decrease as the temperature was lowered to 0.5K (crossover temperature T*), but the decrement was saturated below 0.5K. This saturation corresponded to the transition of switching mechanism from the thermal activation (TA) to MQT. The Q factors of about 70 were obtained in this experiment.
The other is the detection of the energy level quantization above the T*[11]. The switching current distribution had a single peak when the sweep rate was lower than 10A/sec. While as the sweep rate was increased up to 121.4 A/sec, the SCD tended to be wider and periodic peaks become clearly apparent[11] as shown in Fig. 3. The appearance of multiple peaks directly reflects the formation of the discretized bound energy levels in the washboard type potential. It is important that such a behavior can be expected only when the damping of the junction is sufficiently low. The estimated Q value is larger than 1000 in this experiment.
We can also evaluate Q factor from the I-V curve shown in Fig. 1. In addition to the sequential switching from zero to third branches, we find anomalous feature, i.e. the lack of retrapping down to zero voltage within our system resolution. The ratio of amplitude of retrapping current to the critical current reflects the damping in the dynamics of Josephson junctions. The Q value evaluated from the experimental data is more than 100.
All these estimations indicate that IJJs keep high Q value enough to observe the quantum coherence such as the Rabi oscillation in the qubit.

5. Strong coupling effect

Since thickness of the superconducting layer of HTSC is atomic scale, the electromagnetic coupling effect between the adjacent junctions cannot be avoided. One of the peculiarity induced due to such strong coupling is the collective switching reported as the uniform switching[5,12] in Bi2212. We found another

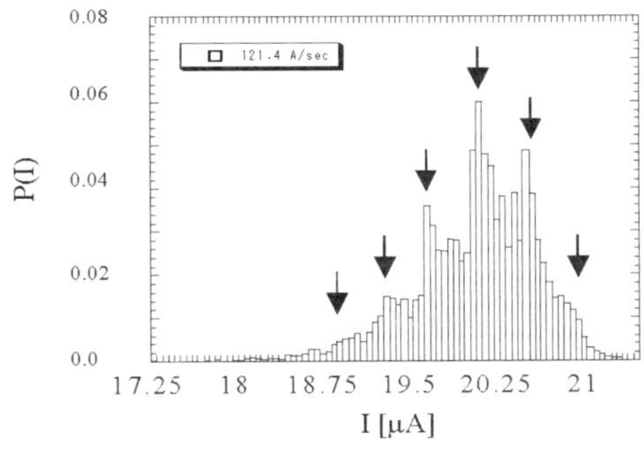

Fig. 3 Experimental data showing probability P(I) of SCD for dI/dt=121.4 A/sec at 4.2 K. The appearance of multiple peaks in the SCD indicates the quantization of the energy levels in the Bi2212 IJJ.

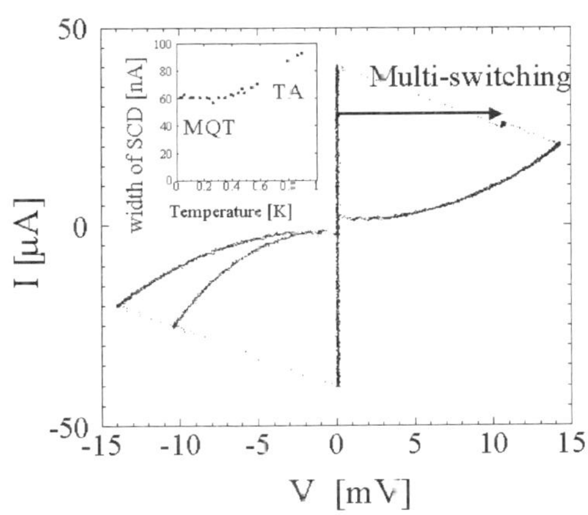

Fig. 4 I-V of multi-switching observed in a Bi2201 IJJ. The inset shows the temperature dependence of the width of SCD.

type of the collective switching. Figure 4 exhibits an I-V curve of Bi2201 IJJ showing simultaneous two junctions multi-switching and the temperature dependence of the width of the SCD. It is clear that the width of the SCD becomes constant at below 0.4K indicating the transition of switching mechanism from TA to the MQT. Basically, this behavior seems quite similar to that observed in isolated switching junctions. We are now trying to figure out whether such coupling effect modifies the switching dynamics at the MQT region.

6. Summary

The quantum dynamics of high-T_c Josephson junctions has been analyzed using Bi2212 and Bi2201 IJJs. Most of the switching current distribution measurements showed quite similar features to those observed in metal superconductor junctions. Is spite of *d*-wave nodal quasiparticles, we found that Q values of IJJ were high enough to observe the quantum coherence. On the other hand, the peculiarities in the switching dynamics, such as multi-junction switching effects, need to be clarified in future works.

Acknowledgments

We would like to thank H. Eisaki and Y. Yoshida for supplying single crystal samples.

References

1. R. Kleiner, F. Steinmeyer, G. Kunkel, P. Muller, Phys. Rev. Lett., 68 (1992) 2394.
2. S. Kawabata, S. Kashiwaya, Y. Asano, Y. Tanaka, Phys. Rev. B 70 (2004) 132505.
3. K. Inomata, S. Sato, K. Nakajima, A. Tanaka, Y. Takano, H. B. Wang, M. Nagao, H. Hatano, S. Kawabata, Phys. Rev. Lett. 95 (2005) 107005.
4. T. Bauch, F. Lombardi, F. Tafuri, A. Barone, G. Rotoli, P. Delsing and T. Claeson, Phys. Rev. Lett. 94 (2005), p. 087003.
5. X. Y. Jin, J. Lisenfeld, Y. Koval, A. Lukashenko, A. V. Ustinov, P. Muller, Phys. Rev. Lett. 96 (2006) 177003.
6. H. Kashiwaya, T. Matsumoto, H. Shibata, S. Kashiwaya, H. Eisaki, Y. Yoshida, S. Kawabata, and Y. Tanaka, J. Phys. Soc. Jpn. 77, (2008) 104708.
7. T. Koyama and M. Tachiki, Phys. Rev. B 54 (1996) 16183.
8. S. Sakai, P Budin and N. F. Pedersen, J. Appl. Phys. 73 (1993) 2411.
9. T. Matsumoto, H. Kashiwaya, H. Shibata, H. Eisaki, Y. Yoshida, H. Yamaguchi, S. Kashiwaya, Physica C 468 (2008) 1916
10. H. Kashiwaya, T. Matsumoto, S. Kashiwaya, H. Shibata, H. Eisaki, Y. Yoshida, S. Kawabata, Y. Tanaka, Physica C 468 (2008) 1919.
11. P. Silvestrini, V.G. Palmieri, B. Ruggiero and M. Russo, Phys. Rev. Lett. 79 (1997) 3046.

TOWARDS QUANTUM SIMULATION WITH PLANAR COULOMB CRYSTALS

I. M. BULUTA and S. HASEGAWA*

Department of Quantum Engineering and Systems Science
The University of Tokyo, 7-3-1 Hongo, Bunkyo-ku, Tokyo 113-8656, Japan
**E-mail: hasegawa@q.t.u-tokyo.ac.jp*

Motivated by the prospect of the experimental realization of applications like quantum simulation and scalable quantum computation using planar Coulomb crystals, we investigated planar Coulomb crystals with one or multicomponents and analyzed in detail, for the first time, their structure and properties. We also studied the RF-heating of planar crystals and found out in which conditions it can be minimized. These results together with the estimation of the error in the implementation of quantum simulation allowed us to find the experimental requirements for realizing quantum simulation with planar crystals. Trying to fulfill these requirements we designed an RF trap system for the realization of planar Coulomb crystals and quantum simulation.

Keywords: Quantum simulation; ion traps; planar Coulomb crystals.

1. Introduction

The simulation of quantum systems is known to be a very difficult computational problem. In the early 1980s, Richard Feynman suggested that only a quantum system would be able to efficiently simulate another quantum system. After 25 years Feyman's quantum simulator has not yet been put into practice although some attempts have been made. Even though the "proof-of-principle" has been realized the scaling up for practical applications is not an easy problem. Things are about to change due to the proposal for quantum simulation with planar Coulomb crystals[1]. This challenging idea may also be the solution for scalable ion trap quantum computation. As the first step towards putting this proposal into practice we investigated planar Coulomb crystals in the view of realizing quantum simulation and we designed an RF-trap system for the realization of planar Coulomb crystals in experiment. In the following sections we review our results and discuss the next steps towards the implementation of quantum simulation with planar Coulomb crystals.

2. Study of planar Coulomb crystals

2.1. *Molecular dynamics simulations*

Molecular dynamics simulations have long been used in the study of Coulomb crystals in both Penning and RF traps. Our simulations are designed using the ProtoMol framework[2], which has been already extensively tested for the simulation of single and multi-component Coulomb crystals in RF traps. We ran the large simulations on a 16 node PC-cluster and the smaller ones on a single machine. At each time step the forces are estimated and the equations of motion are integrated using the Leapfrog algorithm. The Coulomb force between the ions was calculated directly and only when the number of ions exceeds 10,000 the Multi-grid algorithm was used. Fig. 1 shows the simulation results for a 10,000 $^{40}Ca^{+}$ ions planar crystal.

2.2. *One component planar crystals*

The structure of planar crystals has been

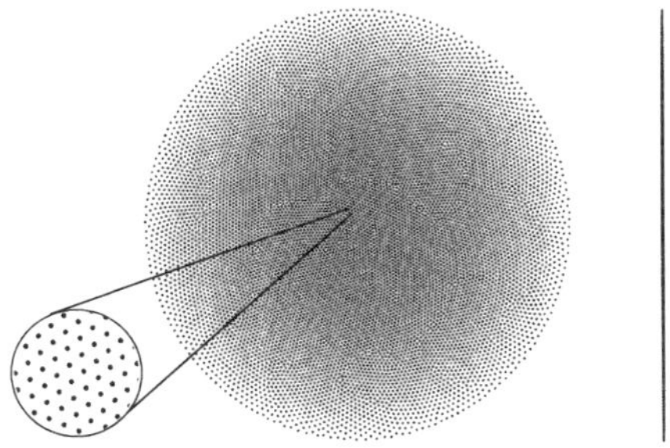

Fig. 1. 10,000 $^{40}Ca^+$ ions in a planar crystal. Radial and axial plane views. Inset: magnification of the center of the crystal. The simulation is performed using static potential. The radius of the crystal is $R \sim 265$ μm.

considered before, however, the particular features due to the RF confinement have not been analyzed in detail so far. We investigated the RF-heating of planar crystals[3] and the effect of spatial separation between components on the sympathetic cooling and RF-heating in two-component planar crystals (bicrystals)[4].

For implementing quantum simulation we are most interested in the triangular lattice structure of the planar crystals. In planar crystals there is a competition between two types of ordering: the triangular lattice ordering characteristic to infinite plasma and the shell structure due to the confining potential. For large crystals ($N > 200$, N is the number of ions) the inner part of the crystal exhibits an almost perfect triangular lattice structure and a circular shell structure near the edges (see Fig. 1). We analyzed in detail the structure of both large and small planar crystals. For small crystals the shell structure is dominant, however, there exist shell structures that are close to an ideal triangular lattice structure and such crystals could be used for implementing quantum simulation and computation. Because only crystals with specific N are needed for our purpose, the precise control of the number of trapped ions is required. This could be achieved in experiment using selective photoionization loading and selective cooling.

2.3. *RF heating*

RF-heating results from the chaotic motion of the ions and manifests itself as an increase in the kinetic energy. Several studies have considered the effects of RF-heating and it was shown that at low temperatures RF-heating is quite small. The dependence of the RF-heating rates on the trapping voltages, temperature and number of ions has never before been studied for planar crystals. From our analysis we found that larger crystals heat faster and in general the heating rates are discouragingly large, meaning that planar crystal may not be maintained without continuous cooling. However, we also found that the heating rate strongly depends on the trapping conditions and can be significantly decreased by choosing the appropriate trapping parameters. In general, in order to obtain reduced RF-heating rates low voltages are required. Nevertheless, planar crystals with more than $N \sim 100$ ions could not be maintained without continuous cooling. Together with the results on the structure of small planar crystals this shows that there is a serious limitation on the number of ions that could be used in experiment. Moreover, it turns out that in RF traps the optimization of the operating parameters is necessary.

2.4. *Multicomponent planar crystals*

In multicomponent Coulomb crystals in RF traps the light ions gather at the center, while the heavy ions form an outer shell (see Fig. 2 which shows a planar bicrystal with 30 $^{40}Ca^+$ and 40 $^{44}Ca^+$ ions). We investigated multicomponent planar crystals in RF traps and found that because of the mass dependence of the radial potential and the application of large axial voltages interesting features occur. First, the number of ions in the inner shell is limited by the mass difference

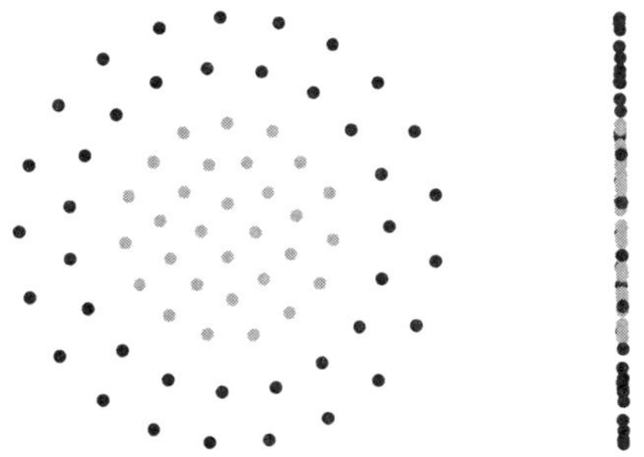

Fig. 2. Planar bicrystal with 30 $^{40}Ca^+$ (gray) and 40 $^{44}Ca^+$ (black) ions. Radial and axial plane views. The simulation is performed using static potential.

between the components. We calculated the maximum number of ions in the inner shell as a function of the ion masses. Second, the spatial separation between the components depends on the trapping parameters as well as the ion masses and charges. We investigated the effect of spatial separation on sympathetic cooling and RF-heating.

The limitation on the number of ions due to the RF heating could be overcome by using the continuous cooling of the outer shell of a planar bicrystal as we suggested in Ref.[4]. The inner shell of a planar bicrystal could be used for implementing quantum simulation and computation, while the continuously cooled outer shell would help maintaining the inner shell at low temperatures via sympathetic cooling. However, the following issues should be taken into consideration: first, the number of ions in the inner shell is restricted by the mass ratio of the components, so for large mass differences we can only have few ions in the inner shell and hence, the limitation problem is not solved. Second, for the sympathetic cooling to be efficient the spatial separation needs to be small (i.e. the mass difference needs to be small). Then, bicrystals with a small mass difference (perhaps isotopes of the same element like in Fig. 2) would be desirable because the number of ions in the inner shell would be large and the small spatial separation would ensure the efficiency of sympathetic cooling.

3. Quantum simulation with planar Coulomb crystals

In planar crystals, in certain conditions, the ions moving in the axial direction can be considered as independent harmonic oscillators weakly coupled by the Coulomb interaction. Therefore, a planar crystal is similar to a microtrap array. Two-qubit interactions necessary for the implementation of quantum simulation and computation are realized by the means of a spin-dependent "pushing force". There are several beam arrangements which implement this "pushing force" (travelling, standing or walking waves). We studied the "push-gate" (for quantum computation and simulation) and the effective spin-spin interaction (for quantum simulation) and estimated the error as a function of the trap and laser parameters. We found that besides the need for minimizing the movement of the ions and the fluctuations of the temperature (i.e. minimize the heating of the planar crystal) we also need very high axial frequencies. The need for very high axial frequencies could be made less stringent by adjusting the amplitude of the "pushing force".

4. Design of the RF trap system

There are various RF trap geometries but the linear segmented quadrupole (like the one in Ref.[5]) type seems most suitable for our purpose as it can provide large axial frequencies and has good optical access. The requirements that our trap should meet are: (a) stable trapping; (b) large anisotropy parameter (the ratio of the radial and axial frequencies) to ensure the planar crystal condition: $\alpha = \omega_z^2/\omega_{xy}^2 > \alpha_3(N)$, where $\alpha_3(N) = (96N/\pi^3 w_1^3)^{1/2}$ and $w_1 = 1.11...$; (c) reduced heating (RF heating and other heating sources). When applying large axial voltages the radial frequency is shifted so the dependence of the anisotropy parameter on the applied voltages is not a simple one, because increasing the axial voltage weakens the radial confinement, meaning that the axial voltage cannot be made arbitrarily large. We know that z_0 (half the length of the inner segments) must be small in order to obtain large axial frequencies and that r_0 (the distance from the trap center to the electrodes' surface) must be large in order to reduce heating (i.e. the ions should be far from the electrodes). However, for large r_0 we require larger RF voltages which leads to increased RF-heating. After taking all these into consideration and performing SIMION simulations we found that the dimensions most suitable for our purposes are $r_0 = 3$ mm, $z_0 = 1.5$ mm and the separation between segments $d = 0.5$ mm. The trap electrodes and the experimental setup are shown in Fig. 3. As compared to the existing RF traps (for example Ref.[5]), our trap provides large axial frequencies for low operating voltages and therefore has reduced RF-heating.

Fig. 3. Our designed RF trap and the experimental setup.

As discussed in Section 3, for implementing quantum simulation very large axial frequencies are required. However, large axial frequencies come at the cost of increased RF-heating so the amplitude of the "pushing force" must be adjusted such that the error is minimized even for smaller axial frequencies. However, this depends on the particular beam setup so it makes the subject of a separate study.

5. Summary

The first step towards the implementation of quantum simulation with planar Coulomb crystals is the experimental realization of planar Coulomb crystals and from our study we found the necessary requirements. The important thing to note is that in the experimental implementation of quantum simulation a trade-off between the axial frequency and the amplitude of the "pushing force" must be reached and therefore, a good trap alone does not ensure the success of the quantum simulation experiments. The laser beam setup for realizing the "pushing force" is particularly important and it has to be thoroughly investigated. Therefore, the next step is the optimization of the laser beam setup. We believe that our RF trap system would make easy-to-build and cheap quantum simulator for practical applications of quantum simulation which would open the door to the exploration of new physical phenomena.

Acknowledgments

The first author acknowledges the financial support of MEXT. We wish to thank our collaborators: M. Todoriki, Y. Hashimoto and M. Kitaoka.

References

1. D. Porras and J.I. Cirac, *Phys. Rev. Lett.* **96**, 250501 (2006).
2. T. Matthey *et al.*, *ACM Trans. Math. Softw.* **30**, 237 (()2004).
3. I.M. Buluta *et al.*, *Phys. Rev. A* **77**, 062320 (2008).
4. I.M. Buluta and S. Hasegawa, *submitted to Phys. Rev. A*, (2008).
5. M. Block *et al.*, *J. Phys. B: At. Mol. Opt. Phys.* **33**, L375 (2000).

THE NEGATIVE BINOMIAL DISTRIBUTION IN QUANTUM PHYSICS

JONAS SÖDERHOLM* and SHUICHIRO INOUE
Institute of Quantum Science, Nihon University,
1-8-14 Kanda-Surugadai, Chiyoda-ku, Tokyo 101-8308, Japan
**E-mail: jonas@phys.cst.nihon-u.ac.jp*

We give examples of situations where the negative binomial distribution has appeared in quantum physics since its debut in the work of Planck. Several of its properties are reviewed, and Mandel's Q-parameter is shown to play an interesting role. The photon-pair distributions of squeezed vacuum and squeezed single-photon states are identified as negative binomial.

Keywords: Negative binomial distribution; Mandel's Q-parameter; Squeezed number states.

1. Planck and the negative binomial distribution

ISQM-TOKYO '08 offers us a suitable occasion to highlight the 150th anniversary of Max Planck's birth in Kiel, Germany on April 23, 1858. His introduction of energy quanta in a desperate attempt to explain the black-body radiation made him the father of quantum mechanics, even though he at first was reluctant to accept its consequences.

In his ground-breaking paper of 1900,[1] Planck derives the expression $\binom{n+m-1}{n}$ for the number of ways n indistinguishable particles can be distributed among m cells. This we recognize as the basis of what later became known as Bose–Einstein statistics.

The negative binomial distribution can be written as

$$\mathcal{P}_{\rm nb}(s,p,n) = \binom{-s}{n} p^s (p-1)^n, \qquad (1)$$

where $s > 0$ and $0 < p < 1$. The negative binomial coefficient

$$\binom{-s}{n} = \frac{(-1)^n\,\Gamma(s+n)}{\Gamma(s)\,n!} \qquad (2)$$

is defined by extending the binomial theorem to negative powers

$$(z+w)^{-s} = \sum_{k=0}^{\infty} \binom{-s}{k} z^k w^{-s-k}, \qquad (3)$$

where $|z| < |w|$. Since the gamma function simplifies to $\Gamma(j) = (j-1)!$ for positive integers j, the relation between Planck's expression and the negative binomial distribution is seen by setting $s = m$.

Denoting the expectation value of n as $\mu = s(1-p)/p$, we can parameterize the negative binomial distribution as

$$P_{\rm nb}(s,\mu,n) = \frac{\Gamma(s+n)\,s^s\mu^n}{\Gamma(s)\,n!\,(s+\mu)^{s+n}}, \qquad (4)$$

which we shall use from here on. The thermal distribution, corresponding to a single cell (mode) in Planck's derivation, is then given by

$$P_{\rm th}(\mu,n) = P_{\rm nb}(1,\mu,n) = \frac{\mu^n}{(\mu+1)^{n+1}}. \qquad (5)$$

2. Mandel's Q-parameter

In quantum optics, Mandel's Q-parameter[2] is frequently used to compare a distribution's variance σ^2 and average value μ. The parameter vanishes for the Poisson distribution, in which case we have $\sigma^2 = \mu$. The negative binomial distribution is seen to be super-Poissonian, since Mandel's Q-parameter is positive

$$Q = \frac{\sigma^2 - \mu}{\mu} = \frac{\mu}{s} > 0. \qquad (6)$$

Hence, the parameter vanishes only in the limit $s \to \infty$, where the negative binomial

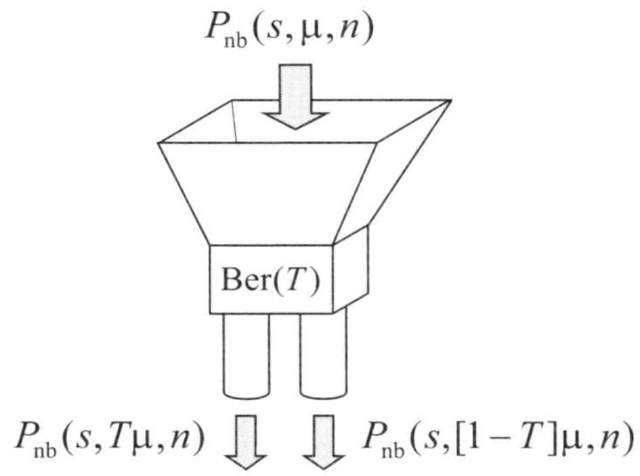

Fig. 1. The negative binomial distribution keeps its form under splitting described by a Bernoulli test.

becomes the Poisson distribution

$$P_{\text{poi}}(\mu,n) = \lim_{s\to\infty} P_{\text{nb}}(s,\mu,n) = \frac{e^{-\mu}\mu^n}{n!}. \tag{7}$$

As is well known, Glauber's coherent state

$$|\alpha\rangle = e^{\alpha\hat{a}^\dagger - \alpha^*\hat{a}}\,|0\rangle = e^{-|\alpha|^2/2}\sum_{n=0}^{\infty}\frac{\alpha^n}{\sqrt{n!}}\,|n\rangle, \tag{8}$$

whose photon distribution is Poissonian with $\mu = |\alpha|^2$, and the thermal state

$$\hat{\rho}_{\text{th}}(\mu) = \sum_{n=0}^{\infty} P_{\text{th}}(\mu,n)\,|n\rangle\langle n| \tag{9}$$

both keep their respective form when passing through a beam splitter. Modeling the beam splitter as a Bernoulli test, which lets a particle be transmitted with probability T, we find that this is a general property of the negative binomial distribution, as illustrated in Fig. 1.

The convolution of two negative binomial distributions is in general not another negative binomial distribution. However, it is true if the distributions have the same Q-parameter. The parameters are then related according to Fig. 2. From repeated use of this fact, we find that convolution of m thermal distributions, corresponding to m cells in Planck's paper,[1] is the so-called multithermal distribution[3]

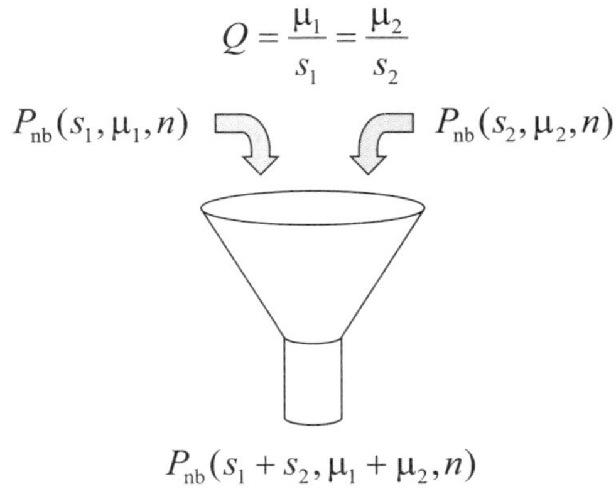

Fig. 2. The convolution of two negative binomial distributions that have the same value of Mandel's Q-parameter is another negative binomial distribution.

$$P_{\text{nb}}(m, m\mu_{\text{th}}, n) = \frac{(m+n-1)!}{(m-1)!\,n!\,(1+m/\mu_{\text{th}})^n(1+\mu_{\text{th}}/m)^m}. \tag{10}$$

3. Down-conversion, SU(1,1), and squeezing

The multithermal distribution also appears naturally when considering nondegenerate down-conversion. When one mode is seeded with a Fock state, the number of generated photon pairs is described by a multithermal distribution. Using the notation $\xi = re^{i\theta}$, the generated SU(1,1) coherent states can be written as

$$\begin{aligned}|\xi;k,0\rangle &= e^{\xi\hat{a}^\dagger\hat{b}^\dagger - \xi^*\hat{a}\hat{b}}|k,0\rangle\\ &= \sum_{n=0}^{\infty} e^{in\theta}\sqrt{P_{\text{nb}}\left(k+1,[k+1]\sinh^2 r,n\right)}\\ &\quad\times|n+k,n\rangle.\end{aligned} \tag{11}$$

The states obtained when the first mode is ignored have an interesting relation to the thermal states[4]

$$\begin{aligned}
&\mathrm{Tr}_1|\xi;k,0\rangle\langle\xi;k,0| \\
&= \sum_{n=0}^{\infty} P_{\mathrm{nb}}(k+1,[k+1]\mu_{\mathrm{th}},n)\,|n\rangle\langle n| \\
&= \frac{\hat{a}^k\hat{\rho}_{\mathrm{th}}(\mu_{\mathrm{th}})\,(\hat{a}^\dagger)^k}{k!\,\mu_{\mathrm{th}}^k}, \qquad (12)
\end{aligned}$$

where $\mu_{\mathrm{th}} = \sinh^2 r$. In the case where $k = 0$, the two-mode squeezed vacuum is generated and Eq. (12) reduces to the thermal state, which is a well-known fact.[5]

Apart from the single-mode mixed states above, coherent single-mode superpositions with negative binomial photon distributions have also been considered.[6,7] This class of states includes the Glauber's coherent states (8), and the phase states can be obtained in the appropriate limit.[7]

Also degenerate down-conversion generate photon pairs with negative binomial distributions. Although the Fock-basis expansion of the squeezed number states

$$|\xi;k\rangle = \exp\left(\frac{\xi\hat{a}^\dagger\hat{a}^\dagger - \xi^*\hat{a}\hat{a}}{2}\right)|k\rangle \qquad (13)$$

has been know for a long time,[8] no one seems to have recognized that the squeezed vacuum and squeezed single-photon states can be expressed as

$$|\xi;0\rangle = \sum_{n=0}^{\infty} e^{in\theta}\sqrt{P_{\mathrm{nb}}\left(\frac{1}{2},\frac{\sinh^2 r}{2},n\right)}\,|2n\rangle \qquad (14)$$

and

$$\begin{aligned}
|\xi;1\rangle = &\sum_{n=0}^{\infty} e^{in\theta}\sqrt{P_{\mathrm{nb}}\left(\frac{3}{2},\frac{3\sinh^2 r}{2},n\right)} \\
&\times|2n+1\rangle, \qquad (15)
\end{aligned}$$

respectively.

4. Related distributions

The negative binomial distribution can be written as a compound Poisson distribution[9]

$$P_{\mathrm{nb}}(s,s\beta,n) = \int_0^\infty P_{\mathrm{poi}}(\mu_{\mathrm{p}},n)P_{\mathrm{gamma}}(s,\beta,\mu_{\mathrm{p}})\,\mathrm{d}\mu_{\mathrm{p}}, \qquad (16)$$

where the gamma distribution is given by

$$P_{\mathrm{gamma}}(\alpha,\beta,\theta) = \frac{\theta^{\alpha-1}e^{-\theta/\beta}}{\beta^\alpha\Gamma(\alpha)}, \quad \alpha,\beta,\theta > 0. \qquad (17)$$

Interestingly, Mandel's photoelectric equation for detection times τ much shorter than the coherence time takes the form[10]

$$P_{\mathrm{det}}(n,\tau) = \int_0^\infty \frac{(\eta\tau I)^\tau\, e^{-\eta\tau I}P(I)}{n!}\,\mathrm{d}I, \qquad (18)$$

where η is the detection efficiency and the optical intensity I is distributed according to P. It hence follows from Eq. (16) that the number of detected photons is given by a negative binomial distribution for any impinging light with a gamma-distributed intensity.[11]

The negative binomial distribution can also be expressed using the logarithm distribution defined by

$$P_{\log}(p,n) = -\frac{(1-p)^n}{n\ln p}, \quad n = 1,2,3,\ldots \qquad (19)$$

That is, $P_{\log}(p,0) = 0$. Now, consider the convolution of N independent identical random variables with probability distribution (19). If N has a Poissonian probability distribution with average μ_{p}, the resulting probability distribution is[12]

$$P_{\mathrm{nb}}\left(-\frac{\mu_{\mathrm{p}}}{\ln p},-\frac{\mu_{\mathrm{p}}[1-p]}{p\ln p},n\right). \qquad (20)$$

This decomposition has been used in high-energy physics to interpret measured multiplicity distribution as arising from Poisson-distributed clusters decaying into daughter particles according to the logarithm distribution.[13]

In addition to the decompositions above, several others can be found in Ref. 14. The fact that the negative binomial distribution also arises from simple birth-and-death processes is also discussed there.

5. Conclusions

As we have seen, the negative binomial distribution appears in a variety of situations. Despite its long history in quantum mechanics, we believe that it deserves more attention than it has received. In particular, since it generalizes and share several properties with the more commonly used thermal and Poisson distributions.

Acknowledgments

This work was supported by the National Institute of Information and Communications Technology.

References

1. M. Planck, *Verh. d. Deutsch. Phys. Ges.* **2**, 237 (1900).
2. L. Mandel, *Opt. Lett.* **4**, 205 (1979).
3. L. Mandel, *Proc. Phys. Soc.* **74**, 233 (1959).
4. G. S. Agarwal, *Phys. Rev.* **A45**, 1787 (1992).
5. B. R. Mollow and R. J. Glauber, *Phys. Rev.* **160**, 1076 (1967).
6. H.-C. Fu and R. Sasaki, *J. Math. Phys.* **38**, 3968 (1997).
7. H.-C. Fu and R. Sasaki, *J. Phys. Soc. Jpn.* **41**, 519 (1997).
8. M. S. Kim, F. A. M. de Oliveira, and P. L. Knight, *Phys. Rev.* **A40**, 2494 (1989).
9. M. Greenwood and G. U. Yule, *J. Roy. Stat. Soc.* **A83**, 255 (1920).
10. L. Mandel and E. Wolf, *Optical Coherence and Quantum Optics* (Cambridge University Press, 1995), Sec. 14.8.
11. G. J. Troup, *Proc. Phys. Soc.* **86**, 39 (1965).
12. R. Lüders, *Biometrika* **26**, 108 (1934).
13. A. Giovannini and L. Van Hove, *Z. Phys.* **C30**, 391 (1986).
14. N. L. Johnson, A. W. Kemp, and S. Kotz, *Univariate Discrete Distributions*, 3rd ed. (Wiley, Hoboken, NJ, 2005).

ON THE ELEMENTARY DECAY PROCESS

D. KOUZNETSOV

Institute for Laser Science, University of Electro Communications, Chofu, Tokyo 182-8585, Japan
E-mail: dima@ils.uec.ac.jp

The quantum mechanical system with continuum of modes is considered as model of the decaying atom. The exponential decay and the Lorentian shape of distribution of the emitted photon are deduced as the non-perturbative solution of the time-dependent Schrödinger equation. The possible application of the formalism to the decay of trapped atoms is discussed.

Keywords: Exponential decay; non-perturbative approach

1. Introduction

The model of exponential decay was suggested a century ago [1], with the discrete momentum representation for photons and perturbation theory for the interaction. Such an approach allows to guess the exponential decay and the Lorentian profile of the emitted photons. Then, the perturbative deduction was used by many authors [2,3,4]. The discrete representation becomes cumbersome at consideration of intermediate level(s) [5,6,7]. In addition, the only experiments serve as verification of the theory of perturbation used.

In this paper, I suggest the non-perturbative description of the spontaneous decay for the simplest possible model (fig.1), that still allows the irreversible decay. The decay solution is verified by the substitution into the Schrödinger equation. Up to my knowledge, it is first case of such verification, ever reported.

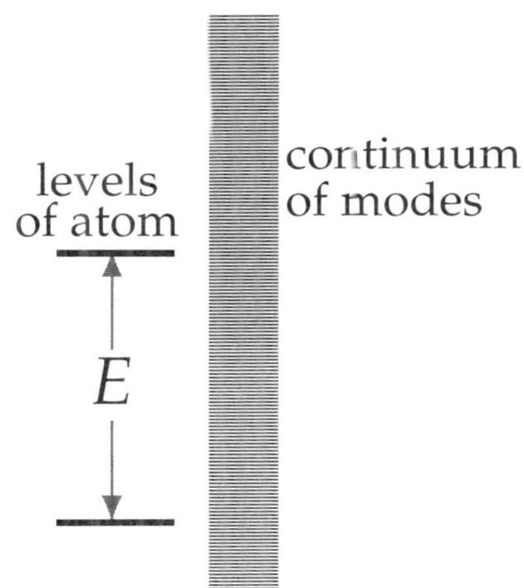

Fig. 1. Energy levels of the two-level atom and levels of the filed. Within this model, the spectrum of energies of photons goes from $-\infty$ to ∞, i.e. photons may have negative energy.

2. Hamiltonian

Consider the model Hamiltonian

$$H = H_0 + H_1 \,, \qquad (1)$$

where

$$H_0 = \int_{-\infty}^{\infty} a_k^\dagger a_k k \,\mathrm{d}k + E u^\dagger u \qquad (2)$$

is "interaction-free" Hamiltonian, and

$$H_1 = \int_{-\infty}^{\infty} \left(a_k^\dagger u + a_k u^\dagger\right) \gamma \,\mathrm{d}k \qquad (3)$$

describes interaction between the atom and the continuum of modes of the field a in the resonant approximation. Parameter E has sense of energy of the atom (fig.1). I use the Planck system of units; so, both, speed of light and the constant of Planck are unity; k can be interpreted as wavenumber, as well as frequency of the mode, or energy of excitation of a single quantum in this mode.

Within this model, the atom has only two levels, ground state $|0\rangle$ and the excited state $u^\dagger|0\rangle$. The atom operator u has following properties:

$$u^2 = 0\,, \quad u^\dagger u + u u^\dagger = 1\,, \quad u|0\rangle = 0\,. \tag{4}$$

The filed has continuum of modes. Operators a_k can be considered as the linear superposition of the field operators of the plane waves with the same wavenumber, at least while k is positive. These operators are supposed to be normalized in such a way, that

$$a_k a_p^\dagger - a_p^\dagger a_k = \delta(k-p)\,. \tag{5}$$

Parameter η describes the coupling. In general, this parameter could depend on wavenumber, but in the resonant approximation this dependence is assumed to be slow; so, I treat η as positive constant. For the same reason, I extend the integration in Eqs. (2),(3) to the $(-\infty, \infty)$ range. In the similar way, I assume, that the integrals with respect to k have sense of the principal value. Such a model allows the exact solution in the compact analytical form in terms of elementary functions.

For the time-dependent Schrödinger equation

$$\mathrm{i}\, \mathrm{d}\Psi/\mathrm{d}t = H\Psi\,, \tag{6}$$

the solution can be presented in the following form:

$$\Psi = h(t)\mathrm{e}^{-\mathrm{i}Et}u^\dagger|0\rangle + \fint_{-\infty}^{\infty} \phi_k(t)\mathrm{e}^{\mathrm{i}kt}a_k^\dagger|0\rangle \mathrm{d}k, \tag{7}$$

where h and ϕ_k for real values of subscript k are c-number functions of real argument. The spontaneous decay from excited state corresponds to the initial conditions

$$h(0) = 1\,, \quad \phi_q(0) = 0\,. \tag{8}$$

In order to eliminate parameter E, let $f_{k-E}(t) = \phi_k(t)$; then

$$\mathrm{i}\frac{\mathrm{d}h(t)}{\mathrm{d}t} = \fint_{-\infty}^{\infty} \eta\, f_q(t)\mathrm{e}^{-\mathrm{i}qt}\,\mathrm{d}q \tag{9}$$

$$\mathrm{i}\frac{\mathrm{d}f_q(t)}{\mathrm{d}t} = \eta h(t)\mathrm{e}^{\mathrm{i}qt} \tag{10}$$

Equations (9) and (10) have sense of the Schrödinger equation in the representation of the distribution $\{h, f\}$ of excitations. Formally, the use of the same principal value integrating by the new variable $q = k - E$ is not correct; such a "principal value" is justified by the slow divergence of integrals in Eq. (23) below.

In order to express the solution, rewrite equation (10) as

$$f_q(t) = -\mathrm{i}\eta \int_0^t h(\tau)\mathrm{e}^{iq\tau}\mathrm{d}\tau\,. \tag{11}$$

Substitution of Eq. (11) into (9) gives

$$\frac{\mathrm{d}h(t)}{\mathrm{d}t} = -\eta^2 \fint_{-\infty}^{\infty}\mathrm{d}q \int_0^t \mathrm{d}\tau\, h(\tau)\mathrm{e}^{iq\tau - \mathrm{i}qt}\,. \tag{12}$$

With new variable of integration $x = \tau - t$, equation (12) can be rewritten as

$$\frac{\mathrm{d}h(t)}{\mathrm{d}t} = -\eta^2 \fint_{-\infty}^{\infty}\mathrm{d}q \int_{-t}^{0} \mathrm{d}x\, h(t+x)\mathrm{e}^{iqx}\,. \tag{13}$$

With use of the generalized signum [8] function ϵ, the integration with respect to x can be extended to the symmetric range:

$$\frac{\mathrm{d}h(t)}{\mathrm{d}t} = -\eta^2 \fint_{-\infty}^{\infty}\mathrm{d}q \int_{-t}^{t} \mathrm{d}x\, h(t+x)\frac{1-\epsilon(x)}{2}\mathrm{e}^{iqx}\,; \tag{14}$$

then

$$\frac{\mathrm{d}h(t)}{\mathrm{d}t} = -\eta^2 \fint_{-\infty}^{\infty}\mathrm{d}q \int_{-t}^{t} \mathrm{d}x\, h(t+x)\mathrm{e}^{iqx} + \mathcal{E}(t); \tag{15}$$

where

$$\mathcal{E}(t) = \fint_{-\infty}^{\infty}\mathrm{d}q \int_{-t}^{t} \mathrm{d}x\, \epsilon(x) h(t+x)\mathrm{e}^{iqx}. \tag{16}$$

It is possible to show, that

$$\mathcal{E}(t) = 0\,; \tag{17}$$

it follows from the anticommutativity [8] of generalized functions $\epsilon(x)$ and $\delta(x)$.

3. The law of decay and the spectrum

Substituting Eq.(17) into Eq. (15), I get

$$\frac{\mathrm{d}h(t)}{\mathrm{d}t} = -\pi\eta^2 h(t)\,, \tag{18}$$

that gives exponential decay

$$h(t) = \exp(-\pi\eta^2 t) \tag{19}$$

of amplitude of probability of the atom to be still excited at time t.

Substitution of the exponential decay (19) into Eq.(11), I get

$$\begin{aligned} f_q(t) &= -\mathrm{i}\eta \int_0^t \mathrm{e}^{-\pi\eta^2\tau+\mathrm{i}q\tau}\mathrm{d}\tau \\ &= -\mathrm{i}\eta \frac{\mathrm{e}^{-\pi\eta^2 t+\mathrm{i}qt} - 1}{-\pi\eta^2 + \mathrm{i}q} \\ &= \eta \frac{1-\mathrm{e}^{-\pi\eta^2 t+\mathrm{i}qt}}{q+\mathrm{i}\pi\eta^2}\,. \end{aligned} \tag{20}$$

In such a way, Eqs. (19) and (20) express the non-perturbative solution of the Schrödinger equation (9),(10).

4. Check

In this section, I show the main difference between the solution above and the perturbative solutions [1,2,3,5,4]: the solution (19),(20) satisfies the Schrödinger equation (9),(10) exactly, and this can be checked with the direct substitution. I trace this check below.

Substituting the solution f_k by Eq.(20) into right hand side of (9), I get

$$\eta \int_{-\infty}^{\infty} f_k \mathrm{e}^{-\mathrm{i}kt}\mathrm{d}k = \eta^2 \int_{-\infty}^{\infty} \mathrm{d}k \frac{\mathrm{e}^{-\mathrm{i}kt} - \mathrm{e}^{-\pi\eta^2 t}}{k+\mathrm{i}\pi\eta^2}\,. \tag{21}$$

Consider integrals I_1 and I_2 for each of terms in the numerator of the integrand separately. For the first term, I close the contour of integration in the lower complex half-plane. This gives

$$\begin{aligned} I_1 &= \eta^2 \int_{-\infty}^{\infty} \mathrm{d}q \frac{\mathrm{e}^{-\mathrm{i}qt}}{q+\mathrm{i}\pi\eta^2} \\ &= -2\pi\mathrm{i}\eta^2 \operatorname{Rez}\left(\frac{\mathrm{e}^{-\mathrm{i}qt}}{k+\mathrm{i}\pi\eta^2}, q=-\mathrm{i}\eta^2 t\right) \\ &= -2\pi\mathrm{i}\eta^2 \exp(-\pi\eta^2 \mathrm{t})\,. \end{aligned} \tag{22}$$

For the second term in the numerator in the integrand in (21), the integral can be expressed through function arctan, giving

$$\begin{aligned} I_2 &= -\eta \mathrm{e}^{-\pi\eta^2 t} \int_{-\infty}^{\infty} \frac{\mathrm{d}k}{k+\mathrm{i}\pi\eta^2} \\ &= -\eta \mathrm{e}^{-\pi\eta^2 t} \int_{-\infty}^{\infty} \mathrm{d}k \frac{k - \mathrm{i}\pi\eta}{k^2 + \left(\pi\eta^2\right)^2} \\ &= \mathrm{i}\pi\eta^2 \mathrm{e}^{-\pi\eta^2 t}\,. \end{aligned} \tag{23}$$

Combining Eqs. (22) and (23), I express the right hand side of (9) as $-\mathrm{i}\pi\eta^2 \exp(-\pi\eta^2 t)$. Substitution of $f(t)$ by Eq (19) into the left hand side of Eq. (9) gives the same expression; i.e., h and f by (19) and (20) are solution of Eq.(9),(10).

Differentiation of expression (20) with respect to t gives

$$\begin{aligned} \frac{\mathrm{d}f_q(t)}{\mathrm{d}t} &= -\eta \frac{-\pi\eta^2 + \mathrm{i}q}{q+\mathrm{i}\pi\eta^2} \mathrm{e}^{-\pi\eta^2 t+\mathrm{i}qt} \\ &= -\mathrm{i}\eta \mathrm{e}^{-\pi\eta^2 t+\mathrm{i}qt} \end{aligned} \tag{24}$$

As in the left hand side of Eq.(10), as in the right hand side, I get the same quantity $\eta \mathrm{e}^{-\pi\eta^2 t+\mathrm{i}qt}$. This confirms, that the second part of the Schrödinger equation (9) is also satisfied.

In such a way, the non-perturbative solution is confirmed to satisfy the Schrödinger equation (9),(10).

5. Interpretation

The solution (19)(20) allows the conventional interpretation. In the limit $t \to \infty$,

$$f_k(\infty) = \frac{\mathrm{i}\eta}{-\pi\eta^2 + \mathrm{i}k}\,. \tag{25}$$

In this approximation, the spectrum of the emitted photons is perfect Lorentian:

$$|\phi_k(\infty)|^2 = \frac{\eta^2}{\left(\pi\eta^2\right)^2 + \left(k-E\right)^2}\,. \tag{26}$$

The distribution of emitted photons by the wavenumbers at finite value of t may correspond to the "kicked" atom [9]. If an atom is excited at time $t = 0$, and at time $t \le \tau = 1/(2\pi\eta^2)$, the atom is destroyed, the state of photons remains the same as it was at time t; then, the spectrum

$$|f_q(t)|^2 = \eta^2 \frac{1+e^{-2\pi\eta^2 t} + 2e^{-\pi\eta^2 t}\cos(qt)}{q^2 + \left(\pi\eta^2\right)^2} \quad (27)$$

could be observed (q is detuning from the center of the spectral line). Similar spectrum appears at the time-frequency oscillations [6,7]; however, the "destruction" of atom at some fixed moment allows to avoid need to deal with the photon correlation. Actual destruction of atoms is not required: the atom could be just ionized, or kicked out from the view of the focusing system, of just suddenly accelerated, moving its spectral line out from the range of the spectrum analyzer due to the Doppler effect.

Possible generalization of the non-perturbative solution to the case of non-trivial distribution of density of bosonic states is straightforward. For example, the spectrum of bosonic states is deformed for the atoms trapped in vicinity of a nanovire [10]. For non-constant coupling η, the numerical evaluation of the integrals in the analogy of equation (12) may be necessary, but even the numerical solution can be non-perturbative.

6. Conclusions

The non-perturbative analysis of the decay process in quantum mechanics is suggested. Example (9),(10) with exponential decay (9) and Lorentian (26) profile of the emitted photon is considered. The solution allows verification by substitution into the Schrödinger equation in the representation (9),(10).

Acknowledgments

This work was supported by COE, Japan.

References

1. V. Weisscopf, E. Wigner. Berechnung der naturlichen Lichtteorie, Zeitschrift für Physik **63** , 54-73 (1930).
2. J. J. Sakurai. Advanced quantum mechanics. Reading, Mass. : Addison-Wesley Pub. 1967.
3. L.D.Landau and E.M.Lifshitz. Quantum mechanics : non-relativistic theory. Oxford : Butterworth-Heinemann, 1991.
4. V. P. Bykov. Radiation of Atoms in a Resonant Environment. Worls Scientific, 1994.
5. J. Kimble, A. Mezzacappa. Time dependence of photon correlations in a 3-level atomic cascade. *Phys. Rev.* **A31**, 3686 (1985).
6. D. Kuznetsov. The decay of a three-level atom and time-frequency oscillations, J. of Modern Optics **38**, No.1, 65-68 (1991)
7. D. Yu. Kuznetsov. On the elementary decay process of three-level systems. (in book: Dynamics of elementary atomic-molecular processes in gas and plasma. Nova Science publisher, NY, 1996)
8. Yu. M. Shirokov. Algebra of generalized functions. Theoretical and Mathematical Physics **39**, 291-301 (1979)
9. M. Sadgrove et all. Deviations from early-time quasilinear behavior for the atom-optics kicked rotor near the classical limit. Phys. Rev. E **71**, 027201 (2005)
10. K. P. Nayak et all. Optical nanofiber as an efficient tool for manipulating and probing atomic fluorescence. Optics Express **15**, 5431-5438 (2007)

LIST OF PARTICIPANTS

Abe, Eisuke
Institute for Solid State Physics
The University of Tokyo

Abe, Nobuo
Advanced Research Laboratory
Hitachi, Ltd.

Akima, Hisanao
Advanced Research Laboratory
Hitachi, Ltd.

Amaha, Shinichi
Quantum Spin Information Project
ICORP- JST
NTT Atsugi Research & Development Center

Ando, Tsuneya
Department of Physics
Tokyo Institute of Technology

Ando, Masahiko
Hitachi Cambridge Laboratory
Hitachi Europe Ltd.

Ashhab, Sahel
Digital Materials Laboratory, ASI
RIKEN

Bauer, Gerrit E.W.
Kavli Institute of NanoScience
Delft University of Technology

Blatt, Rainer
Institute for Experimental Physics
University of Innsbruck

Buluta, Iulia M.
Department of Quantum Engineering and
Systems Science
The University of Tokyo

Ferrus, Thierry A.
Hitachi Cambridge Laboratory
Hitachi Europe Ltd.

Fujikawa, Kazuo
Institute of Quantum Science
Nihon University

Fujita Kazuhiro
Institute for Solid State Physics
The University of Tokyo

Fukui, Nobuyuki
WPI Advanced Institute for Materials
Research
Tohoku University

Fukuyama, Hidetoshi
Department of Applied Physics
Faculty of Science
Tokyo University of Science

Furukawa, Shunsuke
Condensed Matter Theory Laboratory
RIKEN

Fuseya, Yuki
Department of Physics
The University of Tokyo

Goto, Kohei
Department of Applied Physics
Faculty of Engineering
The University of Tokyo

Hamada, Tomoyuki
Advanced Research Laboratory,
Hitachi, Ltd.

Hamamoto, Yuji
Institute for Solid State Physics
The University of Tokyo

Hammura, Kiyotaka
Hitachi Cambridge Laboratory
Hitachi Europe Ltd.

Harada, Ken
Advanced Research Laboratory
Hitachi, Ltd.

Hashizume, Tomihiro
Advanced Research Laboratory
Hitachi, Ltd.

Hatano, Tsuyoshi
Quantum Spin Information Project
ICORP-JST
NTT Atsugi Research & Development Center

Heike, Seiji
Advanced Research Laboratory
Hitachi, Ltd.

Hiroi, Zenji
Institute for Solid State Physics
The University of Tokyo

Hofmann, Holger F.
Graduate School of Advanced Sciences of Matter
Hiroshima University

Hosono, Kazuhiro
Department of Physics
Tokyo Metropolitan University

Houck, Andrew
Departments of Applied Physics
Yale University

Ichimura, Masahiko
Advanced Research Laboratory
Hitachi, Ltd.

Ichinose, Hideki
Advanced Science Institute
Single Quantum Dynamics Research Group
RIKEN

Ieda, Jun'ichi
Institute for Materials Research
Tohoku University

Iitaka, Toshiaki
Advanced Science Institute
RIKEN

Imai, Hiromitsu
Department of Physics
Faculty of Science & Technology
Tokyo University of Science

Ishioka, Sachio
Advanced Research Laboratory
Hitachi, Ltd.

Itano, Wayne M
Time and Frequency Division
National Institute of Standards and Technology

Ito, Kenchi
Advanced Research Laboratory
Hitachi, Ltd.

Iye, Yasuhiro
The Institute for Solid State Physics
The University of Tokyo

Johansson, Robert. J
Digital Materials Laboratory, ASI
RIKEN

Kane, Charles L
Department of Physics and Astronomy
University of Pennsylvania

Kasai, Hiroto
Advanced Research Laboratory
Hitachi, Ltd.

Kashiwaya, Satoshi
Nanoelectronics
Advanced Industrial Science and Technology

Kawabata, Shiro
Nanotechnology Recearch Insititue
Advanced Industrial Science and Technology

Kawakami, Junzo
Hitachi, Ltd.

Kim, Joong Jung
Electron Holography Unit
Okinawa Institute of Science and Technology

Kitagawa, Isao
Advanced Research Laboratory
Hitachi, Ltd.

Kobayashi, Akito
Institute for Advanced Research
Nagoya University

Kodama, Takaaki
Institute for Solid State Physics
The University of Tokyo

Kohashi Teruo
Advanced Research Laboratory
Hitachi, Ltd.

Kohno, Hiroshi
Department of Materials Engineering Science
Graduate School of Engineering Science
Osaka University

Koizumi, Hideaki
Advanced Research Laboratory
Hitachi, Ltd.

Kokado, Satoshi
Faculty of Engineering
Shizuoka University

Kouznetsov, Dmitrii
Institute for Laser Science
University of Electro-Communications

Kubo, Toshihiro
Quantum Spin Information Project
ICORP-JST
NTT Atsugi Research & Development Center

Lambert, Neill
Digital Materials Laboratory, ASI
RIKEN

Liu, Yu-xi
Digital Materials Laboratory, ASI
RIKEN

Mamishin, Shuichi
Advanced Research Laboratory
Hitachi,Ltd.

Marcus, Charles M
Department of Physics
Harvard University

Maruyama, Koji
Digital Materials Laboratory, ASI
RIKEN

Maruyama, Eiichi
RIKEN

Matsuo, Mari
Graduate School of Humanities and Sciences
Ochanomizu University

Matsuoka, Hideyuki
Advanced Research Laboratory
Hitachi, Ltd.

Mooij, J E
Kavli Institute of Nanoscience
Delft University of Technology

Morinaga, Atsuo
Department of Physics
Faculty of Science & Technology
Tokyo University of Science

Morota, Misako
Institute for Solid State Physics
The University of Tokyo

Murakami, Shuichi
Department of Physics
Tokyo Institute of Technology

Nagaosa, Naoto
Department of Applied Physics
Faculty of Engineering
The University of Tokyo

Nakajima, Ryota
Advanced Research Laboratory
Hitachi, Ltd.

Nakamura, Michiharu
Hitachi, Ltd.

Nakano, Hayato
Superconducting Quantum Research Group
NTT Basic Research Laboratories
NTT Corporation

Nambu, Akira
Advanced Research Laboratory
Hitachi, Ltd.

Nitta, Junsaku
Department of Materials Science
Graduate School of Engineering
Tohoku University

Nori, Franco
Digital Materials Laboratory, ASI
RIKEN
Department of Physics,
Center for Theoretical Physics
The University of Michigan

Obuse, Hideaki
Department of Physics
Kyoto University

Ohashi, Yoji
Faculty of Science and Technology
Keio University

Ohnishi, Kohei
Institute for Solid State Physics
The University of Tokyo

Ohno, Hideo
Laboratory for Nanoelectronics and Spintronics
Research Institute of Electrical Communication
Tohoku University

Okamoto, Kiyomi
Department of Physics
Tokyo Institute of Technology

Ono, Takafumi
Graduate School of Advanced Sciences of Matter
Hiroshima University

Ono, Yoshimasa A.
Cernter for Innovation in Engineering Education
School of Engineering
The University of Tokyo

Onoda, Shigeki
Condensed Matter Theory Laboratory
RIKEN

Osakabe, Nobuyuki
Advanced Research Laboratory
Hitachi, Ltd.

Otani, Yoshichika
Institute for Solid State Physics
The University of Tokyo

Parkin, Stuart S.P.
IBM Almaden Research Center
IBM

Saikawa, Kazuhiko

Saito, Hiroki
Department of Applied Physics and Chemistry
The University of Electro-Communications

Saitoh, Eiji
Department of Applied Physics & Physico-Informatics
Faculty of Science and Technology
Keio University

Saitoh, Kazuo
Advanced Research Laboratory
Hitachi, Ltd.

Sakurai, Akio
Department of Physics
Kyoto Sangyo University

Sato, Masahiro
Condensed Matter Theory Laboratory
RIKEN

Sato, Takahiro
Institute for Solid State Physics
The University of Tokyo

Shibata, Junya
Kanagawa Institute of Technology

Shimazu, Yoshihiro
Department of Physics
Yokohama National University

Shimizu, Fujio
Institute for Laser Science
University of Electro-Communications

Shiokawa, Kazutomu
Physics
Academia Sinica

Shirasaki, Ryoen
Department of Physics
Faculty of Engineering
Yokohama National University

Shitade Atsuo
Department of Applied Physics
Faculty of Engineering
The University of Tokyo

Söderholm, Jonas
Institute of Quantum Science
Nihon University

Sugano, Ryoko
Advanced Research Laboratory
Hitachi, Ltd.

Sugawara, Akira
Advanced Research Laboratory
Hitachi, Ltd.

Sugimoto, Naoyuki
Department of Applied Physics
Faculty of Engineering
The University of Tokyo

Suwa, Yuji
Advanced Research Laboratory
Hitachi, Ltd.

Taguchi, Gen
Graduate School of Advanced Sciences of Matter
Hiroshima University

Taguchi, Katsuhisa
Department of Physics
Tokyo Metropolitan University

Tajima, Naoya
Condenced Molecular Materials Lab.
RIKEN

Takeda, Haruo
Hitachi, Ltd.

Takahashi, Hiromasa
Advanced Research Laboratory
Hitachi, Ltd.

Takahashi, Yoshiro
Quantum Optics Lab.
Department of Physics
Kyoto University

Takayanagi, Hideaki
Research Institute for Science and Technology
Tokyo University of Science

Takeuchi, Akihito
Department of Physics
Tokyo Metropolitan University

Tanamoto, Tetsufumi
Advanced LSI Technology Laboratory
Corporate R & D Center
Toshiba Corporation

Tanner, Mike G.
Hitachi Cambridge Laboratory
Hitachi Europe Ltd.

Tarucha, Seigo
Department of Applied Physics
The University of Tokyo

Tatara, Gen
Department of Physics
Tokyo Metropolitan University

Togawa, Yoshihiko
Advanced Science Institute
Quantum Phenomena Observation Technology Team
RIKEN

Tokura, Yoshinori
Department of Applied Physics
The University of Tokyo

Tomaru, Tatsuya
Advanced Research Laboratory
Hitachi, Ltd.

Tonomura, Akira
Advanced Research Laboratory
Hitachi, Ltd.
Initial Research Project
Okinawa Institute of Science and Technology
Advanced Science Institute
RIKEN

Tsubota, Makoto
Department of Physics
Osaka City University

Tsuchiya, Shunji
Department of Physics
Keio University

Uchida, Masaya
Advanced Science Institute
Quantum Phenomena Observation
Technology Team
RIKEN

Uchino, Shun
Department of Physics
The University of Tokyo

Ueda, Masahito
Department of Physics
The University of Tokyo

Utsumi, Yasuhiro
Division of Condensed Matter Theory
Institute for Solid State Physic
The University of Tokyo

Watanabe, Keiji
Central Research Laboratory
Hitachi, Ltd.

Williams, David A.
Hitachi Cambridge Laboratory
Hitachi Europe Ltd.

Wunderlich, Jörg
Hitachi Cambridge Laboratory
Hitachi Europe Ltd.

Xu, Xiulai
Hitachi Cambridge Laboratory
Hitachi Europe Ltd.

Yamada, Eizaburo
Meisei University

Yamada, Masaki
Advanced Reserch Laboratory
Hitachi, Ltd.

Yamaguchi, Hiroshi
NTT Basic Research Laboratories
NTT Corporation

Yang, Chen Ning
Center for Advanced Study
Tsinghua University

Yorozu, Shinichi
Nano Electronics Research Laboratories
NEC Corporation

Yoshida, Takaho
Advanced Research Laboratory
Hitachi, Ltd.

Yunomae, Yoshihiro
Department of Physics
Waseda University

Zeilinger, Anton
Institute of Quantum Optics and Quantum
Information
Austrian Academy of Sciences & Quantum
Optics
Quantum Nanophysics and Quantum
Information
Faculty of Physics
University of Vienna

Zwierlein, Martin W
Department of Physics
Research Laboratory of Electronics, and
Center for Ultracold Atoms
Massachusetts Institute of Technology

AUTHOR INDEX